Lectures on Linear Partial Differential Equations

Lectures on Linear Partial Differential Equations

Gregory Eskin

Graduate Studies
in Mathematics
Volume 123

American Mathematical Society
Providence, Rhode Island

2010 *Mathematics Subject Classification.* Primary 35J25, 35L40, 35K30, 35L05, 35L30, 35P20, 35P25, 35S05, 35S30.

For additional information and updates on this book, visit
www.ams.org/bookpages/gsm-123

Library of Congress Cataloging-in-Publication Data

Eskin, G. I. (Grigorii Il′ich)
Lectures on linear partial differential equations / Gregory Eskin.
p. cm. — (Graduate studies in mathematics ; v. 123)
Includes bibliographical references.
ISBN 978-0-8218-5284-2 (alk. paper)
1. Differential equations, Elliptic. 2. Differential equations, Partial. I. Title.

QA372.E78 2011
515′.3533—dc22

2010048243

∞ The paper used in this book is acid-free and falls within the guidelines
established to ensure permanence and durability.
Visit the AMS home page at http://www.ams.org/

10 9 8 7 6 5 4 3 2 1 16 15 14 13 12 11

In memory of my brother Michael Eskin

Contents

Preface

This book is based on the lectures on partial differential equations that I have given for many years at UCLA. It does not assume any knowledge of partial differential equations and can be considered as a first graduate course in linear PDE. However, some basic knowledge of the Fourier transform, Lebesgue integrals and elementary functional analysis is required. It is organized as lecture notes with emphasis on clarity and accessibility.

We shall briefly describe the content of the book. The first three chapters are the elementary theory of distributions and Fourier transforms of distributions with applications to partial differential equations with constant coefficients. It is similar to the first chapters of the books by Gelfand and Shilov [**GSh**] and Shilov [**Sh**]. Additional material includes the wave front sets of distributions, Sobolev spaces, the stationary phase lemma, the radiation conditions, and potential theory.

In Chapter IV the Dirichlet and the Neumann boundary value problems are considered for second order elliptic equations in a smooth bounded domain. The existence, uniqueness, and regularity of solutions are proven. A nontraditional topic of this chapter is the proof of the existence and uniqueness of the solutions of the Neumann and Dirichlet problems for homogeneous equations in Sobolev spaces of negative order on the boundary.

Chapter V is devoted to scattering theory including inverse scattering, inverse boundary value problem, and the obstacle problem.

Chapter VI starts with the theory of pseudodifferential operators with classical symbols. It is followed by the theory of parabolic Cauchy problems based on pseudodifferential operators with symbols analytic in the half-plane and heat kernel asymptotics.

The next topic of Chapter VI is the Cauchy problem for hyperbolic equations of order $m \geq 2$, the domains of dependence of solutions to hyperbolic equations, and Hörmander's theory [**H1**] of propagation of singularities for the equations of real principal type with applications to hyperbolic equations.

In Chapter VII the Fredholm property for elliptic boundary value problems and parametrices in smooth domains are studied following the approach of the author's book [**E1**]. The main application of the parametrix is the study of heat trace asymptotics as $t \to 0$. The parametrix construction allows one to compute explicitly two leading terms of the heat trace asymptotics for the cases of Dirichlet and Neumann boundary conditions. Chapter VII concludes with elements of the spectral theory of elliptic operators and the proof of the index theorem for elliptic operators in $\mathbb{R}^n$ following the works of Atiyah-Singer [**AtS1**], [**AtS2**] and Seeley [**Se3**].

The last Chapter VIII is devoted to the theory of Fourier integral operators. Starting with the local theory of FIO, we proceed to the global theory. We consider only a subclass of Hörmander's FIOs (see [**H1**]), assuming that the Lagrangian manifold of the FIO corresponds to the graph of a canonical transformation. In particular, having a global canonical transformation, we construct a global FIO corresponding to this canonical transformation. Next, following Maslov [**M1**], [**M2**], [**MF**], we construct a global geometric optic solution for a second order hyperbolic equation on arbitrary time interval $[0, T]$.

Chapter VIII concludes with a section on the oblique derivative problem. The oblique derivative problem is a good example of nonelliptic boundary value problem, and it attracted the attention of many mathematicians: Egorov-Kondrat'ev [**EgK**], Malutin [**Mal**], Mazya-Paneah [**MaP**], Mazya [**Ma**], and others. The section is based on the author's paper [**E3**], and it uses the FIOs to greatly simplify the problem. Similar results are obtained independently by Sjöstrand [**Sj**] and Duistermaat-Sjöstrand [**DSj**].

At the end of each chapter there is a problem section. Some problems are relatively simple exercises that help to study the material. Others are more difficult problems that cover additional topics not included in the book. In those cases hints or references to the original sources are given.

Acknowledgments

I want to thank my friend and collaborator Jim Ralston for many fruitful discussions and advice. I am very grateful to my former students Joe Bennish, Brian Sako, Carol Shubin, Borislava Gutarz, Xiaosheng Li and others who took notes during my classes. These notes were the starting point of this book. I express my deep gratitude to the anonymous referees whose remarks and suggestions greatly improved the book. I am very thankful to Neelesh Tiruviluamala for pointing out a mistake in an earlier draft. I am also grateful to Bernard Lascar and Ciprian Manolescu for valuable comments. I am especially grateful to my wife Marina who prepared the manuscript. Without her help this book would not be written.

My deepest gratitude to the editor Natalya Pluzhnikov for the outstanding work of improving and correcting the manuscript. It is impossible to overestimate her contribution.

Finally, I am very grateful to Sergei Gelfand for encouragement to write a book.

Chapter I

Theory of Distributions

Introduction to Chapters I, II, III

Distributions and their Fourier transforms are the basis of the theory of linear partial differential equations. We present the main elements of the theory of distributions in the first twelve sections accompanied by many examples. The Sobolev spaces in $\mathbb{R}^n$ are studied in §13 using the theory of distributions. In §14 we introduce the notion of wave front sets of distributions and give some of its applications. In §16 we study the Cauchy problem for the heat, Schrödinger, and wave equations. We also consider the Dirichlet problem for the Laplace and Helmholtz (the reduced wave) equations. The fundamental solutions for all these equations are constructed. In §17 we demonstrate the power of distribution theory by constructing a fundamental solution for an arbitrary linear partial differential equation with constant coefficients. In §18 we describe the class of hypoelliptic equations, i.e., linear partial differential equations whose distribution solutions are C^∞ functions. In §19 the existence and uniqueness of the solution of the nonhomogeneous Helmholtz's equation is proved in the class of solutions satisfying the radiation conditions. For this purpose we prove and use in §19 the stationary phase lemma.

The last §20 is devoted to the study of simple and double layer potentials in bounded domains. These potentials will be used in Chapter V (scattering theory).

Sections 6, 15 and 21 are the problem sets. We use some problems to supplement the content of the book. For example, we always construct fundamental solutions using the Fourier transform because this approach can be applied to more general equations. However, for the wave and the

Laplace equations the traditional approach is simpler, and we offer this approach as problems (with some hints).

1. Spaces of infinitely differentiable functions

Denote by $C_0^\infty(\mathbb{R}^n)$ the space of infinitely differentiable complex-valued functions in $\mathbb{R}^n$ with compact supports, i.e., $\varphi(x) \in C_0^\infty(\mathbb{R}^n)$ if $\varphi(x)$ has continuous partial derivatives of every order and $\varphi(x) = 0$ when $|x| > R$ for some R depending of $\varphi(x)$. For example,

$$\chi(x) = \begin{cases} e^{-\frac{1}{1-|x|^2}} & \text{if} \quad |x| < 1, \\ 0 & \text{if} \quad |x| \geq 1, \ |x| = \sqrt{x_1^2 + \cdots + x_n^2}, \end{cases} \tag{1.1}$$

is a C_0^∞ function.

Definition 1.1. Let $f(x)$ be a continuous function in $\mathbb{R}^n$. The support of $f(x)$ is the closure of the set where $f(x) \neq 0$.

We denote the support of $f(x)$ by supp f.

Example 1.1.

a) The support of $\chi(x)$ is the closed ball of radius 1.

b) The support of $f(x_1) = x_1^2 - 1$ is $\mathbb{R}^1$.

Definition 1.2. Let $\varphi(x)$ be a measurable bounded and $\psi(x)$ a Lebesgue integrable functions in $\mathbb{R}^n$, $\|\psi\|_{L_1} = \int_{\mathbb{R}^n} |\psi(x)|dx < +\infty$. Then the convolution of $\varphi(x)$ and $\psi(x)$ (denoted by $(\varphi * \psi)(x)$) is the following integral:

$$(\phi * \psi)(x) = \int_{\mathbb{R}^n} \phi(x-y)\psi(y)dy. \tag{1.2}$$

Proposition 1.1. *If $\varphi(x) \in L_1$ and $\psi(x) \in L_1$, then the integral* (1.2) *exists for almost all $x \in \mathbb{R}^n$, $\varphi * \psi \in L_1(\mathbb{R}^n)$ and*

$$\|\varphi * \psi\|_{L_1} \leq \|\varphi\|_{L_1} \|\psi\|_{L_1}. \tag{1.3}$$

The proof of Proposition 1.1 is given at the end of this section.

1.1. Properties of the convolution.

a) $\varphi * \psi = \psi * \varphi$.

Proof: We have

$$(\varphi * \psi)(x) = \int_{\mathbb{R}^n} \varphi(x-y)\psi(y)dy. \tag{1.4}$$

Changing the variables $x - y = t$, we get

$$\int_{\mathbb{R}^n} \varphi(x-y)\psi(y)dy = \int_{R^n} \varphi(t)\psi(x-t)dt = \psi * \varphi.$$

With more details in the case of one variable we have

$$x_1 - y_1 = t_1, \quad dt_1 = -dy_1.$$

Therefore

$$\begin{aligned}\int_{-\infty}^{\infty} \varphi(x_1 - y_1)\psi(y_1)dy_1 &= -\int_{+\infty}^{-\infty} \varphi(t_1)\psi(x_1 - t_1)dt_1 \\ &= \int_{-\infty}^{\infty} \psi(x_1 - t_1)\varphi(t_1)dt_1.\end{aligned}$$

b) Assume that $\psi \in L_1(\mathbb{R}^n)$ and $\varphi(x)$ and $\frac{\partial^k \varphi(x)}{\partial x^k}$, $|k| \le m$, are continuous and bounded. Then $\varphi * \psi$ and $\frac{\partial^k}{\partial x^k}(\varphi * \psi)$, $|k| \le m$, are also continuous and bounded.

Here we use the standard notation : $k = (k_1, \dots, k_n)$, $|k| = k_1 + \dots + k_n$, $\frac{\partial^k}{\partial x^k} = \frac{\partial^{k_1+k_2+\dots+k_n}}{\partial x_1^{k_1} \partial x_2^{k_2} \dots \partial x_n^{k_n}}$.

The proof of property b) is given at the end of the section.

Property b) shows that for the convolution to be smooth it is enough that one of the two functions (φ or ψ) is smooth.

Proposition 1.2. *Let $\varphi(x)$ and $\psi(x)$ be continuous functions with compact supports, $\operatorname{supp} \varphi \subset U_1$, $\operatorname{supp} \psi \subset U_2$, where U_1 and U_2 are open sets in $\mathbb{R}^n$. Then*

$$\operatorname{supp}(\varphi * \psi) \subset U_1 + U_2,$$

where $U_1 + U_2$ is the open set in $\mathbb{R}^n$ consisting of all sums $x + y$, where $x \in U_1$ and $y \in U_2$.

Proof: We have

$$(\varphi * \psi)(x) = \int_{\mathbb{R}^n} \varphi(x - y)\psi(y)dy = \int_{U_2} \varphi(x - y)\psi(y)dy,$$

since $\psi(y) = 0$ for $y \notin U_2$. If $x_0 \notin U_1 + U_2$, then $x_0 - y \notin U_1$ for all $y \in U_2$. Then $\varphi(x_0 - y) = 0$ for all $y \in U_2$ and, therefore, $(\varphi * \psi)(x_0) = 0$. Hence $\operatorname{supp}(\varphi * \psi) \subset U_1 + U_2$. □

1.2. Approximation by C_0^∞-functions.

Proposition 1.3. *The C_0^∞-functions are dense in the space $C_c(\mathbb{R}^n)$ of continuous functions with compact supports: for any continuous $\varphi(x)$ with support in B_R, there exists a sequence of C_0^∞-functions $\varphi_m(x)$ with support in B_R such that*

$$\max_{x \in \mathbb{R}^n} |\varphi(x) - \varphi_m(x)| \to 0$$

as $m \to \infty$. Here $B_R = \{x : |x| < R\}$.

Proof: Let $c_n = \int_{\mathbb{R}^n} \chi(x)dx$, where $\chi(x)$ is the same as in (1.1). Set

$$\chi_\delta(x) = \frac{1}{c_n\delta^n}\chi\Big(\frac{x}{\delta}\Big).$$

Then $\chi_\delta(x) \geq 0$, and changing the variables $y = \frac{x}{\delta}$, we get

$$\int_{\mathbb{R}^n} \chi_\delta(x)dx = \frac{1}{c_n\delta^n}\int_{\mathbb{R}^n} \chi\Big(\frac{x}{\delta}\Big)dx = \frac{1}{c_n}\int_{\mathbb{R}^n} \chi(y)dy = 1.$$

Given a continuous function $\varphi(x)$ with support in B_R, define $\varphi_\delta(x) = \chi_\delta * \varphi$. Since $\operatorname{supp}\varphi$ is a closed set and B_R is open, there exists $R_1 < R$ such that $\operatorname{supp}\varphi \subset B_{R_1}$. If $0 < \delta < R - R_1$, then, by Proposition 1.2, $\operatorname{supp}\varphi_\delta \subset B_R$, since $\operatorname{supp}\chi_\delta \subset B_\delta$ and $B_{R_1} + B_\delta \subset B_R$.

By property b) of the convolution $\varphi_\delta(x) \in C^\infty$. Therefore it remains to proof that $\varphi_\delta(x)$ converges to $\varphi(x)$ uniformly as $\delta \to 0$. We have

$$|\varphi_\delta(x) - \varphi(x)| = \Big|\int_{\mathbb{R}^n} \varphi(x-y)\chi_\delta(y)dy - \varphi(x)\int_{\mathbb{R}^n} \chi_\delta(y)dy\Big|$$

$$\leq \int_{|y|\leq\delta} \chi_\delta(y)|\varphi(x-y) - \varphi(x)|dx.$$

Since $\varphi(x)$ is uniformly continuous, for any $\varepsilon > 0$ there exists $\delta_0 > 0$ such that $|\varphi(x-y) - \varphi(x)| < \varepsilon$ for all $x \in \mathbb{R}^n$ and all $|y| < \delta_0$. Thus

$$|\varphi_\delta(x) - \varphi(x)| < \varepsilon \int_{|y|\leq\delta} \chi_\delta(y)dy = \varepsilon$$

when $0 < \delta < \delta_0$. □

Finally, we introduce a sequential topology in the space $C_0^\infty(\mathbb{R}^n)$.

Definition 1.3. We call $\mathcal{D}$ the space of all C_0^∞ functions with the following notion of convergence: a sequence $\varphi_m \in C_0^\infty(\mathbb{R}^n)$ converges to $\varphi(x) \in C_0^\infty(\mathbb{R}^n)$ if

a) there exists $R > 0$ such that $\operatorname{supp}\varphi_m \subset B_R$ for all $m = 1, 2, \ldots$;

b) $\max_{x\in\mathbb{R}^n} |\varphi_m(x) - \varphi(x)| \to 0$ and $\max_{x\in\mathbb{R}^n} |\frac{\partial^k\varphi_m(x)}{\partial x^k} - \frac{\partial^k\varphi(x)}{\partial x^k}| \to 0$ as $m \to \infty$ for all $k = (k_1, \ldots, k_n)$.

Remark 1.1. In Definition 1.3 we have defined the convergence of sequences in $\mathcal{D}$, i.e., the sequential topology in $\mathcal{D}$. We will not define open sets in $\mathcal{D}$, i.e., we do not describe the topological structure of $\mathcal{D}$, since we will not use it in this book.

Example 1.2.

a) $\varphi_m(x) = e^{-m}\chi(x)\sin mx \to 0$ in $\mathcal{D}$, since $\operatorname{supp}\varphi_m \subset \operatorname{supp}\chi \subset \overline{B}_1$ and $|\frac{\partial^k\varphi_m}{\partial x^k}| \leq C_k e^{-m} m^{|k|} \to 0$ as $m \to \infty$.

b) $\varphi_m(x) = \frac{1}{m}\chi(\frac{x}{m})$ does not converge to 0 in $\mathcal{D}$: although $|\frac{\partial^k \varphi_m}{\partial x^k}| \leq \frac{c_k}{m^{|k|+1}} \to 0$ as $m \to \infty$, since $\operatorname{supp}\varphi_m = \operatorname{supp}\chi(\frac{x}{m}) = \overline{B}_m$ spreads as $m \to \infty$.

1.3. Proof of Proposition 1.1.

Since the repeated integral

$$\int_{\mathbb{R}^n}\left(\int_{\mathbb{R}^n}|\varphi(x-y)|\,|\psi(y)|dx\right)dy = \int_{\mathbb{R}^n}\left(\int_{\mathbb{R}^n}|\varphi(t)|dt\right)|\psi(y)|dy$$
$$= \int_{\mathbb{R}^n}|\varphi(t)|dt\int_{\mathbb{R}^n}|\psi(y)|dy$$

is finite, Fubini's theorem implies that the repeated integral

$$\int_{\mathbb{R}^n}\left(\int_{\mathbb{R}^n}|\varphi(x-y)||\psi(y)|dy\right)dx$$

and the double integral

$$\int_{\mathbb{R}^n}\int_{\mathbb{R}^n}|\varphi(x-y)||\psi(y)|dxdy$$

are finite and all three integrals are equal. Moreover, Fubini's theorem (see, for example, [**R**]) implies that the double integral

$$\int_{\mathbb{R}^n\times\mathbb{R}^n}\varphi(x-y)\psi(y)dydx$$

exists and for almost all $x \in \mathbb{R}^n$ the integral $\int_{\mathbb{R}^n}\varphi(x-y)\psi(y)dy$ exists. Thus

$$\|\varphi * \psi\|_{L_1} = \int_{\mathbb{R}^n}\left|\int_{\mathbb{R}^n}\varphi(x-y)\psi(y)dy\right|dx$$
$$\leq \int_{\mathbb{R}^n}\int_{\mathbb{R}^n}|\varphi(x-y)||\psi(y)|dydx = \|\varphi\|_{L_1}\|\psi\|_{L_1}.$$

1.4. Proof of property b) of the convolution.

We shall need the following theorem [**R**]:

Theorem 1.4 (Lebesgue convergence theorem). *If $f_m(x) \to f(x)$ almost everywhere as $m \to \infty$ and $|f_m(x)| \leq g(x)$, where $\int_{\mathbb{R}^n} g(x)dx < +\infty$, then $\int_{\mathbb{R}^n} f_m(x)dx \to \int_{\mathbb{R}^n} f(x)dx$ as $n \to \infty$.*

Since $\varphi(x)$ is continuous and bounded, we have that $|\varphi(x)| \leq M$ and $\varphi(x_m - y) \to \varphi(x_0 - y)$ as $x_m \to x_0$. Therefore, by the Lebesgue convergence theorem,

$$\int_{\mathbb{R}^n}\varphi(x_m-y)\psi(y)dy \to \int_{\mathbb{R}^n}\varphi(x_0-y)\psi(y)dy,$$

since $|\varphi(x_m - y)\psi(y)| \leq M|\psi(y)|$ and $\psi \in L_1$.

Hence $\varphi * \psi$ is a continuous function, and $|(\varphi * \psi)(x)| \leq \int_{\mathbb{R}^n} M|\psi(y)|dy$.

Let $h(x) = (\varphi * \psi)(x)$. Then

$$\frac{\partial h(x', x_n)}{\partial x_n} = \lim_{\Delta x_n \to 0} \frac{h(x', x_n + \Delta x_n) - h(x', x_n)}{\Delta x_n}$$

$$= \lim_{\Delta x_n \to 0} \int_{\mathbb{R}^n} \frac{\varphi(x' - y', x_n + \Delta x_n - y_n) - \varphi(x' - y', x_n - y_n)}{\Delta x_n} \psi(y', y_n) dy' dy_n,$$

where $x' = (x_1, \ldots, x_{n-1})$, $y' = (y_1, \ldots, y_{n-1})$.

By the mean value theorem,

$$\frac{\varphi(x' - y', x_n + \Delta x_n - y_n) - \varphi(x' - y', x_n - y_n)}{\Delta x_n}$$
$$= \frac{\partial \varphi(x' - y', x_n + \theta \Delta x_n - y_n)}{\partial x_n},$$

where $0 < \theta < 1$. Since $\frac{\partial \varphi}{\partial x_n}$ is bounded and continuous, we have that $|\frac{\partial \varphi(x' - y', x_n + \theta \Delta x_n - y_n)}{\partial x_n}| \leq M$ and

$$\lim_{\Delta x_n \to 0} \frac{\partial \varphi(x' - y', x_n + \theta \Delta x_n - y_n)}{\partial x_n} = \frac{\partial \varphi(x' - y', x_n - y_n)}{\partial x_n}.$$

Therefore, by the Lebesgue convergence theorem,

$$\frac{\partial h(x', x_n)}{\partial x_n} = \frac{\partial \varphi}{\partial x_n} * \psi$$

and, as was already proven, $\frac{\partial h}{\partial x_n}$ is continuous since $\frac{\partial \varphi}{\partial x_n}$ is continuous.

Analogously one treats partial derivatives of any order.

2. Definition of a distribution

Definition 2.1. A linear continuous functional f on $\mathcal{D}$ is called a distribution. Thus $f(\varphi) : \mathcal{D} \to \mathbb{C}$ is a distribution if the following two properties hold:

1. Linearity: $f(\alpha_1 \varphi_1 + \alpha_2 \varphi_2) = \alpha_1 f(\varphi_1) + \alpha_2 f(\varphi_2)$ for any $\alpha_k \in \mathbb{C}$, $\varphi_k \in \mathcal{D}$, $k = 1, 2$.

2. Continuity: If $\varphi_n \to \varphi$ in $\mathcal{D}$, then $f(\varphi_n) \to f(\varphi)$.

2.1. Examples of distributions.

a) Let $f(x)$ be a locally absolutely integrable function, i.e.,

$$\int_{B_R} |f(x)| dx < +\infty$$

for any $R > 0$. (We assume that $f(x)$ is Lebesgue integrable.) Define

$$f(\varphi) = \int_{\mathbb{R}^n} f(x) \varphi(x) dx, \ \forall \varphi \in \mathcal{D}. \tag{2.1}$$

Then $f(x)$ is a linear functional. If $\varphi_m(x) \to \varphi$ in $\mathcal{D}$, then $\operatorname{supp}\varphi_m \subset B_{R_0}$ for some $R_0 > 0$ and $\max_{x\in\mathbb{R}^n} |\varphi_m(x)-\varphi(x)| \to 0$ as $m \to \infty$. Therefore

$$|f(\varphi_m) - f(\varphi)| \le \int_{B_{R_0}} |f(x)||\varphi_m(x) - \varphi(x)|dx$$
$$\le \max_{x\in\mathbb{R}^n} |\varphi_m(x) - \varphi(x)| \int_{B_{R_0}} |f(x)|dx \to 0$$

as $m \to \infty$.

b) Let $f(x)$ be as in **a)**. Denote by $f_k(\varphi)$ the following linear functional:

$$f_k(\varphi) = \int_{\mathbb{R}^n} f(x)\frac{\partial^k \varphi(x)}{\partial x^k} dx.$$

Then $f_k(\varphi)$ is also a distribution since $\varphi_m \to \varphi$ in $\mathcal{D}$ implies that $\max_{x\in\mathbb{R}^n} |\frac{\partial^k\varphi(x)}{\partial x^k} - \frac{\partial^k\varphi_m(x)}{\partial x^k}| \to 0$.

c) Delta-function is a distribution defined by the formula

$$\delta(\varphi) = \varphi(0).$$

It is clear that $\delta(\varphi)$ is a linear continuous functional since $\varphi_m(x) \to \varphi(x)$ implies that $\varphi_m(0) \to \varphi(0)$.

d) For $n = 2$ we introduce polar coordinates

$$x_1 = r\cos\theta, \quad x_2 = r\sin\theta, \quad r = \sqrt{x_1^2 + x_2^2}$$

and define a linear functional $f(\varphi)$ by the formula

$$f(\varphi) = \int_0^{2\pi} \varphi(\cos\theta, \sin\theta)\, d\theta,$$

i.e., $\varphi(x_1, x_2)$ is integrated over the unit circle. It is clear that $f(\varphi)$ is a distribution. □

2.2. Regular functionals.

We call the distribution defined by formula (2.1) the regular functional corresponding to the locally integrable function $f(x)$. Two distributions $f_1(\varphi)$ and $f_2(\varphi)$ are called equal if $f_1(\varphi) = f_2(\varphi)$ for all $\varphi \in \mathcal{D}$.

Proposition 2.1. *Let $f_1(\varphi)$ and $f_2(\varphi)$ be two regular functionals corresponding to $f_1(x)$ and $f_2(x)$. Then $f_1(\varphi) = f_2(\varphi)$, $\forall \varphi \in \mathcal{D}$, iff $f_1(x) = f_2(x)$ almost everywhere.*

Proof: We have

$$f_1(\varphi) - f_2(\varphi) = \int_{\mathbb{R}^n} (f_1(x) - f_2(x))\varphi(x)dx = 0, \quad \forall\varphi \in \mathcal{D}.$$

Taking real-valued $\varphi \in \mathcal{D}$ and separating the real and the imaginary part of the integral, we can assume without loss of generality that $f_1(x) - f_2(x)$ is real-valued. Let

$$\operatorname{sgn}(f_1(x) - f_2(x)) = \begin{cases} 1 & \text{if} \quad f_1(x) - f_2(x) > 0, \\ 0 & \text{if} \quad f_1(x) - f_2(x) = 0, \\ -1 & \text{if} \quad f_1(x) - f_2(x) < 0. \end{cases}$$

Then $\operatorname{sgn}(f_1 - f_2)$ is a bounded measurable function.

Fix an arbitrary $R > 0$. It is known from Lebesgue measure theory (see, for example, [**R**]) that there exists a sequence of step-functions $\{s_m(x)\}$ such that $s_m(x) \to \operatorname{sgn}(f_1 - f_2)$ almost everywhere in B_R as $m \to \infty$ and $|s_m(x)| \leq 1$. For any step-function $s_m(x)$, there exists a sequence of continuous functions $\{c_p(x)\}$ with supports in B_R such that $c_p(x) \to s_m(x)$ almost everywhere (a.e.) in B_R as $p \to \infty$ and $|c_p(x)| \leq 1$. Using these two facts and the Proposition 1.3 we can find a sequence $\varphi_m(x)$ of C_0^∞-functions with supports in B_R such that $\varphi_m(x) \to \operatorname{sgn}(f_1 - f_2)$ a.e. in B_R and $|\varphi_m(x)| \leq 1$. Then the Lebesgue convergence theorem implies that

$$\int (f_1(x) - f_2(x))\varphi_m(x)dx \to \int_{B_R} |f_1(x) - f_2(x)|dx \qquad \text{as } m \to \infty.$$

Thus $\int_{B_R} |f_1(x) - f_2(x)|dx = 0$. Therefore $f_1(x) = f_2(x)$ a.e. in B_R. Since R is arbitrary, we get $f_1(x) = f_2(x)$ a.e. in $\mathbb{R}^n$. □

The distributions form a linear space since a linear combination $\alpha_1 f_1 + \alpha_2 f_2$ of distributions is also a distribution:

$$(\alpha_1 f_1 + \alpha_2 f_2)(\varphi) = \alpha_1 f_1(\varphi) + \alpha_2 f_2(\varphi).$$

The linear space of distributions is denoted by $\mathcal{D}'$.

2.3. Distributions in a domain.

Let $\Omega \subset \mathbb{R}^n$ be an arbitrary domain. The space $\mathcal{D}(\Omega)$ is the space $C_0^\infty(\Omega)$ with the following notion of convergence: $\varphi_n(x) \in C_0^\infty(\Omega)$ converges to $\varphi(x) \in C_0^\infty(\Omega)$ in $\mathcal{D}(\Omega)$ if there exists a compact subdomain $\overline{B} \subset \Omega$ such that $\operatorname{supp} \varphi_n \subset \overline{B}$ for all $n \geq 1$ and $\frac{\partial^k \varphi_n(x)}{\partial x^k} \to \frac{\partial^k \varphi}{\partial x^k}$ uniformly on $\overline{B}$ for all $0 \leq |k| < +\infty$.

The space of distribution $\mathcal{D}'(\Omega)$ is the space of all linear continuous functionals on $\mathcal{D}(\Omega)$.

Let $U \subset \mathbb{R}^n$ be an open set and let $f \in \mathcal{D}'(\mathbb{R}^n)$. We shall define the restriction $f|_U$ of f to U as $f|_U(\varphi) = f(\varphi)$ for all $\varphi \in \mathcal{D}(U)$. It is obvious that $f|_U \in \mathcal{D}'(U)$.

Example 2.1 (A nonextendible distribution). Let $f(x_1) = e^{x_1^2}$. Consider the regular functional in $\mathcal{D}'(0, +\infty)$ defined by $f(\varphi) = \int_0^\infty e^{\frac{1}{x_1^2}} \varphi(x_1)dx_1$,

$\forall \varphi \in \mathcal{D}(0,+\infty)$. One can prove (cf. Problem 8.12) that this functional is not a restriction to $(0,+\infty)$ of any $f \in \mathcal{D}'(\mathbb{R}^1)$.

3. Operations with distributions

3.1. Derivative of a distribution.

Let $f(x_1)$ be an absolutely continuous function in $\mathbb{R}^1$. Then the formula of integration by parts holds:

$$\int_{-\infty}^{\infty} \frac{df(x_1)}{dx_1}\varphi(x_1)dx_1 = -\int_{-\infty}^{\infty} f(x_1)\frac{d\varphi(x_1)}{dx_1}dx_1, \quad \forall \varphi \in \mathcal{D}. \tag{3.1}$$

Thus, for the regular functionals corresponding to $\frac{df}{dx_1}$ and $f(x_1)$ we get

$$\frac{df}{dx_1}(\varphi) = -f\Big(\frac{d\varphi}{dx_1}\Big). \tag{3.2}$$

Formula (3.2) leads to the following definition of a derivative of a distribution:

Definition 3.1. For any $f \in \mathcal{D}'$ and any $k = (k_1, \dots, k_n)$,

$$\frac{\partial^k f}{\partial x^k}(\varphi) = (-1)^{|k|} f\Big(\frac{\partial^k \varphi}{\partial x^k}\Big). \tag{3.3}$$

The right hand side of (3.3) is well defined since $\frac{\partial^k \varphi}{\partial x^k} \in \mathcal{D}$ and $\varphi_m \to \varphi$ in $\mathcal{D}$ implies that $\frac{\partial^k \varphi_m}{\partial x^k} \to \frac{\partial^k \varphi}{\partial x^k}$ in $\mathcal{D}$. Therefore $(-1)^{|k|} f(\frac{\partial^k \varphi}{\partial x^k})$ is a linear continuous functional on $\mathcal{D}$ and this functional, by definition, is $\frac{\partial^k f}{\partial x^k}(\varphi)$.

Example 3.1.

i) Consider the regular functional θ corresponding to the function $\theta(x_1) = 1$ for $x_1 > 0$, $\theta(x_1) = 0$ for $x_1 < 0$. Then

$$\frac{d}{dx_1}\theta(\varphi) = -\theta\Big(\frac{d\varphi}{dx_1}\Big) = -\int_0^{\infty} \frac{d\varphi}{dx_1}(x_1)dx_1 = \varphi(0) = \delta(\varphi).$$

ii)

$$\frac{d^k \delta}{dx_1^k}(\varphi) = (-1)^{|k|}\delta\Big(\frac{d^k \varphi}{dx_1^k}\Big) = (-1)^{|k|}\frac{d^k \varphi(0)}{dx_1^k}.$$

3.2. Multiplication of a distribution by a C^∞-function.

Again, if $f(\varphi)$ is a regular functional corresponding to $f(x)$ and if $a(x) \in C^\infty(\mathbb{R}^n)$, then, obviously,

$$\int_{\mathbb{R}^n} (a(x)f(x))\varphi(x)dx = \int_{\mathbb{R}^n} f(x)(a(x)\varphi(x))dx. \tag{3.4}$$

This identity leads to the following definition.

Definition 3.2. For any $f \in \mathcal{D}'$ and any $a(x) \in C^\infty$, af is the distribution defined by the formula:

$$af(\varphi) = f(a\varphi), \quad \forall \varphi \in \mathcal{D}. \tag{3.5}$$

Note that the right hand side of (3.5) defines a linear continuous functional on $\mathcal{D}$. Indeed, $a\varphi \in \mathcal{D}$, $\operatorname{supp}(a\varphi) \subset \operatorname{supp}\varphi$, and the Leibniz formula gives that $\varphi_m \to \varphi$ in $\mathcal{D}$ implies that $a\varphi_m \to a\varphi$ in $\mathcal{D}$.

Example 3.2. $x_1\delta = 0$. Indeed, $x_1\delta(\varphi) = \delta(x_1\varphi) = 0\varphi(0) = 0$ for $\forall \varphi \in \mathcal{D}$.

3.3. Change of variables for distributions.

Let $x = s(y)$ be a one-to-one C^∞ map of $\mathbb{R}^n$ onto $\mathbb{R}^n$:

$$x_k = s_k(y_1, \dots, y_n), \quad 1 \le k \le n. \tag{3.6}$$

We assume that the inverse map $y = s^{-1}(x)$ is also C^∞.

Denote by $J(x) = \det\left[\frac{\partial s_i^{-1}(x)}{\partial x_j}\right]_{i,j=1}^n$ the Jacobian of the inverse map $y = s^{-1}(x)$. For a regular functional, after changing the variable $y = s^{-1}(x)$ we get:

$$\int_{\mathbb{R}^n} f(s(y))\varphi(y)dy = \int_{\mathbb{R}^n} f(x)\varphi(s^{-1}(x))|J(x)|dx. \tag{3.7}$$

Let $(f \circ s)(y) = f(s(y))$. Then we can rewrite (3.7) in the following form:

$$f \circ s\ (\varphi) = f(\varphi(s^{-1}(x))|J(x)|). \tag{3.8}$$

Note that if $\varphi \in \mathcal{D}$, then $\psi(x) = \varphi(s^{-1}(x))J(x)$ is also in $\mathcal{D}$. Moreover, if $\varphi_n \to \varphi$ in $\mathcal{D}$, then $\psi_n \to \psi$ in $\mathcal{D}$. Therefore the right hand side of (3.8) is a linear continuous functional on $\mathcal{D}$.

Hence for any $f \in \mathcal{D}'$ we can define the change of variable $x = s(y)$ in f by formula (3.8).

Example 3.3. If $x_1 = s(y_1) = ky_1 + b$ is a linear map in $\mathbb{R}^1$, we get

$$f \circ s\ (\varphi) = f\left(\varphi\Big(\frac{x_1 - b}{k}\Big)\frac{1}{|k|}\right).$$

4. Convergence of distributions

Definition 4.1. We say that a sequence of distributions $f_n \in \mathcal{D}'$ converges to a distribution f if

$$f_n(\varphi) \to f(\varphi) \quad \text{for any} \quad \varphi \in D, \tag{4.1}$$

i.e., $f_n \to f$ in $\mathcal{D}'$ if (4.1) holds.

Example 4.1. If $f_n(x) \to f(x)$ a.e. and $|f_n(x)| \leq g(x)$ for all n, where $g(x)$ is a locally integrable function, then, by the Lebesgue convergence theorem,

$$f_n(\varphi) = \int_{\mathbb{R}^n} f_n(x)\varphi(x)dx \to \int_{\mathbb{R}^n} f(x)\varphi(x)dx = f(\varphi), \quad \forall \varphi \in \mathcal{D}.$$

In particular, consider

$$\ln(x_1 + i\varepsilon) = \ln|x_1 + i\varepsilon| + i\arg(x_1 + i\varepsilon), \tag{4.2}$$

where $\varepsilon > 0$ and $0 < \arg(x_1 + i\varepsilon) < \pi$ for all $x_1 \in \mathbb{R}^1$.

Denote

$$\ln(x_1 + i0) = \lim_{\varepsilon \to +0} \ln(x_1 + i\varepsilon) = \ln|x_1| + i\pi\theta(-x_1), \tag{4.3}$$

where $\theta(-x_1) = 0$ for $x_1 > 0$ and $\theta(-x_1) = 1$ for $x_1 < 0$. Since $|\ln(x_1+i\varepsilon)| \leq \pi + 1 + |\ln|x_1||$ for $|x_1| \leq 1$, and $|\ln(x_1 + i\varepsilon)| \leq \pi + \ln||x_1| + 1|$ for $|x_1| \geq 1$, $0 < \varepsilon < 1$, we get by the Lebesgue convergence theorem that the regular functionals, corresponding to $\ln(x_1 + i\varepsilon)$, converge as $\varepsilon \to 0$ to the regular functional, corresponding to $\ln(x_1 + i0)$.

Example 4.2. Denote $f_n(x_1) = n$ if $0 < x_1 < \frac{1}{n}$, and $f_n(x_1) = 0$ if x_1 does not belong to $[0, \frac{1}{n}]$. Then $\int_{-\infty}^{\infty} f_n(x_1)dx_1 = 1$ and

$$\left|\int_{-\infty}^{\infty} f_n(x_1)\varphi(x_1)dx_1 - \varphi(0)\right| = \left|n\int_0^{\frac{1}{n}} (\varphi(x_1) - \varphi(0))dx_1\right|$$
$$\leq \max_{0 \leq x_1 \leq \frac{1}{n}} |\varphi(x_1) - \varphi(0)| \to 0$$

as $n \to \infty$. Therefore

$$f_n(\varphi) \to \delta(\varphi)$$

as $n \to \infty$.

Theorem 4.1. *If $f_m \to f$ in $\mathcal{D}'$, then for any $k = (k_1, \ldots, k_n)$,*

$$\frac{\partial^k f_m}{\partial x^k} \to \frac{\partial^k f}{\partial x^k} \quad \textit{in} \quad \mathcal{D}'.$$

Proof: We have, by the definition of the derivative of a distribution:

$$\frac{\partial^k f_m}{\partial x^k}(\varphi) = (-1)^{|k|} f_m\Big(\frac{\partial^k \varphi}{\partial x^k}\Big) \to (-1)^{|k|} f\Big(\frac{\partial^k \varphi}{\partial x^k}\Big) = \frac{\partial^k f}{\partial x^k}(\varphi).$$ □

Example 4.3. Let λ be a complex number. Consider $(x_1+i\varepsilon)^\lambda = e^{\lambda \ln(x_1+i\varepsilon)}$, where $\varepsilon > 0$ and $\ln(x_1 + i\varepsilon)$ is defined as in Example 4.1. Thus

$$(x_1 + i\varepsilon)^\lambda = |x_1 + i\varepsilon|^\lambda e^{i\lambda \arg(x_1+i\varepsilon)}, \quad 0 < \arg(x_1 + i\varepsilon) < \pi.$$

Denote

$$(x_1 + i0)^\lambda = \lim_{\varepsilon \to +0} (x_1 + i\varepsilon)^\lambda = |x_1|^\lambda e^{i\lambda\pi\theta(-x_1)}. \tag{4.4}$$

Note that $(x_1 + i0)^\lambda$ defines a regular functional if $\operatorname{Re}\lambda > -1$. By the Lebesgue convergence theorem,

$$\int_{\mathbb{R}^1} (x_1 + i\varepsilon)^\lambda \varphi(x_1)dx_1 \to \int_{\mathbb{R}^1} (x_1 + i0)^\lambda \varphi(x_1)dx_1$$

if $\varepsilon \to 0$, $\operatorname{Re}\lambda > -1$, $\forall\varphi \in \mathcal{D}$.

We denote by $(x_1 + i\varepsilon)^\lambda$ and $(x_1 + i0)^\lambda$ the regular functionals corresponding to integrable functions $(x_1 + i\varepsilon)^\lambda$ and $(x_1 + i0)^\lambda$. For any integer $m > 0$ and noninteger λ such that $\operatorname{Re}\lambda > -1$, we have

$$(x_1 + i\varepsilon)^{\lambda-m} = \frac{1}{\lambda(\lambda-1)\cdots(\lambda-m+1)} \frac{d^m}{dx_1^m}(x_1 + i\varepsilon)^\lambda.$$

By Theorem 4.1, the distributions $(x_1 + i\varepsilon)^{\lambda-m}$ converge as $\varepsilon \to 0$ to the distribution $\frac{1}{\lambda(\lambda-1)\cdots(\lambda-m+1)} \frac{d^m}{dx_1^m}(x_1 + i0)^\lambda$.

We shall denote the limiting distribution by $(x_1 + i0)^{\lambda-m}$. Therefore

$$(x_1 + i0)^{\lambda-m} = \frac{1}{\lambda(\lambda-1)\cdots(\lambda-m+1)} \frac{d^m}{dx_1^m}(x_1 + i0)^\lambda. \tag{4.5}$$

If $\lambda = -k$ is an integer, $k > 0$, then, starting with the identity

$$\frac{d^k}{dx_1^k}\ln(x_1 + i\varepsilon) = \frac{(-1)^{(k-1)}(k-1)!}{(x_1 + i\varepsilon)^k}, \quad \varepsilon > 0,$$

and using Theorem 4.1 and Example 4.1, we get that

$$\lim_{\varepsilon\to+0}(x_1 + i\varepsilon)^{-k} = \frac{(-1)^{k-1}}{(k-1)!}\frac{d^k}{dx_1^k}\ln(x_1 + i0) \quad \text{in} \quad \mathcal{D}'. \tag{4.6}$$

Therefore, for any $\lambda \in \mathbb{C}$, there exists a limit of regular functionals $(x_1 + i\varepsilon)^\lambda$ as $\varepsilon \to +0$, and we denote this limiting distribution by $(x_1 + i0)^\lambda$:

$$\lim_{\varepsilon\to+0}(x_1 + i\varepsilon)^\lambda = (x_1 + i0)^\lambda \quad \text{in} \quad \mathcal{D}', \quad \forall\lambda \in \mathbb{C}. \tag{4.7}$$

Analogously one defines $(x_1 - i0)^\lambda$ as $\lim_{\varepsilon\to+0}(x_1 - i\varepsilon)^\lambda$ in D', $\forall\lambda \in \mathbb{C}$, where $(x_1 - i\varepsilon)^\lambda = |x_1 - i\varepsilon|^\lambda e^{i\lambda\arg(x_1 - i\varepsilon)}$ and $-\pi < \arg(x_1 - i\varepsilon) < 0$. Thus $(x_1 - i0)^\lambda = |x_1|^\lambda e^{-i\lambda\pi\theta(-x_1)}$.

4.1. Delta-like sequences.

Example 4.4.

a) Let $\beta(x)$ be nonnegative and $\int_{\mathbb{R}^n}\beta(x)dx = 1$. Let

$$\beta_\varepsilon(x) = \frac{1}{\varepsilon^n}\beta\Big(\frac{x}{\varepsilon}\Big). \tag{4.8}$$

Then

$$\beta_\varepsilon(\varphi) \to \delta(\varphi) = \varphi(0) \quad \text{as} \quad \varepsilon \to 0, \ \forall\varphi \in D. \tag{4.9}$$

Indeed, changing the variables $y = \frac{x}{\varepsilon}$, $dy = \frac{1}{\varepsilon^n}dx$, we get

$$\beta_\varepsilon(\varphi) - \varphi(0) = \frac{1}{\varepsilon^n}\int_{\mathbb{R}^n} \beta\Big(\frac{x}{\varepsilon}\Big)\varphi(x)dx - \varphi(0) = \int_{\mathbb{R}^n} \beta(y)(\varphi(\varepsilon y) - \varphi(0))dy.$$

Since $|\varphi(\varepsilon y) - \varphi(0)| \leq 2C$ and $\varphi(\varepsilon y) \to \varphi(0)$ for $\forall y \in \mathbb{R}^n$, we obtain, by the Lebesgue convergence theorem, that

$$\Big|\int_{\mathbb{R}^n} \beta(y)(\varphi(\varepsilon y) - \varphi(0))dy\Big| \leq \int_{\mathbb{R}^n} \beta(y)|\varphi(\varepsilon y) - \varphi(0)|dy \to 0$$

as $\varepsilon \to 0$. The sequence of regular functionals β_ε is called a delta-like sequence.

Consider some particular cases of delta-like sequences.

b) The Poisson kernel for the Laplace equation:

$$P(x_1, x_2) = \frac{x_2}{\pi(x_1^2 + x_2^2)}$$

satisfies the Laplace equation

$$\frac{\partial^2 P}{\partial x_1^2} + \frac{\partial^2 P}{\partial x_2^2} = 0$$

in the half-plane $x_2 > 0$. Here $\beta(x_1) = \frac{1}{\pi(1+x_1^2)}$, $\varepsilon = x_2$. Note that

$$\int_{-\infty}^{\infty} \frac{dx_1}{\pi(1 + x_1^2)} = 1.$$

Thus

$$P(x_1, x_2) = \frac{1}{x_2\pi(1 + \frac{x_1^2}{x_2^2})} = \frac{1}{x_2}\beta\Big(\frac{x_1}{x_2}\Big)$$

and

$$P(x_1, x_2) \to \delta \text{ as } x_2 \to +0.$$

c) The heat kernel:

$$E(x, t) = \frac{1}{(4\pi t)^{\frac{n}{2}}} e^{-\frac{|x|^2}{4t}}$$

satisfies the heat equation

$$\frac{\partial}{\partial t}E - \sum_{k=1}^{n} \frac{\partial^2}{\partial x_k^2}E = 0$$

for $t > 0$, $\forall x \in \mathbb{R}^n$. Here $\beta(x) = \frac{1}{\pi^{\frac{n}{2}}}e^{-|x|^2}$, $\varepsilon = \sqrt{4t}$. Note that

$$\frac{1}{\pi^{\frac{n}{2}}}\int_{\mathbb{R}^n} e^{-|x|^2}dx = \Big(\frac{1}{\sqrt{\pi}}\int_{\infty}^{\infty} e^{-x_1^2}dx_1\Big)^n = 1.$$

Thus

$$E(x,t) = \frac{1}{(4t)^{\frac{n}{2}}}\beta\Big(\frac{x}{2\sqrt{t}}\Big) \to \delta \quad \text{as} \quad t \to +0.$$

5. Regularizations of nonintegrable functions

If $f(x)$ is not a locally integrable function, the problem is how to define a distribution naturally related to $f(x)$.

Consider, for example, the case where $f(x_1)$ is locally integrable for $x_1 \in \mathbb{R}^1$, $|x_1| > \delta, \forall \delta > 0$,

$$\int_{|x_1|<\delta} |f(x_1)|dx_1 = +\infty \quad \text{and} \quad |f(x_1)| \leq \frac{C}{|x_1|^m}, \qquad m \geq 1.$$

It is natural to assume that f is a distribution that satisfies

$$f(\varphi) = \int_{\mathbb{R}^1} f(x_1)\varphi(x_1)dx_1 \quad \text{if} \quad \varphi \in \mathcal{D},\ \varphi(x_1) = 0 \quad \text{for} \quad |x_1| < \delta. \tag{5.1}$$

The problem of defining $f \in \mathcal{D}'$ such that (5.1) holds for any $\varphi(x_1) = 0$ in a neighborhood of $x_1 = 0$ is called the regularization problem for $f(x_1)$. One of the distributions $f_1 \in \mathcal{D}'$ that solves the problem of the regularization of $f(x_1)$ is given by the formula:

$$f_1(\varphi) = \int_{|x_1|\leq 1} f(x_1)\left(\varphi(x_1) - \sum_{k=0}^{m-1} \frac{\varphi^{(k)}(0)}{k!}x_1^k\right)dx_1 + \int_{|x_1|>1} f(x_1)\varphi(x_1)dx_1. \tag{5.2}$$

Since

$$\left|\varphi(x_1) - \sum_{k=0}^{m-1}\frac{\varphi^{(k)}(0)}{k!}x_1^k\right| = \left|\frac{\varphi^{(m)}(\theta)}{m!}x_1^m\right| \leq C|x_1|^m,$$

(5.2) defines a linear continuous functional f_1 on $\mathcal{D}$ that coincides with (5.1) when $\varphi(x) = 0$ in a neighborhood of $x_1 = 0$. However, the solution of the regularization problem is not unique since any distribution of the form

$$f_2(\varphi) = f_1(\varphi) + \sum_{k=1}^{N} c_k \frac{d^k\delta}{dx_1^k}(\varphi), \tag{5.3}$$

where $c_k \in \mathbb{C}$ are arbitrary, also coincides with (5.1) when $\varphi(x_1) = 0$ in a neighborhood of $x_1 = 0$.

In applications one tries to define a regularization $f \in \mathcal{D}'$ of a nonintegrable function $f(x)$ in such a way that $f \in \mathcal{D}'$ retains some properties of $f(x)$ in addition to (5.1) such as symmetry, homogeneity, etc.

5.1. Regularization in $\mathbb{R}^1$.

Example 5.1 (Principal value integrals. Distribution p.v. $\frac{1}{x_1}$). If $g(x_1)$ is absolutely integrable on $\delta < |x_1| < +\infty$ for any $\delta > 0$, we define the principal value integral, p.v. $\int_{-\infty}^{\infty} g(x_1)dx_1$, as the limit of $\int_{-\infty}^{-\delta} g(x_1)dx_1 + \int_{\delta}^{\infty} g(x_1)dx_1$ as $\delta \to 0$, assuming that this limit exists:

$$\text{p.v.} \int_{-\infty}^{\infty} g(x_1)dx_1 = \lim_{\delta\to 0} \int_{|x_1|>\delta} g(x_1)dx_1. \tag{5.4}$$

We show that for any $\varphi \in \mathcal{D}$ the principal value integral

$$\text{p.v.} \int_{-\infty}^{\infty} \frac{1}{x_1}\varphi(x_1)dx_1$$

exists and defines a distribution that we denote by p.v. $\frac{1}{x_1}$.

Indeed,

$$\begin{aligned}\int_{|x_1|>\delta} \frac{\varphi(x_1)}{x_1}dx_1 &= \int_{\delta<|x_1|<1} \frac{\varphi(x_1)-\varphi(0)}{x_1}dx_1 \\ &+ \int_{\delta<|x_1|<1} \frac{\varphi(0)}{x_1}dx_1 + \int_{|x_1|>1} \frac{\varphi(x_1)}{x_1}dx_1.\end{aligned} \tag{5.5}$$

In the right hand side of (5.5) the second integral is equal to zero because $\frac{\varphi(0)}{x_1}$ is an odd function, and the first integral converges as $\delta \to 0$ since $|\frac{\varphi(x_1)-\varphi(0)}{x_1}| \leq C$.

Therefore the principal value integral exists and

$$\text{p.v.} \int_{-\infty}^{\infty} \frac{\varphi(x_1)}{x_1}dx = \int_{|x_1|<1} \frac{\varphi(x_1)-\varphi(0)}{x_1}dx + \int_{|x_1|>1} \frac{\varphi(x_1)}{x_1}dx_1. \tag{5.6}$$

It is clear that (5.5) defines a linear continuous functional on $\mathcal{D}$. Note that if $\varphi(0) \neq 0$ then the integrals

$$\int_0^{\infty} \frac{\varphi(x_1)}{x_1}dx_1 \quad \text{and} \quad \int_{-\infty}^{0} \frac{\varphi(x_1)}{x_1}dx_1$$

diverge, but the principal value integral

$$\text{p.v.} \int_{-\infty}^{\infty} \frac{\varphi(x_1)}{x_1}dx_1$$

exists.

Example 5.2 (Distributions $\frac{1}{x_1+i0}$ and $\frac{1}{x_1-i0}$). These distributions were already defined in Example 4.3, and they are derivatives in the sense of distributions of regular functionals $\ln(x_1 + i0)$ and $\ln(x_1 - i0)$:

$$(x_1 \pm i0)^{-1} = \frac{d}{dx_1} \ln(x_1 \pm i0). \tag{5.7}$$

We find the relation between $(x_1 \pm i0)^{-1}$ and p.v. $\frac{1}{x_1}$. We have:

(5.8)
$$\int_{-\infty}^{\infty} \frac{\varphi(x_1)}{x_1 \pm i\varepsilon} dx_1 = \int_{-\infty}^{\infty} \frac{x_1 \mp i\varepsilon}{x_1^2+\varepsilon^2}\varphi(x_1)dx_1 = \int_{|x_1|<1} \frac{x_1(\varphi(x_1)-\varphi(0))}{x_1^2+\varepsilon^2}dx_1$$
$$+ \int_{|x_1|<1} \frac{x_1\varphi(0)}{x_1^2+\varepsilon^2}dx_1 + \int_{|x_1|>1} \frac{x_1}{x_1^2+\varepsilon^2}\varphi(x_1)dx_1 \mp i \int_{-\infty}^{\infty} \frac{\varepsilon\varphi(x_1)dx_1}{x_1^2+\varepsilon^2}.$$

In the right hand side of (5.8) the second integral is equal to zero because $\frac{x_1\varphi(0)}{x_1^2+\varepsilon^2}$ is an odd function, the first and the third integrals converge as $\varepsilon \to 0$ by the Lebesgue convergence theorem, and the last integral in (5.8) is a delta-like sequence (cf. Example 4.4, part b)).

Since
$$\int_{-\infty}^{\infty} \frac{\varepsilon\varphi(x_1)}{x_1^2+\varepsilon^2}dx_1 = \pi\int_{-\infty}^{\infty} \frac{1}{\varepsilon}\beta\Big(\frac{x_1}{\varepsilon}\Big)\varphi(x_1)dx_1 \to \pi\varphi(0),$$
where $\beta(x_1) = \frac{1}{\pi(x_1^2+1)}$, we get

(5.9)
$$\lim_{\varepsilon\to+0} \int_{-\infty}^{\infty} \frac{\varphi(x_1)dx_1}{x_1 \pm i\varepsilon} = \int_{|x_1|<1} \frac{\varphi(x_1)-\varphi(0)}{x_1}dx_1 + \int_{|x_1|>1} \frac{\varphi(x_1)}{x_1}dx_1 \mp i\pi\varphi(0).$$

Comparing (5.9) and (5.6) we have

(5.10)
$$\frac{1}{x_1 \pm i0} = \text{p.v.}\frac{1}{x_1} \mp i\pi\delta.$$

Note that all three distributions $\frac{1}{x_1+i0}$, $\frac{1}{x_1-i0}$ and p.v. $\frac{1}{x_1}$ are different regularization of the nonintegrable function $\frac{1}{x}$.

Example 5.3 (Distributions x_+^λ and x_-^λ). For an arbitrary $\lambda \in \mathbb{C}$, denote

(5.11)
$$x_+^\lambda = \begin{cases} x_1^\lambda = e^{\lambda \ln x_1} & \text{for } x_1 > 0, \\ 0 & \text{for } x_1 < 0, \end{cases}$$
$$x_-^\lambda = \begin{cases} 0 & \text{for } x_1 > 0, \\ |x_1|^\lambda = e^{\lambda \ln |x_1|} & \text{for } x_1 < 0. \end{cases}$$

If $\operatorname{Re}\lambda > -1$, x_+^λ and x_-^λ define regular functionals. For $\operatorname{Re}\lambda \le -1$, $x_\pm^\lambda$ are not integrable in a neighborhood of $x_1 = 0$.

We define distributions regularizing $x_\pm^\lambda$ for $\operatorname{Re}\lambda \le -1$ by using the method of analytic continuation in λ. The same method was originally used to get the analytic continuation of the Gamma-function:
$$\Gamma(\lambda) = \int_0^\infty x_1^{\lambda-1}e^{-x_1}dx_1.$$

Assuming that $\operatorname{Re}\lambda > -1$, we have for an arbitrary positive integer m:

$$(5.12)\quad \int_0^\infty x_1^\lambda \varphi(x_1)dx_1 = \int_0^1 x_1^\lambda \left(\varphi(x_1) - \sum_{k=0}^{m-1} \frac{\varphi^{(k)}(0)}{k!} x_1^k\right) dx_1$$

$$+ \int_0^1 \sum_{k=0}^{m-1} \frac{\varphi^{(k)}(0)}{k!} x_1^{\lambda+k} dx_1 + \int_1^\infty x_1^\lambda \varphi(x_1)dx_1$$

$$= \int_0^1 x_1^\lambda \left(\varphi(x_1) - \sum_{k=0}^{m-1} \frac{\varphi^{(k)}(0)}{k!} x_1^k\right) dx_1$$

$$+ \int_1^\infty x_1^\lambda \varphi(x_1)dx_1 + \sum_{k=0}^{m-1} \frac{\varphi^{(k)}(0)}{k!} \frac{1}{\lambda+k+1}.$$

The right hand side of (5.12) is meromorphic in λ for $\operatorname{Re}\lambda > -m-1$ with simple poles at the points $-1, -2, \dots, -m$, with residues $\frac{\varphi^{(k)}(0)}{k!}, k = 0, \dots, m-1$, and it defines a linear continuous functional on $\mathcal{D}$ for $\operatorname{Re}\lambda > -m-1$. This functional is an analytic continuation of $x_+^\lambda(\varphi)$ defined originally for $\operatorname{Re}\lambda > -1$. We denote this functional also by x_+^λ. Since m in (5.12) is arbitrary, the distribution x_+^λ is defined for any complex λ except negative integers. Therefore $x_+^\lambda(\varphi)$ is a meromorphic function of λ for each $\varphi \in D$ with simple poles at $-1, -2, \dots, -n, \dots$ and x_+^λ is a regularization of a nonintegrable function x_+^λ for $\operatorname{Re}\lambda \le -1$. For $\lambda \in \mathbb{C}$ such that $\operatorname{Re}\lambda > -1$ and for any integer $m > 0$, we get using integration by parts

$$(5.13)\quad \int_0^\infty x_+^\lambda \varphi(x_1)dx_1 = \frac{1}{(\lambda+m)(\lambda+m-1)\cdots(\lambda+1)} \frac{d^m}{dx_1^m} x_+^{\lambda+m}(\varphi).$$

Since the left and the right hand sides (5.13) have analytic extension to $\mathbb{C}\setminus N_-$, where N_- are negative integers, we get (by the uniqueness of analytic continuation) that (5.13) holds for all $\lambda \in \mathbb{C}\setminus N_-$. In particular, if $-m-1 < \operatorname{Re}\lambda < -m$, let $\mu = \lambda + m$. Then x_+^μ is a regular functional: $x_+^\mu(\varphi) = \int_0^\infty x^\mu \varphi(x)dx$ and we obtain from (5.13):

$$(5.14)\quad x_+^\lambda(\varphi) = \frac{1}{(\lambda+m)\cdots(\lambda+1)} \frac{d^m}{dx_1^m} x_+^\mu(\varphi)$$

$$= \frac{(-1)^m}{(\lambda+m)\cdots(\lambda+1)} \int_0^\infty x_1^{\lambda+m} \frac{d^m\varphi(x_1)}{dx_1^m} dx_1,$$

where $-m-1 < \operatorname{Re}\lambda < -m$.

5.2. Regularization in $\mathbb{R}^n$.

Example 5.4 (Distributions p.v. $\frac{1}{|x|^2-k^2}$, $\frac{1}{|x|^2-(k\pm i0)^2}$, $k > 0$). Introduce polar coordinates in $\mathbb{R}^n$: $x = r\omega$, where $r = |x|$, $|\omega| = 1$ and $d\omega$ is the surface area element of the unit sphere in $\mathbb{R}^n$.

For any $\varepsilon > 0$ we have

$$\int_{\mathbb{R}^n} \frac{\varphi(x)dx}{|x|^2 - (k \pm i\varepsilon)^2} = \int_0^\infty \int_{|\omega|=1} \frac{\varphi(r\omega)r^{n-1}dr d\omega}{r^2 - (k \pm i\varepsilon)^2}. \tag{5.15}$$

Make the change of variables

$$r - k = t, \quad r = k + t.$$

Then

$$\int_{\mathbb{R}^n} \frac{\varphi(x)dx}{|x|^2 - (k \pm i\varepsilon)^2} = \int_{-k}^\infty \frac{\psi_\varepsilon(t)dt}{t \mp i\varepsilon}, \tag{5.16}$$

where

$$\psi_\varepsilon(t) = \frac{(k+t)^{n-1}}{2k + t \pm i\varepsilon} \int_{|\omega|=1} \varphi((k+t)\omega)d\omega. \tag{5.17}$$

Analogously to Example 5.2 we get that the limit of (5.16) exists and

$$\begin{aligned} \lim_{\varepsilon \to +0} \int_{\mathbb{R}^n} \frac{\varphi(x)dx}{|x|^2 - (k \pm i\varepsilon)^2} &= \text{p.v.} \int_{-k}^\infty \frac{\psi_0(t)}{t} dt \pm i\pi\psi_0(0) \\ &= \text{p.v.} \int_{\mathbb{R}^n} \frac{\varphi(x)dx}{|x|^2 - k^2} \pm i\pi \frac{k^{n-2}}{2} \int_{|\omega|=1} \varphi(k\omega)d\omega, \end{aligned} \tag{5.18}$$

where by definition

$$\text{p.v.} \int_{\mathbb{R}^n} \frac{\varphi(x)}{|x|^2 - k^2} dx = \lim_{\delta \to +0} \int_{||x|^2 - k^2| > \delta} \frac{\varphi(x)dx}{|x|^2 - k^2}. \tag{5.19}$$

As in (5.6) we have:

$$\text{p.v.} \int_{\mathbb{R}^n} \frac{\varphi(x)dx}{|x|^2 - k^2} = \int_{-k}^k \frac{\psi_0(t) - \psi_0(0)}{t} dt + \int_k^\infty \frac{\psi_0(t)}{t} dt.$$

Since $k > 0$, we have

$$\frac{1}{|x|^2 - (k \pm i0)^2} = \frac{1}{|x|^2 - k^2 \mp i0},$$

where

$$\frac{1}{|x|^2 - k^2 \mp i0} = \lim_{\varepsilon \to 0} \frac{1}{|x|^2 - k^2 \mp i\varepsilon}.$$

Let $d\sigma$ be the surface area element on the sphere $|x|^2 = k^2$. Note that $d\sigma = k^{n-1}d\omega$.

Denote by $\delta(|x|^2 - k^2)$ the distribution that acts on $\varphi \in \mathcal{D}$ by the formula:

$$\int_{|x|=k} \frac{1}{2k} \varphi(x) d\sigma. \tag{5.20}$$

Then (5.18) can be rewritten in the form

$$\frac{1}{|x|^2 - k^2 \pm i0} = \text{p.v.}\, \frac{1}{|x|^2 - k^2} \mp i\pi\delta(|x|^2 - k^2). \tag{5.21}$$

Example 5.5 (Distributions $\frac{1}{S\pm i0}$, p.v. $\frac{1}{S}$, $\delta(S)$, where $S(x) = 0$ is a smooth surface)**.** A surface $S(x) = 0$ is called smooth if $S(x) \in C^\infty(\mathbb{R}^n)$, $S(x)$ is real-valued, and $\nabla S(x_0) = (\frac{\partial S(x_0)}{\partial x_1}, \dots, \frac{\partial S(x_0)}{\partial x_n}) \neq 0$ when $S(x_0) = 0$. We have

$$\frac{1}{S \pm i\varepsilon}(\varphi) = \int_{\mathbb{R}^n} \frac{\varphi(x)}{S(x) \pm i\varepsilon} dx.$$

Let $\{\alpha_j(x)\}$ be a partition of unity in $\mathbb{R}^n$, i.e., $\alpha_j(x) \in C_0^\infty(\mathbb{R}^n)$ and $\sum_j \alpha_j(x) \equiv 1$. We choose a partition of unity such that if $\operatorname{supp} \alpha_j \cap \{S(x) = 0\} \neq \emptyset$, then $\frac{\partial S}{\partial x_{n_j}} \neq 0$ on $\operatorname{supp} \alpha_j$ for some n_j, and the change of variables

$$S(x) = t, \quad x'_{n_j} = x'_{n_j} \tag{5.22}$$

has a C^∞ inverse on $\operatorname{supp} \alpha_j$. Here x'_{n_j} are $n-1$ remaining coordinates after x_{n_j} is deleted.

Denote by

$$x_{n_j} = h(t, x'_{n_j}), \quad x'_{n_j} = x'_{n_j} \tag{5.23}$$

the inverse map to (5.22) . Making the change of variables (5.23), we get

$$\int_{\mathbb{R}^n} \frac{\alpha_j(x)\varphi(x)dx}{S(x) \pm i\varepsilon} = \int_{\mathbb{R}^n} \frac{\alpha_j(h(t, x'_{n_j}), x'_{n_j})\varphi(h(t, x'_{n_j}), x'_{n_j})|\frac{\partial h(t,x'_{n_j})}{\partial t}|}{t \pm i\varepsilon} dt dx'_{n_j}.$$

Taking the limit as $\varepsilon \to 0$ we obtain, as in (5.18):

$$\begin{aligned} \lim_{\varepsilon\to 0} \int_{\mathbb{R}^n} \frac{\alpha_j(x)\varphi(x)dx}{S(x) \pm i\varepsilon} &= \text{p.v.} \int_{\mathbb{R}^n} \frac{\alpha_j\varphi|\frac{\partial h(t,x'_{n_j})}{\partial t}|}{t} dt dx'_{n_j} \\ &\mp i\pi \int_{\mathbb{R}^{n-1}} \alpha_j(h(0, x'_{n_j}), x'_{n_j})\varphi(h(0, x'_{n_j}), x'_{n_j})\left|\frac{\partial h(0, x'_{n_j})}{\partial t}\right| dx'_{n_j}. \end{aligned} \tag{5.24}$$

Denote by $d\sigma$ the surface area element of the surface $S(x) = 0$. On $\operatorname{supp} \alpha_j$ we have:

$$d\sigma = \sqrt{1 + |\nabla_{x'_{n_j}} h(0, x'_{n_j})|^2} dx'_{n_j}. \tag{5.25}$$

Since $S(x'_{n_j}, h(t, x'_{n_j})) \equiv t$, we have:

$$\frac{\partial S}{\partial x_{n_j}} \frac{\partial h}{\partial t} = 1, \qquad \frac{\partial S}{\partial x_k} + \frac{\partial S}{\partial x_{n_j}} \frac{\partial h}{\partial x_k} = 0 \quad \text{for} \quad k \neq n_j. \tag{5.26}$$

Therefore

$$d\sigma = \sqrt{1 + \left(\frac{\partial S}{\partial x_{n_j}}\right)^{-2} |\nabla_{x'_{n_j}} S|^2} dx'_{n_j} = |\nabla_x S| \left|\frac{\partial h}{\partial t}\right| dx'_{n_j}. \tag{5.27}$$

Substituting (5.27) in (5.24) we get in the original system of coordinates

$$\int_{\mathbb{R}^n} \frac{\alpha_j(x)\varphi(x)dx}{S(x) \pm i0} = \text{p.v.} \int_{\mathbb{R}^n} \frac{\alpha_j(x)\varphi(x)dx}{S(x)} \mp i\pi \int_{S(x)=0} \frac{\alpha_j(x)\varphi(x)d\sigma}{|\nabla S(x)|}.$$

Summing over j, we obtain

$$\frac{1}{S(x) \pm i0} = \text{p.v.} \frac{1}{S(x)} \mp i\pi\delta(S),$$

where $\delta(S)$ is the delta-function of the surface $S(x) = 0$:

$$\delta(S)(\varphi) = \int_{S(x)=0} \frac{\varphi(x)}{|\nabla S(x)|} d\sigma.$$

Note, in particular, that

$$\delta(|x|^2 - k^2) = \frac{1}{2k}\delta(|x| - k),$$

since $|\nabla_x(|x|^2 - k^2)| = 2|x| = 2k$ for $|x|^2 - k^2 = 0$ and $|\nabla_x(|x| - k)| = 1$ on $|x| = k$.

6. Supports of distributions

Let f be a distribution. We say that $f = 0$ on an open set $U \subset \mathbb{R}^n$ if $f(\varphi) = 0$ for any $\varphi \in C_0^\infty$ with support in U. Let $U_{\max}$ be the largest open set where $f = 0$. Then the complement of U is called the support of f:

(6.1) $$\operatorname{supp} f = \mathbb{R}^n \setminus U_{\max}.$$

Example 6.1.

a) $\operatorname{supp} \delta = \{0\}$ as the support of $\frac{\partial^k \delta}{\partial x^k}$ for any k.

b) $\operatorname{supp} \delta(S) = \{x : S(x) = 0\}$, where $\delta(S)$ is defined as in §5.

c) $\operatorname{supp} x_+^\lambda = [0, +\infty)$. □

For any ball B_R and any m, we introduce the norms:

(6.2) $$|[\varphi]|_{m,B_R} = \sum_{|k|=0}^{m} \max_{x \in \bar{B}_R} \left| \frac{\partial^k \varphi}{\partial x^k} \right|,$$

where $\varphi \in C_0^\infty(B_R)$, i.e., $\operatorname{supp} \varphi \subset B_R$ and $\bar{B}_R$ is the closure of B_R.

6.1. General form of a distribution with support at 0.

Proposition 6.1. *For any distribution $f \in \mathcal{D}'$ and for any ball B_R, there exist m and C depending on f and B_R such that*

(6.3) $$|f(\varphi)| \le C|[\varphi]|_{m,B_R}, \quad \forall \varphi \in C_0^\infty(B_R).$$

Proof: If (6.3) does not hold for any m and C, then there exists a sequence $\{\varphi_p\}$, $\varphi_p \in C_0^\infty(B_R)$, such that

$$|f(\varphi_p)| > p|[\varphi_p]|_{p,B_R}, \quad p = 1, 2, \ldots . \tag{6.4}$$

Let

$$\psi_p = \frac{\varphi_p}{p|[\varphi_p]|_{p,B_R}}. \tag{6.5}$$

Then $\psi_p \in C_0^\infty(B_R)$, $|[\psi_p]|_{p,B_k} = \frac{1}{p} \to 0$, and $|f(\psi_p)| \geq 1$. Note that $\psi_p \to 0$ in $\mathcal{D}$ since for any k, $|[\psi_p]|_{k,B_k} \leq |[\psi_p]|_{p,B_k} = \frac{1}{p}$ if $k < p$ and therefore $\lim_{p\to\infty} |[\psi_p]|_{k,B_R} = 0$ for $\forall k$, i.e., $\psi_p \to 0$ in $\mathcal{D}$. However, $|f(\psi_p)| \geq 1$, and this is a contradiction. □

Theorem 6.2. *Any distribution f with support at 0 has the form*

$$f = \sum_{|k|=0}^{m} c_k \frac{\partial^k \delta}{\partial x^k}. \tag{6.6}$$

Proof: Denote by $\beta(x)$ a C_0^∞ function such that $\beta(x) = 1$ for $|x| < \frac{1}{2}$, $\beta(x) = 0$ for $|x| > 1$.

For any $\varepsilon > 0$ we have that $1 - \beta(\frac{x}{\varepsilon}) = 0$ for $|x| < \frac{1}{2}\varepsilon$ and therefore $f\left((1 - \beta(\frac{x}{\varepsilon}))\varphi\right) = 0$ for any $\varphi \in \mathcal{D}$ since $\operatorname{supp} f = \{0\}$. Hence

$$f(\varphi) = f\left(\beta\Big(\frac{x}{\varepsilon}\Big)\varphi\right) \quad \text{for any} \quad \varepsilon > 0. \tag{6.7}$$

Applying Proposition 6.1 to f and the ball B_1, we get

$$|f(\varphi)| \leq C|[\varphi]|_{m,B_1} \quad \text{for some } C \text{ and } m. \tag{6.8}$$

By the Taylor formula,

$$\varphi(x) = \sum_{|k|=0}^{m} \frac{1}{k!} \frac{\partial^k \varphi(0)}{\partial x^k} x^k + R_m(x), \tag{6.9}$$

where $R_m(x) \in C^\infty$ and $|R_m(x)| \leq C|x|^{m+1}$. Therefore

$$f(\varphi) = \sum_{k=0}^{m} d_k \frac{\partial^k \varphi(0)}{\partial x^k} + f\left(\beta\Big(\frac{x}{\varepsilon}\Big) R_m(x)\right), \tag{6.10}$$

where

$$d_k = f\left(\frac{x^k}{k!}\beta\Big(\frac{x}{\varepsilon}\Big)\right). \tag{6.11}$$

Note that d_k does not depend on ε since $\operatorname{supp} f = \{0\}$. Applying the estimate (6.8) to $\beta(\frac{x}{\varepsilon})R_m(x)$, we get

$$\left|f\left(\beta\Big(\frac{x}{\varepsilon}\Big)R_m(x)\right)\right| \leq C\left|\left[\beta\Big(\frac{x}{\varepsilon}\Big)R_m(x)\right]\right|_{m,B_1}.$$

Since $|R_m(x)| \le C|x|^{m+1}$ and $\beta(\frac{x}{\varepsilon}) = 0$ for $|x| > \varepsilon$, we have $|\beta(\frac{x}{\varepsilon})R_m(x)| \le C\varepsilon^{m+1}$. Analogously, for any derivative of $\beta(\frac{x}{\varepsilon})R_m(x)$ we get

$$\left|\frac{\partial^k}{\partial x^k}\left(\beta\left(\frac{x}{\varepsilon}\right)R_m(x)\right)\right| \le C\varepsilon^{m+1-|k|}, \quad 0 < |k| \le m.$$

Thus

$$\left|\left[\beta\left(\frac{x}{\varepsilon}\right)R_m(x)\right]\right|_{m,B_1} \le C\varepsilon.$$

Since $f(\beta(\frac{x}{\varepsilon})R_m(x))$ does not depend on ε, we have that $f(\beta(\frac{x}{\varepsilon})R_m(x)) = 0$. Therefore

$$f(\varphi) = \sum_{|k|=0}^{m} d_k \frac{\partial^k \varphi(0)}{\partial x^k} = \sum_{|k|=0}^{m} (-1)^{|k|} d_k \frac{\partial^k \delta}{\partial x^k}(\varphi).$$

□

6.2. Distributions with compact supports.

Denote by $\mathcal{E}' \subset \mathcal{D}'$ the space of distributions with compact supports.

Proposition 6.3. *Any distribution f can be represented as a limit in $\mathcal{D}'$ of distributions $f_n \in \mathcal{E}'$.*

Proof: Let $\beta(x)$ be the same as in the proof of Theorem 6.2, and let $f_n = \beta(\frac{x}{n})f$. Then $\operatorname{supp} f_n \subset B_n$ and $f_n(\varphi) = f(\beta(\frac{x}{n})\varphi) \to f(\varphi)$, $\forall\varphi$, as $n \to \infty$. □

Proposition 6.1 combined with the Hahn-Banach theorem allows one to prove the following theorem:

Theorem 6.4. *Any distribution with a compact support can be represented in the form*

$$f = \sum_{|k|=0}^{m} \frac{\partial^k f_k}{\partial x^k},$$

where f_k are regular functionals corresponding to the continuous functions $f_k(x)$ with compact supports.

Proof: Let $\operatorname{supp} f \subset B_R$. By Proposition 6.1,

$$|f(\varphi)| \le C|[\varphi]|_{m_1,B_R}, \quad \forall\varphi \in C_0^\infty(B_R) \tag{6.12}$$

for some $m_1 \ge 0$. Denote by $\|\varphi\|_p$ the following norm:

$$\|\varphi\|_p^2 = \sum_{|k|=0}^{p} \int_{B_R} \left|\frac{\partial^k \varphi}{\partial x^k}\right|^2 dx. \tag{6.13}$$

Note that

$$\frac{\partial^k \varphi(x)}{\partial x^k} = \int_{-\infty}^{x_1} \cdots \int_{-\infty}^{x_n} \frac{\partial^n}{\partial y_1 \partial y_2 \cdots \partial y_n} \frac{\partial^k \varphi(y)}{\partial y^k} dy.$$

Therefore, by the Cauchy-Schwartz inequality,

$$\|[\varphi]\|^2_{m_1,B_R} \leq C_1\|\varphi\|^2_{m_1+n}, \quad \forall\varphi \in C_0^\infty(B_R), \tag{6.14}$$

i.e.,

$$|f(\varphi)| \leq C_2\|\varphi\|_{m_1+n}, \quad \forall\varphi \in C_0^\infty(B_R). \tag{6.15}$$

Denote by $\vec{L}_2(B_R)$ the space of vector-valued functions $\vec{g}(x) = \{g_k(x) \in L_2(B_R),\ k = (k_1,\dots,k_n),\ 0 \leq |k| \leq m_1+n\}$, with the norm $\|\vec{g}\|_0^2 = \sum_{|k|=0}^{m_1+n}\|g_k\|_0^2$. By the Riesz theorem (see, for example, [**R**]), any bounded linear functional Φ on $\vec{L}_2(B_R)$ has the form $\Phi(g) = \sum_{|k|=0}^{m_1+n}(g_k,h_k)$, where $h_k \in L_2(B_R)$ and (g_k,h_k) is the scalar product in $L_2(B_R)$. Let $\mathcal{L}$ be a linear (not closed) subspace of $\vec{L}_2(B_R)$ consisting of all vector-valued functions of the form $\{\frac{\partial^k\varphi}{\partial x^k}, \varphi \subset C_0^\infty(B_k), 0 \leq |k| \leq m_1+n\}$. It follows from (6.15) that $f(\varphi)$ defines a bounded linear functional Ψ on $\mathcal{L} \subset \vec{L}(B_R)$. Extending Ψ as a bounded linear functional to $\vec{L}_2(B_R)$ (the Hahn-Banach theorem; see, for example, [**R**]) we get $\{\psi_k(x),\ \psi_k(x) \in L_2(B_R), 0 \leq |k| \leq m_1+n\}$ so that $\Psi(g) = \sum_{|k|=0}^{m_1+n}(g_k,\psi_k)$. In particular,

$$f(\varphi) = \sum_{|k|=0}^{m_1+n}\left(\frac{\partial^k\varphi}{\partial x^k},\psi_k\right) = \sum_{|k|=0}^{m_1+n}\int_{B_R}\frac{\partial^k\varphi(x)}{\partial x^k}\overline{\psi_k(x)}dx. \tag{6.16}$$

Let

$$\psi_{k1}(x) = \int_{a-1}^{x_1}\cdots\int_{a_n}^{x_n}\psi_k(y)dy_1\cdots dy_n,$$

where $(a_1,\dots,a_n)$ is a point in B_R. Then $\psi_{k1}(x)$ are continuous in $\overline{B_R}$ and

$$\psi_k(x) = \frac{\partial^n\psi_{k1}(x)}{\partial x_1\cdots\partial x_n} \quad \text{in} \quad B_R.$$

Therefore we have

$$f(\varphi) = \sum_{|k|=0}^{m}\int_{B_R}\frac{\partial^k\varphi(x)}{\partial x^k}f_{k1}(x)dx, \quad \forall\varphi \in C_0^\infty(B_R), \tag{6.17}$$

where $f_{k1}(x)$ are continuous in $\overline{B_R}$, $m = m_1+2n$. Since $\operatorname{supp} f \subset B_R$, there exists $\chi(x) \in C_0^\infty(B_R)$ equal to 1 in a neighborhood of $\operatorname{supp} f$. Therefore, for any $\varphi \in C_0^\infty(\mathbb{R}^n)$, we have from (6.17):

(6.18)

$$f(\varphi) = f(\chi\varphi) = \sum_{|k|=0}^{m}f_{k1}\left(\frac{\partial^k(\chi\varphi)}{\partial x^k}\right) = \sum_{|k|=0}^{m}\frac{\partial f_k}{\partial x^k}(\varphi), \quad \forall\varphi \in C_0^\infty(\mathbb{R}^n),$$

where f_k are regular functionals corresponding to continuous functions $f_k(x)$ in $\mathbb{R}^n$ with $\operatorname{supp} f_k \subset B_R$.

Remark 6.1. Let $g \in \mathcal{D}'(B_{R'})$, where $\overline{B_R} \subset B_{R'}$ (cf. the definition of $\mathcal{D}'(\Omega)$ in §2). Then, as in (6.17), the restriction of g to B_R can be represented in the form

$$g(\varphi) = \sum_{|k|=0}^{m} \frac{\partial^k g_k}{\partial x^k}(\varphi), \quad \forall \varphi \in C_0^\infty(B_R),$$

where the g_k are regular functionals in $\mathcal{D}'(B_R)$ corresponding to the functions $g_k(x)$ continuous in $\overline{B_R}$.

7. The convolution of distributions

7.1. Convolution of $f \in \mathcal{D}'$ and $\varphi \in C_0^\infty$.

If $\varphi \in C_0^\infty(\mathbb{R}^n)$, then, obviously, $\varphi(x-y) \in C_0^\infty(\mathbb{R}^n)$ as a function of y for any fixed $x \in \mathbb{R}^n$.

Definition 7.1. We define the convolution of a distribution $f \in \mathcal{D}'$ and a function $\varphi \in \mathcal{D}$ by the formula:

$$(f * \varphi)(x) = f(\varphi(x - \cdot)), \tag{7.1}$$

where f acts on $\varphi(x-y)$ as a function of y.

Since $\varphi(x_m - y) \to \varphi(x_0 - y)$ in $\mathcal{D}$ as $x_m \to x_0$ and since f is a continuous functional, we have that $(f * \varphi)(x)$ is a continuous function of x. Also we have that

$$\begin{aligned} &\frac{\varphi(x'-y', x_n + h_n - y_n) - \varphi(x'-y', x_n - y_n)}{h_n} \\ &\quad = \frac{\partial \varphi(x'-y', x_n + \theta h_n - y_n)}{\partial x_n} \to \frac{\partial \varphi(x'-y', x_n - y_n)}{\partial x_n} \quad \text{in} \quad \mathcal{D}, \end{aligned} \tag{7.2}$$

as $h_n \to 0$ and (x', x_n) is fixed. Here $x' = (x_1, \dots, x_{n-1}), y' = (y_1, \dots, y_{n-1})$. Therefore, the limit

$$\begin{aligned} \frac{\partial (f * \varphi)}{\partial x_n} &= \lim_{\Delta x_n \to 0} \frac{(f * \varphi)(x', x_n + \Delta x_n) - (f * \varphi)(x', x_n)}{\Delta x_n} \\ &= \lim_{\Delta x_n \to 0} f\Big(\frac{\partial \varphi(x' - \cdot, x_n + \theta \Delta x_n - \cdot)}{\partial x_n}\Big) \end{aligned}$$

exists and

$$\frac{\partial (f * \varphi)}{\partial x_n} = f\Big(\frac{\partial \varphi(x - \cdot)}{\partial x_n}\Big).$$

Analogously, one can show that each partial derivative $\frac{\partial^k (f*\varphi)}{\partial x^k}$ exists and is continuous, and

$$\frac{\partial^k (f * \varphi)}{\partial x^k} = f\Big(\frac{\partial^k \varphi(x - \cdot)}{\partial x^k}\Big) = f * \frac{\partial^k \varphi}{\partial x^k}. \tag{7.3}$$

Note that

$$\frac{\partial^k \varphi(x-y)}{\partial x^k} = (-1)^k \frac{\partial^k \varphi(x-y)}{\partial y^k}.$$

Therefore, by the definition of the derivative of a distribution,

$$f\Big(\frac{\partial^k \varphi(x-y)}{\partial x^k}\Big) = (-1)^k f\Big(\frac{\partial^k \varphi(x-y)}{\partial y^k}\Big) = \frac{\partial^k f}{\partial x^k}(\varphi(x-y)) = \frac{\partial^k f}{\partial x^k} * \varphi.$$

Thus

$$\frac{\partial^k}{\partial x^k}(f * \varphi) = \frac{\partial^k f}{\partial x^k} * \varphi = f * \frac{\partial^k \varphi}{\partial x^k}, \quad \forall f \in \mathcal{D}', \quad \forall \varphi \in \mathcal{D}. \tag{7.4}$$

Proposition 7.1. *Let $f \in \mathcal{E}'$, i.e., f has a compact support,* $\operatorname{supp} f \subset U_1$, *and let* $\operatorname{supp} \varphi \subset U_2$, *where U_1 and U_2 are bounded open sets. Then $f * \varphi \in C_0^\infty(\mathbb{R}^n)$, (7.4) holds, and*

$$\operatorname{supp}(f * \varphi) \subset U_1 + U_2, \tag{7.5}$$

where $U_1 + U_2$ is the open set of all sums $x + y,\ x \in U_1,\ y \in U_2$.

Proof: It has already been proven that $f * \varphi \in C^\infty(\mathbb{R}^n)$ and (7.4) holds. Note that $\operatorname{supp} \varphi(x-y)$ in y is contained in $x - U_2$. If $x \notin U_1 + U_2$, then $x - U_2$ does not intersect U_1 and therefore $f(\varphi(x-\cdot)) = 0$ since $\operatorname{supp} f \subset U_1$. Thus $(f * \varphi)(x) \in C_0^\infty(U_1 + U_2)$ if f has a compact support. □

Proposition 7.2. *For any $\psi \in \mathcal{D}$, $\varphi \in \mathcal{D}$, and $f \in \mathcal{D}'$, the following formula holds:*

$$(f * \varphi)(\psi) = \int_{\mathbb{R}^n} (f * \varphi)(x)\psi(x)dx = f(\varphi_1 * \psi), \tag{7.6}$$

where $\varphi_1(x) = \varphi(-x)$, and $(\varphi_1 * \psi)(x) = \int_{\mathbb{R}^n} \varphi_1(x-t)\psi(t)dt = \int_{\mathbb{R}^n} \varphi(t-x)\psi(t)dt$.

Proof: We have

$$(f * \varphi)(\psi) = \int_{\mathbb{R}^n} f(\varphi(x-\cdot))\psi(x)dx. \tag{7.7}$$

Let $S_N(y) = \sum_{|k|=0}^N \varphi(\xi_k - y)\psi(\xi_k)\Delta x_k$ be a Riemann sum of the integral $(\varphi_1 * \psi)(y) = \int_{\mathbb{R}^n} \varphi(x-y)\psi(x)dx$. Since $S_N(y)$ converges to $(\varphi_1 * \psi)(y)$ uniformly in y and the same is true for each partial derivative $\frac{\partial^k}{\partial y^k} S_N(y)$, we have that $S_N(y) \to (\varphi_1 * \psi)(y)$ in $\mathcal{D}$. Therefore

$$f(S_N) = \sum_{k=0}^N f(\varphi(\xi_k - \cdot))\psi(\xi_k)\Delta x_k \to f(\varphi_1 * \psi). \tag{7.8}$$

Since $\sum_{k=0}^N f(\varphi(\xi_k - \cdot))\psi(\xi_k)\Delta x_k$ is the Riemann sum of the integral (7.7), we get (7.6) from (7.8). □

7.2. Convolution of $f \in \mathcal{D}'$ and $g \in \mathcal{E}'$.

Definition 7.2. The convolution of $f \in \mathcal{D}'$ and $g \in \mathcal{E}'$ is defined by the formula:

$$(f * g)(\psi) = f(g_1 * \psi), \quad \forall \psi \in \mathcal{D}, \tag{7.9}$$

where g_1 is the distribution obtained from g by the change of variables $x \to -x$ (cf. (3.8)):

$$g_1(\varphi) = g(\varphi(-x)). \tag{7.10}$$

Since g_1 has a compact support, we have that $g_1 * \psi = g_1(\psi(x - \cdot)) \in \mathcal{D}$ (see Proposition 7.1). It follows from (7.3) and Proposition 6.1 that

$$\left|\frac{\partial^k}{\partial x^k}(g_1 * \psi)\right| \leq C|[\psi]|_{m+|k|,B_R}$$

for some m and C.

Consider a sequence $\psi_n \to \psi$ in $\mathcal{D}$. We assume that $\operatorname{supp}\psi \subset B_R$ and $\operatorname{supp}\psi_n \subset B_R$. Then for any k,

$$\max_{x\in\mathbb{R}^n}\left|\frac{\partial^k}{\partial x^k}(g_1 * \psi_n) - \frac{\partial^k}{\partial x^k}(g_1 * \psi)\right| \leq C|[\psi_n - \psi]|_{m+|k|,B_R} \to 0$$

as $n \to \infty$. By Proposition 7.1, $\operatorname{supp}(g*\psi_n) \subset B_{R_1+R}$, where $\operatorname{supp} g_1 \subset B_{R_1}$. Therefore $g_1 * \psi_n \to g_1 * \psi$ in $\mathcal{D}$ and $f(g_1 * \psi_n) \to f(g_1 * \psi)$.

Thus $f * g$ is a linear continuous functional on $\mathcal{D}$.

It follows from Proposition 7.2 that Definition 7.2 agrees with Definition 7.1 for $g \in C_0^\infty$. □

Proposition 7.3. *Let $f \in \mathcal{D}'$ and $g \in \mathcal{E}'$. Then for any $k = (k_1, \dots, k_n)$,*

$$\frac{\partial^k}{\partial x^k}(f * g) = \frac{\partial^k f}{\partial x^k} * g = f * \frac{\partial^k g}{\partial x^k}. \tag{7.11}$$

Proof: Indeed

$$\begin{aligned}\frac{\partial^k(f*g)}{\partial x^k}(\varphi) &= (-1)^{|k|}(f*g)\Big(\frac{\partial^k\varphi}{\partial x^k}\Big) = (-1)^{|k|}f\Big(g_1 * \frac{\partial^k\varphi}{\partial x^k}\Big)\\ &= (-1)^{|k|}f\Big(\frac{\partial^k}{\partial x^k}(g_1*\varphi)\Big) = \frac{\partial^k f}{\partial x^k}(g_1*\varphi) = \Big(\frac{\partial^k f}{\partial x^k}*g\Big)(\varphi)\end{aligned}$$

and

$$\begin{aligned}\frac{\partial^k}{\partial x^k}(f*g)(\varphi) &= \frac{\partial^k f}{\partial x^k}(g_1*\varphi) = (-1)^{|k|}f\Big(\frac{\partial^k}{\partial x^k}(g_1*\varphi)\Big)\\ &= (-1)^{|k|}f\Big(\frac{\partial^k g_1}{\partial x^k}*\varphi\Big) = \Big(f*\frac{\partial^k g}{\partial x^k}\Big)(\varphi),\end{aligned}$$

where we used (7.4) and that $\frac{\partial^k g_1}{\partial x^k} = (-1)^k\frac{\partial^k g}{\partial x^k}$. □

Proposition 7.4. *For any* $f \in \mathcal{E}'$ *there exists* $f_\varepsilon \in C_0^\infty(\mathbb{R}^n)$ *such that* $f_\varepsilon \to f$ *in* $\mathcal{D}'$ *as* $\varepsilon \to 0$.

Proof: Let β_ε be a delta-like sequence as in Example 4.4, and, in addition, $\beta(x) \in C_0^\infty(B_1)$. Let $f_\varepsilon(x) = f * \beta_\varepsilon = f(\beta_\varepsilon(x - \cdot))$. By (7.6) we have:

$$f_\varepsilon(\varphi) = f(\beta_{\varepsilon 1} * \varphi),$$

where $\beta_{\varepsilon 1}(x) = \beta_\varepsilon(-x)$. For any k we have

$$\frac{\partial^k}{\partial x^k}(\beta_{\varepsilon 1} * \varphi) = \beta_{\varepsilon 1} * \frac{\partial^k \varphi}{\partial x^k} \to \frac{\partial^k \varphi}{\partial x^k}$$

uniformly (see Proposition 1.3). Since $\operatorname{supp}(\beta_{\varepsilon 1} * \varphi) \subset B_{R+1}$ for $\varepsilon < 1$, where $\operatorname{supp}\varphi \subset B_R$ (see Proposition 1.2), we have that $\beta_{\varepsilon 1} * \varphi \to \varphi$ in $\mathcal{D}$. Therefore $f_\varepsilon(\varphi) = f(\beta_{\varepsilon 1} * \varphi) \to f(\varphi)$. □

Example 7.1. For any $f \in \mathcal{D}'$ we have $f * \delta = f$. Indeed, by (7.9), $(f * \delta)(\psi) = f(\delta * \psi)$. We used that $\delta(-x) = \delta(x)$. By (7.1), $\delta * \psi = \delta(\psi(x - \cdot)) = \psi(x)$. Thus $(f * \delta)(\psi) = f(\psi)$ for any $\psi \in \mathcal{D}$. Similarly, we have $f * \frac{\partial^k \delta}{\partial x^k} = \frac{\partial^k f}{\partial x^k}$.

7.3. Direct product of distributions.

Let $f_1 \in \mathcal{D}(\mathbb{R}^{n_1})$ and $f_2 \in \mathcal{D}'(\mathbb{R}^{n_2})$ and let $x = (x', x'')$, where $x' \in \mathbb{R}^{n_1}$, $x'' \in \mathbb{R}^{n_2}$. For any $\varphi(x', x'') \in C_0^\infty(\mathbb{R}^{n_1+n_2})$ the function $\varphi_1(x'') = f_1(\varphi(\cdot, x'')) \in C_0^\infty(\mathbb{R}^{n_2})$ (cf. the proof of Proposition 7.2). If $\varphi_n(x', x'') \to \varphi(x', x'')$ in $\mathcal{D}(\mathbb{R}^{n_1+n_2})$, then $\varphi_{1n}(x'') = f_1(\varphi_n(\cdot, x'')) \to \varphi_1(x'')$ in $\mathcal{D}(\mathbb{R}^{n_2})$. Therefore $f_2(\varphi_1)$ is well defined, and we call $f_2(\varphi_1) = f_2(f_1(\varphi))$ the direct product $(f_1 \times f_2) \in \mathcal{D}'(\mathbb{R}^n)$ of f_1 and f_2:

$$(f_1 \times f_2)(\varphi) = f_2(f_1(\varphi(x', x''))), \quad \forall \varphi \in C_0^\infty(\mathbb{R}^{n_1+n_2}). \tag{7.12}$$

Analogously,

$$(f_2 \times f_1)(\varphi) = f_1(f_2(\varphi(x', x''))) \in \mathcal{D}(\mathbb{R}^{n_1+n_2}).$$

Note that $f_1 \times f_2 = f_2 \times f_1$. This is obvious in the case where $\psi(x', x'') = \sum_{j=1}^N \psi_{j1}(x')\psi_{j2}(x'')$, with $\psi_{j1}(x') \in C_0^\infty(\mathbb{R}^{n_1})$, $\psi_{j2}(x'') \in C_0^\infty(\mathbb{R}^{n_2})$, since

$$(f_1 \times f_2)(\psi) = \sum_{j=1}^N f_1(\psi_{j1}) f_2(\psi_{f2}) = (f_2 \times f_1)(\psi).$$

For any $\varphi \in C_0^\infty(\mathbb{R}^{n_1+n_2})$, there is a sequence of $C_0^\infty(\mathbb{R}^{n_1+n_2})$ functions of the form $\sum_{j=1}^N \psi_{j1}(x')\psi_{j2}(x'')$ converging to $\varphi(x', x'')$ in $\mathcal{D}(\mathbb{R}^{n_1+n_2})$. Therefore we get that $(f_1 \times f_2)(\varphi) = (f_2 \times f_1)(\varphi)$ for all $\varphi \in \mathcal{D}(\mathbb{R}^{n_1+n_2})$.

Proposition 7.5. *Let* $x = (x', x_n)$ *and let* $f \in \mathcal{D}'(\mathbb{R}^n)$. *If* $f(\frac{\partial \varphi}{\partial x_n}) = 0$ *for any* $\varphi(x', x_n) \in C_0^\infty(\mathbb{R}^n)$, *then* $f = f_1 \times 1$, *where* $f_1 \in \mathcal{D}'(\mathbb{R}^{n-1})$ *and* 1 *is the regular functional in* $\mathcal{D}'(\mathbb{R}^1)$ *corresponding to* 1.

Proof: Let $\varphi_0(x_n) \in C_0^\infty(\mathbb{R}^1)$, $\int_{-\infty}^{\infty}\varphi_0(x_n)dx_n = 1$. For an arbitrary $\varphi(x', x_n) \in C_0^\infty(\mathbb{R}^n)$ set $\varphi_1(x') = \int_{-\infty}^{\infty}\varphi(x', x_n)dx_n$. Let $\psi(x', x_n) = \varphi(x', x_n) - \varphi_1(x')\varphi_0(x_n)$. Then $\int_{-\infty}^{\infty}\psi(x', x_n)dx_n = 0$ and therefore $\psi = \frac{\partial\psi_1}{\partial x_n}$, where $\psi_1(x', x_n) = \int_{-\infty}^{x_n}\varphi(x', t)dt - \varphi_1(x')\int_{-\infty}^{x_n}\varphi_0(t)dt \in C_0^\infty(\mathbb{R}^n)$. We have

$$\varphi(x', x_n) = \frac{\partial\psi_1}{\partial x_n} + \varphi_1(x')\varphi_0(x_n).$$

Therefore

$$f(\varphi) = f(\varphi_1(x')\varphi_0(x_n)), \tag{7.13}$$

since $f(\frac{\partial\psi_1}{\partial_n}) = 0$. Denote by f_1 the following distribution in $\mathcal{D}'(\mathbb{R}^{n-1})$:

$$f_1(a(x')) = f(a(x')\varphi_0(x_n)), \quad \forall a(x') \in C_0^\infty(\mathbb{R}^{n-1}). \tag{7.14}$$

Then

$$f(\varphi) = f_1(\varphi_1(x')).$$

Since $\varphi_1(x') = \int_{-\infty}^{\infty}\varphi(x', x_n)dx_n$, we have $f_1(\varphi_1(x')) = \int_{-\infty}^{\infty} f_1(\varphi(x', x_n))dx_n$ (cf. the proof of Proposition 7.2). Therefore

$$f(\varphi) = \int_{-\infty}^{\infty} f_1(\varphi(x', x_n))dx_n, \tag{7.15}$$

i.e., $f = f_1 \times 1$. □

7.4. Partial hypoellipticity.

Let $V \times (a, b) \subset \Omega$, where Ω is a domain in $\mathbb{R}^n$, V is a domain in $\mathbb{R}^{n-1}$, $x = (x', x_n)$, $x' \in V$, $x_n \in (a, b)$. We assume that $\overline{V} \times [a, b]$ is compact and $\overline{V} \times [a, b] \subset \Omega$. Let $u \in \mathcal{D}'(\Omega), f \in C^\infty(\overline{V} \times [a, b])$. Suppose u is a distribution solution of $P(x, \frac{\partial}{\partial x})u = f$ in $V \times (a, b)$, i.e., $u(P(x, -\frac{\partial}{\partial x})\varphi) = f(\varphi)$, $\forall\varphi \in C_0^\infty(V \times (a, b))$, where $P(x, \frac{\partial}{\partial x})$ is a differential operator of the form

$$P\left(x, \frac{\partial}{\partial x}\right) = \frac{\partial^r u}{\partial x_n^r} + \sum_{k=0}^{r-1} p_k\left(x, \frac{\partial}{\partial x'}\right)\frac{\partial^k}{\partial x_n^k}. \tag{7.16}$$

Here $p_k(x, \frac{\partial}{\partial x'})$ are differential operators in $\frac{\partial}{\partial x'}$ with $C^\infty(\overline{V} \times [a, b])$ coefficients, $x' = (x_1, \dots, x_{n-1})$.

Proposition 7.6. *Let $P(x, \frac{\partial}{\partial x})$ has the form (7.16), $u \in \mathcal{D}'(\Omega)$, and $Pu = f$ in $V \times (a, b)$, where $f \in C^\infty(\overline{V} \times [a, b])$. Then u is a C^∞ function of $x_n \in (a, b)$ with values in $\mathcal{D}'(V)$.*

This property of solutions of $P(x, \frac{\partial}{\partial x})u = f$ is called the partial hypoellipticity property.

Proof: We rewrite $Pu = f$ as a system of first order differential operators in $\frac{\partial}{\partial x_n}$. Let $w_1 = u$, $w_2 = \frac{\partial u}{\partial x_n}, \dots, w_r = \frac{\partial^{r-1} u}{\partial x_n}$. Then we have

$\frac{\partial w_r}{\partial x_n} = -\sum_{k=0}^{r-1} p_k(x, \frac{\partial}{\partial x'})w_{k+1} + f$, $\frac{\partial w_k}{\partial x_n} = w_{k+1}$, $1 \leq k \leq r-1$, and the equation $Pu = f$ can be written as a system

$$\frac{\partial w}{\partial x_n} = A\left(x, \frac{\partial}{\partial x'}\right) w + \Phi, \tag{7.17}$$

where $w = (w_1, \dots, w_r)$, $A(x, \frac{\partial}{\partial x'})$ is an $r \times r$ matrix of differential operators in $\frac{\partial}{\partial x'}$, and $\Phi = (f, \dots, 0)$.

It follows from Remark 6.1 that the restriction of the distribution w to $V \times (a, b)$ can be represented in the form

$$w = \sum_{|k|=0}^{m} \frac{\partial^k}{\partial x^k} v_k, \tag{7.18}$$

where v_k are regular functionals corresponding to vector-valued functions $v_k(x)$ continuous in $\overline{V} \times [a, b]$. We rewrite (7.18) in the following form:

$$w = \sum_{j=0}^{m} \frac{\partial^j}{\partial x_n^j} h_j, \tag{7.19}$$

where h_j are linear functionals continuous in $x_n \in [a, b]$ with values in $\mathcal{D}'(V)$.

We call m the order of w. A negative order of w means that w is continuously differentiable in x_n up to the order $|m|$ with values in $\mathcal{D}'(V)$).

Note that $A(x, \frac{\partial}{\partial x'})w$ has the same order as w since $A(x, \frac{\partial}{\partial x'})$ does not contain derivatives with respect to x_n. We denote the right hand side of (7.17) by b:

$$\frac{\partial w}{\partial x_n} = b.$$

Note that the order of b is m, i.e., $b = \sum_{j=0}^{m} \frac{\partial^j b_j}{\partial x_n^j}$. Let

$$b^{(1)} = \sum_{j=0}^{m} \frac{\partial^j}{\partial x_n^j} \int_a^{x_n} b_j dy_n.$$

Then $b^{(1)}$ has order $m - 1$ and $\frac{\partial b^{(1)}}{\partial x_n} = b$ since $\frac{\partial}{\partial x_n} \int_a^{x_n} b_j dy_n = b_j$. Hence $\frac{\partial (w - b^{(1)})}{\partial x_n} = 0$ in $V \times (a, b)$. By Proposition 7.5, $w - b^{(1)} = b^{(0)}$ does not depend on x_n, i.e., the order of $b^{(0)}$ is $-\infty$. Therefore $w = b^{(1)} + b^{(0)}$ has order $m - 1$. Repeating the same argument with w of order $m - 1$, we get that the order of w is $m - 2$. After N steps we get that the order of w is $m - N$, $\forall N$. Therefore $w \in C^\infty$ in $x_n \in (a, b)$ with values in $\mathcal{D}'(V)$. □

8. Problems

1. Let $f(x)$, $x \in \mathbb{R}^1$, be a piecewise absolutely continuous function, i.e., there are finitely many points $c_1, \dots, c_n$ such that $f(x)$ is absolutely continuous on $(-\infty, c_1], [c_1, c_2], \dots, [c_n - 1, c_n], [c_n, +\infty)$ and $f(x)$ has jumps at c_k, $k = 1, \dots, n$:

$$f(c_k + 0) - f(c_k - 0) = h_k.$$

Denote by f the regular functional corresponding to $f(x)$ and by g the regular functional corresponding to $f'(x)$. Prove that

$$\frac{df}{dx} = g + \sum_{k=1}^{n} h_k \delta(x - c_k),$$

where $\frac{df}{dx}$ is the derivative of f in the distribution sense and $\delta(x-c_k)$ is the delta-function at c_k, i.e.,

$$\delta(x - c_k)(\varphi) = \varphi(c_k), \quad k = 1, \dots, n.$$

2. Find the distribution derivatives of the following regular functionals in $\mathbb{R}^1$:

$$\text{a)} \ \frac{d^2}{dx^2}|x|, \quad \text{b)} \ \frac{d^2}{dx^2}|\cos x|, \quad \text{c)} \ \frac{d^2}{dx^2}\sin|x|.$$

3. Find the limits of the following sequences of distributions in $\mathbb{R}^1$:

$$\text{a)} \ \cos nx, \quad \text{b)} \ n^{10}\sin nx, \quad \text{c)} \ \frac{\sin nx}{x}.$$

4. Prove that any trigonometric series $\sum_{n=-\infty}^{\infty} c_n e^{inx}$ such that $|c_n| \le C_1|n|^r + C_2$, converges to a distribution $f \in \mathcal{D}'$.

5. Denote by $T(x)$ a 2π-periodic function such that

$$T(x) = \frac{x(2\pi - x)}{4\pi} \quad \text{for} \quad 0 < x < 2\pi.$$

a) Using the Fourier series

$$T(x) = \frac{\pi}{6} - \frac{1}{2\pi}\sum_{n \ne 0} \frac{1}{n^2} e^{inx},$$

prove that

$$\sum_{n=-\infty}^{\infty} \delta(x - 2\pi n) = \frac{1}{2\pi} \sum_{n=-\infty}^{\infty} e^{inx}.$$

This formula is called the Poisson summation formula.

b) Find $\sum_{n=1}^{\infty} n \sin nx$.

6. Prove that

$$\frac{d}{dx_1}\ln|x_1| = \text{p.v.}\,\frac{1}{x_1}.$$

7. Consider a system of ordinary differential equations with C^∞ coefficients,

$$\frac{du}{dx} = A(x)u,$$

where $u(x) = (u_1(x),\ldots,u_m(x))$ and $A(x)$ is an $m\times m$ C^∞ matrix. Using Proposition 7.5 and the existence of a C^∞ fundamental matrix of solutions (i.e., an $m\times m$ matrix $\Phi(x)$ such that $\frac{d\Phi(x)}{dx} = A(x)\Phi(x)$ and $\Phi(0) = I$, where I is the identity matrix), prove that any distribution solution of $\frac{du}{dx} = A(x)u$ is a regular functional corresponding to a C^∞ function, i.e., any distribution solution is a classical solution.

8. Denote $\theta(x_1,x_2) = \theta(x_1)\theta(x_2)$, where $\theta(t) = 1$ for $t>0$ and $\theta(t) = 0$ for $t<0$. Prove that

$$\frac{\partial^2}{\partial x_1\partial x_2}\theta(x_1,x_2) = \delta,$$

where δ is the delta-function in $\mathbb{R}^2$.

9. A distribution E_y (y is a parameter) is called a fundamental solution to the ordinary differential equation $Lu = f$ if

$$LE_y = \delta(x-y).$$

a) Find the general form of fundamental solution to the second order ordinary differential equation

$$p_0(x)u''(x) + p_1(x)u'(x) + p_2(x)u = f,$$
$$p_k(x)\in C^\infty(\mathbb{R}^1),\ k = 0,1,2,\ p_0(x)\neq 0.$$

b) Find the fundamental solution to

$$\sum_{k=0}^{n} p_k(x)u^{(k)} = f,\quad p_k(x)\in C^\infty,$$
$$k = 0,\ldots,n,\ p_0(x)\neq 0,$$

such that $E_y = 0$ for $x<y$.

c) Find the fundamental solution to $u''+4u = f$ such that $E_y = 0$ for $x<y$.

10. Let $u(x,t)$ be a smooth function in $\mathbb{R}^2$ outside of a curve $x = x(t)$. Suppose that there exist limits

$$u_+(t) = \lim_{(x,t)\to(x(t),t)} u(x,t),\quad x > x(t),$$

and

$$u_-(t) = \lim_{(x,t)\to(x(t),t)} u(x,t), \quad x < x(t).$$

Let $u(x,t)$ be a distribution solution to a nonlinear equation

$$\frac{\partial u}{\partial t} + \frac{\partial}{\partial x} f(u) = 0,$$

i.e.,

$$\int\int_{\mathbb{R}^2} \left(-u(x,t)\frac{\partial \varphi}{\partial t} - f(u)\frac{\partial \varphi}{\partial x} \right) dxdt = 0$$

for any $\varphi(x,t) \in C_0^\infty(\mathbb{R}^2)$. We assume that $f \in C^\infty(\mathbb{R}^1)$. Prove that the following condition holds:

$$\frac{dx}{dt} = \frac{f(u_+(t)) - f(u_-(t))}{u_+(t) - u_-(t)}.$$

This condition is called the Rankine-Hugoniot condition.

11. Let $f(x_1)$ be the Cantor function on $[0,1]$, i.e., $f(x_1)$ is continuous, nondecreasing, $f(0) = 0,\ f(1) = 1$, and $f'(x_1) = 0$ a.e.

Set $f_0(x_1) = f(x_1)$ on $[0,1]$ and $f_0(x_1) = 0$ for x_1 that does not belong to $[0,1]$.

Prove that the distribution derivative of $f_0(x_1)$ is the following functional:

$$\frac{df_0}{dx}(\varphi) = \varphi(1) - \int_0^1 \varphi(x_1)df(x_1),$$

where $\int_0^1 \varphi(x_1)df(x_1)$ is the Stieltjes integral.

12. Prove that there is no distribution on $\mathbb{R}^1$ such that its restriction to $(0,+\infty)$ is a regular functional in $\mathcal{D}'(0,+\infty)$ corresponding to $f(x_1) = e^{\frac{1}{x_1^2}}$.

Chapter II

Fourier Transforms

9. Tempered distributions

Denote by S the Schwartz space of infinitely differentiable functions $\varphi(x)$ such that

$$\left|\frac{\partial^k \varphi(x)}{\partial x^k}\right| \le C_{mk}(1+|x|)^{-m} \tag{9.1}$$

for any k and any positive integer m.

Introduce the norms

$$|[\varphi]|_{m,S} = \sup_{x\in\mathbb{R}^n}(1+|x|)^m \sum_{|k|=0}^{m}\left|\frac{\partial^k \varphi(x)}{\partial x^k}\right|. \tag{9.2}$$

We shall say that $\varphi_n \to \varphi$ in S if

$$|[\varphi_n - \varphi]|_{m,S} \to 0 \quad \text{as} \quad n\to\infty \qquad \text{for all} \quad m = 1, 2, \dots. \tag{9.3}$$

If one introduces the metric $\rho(\varphi, \psi)$, where

$$\rho(\varphi,\psi) = \sum_{m=1}^{\infty} \frac{1}{2^m}\frac{|[\varphi-\psi]|_{m,S}}{1+|[\varphi-\psi]|_{m,S}}, \tag{9.4}$$

then the convergence in S is equivalent to the convergence in the metric: $\varphi_n \to \varphi$ in S iff $\rho(\varphi_n, \varphi) \to 0$. Therefore, the Schwartz space S is a metrizable space with metric (9.4).

Note that if $\varphi_n \in \mathcal{D}$ and $\varphi_n \to \varphi$ in $\mathcal{D}$, then

$$\varphi_n \to \varphi \quad \text{in } S. \tag{9.5}$$

Thus $\mathcal{D}$ is embedded in S: $\mathcal{D} \subset S$. We say that a topological (or normed) space B_1 is embedded in a topological (normed) space B_2 if B_1 can be identified with a subspace of B_2, and the convergence in B_1 implies the convergence in B_2.

Note also that $\mathcal{D}$ is dense in S. Indeed, for any $\varphi(x) \in S$ we have $\varphi_n(x) = \beta(\frac{x}{n})\varphi \in \mathcal{D}$ and $\varphi_n \to \varphi$ in S.

Here $\beta(x) \in C_0^\infty(\mathbb{R}^n)$, $\beta(x) = 1$ for $|x| < \frac{1}{2}$, and $\beta(x) = 0$ for $|x| > 1$.

Definition 9.1. Linear continuous functional on S is called a tempered distribution. The linear space of tempered distributions is denoted by S'.

Thus, if $f \in S'$, then $f(\alpha_1\varphi_1 + \alpha_2\varphi_2) = \alpha_1 f(\varphi_1) + \alpha_2 f(\varphi_2)$ for every $\varphi_1, \varphi_2 \in S$ and $\alpha_1, \alpha_2 \in \mathbb{C}$, and

$$\lim_{n\to\infty} f(\varphi_n) = f(\varphi) \quad \text{if} \quad \varphi_n \to \varphi \quad \text{in} \quad S.$$

We say that $f_n \in S'$ converges to $f \in S'$ if $f_n(\varphi) \to f(\varphi)$ for each $\varphi \in S$. Since $\varphi_n \in \mathcal{D}$ and $\varphi_n \to \varphi$ in $\mathcal{D}$ implies that $\varphi_n \to \varphi$ in S, we have that any tempered distribution is also a linear continuous functional on $\mathcal{D}$. Moreover, $f_n \to f$ in S' implies that $f_n \to f$ in $\mathcal{D}'$. Therefore S' is embedded in $\mathcal{D}'$:

$$S' \subset \mathcal{D}'. \tag{9.6}$$

The definition of the derivative of a tempered distribution is the same as of distribution in $\mathcal{D}$.

We cannot define the product of a tempered distribution and an arbitrary C^∞ function $a(x)$ since there are C^∞ functions $a(x)$ such that the product $a(x)\varphi$ does not belong to S for some $\varphi \in S$.

If $a(x) \in C^\infty$ is such that

$$\left|\frac{\partial^k a(x)}{\partial x^k}\right| \leq C_k(1 + |x|)^{n_k}, \quad \forall k, \tag{9.7}$$

then the Leibniz formula for the derivatives of products implies that $a(x)\varphi \in S$ for each $\varphi \in S$.

Therefore, for each tempered distribution f and each $a(x)$ satisfying (9.7) we can define af by the formula:

$$af(\varphi) = f(a\varphi). \tag{9.8}$$

If $\varphi_n \to \varphi$ in S, then (9.7) implies that $a\varphi_n \to a\varphi$ in S. Therefore $af(\varphi_n) = f(a\varphi_n) \to f(a\varphi) = af(\varphi)$. Thus af is a tempered destribution.

Repeating the proof of Proposition 6.1, we get the following result:

Theorem 9.1. *For any tempered distribution f, there exist an integer m and a constant C such that*

$$|f(\varphi)| \leq C\|[\varphi]\|_{m,S} \quad \forall\varphi \in S. \tag{9.9}$$

Example 9.1.

a) Let $f(x)$ be a Lebesgue measurable function such that

$$\int_{\mathbb{R}^n} |f(x)|(1+|x|)^{-N}dx < +\infty. \tag{9.10}$$

Then $f(x)$ defines a tempered distribution f by the formula

$$f(\varphi) = \int_{\mathbb{R}^n} f(x)\varphi(x)dx. \tag{9.11}$$

We have

$$|f(\varphi)| \leq \int_{\mathbb{R}^n} |f(x)|(1+|x|)^{-N}dx|[\varphi]|_{N,S}.$$

Therefore (9.11) is a linear continuous functional on S. We call such a functional the regular tempered functional corresponding to $f(x)$.

b) $f(x_1) = e^{x_1^2}$ does not define a regular functional in S' since

$$\int_{-\infty}^{\infty} e^{x_1^2}\varphi(x_1)dx_1$$

does not converge for any $\varphi \in S$. However, $f(x_1) = e^{x_1}\sin e^{x_1} = -\frac{d}{dx_1}\cos e^{x_1}$ defines a distribution in S' by the formula $f(\varphi) = \int_{-\infty}^{\infty} \cos e^{x_1}\varphi'(x_1)dx_1$.

Example 9.2. Any $f \in \mathcal{E}'$ is a tempered distribution. Indeed,

$$f(\varphi) = f\left(\beta\left(\frac{x}{R}\right)\varphi\right), \tag{9.12}$$

where $\beta(\frac{x}{R}) = 1$ for $|x| < \frac{1}{2}R$ and R is such that $\operatorname{supp} f \subset B_{\frac{1}{2}R}$.

Formula (9.12) allows us to extend the functional f to the space of all C^∞ functions since for any $\varphi \in C^\infty$, $\beta(\frac{x}{R})\varphi \in C_0^\infty$. In particular, we can consider f on S. If $\varphi_n \to \varphi$ in S, then $\beta(\frac{x}{R})\varphi_n \to \beta(\frac{x}{R})\varphi$ in $\mathcal{D}$. Therefore $f(\varphi_n) \to f(\varphi)$. □

We say that $f_n \in \mathcal{E}'$ converges to $f \in \mathcal{E}'$ if $f_n(\varphi) \to f(\varphi)$, $\forall\varphi \in \mathcal{D}$ and there exists $R > 0$ such that $\operatorname{supp} f_n \subset B_R$ for all n. Example 9.2 shows that

$$\mathcal{E}' \subset S',$$

i.e., $\mathcal{E}'$ is embedded in S'.

9.1. General form of a tempered distribution.

Analogously to Theorem 6.4 we can deduce

Theorem 9.2. *Any $f \in S'$ can be represented in the following form:*

$$f = \sum_{|k|=0}^{m_1} \frac{\partial^k}{\partial x^k} f_k,$$

where f_k are regular functionals in S' corresponding to continuous functions $f_k(x)$ satisfying the estimates

$$|f_k(x)| \leq C_k(1+|x|)^{m_2}.$$

Here m_1, m_2 are integers.

Proof: It follows from Theorem 9.1 that (9.9) holds for some m and C. Since

(9.13)
$$(1+|x|)^m \frac{\partial^k \varphi(x)}{\partial x^k} = \int_{-\infty}^{x_1} \cdots \int_{-\infty}^{x_n} \frac{\partial}{\partial y_1} \cdots \frac{\partial}{\partial y_n} \left[(1+|y|)^m \frac{\partial^k \varphi(y)}{\partial y^k} \right] dy,$$

we have

$$|[\varphi]|_{m,S} \leq C \sum_{|k|=0}^{m+n} \int_{\mathbb{R}^n} (1+|y|)^m \left| \frac{\partial^k \varphi(y)}{\partial y^k} \right| dy.$$

Using the Cauchy-Schwartz inequality we obtain

$$|[\varphi]|_{m,S}^2 \leq C \sum_{|k|=0}^{m+n} \int_{\mathbb{R}^n} (1+|y|)^{2m+n+1} \left| \frac{\partial^k \varphi(y)}{\partial y^k} \right|^2 dy. \tag{9.14}$$

Let $\vec{L}_2(\mathbb{R}^n)$ be the space of vector-valued functions

$$\vec{g}(x) = \{g_k(x) \in L_2(\mathbb{R}^n),\ 0 \leq |k| \leq m+n\},$$

and let $\mathcal{L}$ be a subspace of $\vec{L}_2(\mathbb{R}^n)$ consisting of vector-valued functions of the form

$$\vec{\psi}(x) = \left\{ (1+|y|)^{m+\frac{n+1}{2}} \frac{\partial^k \varphi(y)}{\partial y^k},\ \varphi(y) \in S,\ 0 \leq |k| \leq m+n \right\}.$$

It follows from (9.9) and (9.14) that $f(\varphi)$ defines a bounded functional on $\mathcal{L}$. Therefore, by the Hahn-Banach and Riesz theorems (cf. the proof of Theorem 6.4), we have that

$$f(\varphi) = \sum_{|k|=0}^{m+n} \int_{\mathbb{R}^n} (1+|y|)^{m+\frac{n+1}{2}} \frac{\partial^k \varphi(y)}{\partial y^k} \overline{\psi_k(y)} dy \tag{9.15}$$

for some $\psi_k(y) \in L_2(\mathbb{R}^n)$ and each $\varphi \in S$.

Define

$$f_k(x) = \int_0^{x_1} \cdots \int_0^{x_n} (1+|y|)^{m+\frac{n+1}{2}} \overline{\psi_k(y)} dy. \tag{9.16}$$

Then the $f_k(x)$ are continuous in $\mathbb{R}^n$. Applying the Cauchy-Schwartz inequality, we get:

(9.17)

$$|f_k(x)|^2 \leq \left|\int_0^{x_1} \cdots \int_0^{x_n} (1+|y|)^{2m+n+1} dy\right| \left|\int_0^{x_1} \cdots \int_0^{x_n} |\psi_k(y)|^2 dy\right|$$
$$\leq C(1+|x|)^{2m+2n+1}.$$

Therefore

$$f(\varphi) = \sum_{|k|=0}^{m+n} \int_{\mathbb{R}^n} f_k(x)(-1)^n \frac{\partial^n}{\partial x_1 \cdots \partial x_n} \frac{\partial^k \varphi(x)}{\partial x^k} dx. \tag{9.18}$$

10. Fourier transforms of tempered distributions

Let $f \in L_1(\mathbb{R}^n)$. The Fourier transform of $f(x)$ is defined by

$$\tilde{f}(\xi) = \int_{\mathbb{R}^n} f(x) e^{-ix\cdot\xi} dx, \tag{10.1}$$

where $x \cdot \xi = x_1\xi_1 + \cdots + x_n\xi_n$.

We denote the operator of the Fourier transform by F. Note that $\tilde{f}(\xi)$ is continuous in $\mathbb{R}^n$ since

$$\tilde{f}(\xi^{(m)}) = \int_{\mathbb{R}^n} f(x) e^{-ix\cdot\xi^{(m)}} dx \to \int_{\mathbb{R}^n} f(x) e^{-ix\cdot\xi} dx = \tilde{f}(\xi) \tag{10.2}$$

by the Lebesgue convergence theorem. Note also that $\tilde{f}(\xi)$ is bounded: $|\tilde{f}(\xi)| \leq \int_{\mathbb{R}^n} |f(x)|dx$.

Theorem 10.1 (Fourier inversion formula). *Let $f(x) \in L_1(\mathbb{R}^n)$, $\tilde{f}(\xi) \in L_1(\mathbb{R}^n)$ and let $f(x)$ be continuous and bounded. Then*

$$f(x) = \frac{1}{(2\pi)^n} \int_{\mathbb{R}^n} \tilde{f}(\xi) e^{ix\cdot\xi} d\xi. \tag{10.3}$$

Proof: Let $\beta(x)$ be a continuous function with the following properties: $\beta(0) = 1$, $|\beta(x)| \leq C$, $\beta(x) \in L_1(\mathbb{R}^n), \tilde{\beta}(\xi) \in L_1(\mathbb{R}^n)$, where $\tilde{\beta}(\xi)$ is the Fourier transform of $\beta(x)$ and

$$\int_{\mathbb{R}^n} \tilde{\beta}(\xi) d\xi = (2\pi)^n. \tag{10.4}$$

For example, we can take $\beta(x) = \prod_{j=1}^n \beta_j(x_j)$, where $\beta_j(x_j) = e^{-|x_j|}$. Then

$$F(e^{-|x_j|}) = \int_{-\infty}^0 e^{x_j - ix_j\xi_j} dx_j + \int_0^\infty e^{-x_j - ix_j\xi_j} dx_j$$
$$= \frac{1}{1-i\xi_j} + \frac{1}{1+i\xi_j} = \frac{2}{1+\xi_j^2}.$$

Note that $\int_{-\infty}^{\infty} \frac{d\xi_j}{1+\xi_j^2} = \pi$. Therefore, $\beta(x) = \prod_{j=1}^n e^{-|x_j|}$ has the required properties.

Consider the integral

$$(10.5)\quad f_\varepsilon(x) = \int_{\mathbb{R}^n} \beta(\varepsilon\xi)\tilde{f}(\xi)e^{ix\cdot\xi}d\xi = \int_{\mathbb{R}^n} \beta(\varepsilon\xi)\left(\int_{\mathbb{R}^n} f(y)e^{i(x-y)\cdot\xi}dy\right)d\xi.$$

Since

$$\int_{\mathbb{R}^n}\int_{\mathbb{R}^n} |\beta(\varepsilon\xi)||f(y)|dyd\xi < +\infty,$$

we can change the order of integration (the Fubini theorem). Therefore

$$f_\varepsilon(x) = \int_{\mathbb{R}^n}\left(\int_{\mathbb{R}^n} \beta(\varepsilon\xi)e^{-i(y-x)\cdot\xi}d\xi\right)f(y)dy.$$

Changing the variables $\varepsilon\xi = \eta$, we get

$$f_\varepsilon(x) = \int_{\mathbb{R}^n} \frac{1}{\varepsilon^n}\tilde{\beta}\left(\frac{y-x}{\varepsilon}\right)f(y)dy.$$

Changing the variables $\frac{y-x}{\varepsilon} = t$ again, we obtain

$$(10.6)\quad f_\varepsilon(x) = \int_{\mathbb{R}^n} \tilde{\beta}(t)f(x+\varepsilon t)dt.$$

Since $\tilde{\beta}(t) \in L_1(\mathbb{R}^n)$ and $f(x)$ is continuous and bounded, we get, by the Lebesgue convergence theorem, that

$$\lim_{\varepsilon\to 0} f_\varepsilon(x) = \int_{\mathbb{R}^n} \tilde{\beta}(t)f(x)dt = (2\pi)^n f(x).$$

Also from (10.5) we have, applying the Lebesgue convergence theorem again, that

$$\lim_{\varepsilon\to 0} f_\varepsilon(x) = \int_{\mathbb{R}^n} \tilde{f}(\xi)e^{ix\cdot\xi}d\xi.$$

Therefore (10.3) holds. □

We denote the operator of the inverse Fourier transform by F^{-1}.

10.1. Fourier transforms of functions in S.

Lemma 10.2. *If $\varphi \in S$, then $\tilde{\varphi} = F(\varphi) \in S$. Moreover, F is a continuous one-to-one map of S onto S. Thus, if $\varphi_n \to \varphi$ in S, then $F(\varphi_n) \to F(\varphi)$ in S and $F(S) = S$, $F^{-1}(S) = S$.*

Proof: Using integration by parts we obtain

$$(10.7)\quad F\left(\frac{\partial^k\varphi}{\partial x^k}\right) = \int_{\mathbb{R}^n} \frac{\partial^k\varphi}{\partial x^k}e^{-ix\cdot\xi}dx = (i\xi)^k\tilde{\varphi}(\xi), \qquad \forall k = (k_1,\dots,k_n).$$

Also

$$(10.8)\quad F\Big(x^p\frac{\partial^k\varphi}{\partial x^k}\Big) = i^p\frac{\partial^p}{\partial\xi^p}\int_{\mathbb{R}^n}\frac{\partial^k\varphi(x)}{\partial x^k}e^{-ix\cdot\xi}dx = i^p\frac{\partial^p}{\partial\xi^p}[(i\xi)^k\tilde{\varphi}(\xi)],\quad \forall p=(p_1,\dots,p_n).$$

Therefore

$$(10.9)\quad \Big|\frac{\partial^p}{\partial\xi^p}(i^k\xi^k\tilde{\varphi}(\xi))\Big| \le \int_{\mathbb{R}^n}|x|^{|p|}\Big|\frac{\partial^k\varphi(x)}{\partial x^k}\Big|dx \le \int_{\mathbb{R}^n}\frac{|x|^{|p|}|[\varphi]|_{n+1+m,S}}{(1+|x|)^{n+1+|p|}}dx \le C_m|[\varphi]|_{n+1+m,S},$$

where $m=\max(k,p)$.

The estimate (10.9) shows that $\tilde{\varphi}(\xi)\in S$ if $\varphi\in S$, i.e., $F(S)\subset S$ and

$$(10.10)\qquad |[\tilde{\varphi}]|_{m,S}\le C_m|[\varphi]|_{m+n+1,S},\quad \forall m,$$

so if $\varphi_n\to\varphi$ in S, then $\tilde{\varphi}_n\to\tilde{\varphi}$ in S. Applying the same arguments to F^{-1}, we get that $F^{-1}(S)\subset S$ and, therefore, $F(S)=S$ and F is a homeomorphism of S. □

10.2. Fourier transform of tempered distributions.

For absolutely integrable $f(x)$, $f(x)\in L_1$, we get, using the Fubini theorem:

$$(10.11)\quad \int_{\mathbb{R}^n}f(x)\tilde{\varphi}(x)dx = \int_{\mathbb{R}^n}f(x)\Big(\int_{\mathbb{R}^n}\varphi(\xi)e^{-ix\cdot\xi}d\xi\Big)dx = \int_{\mathbb{R}^n}\Big(\int_{\mathbb{R}^n}f(x)e^{-ix\cdot\xi}dx\Big)\varphi(\xi)d\xi = \int_{\mathbb{R}^n}\tilde{f}(\xi)\varphi(\xi)d\xi$$

for any $\varphi\in S$. Thus we can rewrite (10.11) in the form

$$(10.12)\qquad \tilde{f}(\varphi)=f(\tilde{\varphi}),\quad \forall\varphi\in S.$$

Definition 10.1. For any $f\in S'$ the Fourier transform of f is the functional $\tilde{f}\in S'$ defined by formula (10.12) for all $\varphi\in S$.

We denote the Fourier transform of a distribution f by $\tilde{f}$ or $F(f)$. Lemma 10.2 implies that $\tilde{f}$ is a continuous functional since if $\varphi_n\to\varphi$ in S, then $\tilde{\varphi}_n\to\tilde{\varphi}$ in S and therefore $f(\tilde{\varphi}_n)\to f(\tilde{\varphi})$.

Example 10.1.

a) $F(\delta)=1$ since

$$\tilde{\delta}(\varphi)=\delta(\tilde{\varphi})=\tilde{\varphi}(0)=\int_{\mathbb{R}^n}\varphi(x)dx=1(\varphi).$$

b) $F(\frac{\partial^k \delta}{\partial x^k}) = (i\xi)^k$. Indeed,

$$F\left(\frac{\partial^k \delta}{\partial x^k}\right)(\varphi) = \frac{\partial^k \delta}{\partial x^k}(\tilde{\varphi}) = (-1)^k \frac{\partial^k \tilde{\varphi}(0)}{\partial x^k}$$
$$= (-1)^k \int_{\mathbb{R}^n} (-i\xi)^k \varphi(\xi) d\xi = \int_{\mathbb{R}^n} (i\xi)^k \varphi(\xi) d\xi = (i\xi)^k(\varphi). \qquad \square$$

Note some properties of the Fourier transform of distributions.

Proposition 10.3. *If $f_n \to f$ in S', then $\tilde{f}_n \to \tilde{f}$ in S'.*

Proof: We have $\tilde{f}(\varphi) = f(\tilde{\varphi})$. If $f_m \to f$ in S', then $\tilde{f}_m(\varphi) = f_m(\tilde{\varphi}) \to f(\tilde{\varphi}) = \tilde{f}(\varphi)$ as $m \to \infty$.

Proposition 10.4. *Operator F is a one-to-one map of S' onto S' : $F(S') = S'$.*

Proof: It follows from (10.12) that $\tilde{f} = 0$ implies $f = 0$, i.e., F is a one-to-one map. Take any $g \in S'$. Then, analogously to (10.12), we can define $\Phi(g) \in S'$ by the equality

$$\Phi(g)(\tilde{\varphi}) = g(\varphi). \tag{10.13}$$

Similarly to Proposition 10.3, the map $\Phi(g)$ is a continuous map of S' to S'.

It follows from (10.12) and (10.13) that

$$F\Phi(g)(\varphi) = \Phi(g)(\tilde{\varphi}) = g(\varphi), \quad \forall \varphi \in S.$$

Therefore $F\Phi(g) = g$, $\forall g \in S'$, i.e., F maps S' onto S', and $\Phi = F^{-1}$.

Proposition 10.5. *For any $k = (k_1, \dots, k_n)$ and $f \in S'$,*

$$F\left(\frac{\partial^k f}{\partial x^k}\right) = (i\xi)^k \tilde{f}. \tag{10.14}$$

Proof: We have for any $\varphi \in S$,

$$F\left(\frac{\partial^k f}{\partial x^k}\right)(\varphi) = \frac{\partial^k f}{\partial x^k}(\tilde{\varphi}) = (-1)^k f\left(\frac{\partial^k \tilde{\varphi}}{\partial x^k}\right).$$

It follows from (10.8) that

$$\frac{\partial^k \tilde{\varphi}}{\partial x^k} = \int_{\mathbb{R}^n} (-i\xi)^k \varphi(\xi) e^{-ix\cdot\xi} d\xi = F\Big((-i\xi)^k \varphi(\xi)\Big).$$

Therefore

$$F\left(\frac{\partial^k f}{\partial x^k}\right)(\varphi) = (-1)^k f\left(\frac{\partial^k \tilde{\varphi}}{\partial x^k}\right)$$
$$= f\Big(F((i\xi)^k \varphi(\xi))\Big) = \tilde{f}\Big((i\xi)^k \varphi\Big) = (i\xi)^k \tilde{f}(\varphi).$$

Example 10.2. $F(\theta) = -i(\xi_1 - i0)^{-1}$, where $\theta(x_1) = 1$ for $x_1 > 0$, and $\theta(x_1) = 0$ for $x_1 < 0$. Indeed, consider the following absolutely convergent integral:

$$\int_0^\infty e^{-ix_1(\xi_1 - i\tau)} dx_1 = \int_0^\infty e^{-x_1\tau - ix_1\xi_1} dx_1 = F\left(\theta(x_1)e^{-x_1\tau}\right),$$

where $\tau > 0$. Computing this integral, we get

$$F(\theta(x_1)e^{-x_1\tau}) = \frac{1}{i(\xi_1 - i\tau)} = \frac{-i}{\xi_1 - i\tau}. \tag{10.15}$$

As $\tau \to +0$, $\theta(x_1)e^{-x_1\tau}$ converges to $\theta(x_1)$ in S' by the Lebesgue convergence theorem. Therefore, by Proposition 10.3, $F(\theta e^{-x_1\tau}) \to F(\theta)$ as $\tau \to +0$. The right hand side of (10.15) converges in S' to the distribution $(-i)(\xi_1 - i0)^{-1}$ (see Example 5.2). Thus $F(\theta) = -i(\xi_1 - i0)^{-1} = (-i)\,\text{p.v.}\frac{1}{\xi_1} + \pi\delta$. Analogously $F(1-\theta) = i(\xi_1 + i0)^{-1}$.

Set $\operatorname{sgn} x_1 = 2\theta(x_1) - 1$, i.e., $\operatorname{sgn} x_1 = 1$ if $x_1 > 0$ and $\operatorname{sgn} x_1 = -1$ if $x_1 < 0$. Then we have $F(\operatorname{sgn} x_1) = 2F(\theta) - F(1) = (-2i)\,\text{p.v.}\frac{1}{\xi_1}$, since $F(1) = 2\pi\delta$.

10.3. Generalization of Liouville's theorem.

Example 10.3. A well-known Liouville's theorem of complex analysis states that if $u(x)$ is a harmonic function, i.e., $\Delta u = 0$, and if $|u(x)| \le C(1+|x|)^m$, then $u(x)$ is a polynomial of degree $\le m$. We shall prove

Theorem 10.6. *Let $P(\xi)$ be a polynomial such that $P(0) = 0$ and $P(i\xi) \neq 0$ if $\xi \neq 0$. Let $u \in S'$ be a tempered distribution solution of the partial differential equation $P(\frac{\partial}{\partial x})u = 0$. Then $u(x)$ is a polynomial.*

Proof: Since $u \in S'$, the Fourier transform $\tilde{u}$ of u is defined. By Proposition 10.5 $\tilde{u}$ satisfies the equation $P(i\xi)\tilde{u} = 0$. Since $P(i\xi) \neq 0$ for $\xi \neq 0$, the function $\frac{\psi(\xi)}{P(i\xi)} \in C_0^\infty(\mathbb{R}^n)$ if $\psi(\xi)$ is any C_0^∞ function equal to zero in a neighborhood of $\xi = 0$. Multiplying $P(i\xi)\tilde{u} = 0$ by $\frac{\psi(\xi)}{P(i\xi)}$, we get $\psi(\xi)\tilde{u} = 0$. Take any $\tilde{\varphi} \in C_0^\infty$ such that $\tilde{\varphi}(\xi) = 0$ for $|\xi| < \delta$ for some $\delta > 0$. One can find $\psi(\xi) \in C_0^\infty$ such that $\psi(\xi) = 0$ in a neighborhood of 0 and $\psi = 1$ in a neighborhood of $\operatorname{supp}\tilde{\varphi}$. Thus $\psi\tilde{\varphi} = \tilde{\varphi}$. Therefore $\tilde{u}(\tilde{\varphi}) = \tilde{u}(\psi\tilde{\varphi}) = \psi\tilde{u}(\tilde{\varphi}) = 0$. Hence $\operatorname{supp}\tilde{u} = \{0\}$. Then by Theorem 6.2,

$$\tilde{u} = \sum_{|k| \le m} c_k \frac{\partial^k}{\partial \xi^k}\delta.$$

Taking the inverse Fourier transform, we get

$$u = \sum_{|k| \le m} c_k \frac{1}{(2\pi)^n}(-ix)^k,$$

i.e., u is a polynomial.

11. Fourier transforms of distributions with compact supports

Proposition 11.1. *Let $f \in \mathcal{E}'$, i.e., f has a compact support. Then the Fourier transform $\tilde{f} \in S'$ of f has the form*

$$\tilde{f} = f(e^{-ix\cdot\xi}). \tag{11.1}$$

Moreover, $\tilde{f}$ is a regular functional in S' that can be extended to the complex space $\mathbb{C}^n$ as an entire analytic function of $\xi + i\tau$ satisfying the estimates

$$|\tilde{f}(\xi + i\tau)| \leq C(1 + |\xi| + |\tau|)^m e^{R|\tau|}, \tag{11.2}$$

where R is such that $\operatorname{supp} f \subset B_R$.

Proof: As was explained in Example 9.2, $f(e^{-ix\cdot(\xi+i\tau)})$ makes sense, since f has a compact support and $e^{-ix\cdot(\xi+i\tau)}$ is a C^∞ function of x for any $\xi + i\tau \in \mathbb{C}^n$. In fact, $f(e^{-ix\cdot(\xi+i\tau)}) = f\left(e^{-ix\cdot(\xi+i\tau)}\psi(x)\right)$, where $\psi(x) \in C_0^\infty$, $\operatorname{supp}\psi \subset B_R$, and $\psi \equiv 1$ in a neighborhood of $\operatorname{supp} f$. Let $z_n = \xi_n + i\tau_n$, $z' = \xi' + i\tau' = (\xi_1 + i\tau_1, \dots, \xi_{n-1} + i\tau_{n-1})$. We have that

$$\frac{(e^{-ix'\cdot(\xi'+i\tau')-ix_n(z_n+\Delta z_n)} - e^{-ix'\cdot(\xi'+i\tau')-ix_n z_n})\psi(x)}{\Delta z_n}$$

converges to $-ix_n e^{-ix\cdot(\xi+i\tau)}\psi(x)$ in $\mathcal{D}$ as a function of x if the $\xi + i\tau$ are fixed and $\Delta z_n \to 0$. Since f is a continuous functional, we get that $f(\psi(x)e^{-ix\cdot(\xi+i\tau)})$ is an analytic function of $z_n = \xi_n + i\tau_n$ for any $\xi + i\tau \in \mathbb{C}^n$. Analogously, one can show that $f(\psi e^{-ix\cdot(\xi+i\tau)})$ is an analytic function of z_k, $k = 1, \dots, n$, and

$$\frac{\partial}{\partial z_k} f\left(\psi(x)e^{-ix\cdot(\xi+i\tau)}\right) = f\left(\psi(x)(-ix_k)e^{-ix\cdot(\xi+i\tau)}\right), \quad k = 1, \dots, n.$$

Thus $f\left(\psi e^{-ix\cdot(\xi+i\tau)}\right)$ is an entire function of $\xi + i\tau$.

Applying estimate (6.3), we get

$$|f(\psi e^{-ix\cdot(\xi+i\tau)})| \leq C|[\psi e^{-ix\cdot(\xi+i\tau)}]|_{m,B_R} \leq C_1(1 + |\xi| + |\tau|)^m e^{R|\tau|}. \tag{11.3}$$

It remains to prove that $\tilde{f} = f(\psi e^{-ix\cdot\xi})$.

Denote by $S_N(x) = \sum_k e^{-ix\cdot\xi_k}\varphi(\xi_k)\Delta\xi_k$ the Riemann sum of the integral $\tilde{\varphi}(x) = \int_{\mathbb{R}^n} e^{-ix\cdot\xi}\varphi(\xi)d\xi$. Since $\psi(x)S_N(x) \to \psi(x)\tilde{\varphi}(x)$ in $\mathcal{D}$, we get

$$\sum_k f(\psi(x)e^{-ix\cdot\xi_k})\varphi(\xi_k)\Delta\xi_k = f(\psi(x)S_N(x)) \tag{11.4}$$
$$\to f(\psi(x)\tilde{\varphi}(x)) = \psi f(\tilde{\varphi}) = f(\tilde{\varphi}).$$

The left hand side of (11.4) is the Riemann sum of

$$\int_{\mathbb{R}^n} f(\psi(x)e^{-ix\cdot\xi})\varphi(\xi)d\xi.$$

Therefore

$$\int_{\mathbb{R}^n} f(\psi(x)e^{-ix\cdot\xi})\varphi(\xi)d\xi = f(\tilde{\varphi}) = \tilde{f}(\varphi), \quad \forall\varphi \in S.$$

Thus $\tilde{f}$ is the regular functional corresponding to $f(\psi(x)e^{-ix\cdot\xi})$.

Example 11.1. Let $\delta(|x|^2 - k^2)$ be the same distribution as in Example 5.4. We compute $F(\delta(|x|^2 - k^2))$ for $n = 3$. We have, by (11.1) and (5.20),

$$F(\delta(|x|^2 - k^2)) = \frac{1}{2k}\int_{|x|=k} e^{-ix\cdot\xi}d\sigma.$$

Introducing spherical coordinates (r, θ, φ), we get $d\sigma = k^2 \sin\theta\, d\theta d\varphi$, $x \cdot \xi = |x||\xi| \cos\theta$. Thus

$$F(\delta(|x|^2 - k^2)) = \frac{k}{2}\int_0^{\pi}\int_0^{2\pi} e^{-ik|\xi|\cos\theta} \sin\theta\, d\theta d\varphi.$$

Integrating with respect to φ and changing the variables $t = \cos\theta$, we get

(11.5)

$$F(\delta(|x|^2 - k^2)) = \frac{k}{2}\cdot 2\pi\int_{-1}^{1} e^{-ik|\xi|t}dt = \frac{\pi}{-i|\xi|}e^{-ik|\xi|t}|_{-1}^{1} = 2\pi\frac{\sin k|\xi|}{|\xi|}.$$

Example 11.2. Let f be the regular functional in S' corresponding to $f(x)$ such that $\int_{\mathbb{R}^n} |f(x)|(1 + |x|)^{-N}dx < \infty$.

a) Let $\tilde{f}_m(\xi)$ be the following integral:

$$\tilde{f}_m(\xi) = \int_{|x|\le m} f(x)e^{-ix\cdot\xi}dx. \tag{11.6}$$

Then the sequence of regular functionals $\tilde{f}_m$ corresponding to $\tilde{f}_m(\xi)$ converges in S' to $\tilde{f} = F(f)$. Indeed, let $f_m(x) = f(x)$ for $|x| \le m$, $f_m(x) = 0$ for $|x| > m$ and let f_m be the regular functionals corresponding to $f_m(x)$.

Since $f_m(x) \to f(x)$ as $m \to \infty$ and $|f_m(x)\varphi(x)| \le |f(x)||\varphi(x)|$ for all m and any $\varphi \in S$, the Lebesgue convergence theorem implies that $f_m \to f$ in S'. Then $\tilde{f}_m \to \tilde{f}$ in S'.

b) Let $\beta(x)$ be the same as in (6.7), and let $f_\varepsilon(x) = \beta(\varepsilon x)f(x)$. Then, using the Lebesgue convergence theorem again, we have $f_\varepsilon \to f$ in S' as $\varepsilon \to 0$. Therefore $\tilde{f}_\varepsilon(\xi) = \int_{\mathbb{R}^n} f_\varepsilon(x)e^{-ix\cdot\xi}dx$ converges to $\tilde{f}$ in S'.

c) Set $g(x) = f(x)(1 + |x|^2)^{-k}$, where $k > 0$ is an integer and $2k \ge N$. Then $g(x) \in L_1(\mathbb{R}^n)$ and $f(x) = (1 + |x|^2)^k g(x)$. Taking the Fourier transform we get $\tilde{f} = (1 - \Delta)^k\tilde{g}$, where $\tilde{g}$ is the regular functional corresponding to $\tilde{g}(\xi) = \int_{\mathbb{R}^n} g(x)e^{-ix\cdot\xi}dx$ and $\Delta = \sum_{k=1}^{n}\frac{\partial^2}{\partial\xi_k^2}$ is the Laplacian.

Example 11.3. For any $\lambda \in \mathbb{C} \setminus N_-$, the following formula holds:

$$F(x_+^\lambda) = \Gamma(\lambda+1)e^{-i(\lambda+1)\frac{\pi}{2}}(\xi_1 - i0)^{-\lambda-1}, \tag{11.7}$$

where x_+^λ and $(\xi_1 - i0)^{-\lambda-1}$ are defined in Examples 5.3 and 4.3.

To prove (11.7), consider the following integral for $\operatorname{Re}\lambda > -1$:

$$\int_0^\infty x_1^\lambda e^{-ix_1(\xi_1 - i\tau)}dx_1 = F(x_+^\lambda e^{-x_1\tau}). \tag{11.8}$$

The left hand side of (11.8) is an analytic function of $z = \xi_1 - i\tau$ in the lower half-plane $\operatorname{Im} z = -\tau < 0$. Consider this integral for $\xi_1 = 0,\ z = -i\tau$. Then we get, after changing the variables $y_1 = x_1\tau$:

$$\int_0^\infty x_1^\lambda e^{-x_1\tau}dx_1 = \frac{1}{\tau^{\lambda+1}}\int_0^\infty y_1^\lambda e^{-y_1}dy_1 = \frac{\Gamma(\lambda+1)}{\tau^{\lambda+1}}. \tag{11.9}$$

Since the left and the right hand sides of (11.9) are analytic in $w = \tau + i\xi_1$ for $\tau > 0$, we have that

$$\begin{aligned} F(x_+^\lambda e^{-x_1\tau}) &= \int_0^\infty x_1^\lambda e^{-x_1(\tau+i\xi_1)}dx_1 \\ &= \frac{\Gamma(\lambda+1)}{(\tau+i\xi_1)^{\lambda+1}} = \Gamma(\lambda+1)e^{-(\lambda+1)\ln(\tau+i\xi_1)}, \end{aligned}$$

where $\ln(\tau + i\xi_1) = \ln|\tau + i\xi_1| + i\arg(\tau + i\xi_1)$ and $-\frac{\pi}{2} < \arg(\tau + i\xi_1) < \frac{\pi}{2}$.

We have $\ln(\tau+i\xi_1) = \ln i(\xi_1 - i\tau) = \ln(\xi_1 - i\tau) + i\frac{\pi}{2}$, where $\ln(\xi_1 - i\tau) = \ln|\xi_1 - i\tau| + i\arg(\xi_1 - i\tau)$ and $-\pi < \arg(\xi_1 - i\tau) < 0$. Thus

$$F(x_+^\lambda e^{-x_1\tau}) = \frac{\Gamma(\lambda+1)e^{-i(\lambda+1)\frac{\pi}{2}}}{(\xi_1 - i\tau)^{\lambda+1}}. \tag{11.10}$$

As $\tau \to 0$, $x_+^\lambda e^{-x_1\tau} \to x_+^\lambda$ in S' by the Lebesgue convergence theorem. Therefore, by Proposition 10.3, $F(x_+^\lambda e^{-x_1\tau}) \to F(x_+^\lambda)$ in S'. The right hand side of (11.10) converges to $\Gamma(\lambda+1)e^{-i(\lambda+1)\frac{\pi}{2}}(\xi_1 - i0)^{-\lambda-1}$ by the definition of the distribution $(\xi_1 - i0)^{-\lambda-1}$ (see Example 4.3).

Therefore, (11.7) holds for $\operatorname{Re}\lambda > -1$. Since the right and the left hand sides of (11.10) have analytic continuation to $\mathbb{C} \setminus N_-$, where $N_- = \{-1, -2, -3, \dots\}$, we see that (11.7) holds for any $\lambda \in \mathbb{C} \setminus N_-$.

Example 11.4. Denote $(\ln x_1)_+ = \ln x_1$ for $x_1 > 0$ and $(\ln x_1)_+ = 0$ for $x_1 < 0$. Then

$$F((\ln x_1)_+) = \frac{c_0}{\xi_1 - i0} + \frac{i\ln(\xi_1 - i0)}{\xi_1 - i0}, \tag{11.11}$$

where $c_0 = \frac{d}{d\lambda}(\Gamma(\lambda+1)e^{-i(\lambda+1)\frac{\pi}{2}})\big|_{\lambda=0} = -i\Gamma'(1) - \frac{\pi}{2}$.

To prove (11.11) we differentiate (11.10) with respect to λ. We get

$$F(x_+^\lambda(\ln x_1)_+e^{-x_1\tau}) = \frac{\frac{d}{d\lambda}(\Gamma(\lambda+1)e^{-i(\lambda+1)\frac{\pi}{2}})}{(\xi_1 - i\tau)^{\lambda+1}} - \frac{\Gamma(\lambda+1)e^{-i(\lambda+1)\frac{\pi}{2}}\ln(\xi_1 - i\tau)}{(\xi_1 - i\tau)^{\lambda+1}}. \tag{11.12}$$

Setting $\lambda = 0$ and passing to the limit as $\tau \to +0$, we get (11.11).

12. Fourier transforms of convolutions

Proposition 12.1. *If $f \in S'$ and $g \in \mathcal{E}'$, then $f * g \in S'$ and*

$$F(f * g) = \tilde{g}(\xi)\tilde{f}. \tag{12.1}$$

Proof: Consider $F^{-1}(g_1 * \tilde{\varphi})$, where $\tilde{\varphi} \in C_0^\infty(\mathbb{R}^n)$ and $g_1(x) = g(-x)$. We have

$$g_1 * \tilde{\varphi} = g_1(\tilde{\varphi}(x - \cdot)) \in C_0^\infty(\mathbb{R}^n)$$

(see Proposition 7.1). Therefore

$$F^{-1}(g_1 * \tilde{\varphi}) = \frac{1}{(2\pi)^n}\int_{\mathbb{R}^n} g_1(\tilde{\varphi}(x - \cdot))e^{ix\cdot\xi}dx.$$

Changing the variables $x - y = t$, we get

$$\frac{1}{(2\pi)^n}\int_{\mathbb{R}^n}\tilde{\varphi}(x-y)e^{ix\cdot\xi}dx = \frac{1}{(2\pi)^n}\int_{\mathbb{R}^n}\tilde{\varphi}(t)e^{i(y+t)\cdot\xi}dt = \varphi(\xi)e^{iy\cdot\xi}.$$

Moreover, the integral sum

$$S_N(y) = \frac{1}{(2\pi)^n}\sum_k \tilde{\varphi}(x_k - y)e^{ix_k\cdot\xi}\Delta x_k$$

converges to the integral $\frac{1}{(2\pi)^n}\int_{\mathbb{R}^n}\tilde{\varphi}(x-y)e^{ix\cdot\xi}dx = \varphi(\xi)e^{iy\cdot\xi}$ uniformly in y for $|y| < R_1$, and the same is true for any derivative with respect to y, $|y| < R_1$. Here R_1 is such that $\operatorname{supp} g \subset B_{R_1}$. Therefore

$$\frac{1}{(2\pi)^n}\int_{\mathbb{R}^n} g_1(\tilde{\varphi}(x-\cdot))e^{ix\cdot\xi}dx = \lim_{\max_k|\Delta x_k|\to 0}\frac{1}{(2\pi)^n}\sum_k g_1(\tilde{\varphi}(x_k-\cdot))e^{ix_k\cdot\xi}\Delta x_k$$

$$= \lim_{\max_k|\Delta x_k|\to 0} g_1(S_N) = g_1(\varphi(\xi)e^{iy\cdot\xi}) = \varphi(\xi)g_1(e^{iy\cdot\xi}) = \varphi(\xi)\tilde{g}(\xi).$$

Thus

$$F^{-1}(g_1 * \tilde{\varphi}) = \tilde{g}(\xi)\varphi(\xi). \tag{12.2}$$

We have used that $\tilde{g_1}(\xi) = \tilde{g}(-\xi)$ since $g_1(x) = g(-x)$. Note that $\tilde{g}(\xi)$ is an entire function satisfying

$$\left|\frac{\partial^k \tilde{g}(\xi)}{\partial \xi^k}\right| \le C_k(1 + |\xi|)^m$$

for the real ξ. Indeed (see Proposition 11.1)

$$\tilde{g}(\xi) = g(\psi e^{-ix\cdot\xi}).$$

It follows from the proof of Proposition 11.1 that

$$\frac{\partial^k \tilde{g}(\xi)}{\partial \xi^k} = g((-ix)^k \psi e^{-ix\cdot\xi}).$$

Therefore, applying the estimate (6.3), we get

$$\left|\frac{\partial^k \tilde{g}(\xi)}{\partial \xi^k}\right| \leq C\|[(-ix)^k \psi e^{-ix\cdot\xi}]\|_{m,B_{R_1}} \leq C_k(1+|\xi|)^m.$$

Thus $\tilde{g}(\xi)\varphi(\xi) \in S$.

By the definition of $f * g$ we have

$$(f * g)(\tilde{\varphi}) = f(g_1 * \tilde{\varphi}).$$

It follows from (10.12) that

$$(f * g)(\tilde{\varphi}) = f(g_1 * \tilde{\varphi}) = \tilde{f}(F^{-1}(g_1 * \tilde{\varphi})).$$

Therefore, (12.2) implies that

$$(f * g)(\tilde{\varphi}) = \tilde{f}(\tilde{g}(\xi)\varphi(\xi)) = \tilde{g}(\xi)\tilde{f}(\varphi) \tag{12.3}$$

for any $\varphi \in S$ such that $\tilde{\varphi} \in C_0^\infty(\mathbb{R}^n)$. Since $C_0^\infty(\mathbb{R}^n)$ is dense in S and since $\tilde{g}(\xi)\tilde{f} \in S'$, we see that (12.3) holds for all $\varphi \in S$, $f * g \in S'$. Therefore $F(f * g)(\varphi) = (f * g)(\tilde{\varphi}) = \tilde{g}(\xi)\tilde{f}(\varphi)$, i.e., $F(f * g) = \tilde{g}(\xi)\tilde{f}$.

13. Sobolev spaces

The Sobolev space $H_s(\mathbb{R}^n)$, $s \in \mathbb{R}$, consists of all distributions $f \in S'$ such that $\tilde{f}$ is the regular functional corresponding to a Lebesgue measurable function $\tilde{f}(\xi)$ such that $\int_{\mathbb{R}^n} |\tilde{f}(\xi)|^2(1+|\xi|)^{2s} d\xi < +\infty$. Thus $H_s(\mathbb{R}^n)$ is a complete Hilbert space with the scalar product

$$\langle f, g\rangle_s = \int \tilde{f}(\xi)\overline{\tilde{g}(\xi)}(1+|\xi|)^{2s} d\xi$$

and the norm $\|f\|_s = \sqrt{\langle f, f\rangle_s}$.

It is clear that the convergence in S implies the convergence in H_s. Denote by $\check{\varphi}(\xi)$ the inverse Fourier transform of $\varphi \in S : \check{\varphi}(\xi) = F^{-1}\varphi$. Then

$$f(\varphi) = \tilde{f}(F^{-1}\varphi) = \int_{\mathbb{R}^n} \tilde{f}(\xi)\check{\varphi} d\xi. \tag{13.1}$$

Applying the Cauchy-Schwartz inequality, we get:

$$|f(\varphi)| \leq \int_{\mathbb{R}^n} |\tilde{f}(\xi)|(1+|\xi|)^s \frac{|\check{\varphi}(\xi)|}{(1+|\xi|)^s} d\xi \leq (2\pi)^{-n}\|f\|_s\|\varphi\|_{-s}, \tag{13.2}$$

since $\check{\varphi}(\xi) = (2\pi)^{-n}\tilde{\varphi}(-\xi)$.

It follows from (13.2) that the convergence in H_s implies the convergence in S'. Therefore we have the embedding

$$S \subset H_s \subset S'.$$

Proposition 13.1. *For any $f \in \mathcal{E}'$, where $\mathcal{E}'$ is the space of distributions with compact supports, there exists $s \in \mathbb{R}$ such that $f \in H_s$.*

Proof: It follows from Proposition 11.1 that

$$|\tilde{f}(\xi)| \leq C(1+|\xi|)^m.$$

Take $s < -m - \frac{n}{2}$. Then

$$\int_{\mathbb{R}^n} |\tilde{f}(\xi)|^2 (1+|\xi|)^{2s} d\xi \leq C^2 \int_{\mathbb{R}^n} (1+|\xi|)^{2s+2m} d\xi < +\infty.$$

Thus $f \in H_s$.

Example 13.1. $\delta \in H_{-\frac{n}{2}-\varepsilon}$, $\forall \varepsilon > 0$. Indeed, $F\delta = 1$ and $\int_{\mathbb{R}^n}(1+|\xi|)^{2s} d\xi < +\infty$ for $s < -\frac{n}{2}$. Analogously $\frac{\partial^k \delta}{\partial x^k} \in H_{-\frac{n}{2}-|k|-\varepsilon}$, $\forall \varepsilon > 0$.

Example 13.2.

a) The case $s = 0$. The Plancherel theorem. For any $\varphi \in S$ and $\psi \in S$ we have, similarly to (10.11):

$$\int_{\mathbb{R}^n} \varphi(x)\overline{\psi(x)} dx = \frac{1}{(2\pi)^n} \int_{\mathbb{R}^n} \tilde{\varphi}(\xi)\overline{\tilde{\psi}(\xi)} d\xi. \tag{13.3}$$

The equality (13.3) is called Parseval's equality.

Let $f \in H_0(\mathbb{R}^n)$, $\tilde{f}(\xi) = Ff$. It follows from Lebesgue measure theory that for any $g(\xi) \in L_2(\mathbb{R}^n)$, there exists a sequence of continuous functions with compact supports that converges to $g(\xi)$ in the L_2-norm. Combining this fact with Proposition 1.3, we get that there exists a sequence $\tilde{\varphi}_n(\xi) \in C_0^\infty(\mathbb{R}^n)$ such that

$$\int_{\mathbb{R}^n} |\tilde{f}(\xi) - \tilde{\varphi}_n(\xi)|^2 d\xi \to 0 \quad \text{as} \quad n \to \infty.$$

It follows from (13.3) that

$$\int_{\mathbb{R}^n} |\varphi_m(x) - \varphi_n(x)|^2 dx = \frac{1}{(2\pi)^n} \int_{\mathbb{R}^n} |\tilde{\varphi}_m(\xi) - \tilde{\varphi}_n(\xi)|^2 d\xi,$$

where $\varphi_n = F^{-1}\tilde{\varphi}_n$. Therefore $\{\varphi_n(x)\}$ is a Cauchy sequence in $L_2(\mathbb{R}^n)$. Since $L_2(\mathbb{R}^n)$ is complete, there exists $\Phi(x) \in L_2(\mathbb{R}^n)$ such that $\varphi_n(x) \to \Phi(x)$ in the L_2-norm. Passing to the limit in Parseval's equality

$$\int_{\mathbb{R}^n} \varphi_n(x)\overline{\psi(x)} dx = \frac{1}{(2\pi)^n} \int_{\mathbb{R}^n} \tilde{\varphi}_n(\xi)\overline{\tilde{\psi}(\xi)} d\xi,$$

we get

$$\int_{\mathbb{R}^n} \Phi(x)\overline{\psi(x)}dx = \frac{1}{(2\pi)^n}\int_{\mathbb{R}^n} \tilde{f}(\xi)\overline{\tilde{\psi}(\xi)}d\xi,\ \forall\psi \in S. \tag{13.4}$$

By the definition of the Fourier transform in S', we have

$$f(\overline{\psi(x)}) = \tilde{f}(F^{-1}\overline{\psi(x)}).$$

Since $F^{-1}\overline{\psi(x)} = \frac{1}{(2\pi)^n}\int_{\mathbb{R}^n} e^{ix\cdot\xi}\overline{\psi(x)}dx = \frac{1}{(2\pi)^n}\overline{\tilde{\psi}(\xi)}$, we obtain

$$f(\overline{\psi(x)}) = \frac{1}{(2\pi)^n}\int_{\mathbb{R}^n} \tilde{f}(\xi)\overline{\tilde{\psi}(\xi)}d\xi. \tag{13.5}$$

Comparing (13.4) and (13.5), we see that $f \in H_0(\mathbb{R}^n)$ is the regular functional corresponding to $\Phi(x) \in L_2(\mathbb{R}^n)$. Therefore $H_0(\mathbb{R}^n) = L_2(\mathbb{R}^n)$. This statement is one of the equivalent formulations of Plancherel's theorem. In what follows we will write $f(x)$ instead of $\Phi(x)$ for a regular functional $f \in H_0(\mathbb{R}^n)$.

b) The case where $s = m > 0$ and is integer. Since $F(\frac{\partial^k f}{\partial x^k}) = (i\xi)^k\tilde{f}$ and $(i\xi)^k\tilde{f}(\xi) \in L_2(\mathbb{R}^n)$ for $|k| \le m$ we get by Plancherel's theorem that $\frac{\partial^k f}{\partial x^k} \in L_2(\mathbb{R}^n)$. Therefore, $H_m(\mathbb{R}^n)$ consists of the functions $f(x)$ that belong to $L_2(\mathbb{R}^n)$ and have distribution derivatives $\frac{\partial^k f}{\partial x^k}$ also in $L_2(\mathbb{R}^n)$ for $|k| \le m$. The norm $\|f\|_m$ is equivalent to the following norm:

$$(\|f\|_m')^2 = \sum_{|k|=0}^{m}\int_{\mathbb{R}^n}\left|\frac{\partial^k f}{\partial x^k}\right|^2 dx. \tag{13.6}$$

Indeed, by Plancherel's theorem

$$(\|f\|_m')^2 = \sum_{|k|=0}^{m}\int_{\mathbb{R}^n}\left|\frac{\partial^k f}{\partial x^k}\right|^2 dx = \sum_{|k|=0}^{m}\frac{1}{(2\pi)^n}\int_{\mathbb{R}^n}|(i\xi)^k\tilde{f}(\xi)|^2 d\xi.$$

Since $C_1(1+|\xi|)^{2m} \le \sum_{|k|=0}^{m}|\xi^k|^2 \le C_2(1+|\xi|)^{2m}$, we have

$$C_1'\|f\|_m \le \|f\|_m' \le C_2'\|f\|_m. \tag{13.7}$$

c) The case where $m < 0$ and is an integer. Set

$$\tilde{g}(\xi) = \frac{\tilde{f}(\xi)}{(1+|\xi|^{|m|})}.$$

Then $\tilde{g}(\xi) \in L^2(\mathbb{R}^n)$ and

$$\tilde{f}(\xi) = (1+|\xi|^{|m|})\tilde{g}(\xi).$$

Since $|\xi| = \sum_{k=1}^n \xi_k \frac{\xi_k}{|\xi|}$, we get

$$\tilde{f}(\xi) = \left(1 + \left(\sum_{k=1}^n \frac{\xi_k}{|\xi|}\xi_k\right)^{|m|}\right)\tilde{g}(\xi) = \sum_{|p|=0}^{|m|}(i\xi)^p \tilde{g}_p(\xi),$$

where $\tilde{g}_p(\xi) \in L_2(\mathbb{R}^n)$. Therefore any $f \in H_m$ can be represented in the form:

$$f = \sum_{|p|=0}^{|m|} \frac{\partial^p}{\partial x^p} g_p, \tag{13.8}$$

where $g_p \in L_2(\mathbb{R}^n)$ and the derivatives are understood in the distribution sense.

13.1. Density of $C_0^\infty(\mathbb{R}^n)$ in $H_s(\mathbb{R}^n)$.

Theorem 13.2. *The space $C_0^\infty(\mathbb{R}^n)$ is dense in $H_s(\mathbb{R}^n)$.*

Proof: Let f be any element of H_s and let $\beta(x) \in C_0^\infty(\mathbb{R}^n)$, $\beta(x) \geq 0$, $\beta(x) = 1$ for $|x| < \frac{1}{2}$, $\beta(x) = 0$ for $|x| > 1$, and $\int_{\mathbb{R}^n}\beta(x)dx = 1$. Set $\beta_\delta(x) = \frac{1}{\delta^n}\beta(\frac{x}{\delta})$, $f_\delta(x) = f * \beta_\delta$. Then $f_\delta(x) \in C^\infty$ (see §7) and $\tilde{f}_\delta(\xi) = \tilde{\beta}_\delta(\xi)\tilde{f}(\xi)$ (see Proposition 12.1). Note that

$$\tilde{\beta}_\delta(\xi) = \int_{\mathbb{R}^n}\frac{1}{\delta^n}\beta\left(\frac{x}{\delta}\right)e^{-ix\cdot\xi}dx = \int_{\mathbb{R}^n}\beta(y)e^{-i\delta y\cdot\xi}dy = \tilde{\beta}(\delta\xi).$$

By the Lebesgue convergence theorem,

$$\|f - f_\delta\|_s^2 = \int_{\mathbb{R}^n}|\tilde{f}(\xi)|^2|1 - \tilde{\beta}(\delta\xi)|^2(1 + |\xi|)^{2s}d\xi \to 0$$

as $\delta \to 0$, since $\tilde{\beta}(\delta\xi) \to \tilde{\beta}(0) = \int_{\mathbb{R}^n}\beta(x)dx = 1$ and also $|\tilde{\beta}(\delta\xi)| \leq \int_{\mathbb{R}^n}\beta(x)dx = 1$.

For an arbitrary $\varepsilon > 0$ fix $\delta > 0$ such that

$$\|f - f_\delta\|_s < \frac{\varepsilon}{2}. \tag{13.9}$$

Denote

$$f_{\delta_1}(x) = f_\delta(x)\beta(\delta_1 x).$$

Note that $f_{\delta_1}(x) \in C_0^\infty(\mathbb{R}^n)$.

Choose a positive integer m such that $m \geq s$. Since

$$|\tilde{\beta}(\delta\xi)| \leq C_N(1 + |\delta\xi|)^{-N},\ \forall N,$$

we see that $f_\delta \in H_m$. Thus, using the equivalent norm (13.6), we see

$$\|f_\delta - f_{\delta_1}\|_s^2 \le \|f_\delta - f_{\delta_1}\|_m^2 \le C \sum_{|k|=0}^{m} \int_{\mathbb{R}^n} \left| \frac{\partial^k}{\partial x^k}[(1-\beta(\delta_1 x))f_\delta(x)] \right|^2 dx$$
$$\le C_1 \sum_{|k|=0}^{m} \int_{|x|>\frac{1}{2\delta_1}} \left| \frac{\partial^k}{\partial x^k} f_\delta(x) \right|^2 dx,$$

since $1-\beta(\delta_1 x) = 0$ for $|x| < \frac{1}{2\delta_1}$. Therefore $\|f_\delta - f_{\delta_1}\|_s \to 0$ as $\delta_1 \to 0$. Combining this with (13.9), we obtain that there exists a C_0^∞ function $f_{\delta_1}(x)$ such that $\|f - f_{\delta_1}\|_s < \varepsilon$.

13.2. Multiplication by $a(x) \in S$.

Theorem 13.3. *Let $a(x) \in S$. Then the multiplication by $a(x)$ is a bounded operator in H_s for any $s \in \mathbb{R}$.*

Proof: We have

$$F(af)(\xi) = \frac{1}{(2\pi)^n} \int_{\mathbb{R}^n} \tilde{a}(\xi-\eta)\tilde{f}(\eta)d\eta.$$

Since $a(x) \in S$, we have

$$|\tilde{a}(\xi-\eta)| \le C_N(1+|\xi-\eta|)^{-N}, \ \forall N.$$

The following inequality holds for any s:

$$\frac{(1+|\xi|)^s}{(1+|\eta|)^s} \le C_s(1+|\xi-\eta|)^{|s|}. \tag{13.10}$$

Indeed, for $s>0$ we see that

$$(1+|\xi|)^s \le (1+|\xi-\eta|+|\eta|)^s \le [(1+|\eta|)(1+|\xi-\eta|)]^s.$$

If $s<0$, we reverse the roles of ξ and η. We have

$$|F(af)|(1+|\xi|)^s \le C \int_{\mathbb{R}^n} \frac{(1+|\xi|)^s}{(1+|\eta|)^s(1+|\xi-\eta|)^N}(1+|\eta|)^s|\tilde{f}(\eta)|d\eta.$$

Taking $N > |s|+n+1$ and using (13.10) and the Cauchy-Schwartz inequality we obtain

(13.11)

$$|F(af)|^2(1+|\xi|)^{2s} \le C_1 \left| \int_{\mathbb{R}^n} \frac{1}{(1+|\xi-\eta|)^{\frac{n+1}{2}}} \frac{(1+|\eta|)^s|\tilde{f}(\eta)|}{(1+|\xi-\eta|)^{\frac{n+1}{2}}} d\eta \right|^2$$
$$\le C_1 \int_{\mathbb{R}^n} \frac{d\eta}{(1+|\xi-\eta|)^{n+1}} \int_{\mathbb{R}^n} \frac{(1+|\eta|)^{2s}|\tilde{f}(\eta)|^2}{(1+|\xi-\eta|)^{n+1}} d\eta.$$

Since

$$\int_{\mathbb{R}^n} \frac{d\eta}{(1+|\xi-\eta|)^{n+1}} \le C_2,$$

we conclude, integrating (13.11) with respect to ξ, that

$$\|af\|_s^2 \le C_3 \int_{\mathbb{R}^n} \frac{(1+|\eta|)^{2s}|\tilde{f}(\eta)|^2}{(1+|\xi-\eta|)^{n+1}} d\xi d\eta \le C_4\|f\|_s^2. \tag{13.12}$$ □

13.3. Sobolev's embedding theorem.

Let $\varphi(x)$ be a C_0^∞ function. The following inequality holds:

$$\sum_{|k|=0}^{m} \max_{x\in\mathbb{R}^n} \left|\frac{\partial^k \varphi(x)}{\partial x^k}\right| \le C_s\|\varphi\|_s, \tag{13.13}$$

assuming that $s > m + \frac{n}{2}$. Indeed,

$$\frac{\partial^k \varphi(x)}{\partial x^k} = \frac{1}{(2\pi)^n}\int_{\mathbb{R}^n} (i\xi)^k \tilde{\varphi}(\xi) e^{ix\cdot\xi} d\xi.$$

Applying the Cauchy-Schwartz inequality, we get

$$\left|\frac{\partial^k \varphi(x)}{\partial x^k}\right|^2 \le \left|\frac{1}{(2\pi)^n}\int_{\mathbb{R}^n} \frac{|\xi|^k}{(1+|\xi|)^s}(1+|\xi|)^s|\tilde{\varphi}(\xi)|d\xi\right|^2$$
$$\le \frac{1}{(2\pi)^{2n}}\int_{\mathbb{R}^n} \frac{|\xi|^{2|k|}}{(1+|\xi|)^{2s}} d\xi \|\varphi\|_s^2.$$

Since $2s - 2|k| > n$ for $|k| \le m$, we have

$$\int_{\mathbb{R}^n} \frac{|\xi|^{2|k|}}{(1+|\xi|)^{2s}} d\xi < +\infty.$$

Thus (13.13) holds.

Now we extend (13.13) to an arbitrary $f \in H_s$, $s > m + \frac{n}{2}$. For any $f \in H_s$, there exists a sequence $\varphi_p(x) \in C_0^\infty(\mathbb{R}^n)$ such that $\|f - \varphi_p\|_s \to 0$ as $p \to \infty$ (see Theorem 13.2). Denote by $C_0^m(\mathbb{R}^n)$ the space of continuous functions $v(x)$ having continuous derivatives up to the order m and such that $v(x) \to 0$ as $|x| \to \infty$ and $\frac{\partial^k v(x)}{\partial x^k} \to 0$ as $|x| \to \infty$ and $|k| \le m$.

Denote by $\|v\|_{C_0^m}$ the norm in C_0^m:

$$\|v\|_{C_0^m} = \sum_{|k|=0}^{m} \max_{x\in\mathbb{R}^n} \left|\frac{\partial^k v}{\partial x^k}\right|. \tag{13.14}$$

Since $\varphi_p \to f$ in H_s, the inequality (13.13) implies that $\{\varphi_p\}$ is a Cauchy sequence in $C_0^m(\mathbb{R}^n)$, and since $C_0^m(\mathbb{R}^n)$ is complete, there exists $v(x) \in C_0^m(\mathbb{R}^n)$ such that $\varphi_p \to v$ in $C_0^m(\mathbb{R}^n)$. Since the convergence in $C_0^m(\mathbb{R}^n)$ is obviously stronger than the convergence in S', we have that $\varphi_p \to v$ in S'. Also $\varphi_p \to f$ in S'. Thus $v = f$. Passing to the limit in (13.13), we get for any $f \in H_s$,

$$\|f\|_{C_0^m} \le C_s\|f\|_s, \quad s > m + \frac{n}{2}. \tag{13.15}$$

Therefore we have proven the following theorem.

Theorem 13.4 (Sobolev's embedding theorem). *If $s > m + \frac{n}{2}$, the space H_s is embedded in $C_0^m(\mathbb{R}^n)$: $H_s(\mathbb{R}^n) \subset C_0^m(\mathbb{R}^n)$ and the inequality* (13.15) *holds for any $f \in H_s$.*

13.4. An equivalent norm for noninteger.

Theorem 13.5. *The norm $\|u\|_\gamma$ is equivalent to the norm*

$$(\|u\|'_\gamma)^2 = \int_{\mathbb{R}^n}\int_{\mathbb{R}^n} \frac{|u(x+y)-u(x)|^2}{|y|^{n+2\gamma}} dxdy + \int_{\mathbb{R}^n} |u(x)|^2 dx, \tag{13.16}$$

where $0 < \gamma < 1$.

Proof: Let $u(x)$ be a C_0^∞ function. We have

$$u(x+y) - u(x) = \frac{1}{(2\pi)^n}\int_{\mathbb{R}^n} \tilde{u}(\xi)(e^{iy\cdot\xi}-1)e^{ix\cdot\xi}d\xi.$$

Using Parseval's equality for fixed y, we get

$$\int_{\mathbb{R}^n} |u(x+y)-u(x)|^2 dx = \frac{1}{(2\pi)^n}\int_{\mathbb{R}^n} |\tilde{u}(\xi)|^2|e^{iy\cdot\xi}-1|^2 d\xi.$$

Therefore

$$(\|u\|'_\gamma)^2 = \frac{1}{(2\pi)^n}\int_{\mathbb{R}^n}\int_{\mathbb{R}^n} |\tilde{u}(\xi)|^2\frac{|e^{iy\cdot\xi}-1|^2}{|y|^{n+2\gamma}} d\xi dy + \frac{1}{(2\pi)^n}\int_{\mathbb{R}^n} |\tilde{u}(\xi)|^2 d\xi.$$

We show that

$$\int_{\mathbb{R}^n} \frac{|e^{iy\cdot\xi}-1|^2}{|y|^{n+2\gamma}} dy = C_\gamma|\xi|^{2\gamma}.$$

Making an orthogonal change of variables $y = Oz$ such that $\frac{\xi}{|\xi|}$ will be equal to $(1, 0, \ldots, 0)$ in z-coordinates and then changing z to $\frac{z}{|\xi|}$, we obtain

$$\begin{aligned}\int_{\mathbb{R}^n} \frac{|e^{iy\cdot\xi}-1|^2}{|y|^{n+2\gamma}} dy &= \int_{\mathbb{R}^n} \frac{|e^{iz_1|\xi|}-1|^2}{|z|^{n+2\gamma}} dz \\ &= |\xi|^{2\gamma}\int_{\mathbb{R}^n} \frac{|e^{iz_1}-1|^2}{|z|^{n+2\gamma}} dz = C_\gamma|\xi|^{2\gamma}.\end{aligned} \tag{13.17}$$

Note that the integrals in (13.17) converge since $|e^{iz_1}-1|^2 \le C|z_1|^2$ for $|z_1|$ small and since $0 < \gamma < 1$. Therefore we obtain

$$C_1\|u\|_\gamma \le \|u\|'_\gamma \le C_2\|u\|_\gamma \tag{13.18}$$

for each $u \in C_0^\infty(\mathbb{R}^n)$. Since $C_0^\infty(\mathbb{R}^n)$ is dense in $H_\gamma(\mathbb{R}^n)$ we see that (13.16) is an equivalent norm in $H_\gamma(\mathbb{R}^n)$. □

If $s = m + \gamma$, $0 < \gamma < 1$, $m \in \mathbb{Z}$, then by (13.7) and Theorem 13.5 we get that $\|u\|_{m+\gamma}$ is equivalent to the following norm:

$$\sum_{|k|=m} \int_{\mathbb{R}^n} \int_{\mathbb{R}^n} \frac{\left| \frac{\partial^k u(y)}{\partial y^k} - \frac{\partial^k u(x)}{\partial x^k} \right|^2}{|x-y|^{n+2\gamma}} dx dy + \int_{\mathbb{R}^n} |u(x)|^2 dx. \tag{13.19}$$

13.5. Restrictions to hyperplanes (traces).

Theorem 13.6. *Let $x = (x', x_n)$, where $x' = (x_1, \dots, x_{n-1}) \in \mathbb{R}^{n-1}$. Let $[v]_s$ be the norm in $H_s(\mathbb{R}^{n-1})$. Then for $s > \frac{1}{2}$ the following estimates hold:*

$$\sup_{x_n \in \mathbb{R}^1} [u(x', x_n)]_{s-\frac{1}{2}} \le C\|u\|_s, \tag{13.20}$$

$$[u(x', x_n) - u(x', y_n)]_{s-\alpha-\frac{1}{2}} \le C|x_n - y_n|^\alpha \|u\|_s, \tag{13.20$'$}$$

where $0 < \alpha \le 1, \quad \alpha < s - \frac{1}{2}$.

Proof: Consider $u \in C_0^\infty(\mathbb{R}^n)$. Let $w(\xi', x_n)$ be the Fourier transform of $u(x', x_n)$ in x'. Then

$$w(\xi', x_n) = \frac{1}{2\pi} \int_{-\infty}^{\infty} \tilde{u}(\xi', \xi_n) e^{ix_n \xi_n} d\xi_n,$$

where $\tilde{u}(\xi', \xi_n)$ is the Fourier transform of $u(x', x_n)$ in x' and x_n. By the Cauchy-Schwartz inequality, we get

$$\begin{aligned} |w(\xi', x_n)|^2 \le \frac{1}{(2\pi)^2} \int_{-\infty}^{\infty} |\tilde{u}(\xi', \xi_n)|^2 (1 + |\xi'| + |\xi_n|)^{2s} d\xi_n \\ \times \int_{-\infty}^{\infty} (1 + |\xi'| + |\xi_n|)^{-2s} d\xi_n. \end{aligned} \tag{13.21}$$

Changing the variables $\xi_n = (1 + |\xi'|)\eta_n$, we obtain

$$\begin{aligned} \int_{-\infty}^{\infty} (1 + |\xi'| + |\xi_n|)^{-2s} d\xi_n &= (1 + |\xi'|)^{-2s+1} \int_{-\infty}^{\infty} (1 + |\eta_n|)^{-2s} d\eta_n \\ &= C_1 (1 + |\xi'|)^{-2s+1}. \end{aligned} \tag{13.22}$$

Note that the integrals in (13.22) converge since $s > \frac{1}{2}$. Multiplying (13.21) by $(1 + |\xi'|)^{2s-1}$ and integrating with respect to ξ', we get

$$[u(x', x_n)]_{s-\frac{1}{2}}^2 \le C\|u\|_s^2 \tag{13.23}$$

for any $u \in C_0^\infty(\mathbb{R}^n)$, where C is independent of $u(x', x_n)$ and x_n.

Analogously, we have

$$w(\xi', x_n) - w(\xi', y_n) = \frac{1}{2\pi} \int_{-\infty}^{\infty} \tilde{u}(\xi', \xi_n)(e^{ix_n \xi_n} - e^{iy_n \xi_n}) d\xi_n.$$

Therefore, as in (13.21),

$$(13.21') \qquad |w(\xi', x_n) - w(\xi', y_n)|^2$$

$$\leq \frac{1}{(2\pi)^2} \int_{-\infty}^{\infty} |\tilde{u}(\xi', \xi_n)|^2 (1 + |\xi'| + |\xi_n|)^{2s} d\xi_n \int_{-\infty}^{\infty} \frac{|e^{ix_n\xi_n} - e^{iy_n\xi_n}|^2}{(1 + |\xi'| + |\xi_n|)^{2s}} d\xi_n.$$

Note that $|e^{ix_n\xi_n} - e^{iy_n\xi_n}| \leq C|x_n - y_n|^\alpha |\xi_n|^\alpha$ for any $0 \leq \alpha \leq 1$. Thus, changing the variables $\xi_n = (1 + |\xi'|)\eta_n$, we get, as in (13.22),

$$\int_{-\infty}^{\infty} \frac{|e^{ix_n\xi_n} - e^{iy_n\xi_n}|^2}{(1 + |\xi'| + |\xi_n|)^{2s}} d\xi_n \leq C_1 |x_n - y_n|^{2\alpha} (1 + |\xi'|)^{-2s+1+2\alpha},$$

where

$$C_1 = C \int_{-\infty}^{\infty} \frac{|\eta_n|^{2\alpha}}{(1 + |\eta_n|)^{2s}} d\eta_n < \infty$$

if $2s - 2\alpha > 1$. Therefore, multiplying (13.21′) by $(1 + |\xi'|)^{2s-2\alpha-1}$ and integrating with respect to ξ', we conclude that

$$(13.23') \qquad [u(x', x_n) - u(x', y_n)]^2_{s-\alpha-\frac{1}{2}} \leq C|x_n - y_n|^{2\alpha} \|u\|_s^2,$$

where $0 < \alpha < s - \frac{1}{2}$, C is independent of x_n and y_n, and $u \in C_0^\infty(\mathbb{R}^n)$.

For an arbitrary $u \in H_s(\mathbb{R}^n)$, there exists a sequence $\varphi_m(x', x_n) \in C_0^\infty(\mathbb{R}^n)$ such that $\|\varphi_m - u\|_s \to 0$ as $m \to \infty$. It follows from (13.23) that for a fixed $x_n^{(0)}$, $\{\varphi_m(x', x_n^{(0)})\}$ is a Cauchy sequence in $H_{s-\frac{1}{2}}(\mathbb{R}^{n-1})$. Since $H_{s-\frac{1}{2}}(\mathbb{R}^{n-1})$ is complete, there exists $\lim_{m\to\infty} \varphi_m(x', x_n^{(0)}) \in H_{s-\frac{1}{2}}(\mathbb{R}^{n-1})$. We shall denote this limit by $u(x', x_n^{(0)})$ and we shall say that $u(x', x_n^{(0)})$ is the restriction of $u \in H_s(\mathbb{R}^n)$ to the hyperplane $x_n = x_n^{(0)}$. Thus, (13.20) holds for each $u \in H_s(\mathbb{R}^n)$. Therefore $H_s(\mathbb{R}^n) \subset C(\mathbb{R}^1, H_{s-\frac{1}{2}}(\mathbb{R}^{n-1}))$, where $C(\mathbb{R}^1, H_{s-\frac{1}{2}}(\mathbb{R}^{n-1}))$ is the space of continuous functions of $x_n \in \mathbb{R}^1$ with values in $H_{s-\frac{1}{2}}(\mathbb{R}^{n-1})$. Moreover, (13.23′) shows that $u(x', x_n)$ is a Hölder function of x_n with values in $H_{s-\alpha-\frac{1}{2}}(\mathbb{R}^{n-1})$. □

13.6. Duality of Sobolev spaces.

If $\varphi \in C_0^\infty(\mathbb{R}^n)$, $\psi \in C_0^\infty(\mathbb{R}^n)$, we have

$$|(\varphi, \psi)| = \left| \frac{1}{(2\pi)^n} \int_{\mathbb{R}^n} \tilde{\varphi}(\xi) \overline{\tilde{\psi}(\xi)} d\xi \right| \leq \frac{1}{(2\pi)^n} \|\varphi\|_s \|\psi\|_{-s}$$

for each $s \in \mathbb{R}^1$. Here $(\varphi, \psi) = \int_{\mathbb{R}^n} \varphi(x)\overline{\psi(x)} dx$ is the L_2 scalar product. Taking the closure using Theorem 13.2, we can extend (u, v) to any $u \in H_s(\mathbb{R}^n)$ and any $v \in H_{-s}(\mathbb{R}^n)$ and we obtain that

$$(13.24) \qquad |(u, v)| \leq \frac{1}{(2\pi)^n} \|u\|_s \|v\|_{-s}$$

holds for any such u and v.

Proposition 13.7. *The space $H_{-s}(\mathbb{R}^n)$ is dual to $H_s(\mathbb{R}^n)$ with respect to the extension of the scalar product (φ, ψ) : $H^*_{-s}(\mathbb{R}^n) = H_s(\mathbb{R}^n)$.*

Proof: The dual space means the space of linear continuous functionals. It follows from (13.24) that each $v \in H_{-s}(\mathbb{R}^n)$ defines a linear continuous functional on $H_s(\mathbb{R}^n)$.

Vice versa, let $\Phi(u)$ be a linear continuous functional on $H_s(\mathbb{R}^n)$. Since $H_s(\mathbb{R}^n)$ is a Hilbert space with scalar product $\langle f, g\rangle_s$, the linear functional $\Phi(u)$ can be represented in the following form (this is the Riesz theorem):

$$\Phi(u) = \langle u, g\rangle_s = \int_{\mathbb{R}^n} \tilde{u}(\xi)\overline{\tilde{g}(\xi)}(1 + |\xi|)^{2s} d\xi,$$

where $g \in H_s(\mathbb{R}^n)$. Denote $\tilde{v}(\xi) = (1+|\xi|)^{2s}\tilde{g}(\xi)$. Then $v \in H_{-s}(\mathbb{R}^n)$, $\Phi(u) = (u, v)$ and the norm of Φ is $\|g\|_s = \|v\|_{-s}$. □

13.7. Invariance of $H_s(\mathbb{R}^n)$ under changes of variables.

Let

$$x = s(y) \tag{13.25}$$

be a C^∞ diffeomorphism of $\mathbb{R}^n$ such that $s(y) = y$ for $|y| > R$ and the Jacobian $\det \left[\frac{\partial s_i(y)}{\partial y_j}\right]^n_{i,j=1} \neq 0$ for all $y \in \mathbb{R}^n$. Note that the inverse map $y = s^{-1}(x)$ has the same properties. Let $v(y) = u(s(y))$ and denote by S and S^{-1} the operators $v = Su$ and $S^{-1}v = u$. Changing the variables in (13.6) when $s = m \geq 0$ is an integer, and taking into account that $\frac{\partial^i}{\partial y^i} s(y)$ are bounded for all i, we get

$$(\|u\|'_m)^2 \leq \sum_{|k|=0}^m C \int_{\mathbb{R}^n} \left|\frac{\partial^k v(y)}{\partial y^k}\right|^2 dy = C(\|v\|'_m)^2,$$

i.e., the operator S^{-1} is bounded in $H_m(\mathbb{R}^n)$.

Analogously we have that the operator S is bounded. Therefore

$$C_1\|u\|_s \leq \|v\|_s \leq C_2\|u\|_s, \tag{13.26}$$

where $C_1 > 0, C_2 > 0$, and $s = m \geq 0$ is an integer.

The case $s = m + \gamma$, $m \geq 0$, $0 < \gamma < 1$, can be treated analogously using the equivalent norm (13.19).

Consider $s < 0$. Then by the definition of change of variables for distributions (see (3.8)), we have

$$(Su, \varphi) = (u, (S^{-1}\varphi)J),$$

where $J(x)$ is the Jacobian of the map $y = s^{-1}(x)$. We have, using Theorem 13.3 and (13.26) for $-s > 0$:

$$|(u, (S^{-1}\varphi)J)| \le \|u\|_s \|(S^{-1}\varphi)J\|_{-s} \le C\|u\|_s \|S^{-1}\varphi\|_{-s} \le C'\|u\|_s\|\varphi\|_{-s}.$$

Therefore $|(Su, \varphi)| \le C'\|u\|_s\|\varphi\|_{-s}$ for any $\varphi \in H_{-s}(\mathbb{R}^n)$. This implies that $v = Su$ is a linear continuous functional over $H_{-s}(\mathbb{R}^n)$ with the norm bounded by $C'\|u\|_s$. Hence $v = Su \in H^*_{-s}(\mathbb{R}^n) = H_s(\mathbb{R}^n)$ and

$$\|Su\|_s \le C'\|u\|_s \quad \text{for} \quad s < 0.$$

The inequality $\|S^{-1}v\|_s \le C''\|v\|_s$ is proven analogously. Therefore (13.26) holds for any $s \in \mathbb{R}^1$ and we have proven

Theorem 13.8. *Sobolev's spaces are invariant under the diffeomorphism* (13.25).

The following remark will be used in Chapter V.

Remark 13.1. Consider $u \in H_s(\mathbb{R}^n)$, $s > \frac{1}{2}$. Then the restriction of $u(x)$ to the sphere $|x| = k$ exists and

$$\int_{|x|=k} |u(x)|^2 d\sigma \le C\|u\|_s^2, \tag{13.27}$$

where $d\sigma$ is the surface area element of the sphere $|x| = k$ and C is independent of k for $k \ge \varepsilon_0 > 0$.

Proof: Let $\psi_j(\theta) \in C_0^\infty(S^{n-1})$, $1 \le j \le N$, $\sum_{j=1}^N \psi_j(\theta) = 1$. Set $\varphi_j(x) = \chi(|x| - k)\psi_j(\frac{x}{|x|})$, where $\chi(t) \in C_0^\infty(\mathbb{R}^1)$, $\chi(t) = 1$ for $|t| < \frac{k_1}{2}$, $\chi(t) = 0$ for $|t| > k_1$, $k_1 = \min(k, 1)$. Make the change of variables $s(x) = (r, k\theta)$ on $\operatorname{supp}\varphi_j(x)$, where $r = |x|$, $\theta = \frac{x}{|x|}$, $k\theta$ belongs to the sphere of radius k. Note that the Jacobian of this change of variables is bounded above and below by constants independent of k for $k \ge \varepsilon_0 > 0$, where ε_0 is arbitrary. We have, changing the variables and applying Theorem 13.6:

$$\int_{|x|=k} |u(x)|^2 d\sigma = \sum_{j=1}^N \int_{|x|=k} |\varphi_j u|^2 d\sigma \le C\sum_{j=1}^N (\|\varphi_j u\|'_s)^2,$$

where $\|\varphi_j u\|'_s$ is the Sobolev norm in variables $(r, k\theta)$. Changing the variables back and using (13.26), we get (13.27).

Analogously it follows from (13.20′) that

$$\int_{|\omega|=1} |u(r\omega) - u(k\omega)|^2 k^{n-1} d\omega \le C|r - k|^{2\alpha}\|u\|_s^2, \tag{13.28}$$

where $0 < \alpha \le 1$, $\alpha < s - \frac{1}{2}$, $|r - k| \le 1$. □

Example 13.3. Let $v(x)$ be any function (distribution) in $H_{s-\frac{1}{2}}(\mathbb{R}^{n-1})$, where s is arbitrary. We construct $u(x', x_n) \in H_s(\mathbb{R}^n)$, continuous in x_n, such that $u(x', 0) = v(x')$ and

$$\|u\|_s = C[v]_{s-\frac{1}{2}},$$

where C is independent of v.

Take $\beta(x_n) \in C_0^\infty(\mathbb{R}^1)$ such that $\beta(0) = 1$. Consider $w(\xi', x_n) = \tilde{v}(\xi')\beta(x_n(1+|\xi'|))$. Let $u = F_{\xi'}^{-1}w(\xi', x_n)$, where $F_{\xi'}^{-1}$ is the inverse Fourier transform in ξ'. We have

$$\tilde{u}(\xi', \xi_n) = \tilde{v}(\xi')\tilde{\beta}\left(\frac{\xi_n}{1+|\xi'|}\right)\frac{1}{1+|\xi'|}.$$

Therefore, changing the variables $\xi_n = (1+|\xi'|)\eta_n$, we get

$$\|u\|_s^2 = \int_{\mathbb{R}^n}(1+|\xi'|+|\xi_n|)^{2s}|\tilde{u}(\xi', \xi_n)|^2 d\xi' d\xi_n$$

$$= \int_{\mathbb{R}^n}(1+|\xi'|)^{2s}(1+|\eta_n|)^{2s}|\tilde{v}(\xi')|^2|\tilde{\beta}(\eta_n)|^2(1+|\xi'|)^{-1}d\xi' d\eta_n.$$

$$= [v]_{s-\frac{1}{2}}^2\int_{-\infty}^{\infty}|\tilde{\beta}(\eta_n)|^2(1+|\eta_n|)^{2s}d\eta_n = C^2[v]_{s-\frac{1}{2}}^2.$$

14. Singular supports and wave front sets of distributions

We have defined the restriction of a distribution $g \in \mathcal{D}'(\mathbb{R}^n)$ to an open set $U \subset \mathbb{R}^n$ by considering $g(\varphi)$ only for all $\varphi \in C_0^\infty(U)$, i.e., $\varphi \in C_0^\infty(\mathbb{R}^n)$ and $\operatorname{supp}\varphi \subset U$. In particular we say that $g \in \mathcal{D}'(\mathbb{R}^n)$ is C^∞ in U if there exists $g(x) \in C^\infty(U)$ such that

$$g(\varphi) = \int_U g(x)\varphi(x)dx, \quad \forall\varphi \in C_0^\infty(U). \tag{14.1}$$

Definition 14.1. Let $U_{\max}$ be the maximal open set where $g \in \mathcal{D}'$ is C^∞. Then the complement $\mathbb{R}^n \setminus U_{\max}$ is called the singular support of g:

$$\mathbb{R}^n \setminus U_{\max} = \operatorname{sing\,supp} g.$$

Example 14.1.

a) $\operatorname{sing\,supp}\delta = \operatorname{supp}\delta = \{0\}$.

b) $\operatorname{sing\,supp} x_+^\lambda = \{0\}$ since x_+^λ is C^∞ for $x_1 \in \mathbb{R}^1 \setminus \{0\}$: $x_+^\lambda(\varphi) = \int_0^\infty x_1^\lambda\varphi(x_1)dx_1$ for all $\varphi(x_1) \in C_0^\infty$ such that $\varphi(x_1) = 0$ in a neighborhood of $x_1 = 0$.

c) $\operatorname{sing\,supp}\frac{1}{|x|^2-k^2-i0} = \{x : |x|^2 - k^2 = 0\}$. □

Let $x_0 \notin \operatorname{sing\,supp} g$. Since $\operatorname{sing\,supp} g$ is a closed set, there exist a C_0^∞ function $\varphi(x)$ with $\varphi(x_0) \neq 0$ such that $\varphi g \in C_0^\infty$. Therefore

$$|\widetilde{\varphi g}(\xi)| \leq C_N(1+|\xi|)^{-N}, \quad \forall N. \tag{14.2}$$

The wave front set of a distribution g is a refinement of the notion of singular support. It is a set in $\mathbb{R}^n \times \mathbb{R}^n \setminus \{0\}$, and it will be denoted by $WF(g)$.

Definition 14.2. We say that $(x_0, \xi_0) \in \mathbb{R}^n \times \mathbb{R}^n \setminus \{0\}$ does not belong to the wave front set of $g \in \mathcal{D}'(\mathbb{R}^n)$ if there exists a C_0^∞ function $\varphi(x)$ such that $\varphi(x_0) \neq 0$ and

$$|F(\varphi g)(\xi)| \leq C_N(1+|\xi|)^{-N} \tag{14.3}$$

for all N and all $\xi \neq 0$ satisfying

$$\left|\frac{\xi}{|\xi|} - \frac{\xi_0}{|\xi_0|}\right| < \delta \quad \text{for some} \quad \delta > 0. \tag{14.4}$$

Note that $WF(g)$ is a closed set in $\mathbb{R}^n \times \mathbb{R}^n \setminus \{0\}$ since if $(x_0, \xi_0) \notin WF(g)$, then $(\bar{x}, \bar{\xi})$ for all $|\bar{x} - x_0|$ small and $|\frac{\bar{\xi}}{|\bar{\xi}|} - \frac{\xi_0}{|\xi_0|}|$ small does not belong to $WF(g)$ either. Moreover, $WF(g)$ is a conic set in $\mathbb{R}^n \times \mathbb{R}^n \setminus \{0\}$: if $(x, \xi) \in WF(g)$, then $(x, t\xi) \in WF(g)$ for all $t > 0$.

Therefore $(x_0, \xi_0) \in WF(g)$ if for any $\varphi \in C_0^\infty$, $\varphi(x_0) \neq 0$, $F(\varphi g)(\xi)$ does not decay faster than any negative power of $|\xi|$ in any cone $|\frac{\xi}{|\xi|} - \frac{\xi_0}{|\xi_0|}| < \delta$, $\forall \delta > 0$.

Proposition 14.1. *If $a(x) \in C_0^\infty(\mathbb{R}^n)$, then*

$$WF(ag) \subset WF(g).$$

Proof: If $(x_0, \xi_0) \notin WF(g)$, then there exists $\varphi \in C_0^\infty(\mathbb{R}^n)$, $\varphi(x_0) \neq 0$, such that

$$|F(\varphi g)(\xi)| \leq C_N(1+|\xi|)^{-N}, \quad \forall N, \tag{14.5}$$

if $|\frac{\xi}{|\xi|} - \frac{\xi_0}{|\xi_0|}| < \delta$ for some $\delta > 0$. Set $v = \varphi g$. Then

$$F(\varphi a g)(\xi) = \frac{1}{(2\pi)^n} \int_{\mathbb{R}^n} \tilde{a}(\xi - \eta)\tilde{v}(\eta)d\eta = I_1 + I_2,$$

where

$$I_1(\xi) = \frac{1}{(2\pi)^n} \int_{|\frac{\eta}{|\eta|} - \frac{\xi_0}{|\xi_0|}| < \delta} \tilde{a}(\xi - \eta)\tilde{v}(\eta)d\eta,$$

$$I_2(\xi) = \frac{1}{(2\pi)^n} \int_{|\frac{\eta}{|\eta|} - \frac{\xi_0}{|\xi_0|}| > \delta} \tilde{a}(\xi - \eta)\tilde{v}(\eta)d\eta.$$

Since $|\tilde{a}(\xi-\eta)| \leq C_N(1+|\xi-\eta|)^{-N}$, $\forall N$, we have, using (13.10) with $s = N$, $N > n+1$:

$$
\begin{aligned}
(14.6) \qquad |I_1| &\leq \int C_{N1}(1+|\xi-\eta|)^{-2N}(1+|\eta|)^{-N}d\eta \\
&\leq \int \frac{C_{N1}}{(1+|\xi|)^N}\frac{(1+|\xi|)^N}{(1+|\eta|)^N(1+|\xi-\eta|)^{2N}}d\eta \\
&\leq C_{N2}(1+|\xi|)^{-N}, \quad \forall N, \quad \forall \xi.
\end{aligned}
$$

If

$$
(14.7) \qquad \left|\frac{\xi}{|\xi|} - \frac{\xi_0}{|\xi_0|}\right| < \delta' \quad \text{and} \quad \left|\frac{\eta}{|\eta|} - \frac{\xi_0}{|\xi_0|}\right| > \delta,
$$

where $0 < \delta' < \delta$, then

$$
\left|\frac{\eta}{|\eta|} - \frac{\xi}{|\xi|}\right| > \delta - \delta' > 0,
$$

i.e., the angle between the vectors ξ and η is bounded below. Therefore

$$
|\xi - \eta| > C(|\xi| + |\eta|)
$$

if (14.7) holds. We have for $|\frac{\xi}{|\xi|} - \frac{\xi_0}{|\xi_0|}| < \delta'$:

$$
\begin{aligned}
(14.8) \qquad |I_2(\xi)| &\leq \int_{|\frac{\eta}{|\eta|}-\frac{\xi_0}{|\xi_0|}|>\delta} C_N(1+|\xi-\eta|)^{-N}|\tilde{v}(\eta)|d\eta \\
&\leq \int_{|\frac{\eta}{|\eta|}-\frac{\xi_0}{|\xi_0|}|>\delta} C_{N1}(1+|\xi|+|\eta|)^{-N}(1+|\eta|)^m d\eta \\
&\leq C_{N2}(1+|\xi|)^{-N+m+n+1}, \quad \forall N.
\end{aligned}
$$

In (14.8) we used that $|\tilde{v}(\eta)| \leq C(1+|\eta|)^m$, since $v \in \mathcal{E}'$ (see Proposition 11.1).

It follows from (14.6) and (14.8) that $(x_0, \xi_0) \notin WF(ag)$. Therefore $WF(ag) \subset WF(g)$. □

Proposition 14.2. *The projection $\pi : (x,\xi) \to x$ of $WF(g)$ to $\mathbb{R}^n$ is the singular support of g.*

Proof: If x_0 does not belong to $\pi WF(g)$, then for any $\xi_0 \neq 0$ there exist $\varphi_0(x) \in C_0^\infty(\mathbb{R}^n)$, $\varphi(x_0) \neq 0$, and δ_0 such that

$$
|F(\varphi_0(x)g)(\xi)| \leq C_N(1+|\xi|)^{-N}, \quad \forall N \quad \text{and} \quad \xi \in V_{\xi_0,\delta_0},
$$

where V_{ξ_0,δ_0} is the cone $|\frac{\xi}{|\xi|} - \frac{\xi_0}{|\xi_0|}| < \delta_0$. Such cones form an open cover of $\mathbb{R}^n \setminus \{0\}$. Since the unit sphere S^{n-1} is compact, there exists a finite subcover $\{V_j\}$, $1 \leq j \leq r$, of $\mathbb{R}^n \setminus \{0\}$. Let $\varphi_j(x)$, $1 \leq j \leq r$, be the corresponding C_0^∞ functions, $\varphi_j(x_0) \neq 0$. Denote $\psi_0(x) = \prod_{j=1}^r \varphi_j(x)$. As in the proof of Proposition 14.1 we have that

$$|F(\psi_0 g)(\xi)| \leq C_N(1+|\xi|)^{-N}, \quad \forall N \quad \text{and} \quad \xi \in V_j', \quad 1 \leq j \leq r,$$

where V_j' is a slightly smaller cone than V_j, but $\{V_j'\}$, $1 \leq j \leq r$, still covers $\mathbb{R}^n \setminus \{0\}$. Therefore

$$|F(\psi_0 g)(\xi)| \leq C_N(1+|\xi|)^{-N}, \quad \forall N, \ \forall \xi. \tag{14.9}$$

Thus $\psi_0 g \in C_0^\infty$ and $x_0 \notin \operatorname{sing\,supp} g$. Vice versa, if $x_0 \notin \operatorname{sing\,supp} g$, then (14.9) holds for some $\varphi(x) \in C_0^\infty(\mathbb{R}^n), \varphi(x_0) \neq 0$. Thus x_0 does not belong to $\pi WF(g)$.

Example 14.2.

a) $WF(\delta) = \{(0,\xi), \forall \xi \neq 0\}$. Indeed, $F(\varphi\delta) = \varphi(0)$. If $\varphi(0) \neq 0$, then $F(\varphi\delta)$ does not decay in any direction $\frac{\xi}{|\xi|}$.

b) $WF((x_1 + i0)^{-1}) = \{(0,\xi_1), \xi_1 > 0\}$. Indeed, as in Example 10.2, we have

$$F((x_1+i0)^{-1}) = -2\pi i\theta(\xi_1),$$

where $\theta(\xi_1) = 1$ for $\xi_1 > 0$, $\theta(\xi_1) = 0$ for $\xi_1 < 0$. For any $\varphi(x_1) \in C_0^\infty(\mathbb{R}^1)$, $\varphi(0) \neq 0$, we get

$$\begin{aligned} F(\varphi(x_1)(x_1+i0)^{-1}) &= \frac{1}{2\pi}\tilde{\varphi} * F((x_1+i0)^{-1}) \\ &= -i\int_{-\infty}^{\infty} \theta(\xi_1 - \eta_1)\tilde{\varphi}(\eta_1)d\eta_1 = -i\int_{-\infty}^{\xi_1} \tilde{\varphi}(\eta_1)d\eta_1. \end{aligned}$$

Therefore $F(\varphi(x_1)(x_1+i0)^{-1})$ decays faster than any negative power of $|\xi_1|$ as $\xi_1 \to -\infty$, and it does not decay as $\xi_1 \to +\infty$, since $\varphi(0) = \frac{1}{2\pi}\int_{-\infty}^{\infty}\tilde{\varphi}(\eta_1)d\eta_1 \neq 0$. Therefore,

$$WF((x_1+i0)^{-1}) = \{(0,\xi_1), \xi_1 > 0\}.$$

Example 14.3. Consider the distribution $a(x')\delta(x_n)$, where

$$x = (x', x_n), \ x' = (x_1, \dots, x_{n-1}), \ a(x') \in C_0^\infty(\mathbb{R}^{n-1}).$$

Then $WF(a(x')\delta(x_n)) = \{(x', 0, 0, \xi_n) : x' \in \operatorname{supp} a \text{ and } \xi_n \neq 0\}$, i.e., $x_n = 0$, $\xi' = 0$ if $(x', x_n, \xi', \xi_n) \in WF(a(x')\delta(x_n))$.

Indeed, $\operatorname{sing\,supp}(a(x')\delta(x_n)) = \{(x',0),$ where $x' \in \operatorname{supp} a(x')\}$. If $\varphi(x', x_n) \in C_0^\infty(\mathbb{R}^n)$ and $\varphi(x_0', 0) \neq 0$ for some $x_0' \in \operatorname{supp} a(x')$, we have

$$F(\varphi(x',x_n)a(x')\delta(x_n)) = \int_{\mathbb{R}^{n-1}} \varphi(x',0)a(x')e^{-ix'\cdot\xi'}dx' \tag{14.10}$$

and

$$|F(\varphi(x',x_n)a(x')\delta(x_n))| \leq C_N(1+|\xi'|)^{-N}, \ \forall N, \ \forall \xi' \in \mathbb{R}^{n-1}.$$

If $\xi_0 = (\xi_0', \xi_{n0})$ and $\xi_0' \neq 0$, then

$$\left|\frac{\xi}{|\xi|} - \frac{\xi_0}{|\xi_0|}\right| < \delta, \quad 0 < \delta < \frac{1}{2}\frac{|\xi_0'|}{|\xi_0|}, \tag{14.11}$$

implies that $\frac{|\xi'|}{|\xi|} \geq \frac{|\xi_0'|}{|\xi_0|} - \delta > \frac{1}{2}\frac{|\xi_0'|}{|\xi_0|}$. Indeed,

$$\frac{|\xi_0'|}{|\xi_0|} - \frac{|\xi'|}{|\xi|} \leq \left|\frac{\xi'}{|\xi|} - \frac{\xi_0'}{|\xi_0|}\right| \leq \left|\frac{\xi}{|\xi|} - \frac{\xi_0}{|\xi_0|}\right| < \delta.$$

Therefore $\frac{|\xi'|}{|\xi|} > \frac{|\xi_0'|}{|\xi_0|} - \delta > \frac{1}{2}\frac{|\xi_0'|}{|\xi_0|}$. Since $|\xi'| > C|\xi|$, we have

$$|F(\varphi(x', x_n)a(x')\delta(x_n))| \leq C_{N1}(1 + |\xi|)^{-N}, \ \forall N,$$

if (14.11) holds. Thus $(x_0', 0, \xi_0', \xi_{n0}) \notin WF(a(x')\delta(x_n))$ if $\xi_0' \neq 0$. If $\xi_0' = 0$, then $(x_0', 0, \xi_0', \xi_{n0}) \in WF(a(x')\delta(x_n))$ since (14.10) does not decay when $\xi' = 0$ and $|\xi_n| \to \infty$ for any $\varphi(x', x_n) \in C_0^\infty$ such that $\int_{\mathbb{R}^{n-1}} \varphi(x', 0)a(x')dx' \neq 0$.

14.1. Products of distributions.

Theorem 14.3. *Let g_k, $k = 1, 2$, be two distributions in $\mathcal{D}'$ with the wave fronts sets $\Gamma_k = WF(g_k)$, $k = 1, 2$. Assume that*

$$\Gamma_1 \cap (-\Gamma_2) = \emptyset, \tag{14.12}$$

where $(x, \xi) \in -\Gamma_2$ means that $(x, -\xi) \in \Gamma_2$. Then the product $g_1 g_2 \in \mathcal{D}'$ is defined and

$$WF(g_1 g_2) \subset (\Gamma_1 + \Gamma_2) \cup \Gamma_1 \cup \Gamma_2, \tag{14.13}$$

where $(x, \xi) \in \Gamma_1 + \Gamma_2$ means that $\xi = \xi_1 + \xi_2$ for some $(x, \xi_1) \in \Gamma_1$, $(x, \xi_2) \in \Gamma_2$.

Proof: Arguing as in the proof of Proposition 14.2, for each $x_0 \in \mathbb{R}^n$ we get C_0^∞ functions $\varphi_k(x)$, $k = 1, 2$, such that $\varphi_k(x_0) \neq 0$ and

$$|F(\varphi_k g)(\xi)| \leq C_N(1 + |\xi|)^{-N}, \ \ \forall N, \ \ \forall \xi \in \mathbb{R}^n \setminus V_k, \tag{14.14}$$

where V_k are closed cones slightly larger than $\pi_k^{-1}(x_0)$, and $\pi_k^{-1}(x_0)$ is the conic set in $\mathbb{R}^n \setminus \{0\}$ consisting of all (x_0, ξ) such that $(x_0, \xi) \in WF(g_k)$. Since $\pi_1^{-1}(x_0) \cap (-(\pi_2^{-1}(x_0))) = \emptyset$, we may assume that $V_1 \cap (-V_2) = \emptyset$. Denote by U_{x_0} an open ball centered at x_0 such that $\varphi_k(x) \neq 0$ on $\overline{U_{x_0}}$. Such balls U_{x_0} form an open cover of $\mathbb{R}^n$. We can extract locally finite subcover U_j. Let $\varphi_{jk}(x)$ be C_0^∞ functions, $\varphi_{jk}(x) \neq 0$ on $\overline{U_j}$, and let V_{jk} be the corresponding cones, $k = 1, 2$, $V_{j1} \cap (-V_{j2}) = \emptyset$. Let $\psi_j(x)$ be $C_0^\infty(U_j)$ such that $\sum_j \psi_j^2 \equiv 1$. Then $\psi_j = a_{jk}\varphi_{jk}$, where $a_{jk} \in C_0^\infty(U_j)$. By Proposition 14.1,

$$|\tilde{g}_{jk}(\xi)| \leq C_N(1 + |\xi|)^{-N}, \ \ \forall N, \ \ \forall \xi \in \mathbb{R}^n \setminus V_{jk}', \tag{14.15}$$

where $g_{jk} = \psi_j g_k$, the V'_{jk} is slightly larger than the cone V_{jk} and we assume that $V'_{j1} \cap (-V'_{j2}) = \emptyset$.

It is enough to define $(\psi_j g_1)(\psi_j g_2)$. Then $g_1 g_2 = \sum_j (\psi_j g_1)(\psi_j g_2)$. We show that the integral

$$h_{j3}(\xi) = \frac{1}{(2\pi)^n} \int_{\mathbb{R}^n} \tilde{g}_{j1}(\xi - \eta)\tilde{g}_{j2}(\eta) d\eta \tag{14.16}$$

is absolutely convergent and

$$|h_{j3}(\xi)| \le C(1 + |\xi|)^{m_1 + m_2 + n}, \tag{14.17}$$

where m_k, $k = 1, 2$, are such that $|\tilde{g}_k(\xi)| \le (1 + |\xi|)^{m_k}$. Indeed, since $V'_{j1} \cap (-V'_{j2}) \neq \emptyset$, we can find functions $\alpha_{jk}(\xi)$, homogeneous of degree zero, that are C^∞ for $\xi \neq 0$, equal to 1 on V'_{jk}, and such that $\operatorname{supp} \alpha_{j1}(\xi) \cap \operatorname{supp} \alpha_{j2}(-\xi)$ is empty in $\mathbb{R}^n \setminus \{0\}$. We have

$$h_{j3}(\xi) = \sum_{p=1}^{4} I_p(\xi),$$

where

$$\begin{aligned}
I_1(\xi) &= \frac{1}{(2\pi)^n} \int_{\mathbb{R}^n} \alpha_{j1}(\xi - \eta)\tilde{g}_{j1}(\xi - \eta)\alpha_{j2}(\eta)\tilde{g}_{j2}(\eta) d\eta, \\
I_2(\xi) &= \frac{1}{(2\pi)^n} \int_{\mathbb{R}^n} \alpha_{j1}(\xi - \eta)\tilde{g}_{j1}(\xi - \eta)(1 - \alpha_{j2}(\eta))\tilde{g}_{j2}(\eta) d\eta, \\
I_3(\xi) &= \frac{1}{(2\pi)^n} \int_{\mathbb{R}^n} (1 - \alpha_{j1}(\xi - \eta))\tilde{g}_{j1}(\xi - \eta)\alpha_{j2}(\eta)\tilde{g}_{j2}(\eta) d\eta, \\
I_4(\xi) &= \frac{1}{(2\pi)^n} \int_{\mathbb{R}^n} (1 - \alpha_{j1}(\xi - \eta))\tilde{g}_{j1}(\xi - \eta)(1 - \alpha_{j2}(\eta))\tilde{g}_{j2}(\eta) d\eta.
\end{aligned}$$

Since

$$|(1 - \alpha_{jk}(\xi))\tilde{g}_{jk}(\xi)| \le C_n(1 + |\xi|)^{-N}, \quad \forall N, \quad \forall \xi,$$

we get, using (13.10):

$$\begin{aligned}
|I_4(\xi)| &\le C_N \int_{\mathbb{R}^n} \frac{d\eta}{(1 + |\xi - \eta|)^{2N}(1 + |\eta|)^N} \le \frac{C_{N1}}{(1 + |\xi|)^N}, \\
|I_2(\xi)| &\le C_N \int_{\mathbb{R}^n} (1 + |\xi - \eta|)^{m_1}(1 + |\eta|)^{-N} d\eta \le C_{N1}(1 + |\xi|)^{m_1}, \\
|I_3(\xi)| &\le C_N \int_{\mathbb{R}^n} (1 + |\xi - \eta|)^{-N}(1 + |\eta|)^{m_2} d\eta \le C_{N1}(1 + |\xi|)^{m_2}.
\end{aligned}$$

It remains to estimate $I_1(\xi)$.

Since $\operatorname{supp} \alpha_{j1}(\eta) \cap \operatorname{supp} \alpha_{j2}(-\eta) = \emptyset$, we have that for each ξ, the angle between the vectors $-\eta$ and $\xi - \eta$ is bounded below when $\eta \in \operatorname{supp} \alpha_{j2}(\eta)$

and $\xi - \eta \in \operatorname{supp} \alpha_{j1}(\xi - \eta)$. Therefore $|\xi| \geq C|\eta|$ on $\operatorname{supp} \alpha_{j1}(\xi - \eta)\alpha_{j2}(\eta)$. Thus

$$|I_1(\xi)| \leq C \int_{C|\eta| \leq |\xi|} (1 + |\xi - \eta|)^{m_1}(1 + |\eta|)^{m_2} d\eta \leq C_1(1 + |\xi|)^{m_1+m_2+n}.$$

Therefore the integral (14.16) converges absolutely and satisfies (14.17).

We define $(\psi_j g_1)(\psi_j g_2)$ as the inverse Fourier transform of $h_{j3}(\xi)$:

$$(\psi_j g_1)(\psi_j g_2) = F^{-1}\Big(\frac{1}{(2\pi)^n} \int \tilde{g}_{j1}(\xi - \eta)\tilde{g}_{j2}(\eta) d\eta\Big). \tag{14.18}$$

It remains to prove (14.13). Denote by V''_{j1}, V''_{j2} closed cones in $\mathbb{R}^n \setminus \{0\}$ such that $V''_{j1} \cap (-V''_{j2}) = \emptyset$ and the interior of V''_{jk} contains $\operatorname{supp} \alpha_{jk}(\xi)$, $k = 1, 2$. It is enough to prove that

$$|h_{j3}(\xi)| \leq C_N(1 + |\xi|)^{-N}, \quad \forall N,$$

when $\xi \notin (V''_{j1} + V''_{j2}) \cup V''_{j1} \cup V''_{j2}$. If $\xi \notin V''_{j1} + V''_{j2}$, then $\tilde{\alpha}_{j1}(\xi - \eta)\tilde{\alpha}_{j2}(\eta) = 0$ for all η and, therefore, $I_1(\xi) = 0$. We split $I_2(\xi)$ into two integrals: $I_2(\xi) = I_{21}(\xi) + I_{22}(\xi)$, where

$$I_{21}(\xi) = \int_{|\eta| < \varepsilon|\xi|} \alpha_{j1}(\xi - \eta)\tilde{g}_{j1}(\xi - \eta)(1 - \alpha_{j2}(\eta))\tilde{g}_{j2}(\eta) d\eta$$

and

$$I_{22}(\xi) = I_2(\xi) - I_{21}(\xi).$$

If $\xi \notin \operatorname{supp} V''_{j1}$ and ε is small enough, then $\xi - \eta \notin \operatorname{supp} \alpha_{j1}(\xi - \eta)$ for $|\eta| < \varepsilon|\xi|$, i.e., $I_{21}(\xi) = 0$. We have

$$\begin{aligned}
|I_{22}(\xi)| &\leq \int_{|\eta| > \varepsilon|\xi|} C(1 + |\xi - \eta|)^{m_1} C_N(1 + |\eta|)^{-N} d\eta \\
&\leq C_{N\varepsilon} \int_{|\eta| > \varepsilon|\xi|} \frac{(1 + |\xi| + |\eta|)^{m_1}}{(1 + |\eta|)^{n+1+m_1}(1 + |\xi|)^{N-n-m_1-1}} d\eta \\
&\leq C^{(1)}_{N\varepsilon}(1 + |\xi|)^{-N+n+1+2m_1},
\end{aligned}$$

where N is arbitrary. Analogously one estimates $I_3(\xi)$ for $\xi \notin V''_{j2}$. □

14.2. Restrictions of distributions to a surface.

Example 14.4. Consider $WF(\delta(S))$, where $S(x) = 0$ is a smooth surface (see Example 5.5). We prove that

$$WF(\delta(S)) = \{(x, \xi) : S(x) = 0 \text{ and } \xi = \pm t\nabla S(x), \forall t > 0\}. \tag{14.19}$$

Since $\delta(S) = 0$ outside $S(x) = 0$, it is enough to consider x such that $S(x) = 0$. We show that $(x_0, \xi_0) \notin WF(\delta(S))$ if $S(x_0) = 0$ and ξ_0 is not parallel to $\nabla S(x_0)$, i.e., $\xi_0 \neq \pm t\nabla S(x_0)$, $\forall t > 0$.

Let $\varphi(x)$ be a C_0^∞ function, $\varphi(x_0) \neq 0$ and let $\operatorname{supp}\varphi$ be so small that ξ_0 is not parallel to $\nabla S(x)$ for any $x \in \operatorname{supp}\varphi \cap \{S(x) = 0\}$. We have (see Example 5.5 and Proposition 11.1)

$$F(\varphi\delta(S)) = \int_{S(x)=0} \frac{\varphi(x)e^{-ix\cdot\xi}}{|\nabla S(x)|} d\sigma,$$

where $d\sigma$ is the surface area element of the surface $S(x) = 0$.

Let $x = x(t')$, $t' = (t_1, \dots, t_{n-1})$ be the parametric equations of the surface $S(x) = 0$ on $\operatorname{supp}\varphi$. Then

$$F(\varphi\delta(S)) = \int_{\mathbb{R}^{n-1}} \frac{\varphi(x(t'))e^{-ix(t')\cdot\xi}}{|\nabla S(x(t')|} a(t')dt', \tag{14.20}$$

where $a(t') \in C^\infty(\mathbb{R}^{n-1})$. We have for any t_k, $1 \le k \le n-1$:

$$\frac{\partial}{\partial t_k} e^{-ix(t')\cdot\xi} = (-i)\left(\frac{\partial x(t')}{\partial t_k} \cdot \xi\right) e^{-ix(t')\cdot\xi}.$$

Note that $\frac{\partial x(t')}{\partial t_k}$ is a tangent vector to the surface $S(x) = 0$. Since ξ_0 is not parallel to $\nabla S(x)$ on $\operatorname{supp}\varphi \cap \{x : S(x) = 0\}$ and $\nabla S(x(t'))$ is the only vector (modulo a nonzero constant) that is perpendicular to all tangent vectors $\frac{\partial x(t')}{\partial t_k}$, $1 \le k \le n-1$, we see that there exists k such that $\frac{\partial x(t')}{\partial t_k} \cdot \xi \neq 0$ for all $x(t') \in \operatorname{supp}\varphi \cap \{x : S(x) = 0\}$ and all ξ such that $|\frac{\xi}{|\xi|} - \frac{\xi_0}{|\xi_0|}| < \delta$, $\delta > 0$. Therefore $|\frac{\partial x(t')}{\partial t_k} \cdot \xi| \ge C|\xi|$ for such $x(t')$ and ξ.

For an arbitrary N we have:

$$\begin{aligned} e^{-ix(t')\cdot\xi} &= i\left(\frac{\partial x(t')}{\partial t_k} \cdot \xi\right)^{-1} \frac{\partial}{\partial t_k} e^{-ix(t')\cdot\xi} \\ &= \left[\left(-i\frac{\partial x(t')}{\partial t_k} \cdot \xi\right)^{-1} \frac{\partial}{\partial t_k}\right]^N e^{-ix(t')\cdot\xi}. \end{aligned} \tag{14.21}$$

Substituting (14.21) in (14.20) and integrating by parts N times, we get

$$|F(\varphi\delta(S))(\xi)| \le C_N(1 + |\xi|)^{-N}, \quad \forall N,$$

for all ξ such that $|\frac{\xi}{|\xi|} - \frac{\xi_0}{|\xi_0|}| < \delta$. Therefore

$$WF(\delta(S)) \subset \{(x, \xi) : S(x) = 0 \text{ and } \xi = \pm t\nabla S(x),\ t > 0\}.$$

We show that

$$WF(\delta(S)) = \{(x, \xi) : S(x) = 0 \text{ and } \xi = \pm t\nabla S(x)\}.$$

Note that if $g(x)$ is a real-valued absolutely integrable function, then $\overline{\tilde{g}(\xi)} = \tilde{g}(-\xi)$. The same relation holds for the Fourier transform of real-valued distributions. Therefore, if $(x_0, \xi_0) \in WF(\delta(S))$, then $(x_0, -\xi_0) \in WF(\delta(S))$.

Thus $(x_0, \pm t\nabla S(x_0))$, where $S(x_0) = 0$, belong to $WF(\delta(S))$, since otherwise $\varphi\delta(S)$ will be C_0^∞ in a neighborhood of x_0. □

Denote $N(S) = \{(x, \pm t\nabla S(x)) : S(x) = 0, \forall t > 0\}$.

Corollary 14.4. *One can define the restriction of $f \in \mathcal{D}'$ to a smooth surface $S(x) = 0$ as the product of f and $a\delta(S)$, where $a \in C_0^\infty(S)$:*

$$f\big|_S(a) = (fa\delta(S))(1), \tag{14.22}$$

assuming that $WF(f) \cap N(S) = \emptyset$.

Note that $a\delta(S)$ and therefore $fa\delta(S)$ are distributions with compact supports.

Corollary 14.4 follows from Theorem 14.3 because $f\delta(S)$ is defined if $WF(f) \cap N(S) = \emptyset$, where $N(S) = WF(\delta(S))$.

For example, the restriction of $f \in \mathcal{D}'(\mathbb{R}^n)$ to the hyperplane $x_n = 0$ is the following distribution in $\mathcal{D}'(\mathbb{R}^{n-1})$:

$$f\big|_{x_n=0}(\varphi(x')) = (f\varphi(x')\delta(x_n))(1), \tag{14.23}$$

where $\varphi(x')$ is arbitrary $C_0^\infty(\mathbb{R}^{n-1})$ function. This restriction is well defined if $(x', 0, 0, \pm 1)$ does not belong to $WF(f)$.

15. Problems

1. **a)** Find three linearly independent solutions in $\mathcal{D}'(\mathbb{R}^1)$ of the differential equation $x^2 \frac{du}{dx} = 0$.
   b) Find $k+q$ linearly independent solutions in $\mathcal{D}'(\mathbb{R}^1)$ of $x^q y^{(k)} = 0$.

2. Find the Fourier transform of the following distributions:
   a) $f(x_1) = |x_1|$;
   b) $f(x_1) = \sin ax_1$;
   c) $f(x_1) = x_1 \sin ax_1$;
   d) $f(x_1) = \ln|x_1|$.

3. Prove that any solution $u \in S'$ of $(\Delta + k^2)u = 0$ is an entire function satisfying the estimate

$$|u(x_1 + iy_1, \ldots, x_n + iy_n)| \leq C_{k_1}(1 + |x|)^m e^{k_1|y|}$$

for some m and arbitrary $k_1 > k$.

4. A function $f(x)$, continuous for $x \neq 0$, is called homogeneous of degree λ if $f(tx) = t^\lambda f(x)$ for all $t > 0$ and $x \neq 0$. Analogously, a distribution $f \in \mathcal{D}'$ is called homogeneous of degree λ if $(f \circ s)(\varphi) = t^\lambda f(\varphi)$, where $(f \circ s)(\varphi) = t^{-n} f(\varphi(t^{-1}\cdot))$ and $s(x) = tx$, $t > 0$. Prove that

a) x_+^λ, $\lambda \in \mathbb{C} \setminus N_-$, is homogeneous of degree λ;
b) δ is homogeneous of degree $-n$;
c) $(x_1 \pm i0)^\lambda$ are homogeneous of degree λ for each $\lambda \in \mathbb{C}$.

5. Prove that if $f \in \mathcal{D}'$ is homogeneous of degree λ, then $\frac{\partial^k f}{\partial x^k}$ is homogeneous of degree $\lambda - |k|$.

6. Prove that if f is a homogeneous distribution of degree λ, then its Fourier transform $\tilde{f}$ is a homogeneous distribution of degree $-\lambda-n$.

7. A distribution f is called spherically symmetric if $f \circ s(\varphi) = f(\varphi)$ for each orthogonal transformation $x = sy$ in $\mathbb{R}^n$. Here $(f \circ s)(\varphi) = f(\varphi \circ s^{-1})$. Prove that if a tempered distribution f is spherically symmetric, then its Fourier transform $\tilde{f} \in S'$ is also spherically symmetric.

8. Consider the regular functional $|x|^\lambda$ in $\mathbb{R}^n$ with $\operatorname{Re}\lambda > -n$.
   a) Prove that $|x|^\lambda$ has an analytic extension to $\mathbb{C} \setminus \{-n, -n-2, -n-4, \dots\}$.
   b) Prove that
$$F(|x|^\lambda) = C_\lambda |\xi|^{-\lambda-n},$$
$\lambda \neq 0, 2, 4, \dots$, $\lambda \neq -n, -n-2, -n-4, \dots$.
   c) Find the constant C_λ by using that $|x|^\lambda(\tilde{\varphi}) = C_\lambda |\xi|^{-\lambda-n}(\varphi)$ with $\varphi = e^{-\frac{|\xi|^2}{2}}$.

9. Find the wave front set of x_+^λ, $\lambda \in \mathbb{C} \setminus N_-$.

10. Prove that $WF((S+i0)^{-1}) = \{(x, t\nabla S(x)) : S(x) = 0, \ \forall t > 0\}$ and $WF((S-i0)^{-1}) = \{(x, -t\nabla S(x) : S(x) = 0 \text{ and } t > 0\}$. Here $(S \pm i0)^{-1}$ are the same distributions as in Example 5.5.

11. Let M_{n-r} be the intersection of r smooth surfaces $S_1(x) = 0, \dots, S_r(x) = 0$ such that if $x_0 \in M_{n-r}$, then the gradients $\nabla S_1(x_0), \dots, \nabla S_r(x_0)$ are linearly independent. Thus M_{n-r} is a smooth manifold of dimension $n-r$. Define $\delta(M_{n-r})$ and find $WF(\delta(M_{n-r}))$.

12. Prove that the differential equation
$$-2x^3 y' = y$$
has no nonzero solutions in $\mathcal{D}'(\mathbb{R}^1)$.

13. a) For an arbitrary $f \in \mathcal{D}'(\mathbb{R}^1)$ find $u \in \mathcal{D}'(\mathbb{R}^1)$ such that $xu = f$.
    b) Prove that the differential equation $xy' = f$ has a solution in $\mathcal{D}'(\mathbb{R}^1)$ for each $f \in \mathcal{D}'(\mathbb{R}^1)$.

14. a) Prove that
$$\frac{x_+^{\lambda-1}}{\Gamma(\lambda)} * \frac{x_+^{\mu-1}}{\Gamma(\mu)} = \frac{x_+^{\lambda+\mu-1}}{\Gamma(\lambda+\mu)}.$$

b) Solve the Abel equation,

$$\frac{1}{\Gamma(\lambda)}\int_0^x (x-y)^{\lambda-1}u(y)dy = f(x), \quad x>0, \quad 0<\lambda<1,$$

assuming that $f(x)$ is absolutely continuous. (Hint: Use the Fourier transform formula (11.7).) Answer:

$$u(x) = \frac{1}{\Gamma(1-\alpha)}\int_0^x f'(y)(x-y)^{-\lambda}dy.$$

15. Deduce from Theorem 10.6 the classical Liouville theorem: If $f(z)$ is an entire function and $|f(z)| \leq C(1+|z|)^N$, then $f(z) = \sum_{k=0}^N c_k z^k$.

16. A distribution $g_\lambda \in \mathcal{D}'$ depends analytically on a parameter $\lambda \in \Omega \subset \mathbb{C}$ if $g_\lambda(\varphi)$ is analytic in the domain Ω for every $\varphi \in \mathcal{D}$. Prove that if $g_\lambda \in S'$ depends analytically on λ, then $\tilde{g}_\lambda \in S'$ also depends analytically on $\lambda \in \Omega$.

17. Suppose $a(x) \in C^\infty(\mathbb{R}^n)$ and $|\frac{\partial^k a(x)}{\partial x^k}| \leq C_k$ for any $|k| \geq 0$. Prove that

$$\|au\|_s \leq C_s\|u\|_s, \quad \forall s, \quad \forall u \in C_0^\infty(\mathbb{R}^n).$$

(Hint: For $s \geq 0$ use the norms (13.6) and (13.19). For $s<0$ use the duality arguments as in the proof of Theorem 13.8.)

18. Prove the invariance of the Sobolev spaces $H_{m+\gamma}$, $0<\gamma<1$, under the change of variables $x = s(y)$ using the norm (13.19).

19. Let $\mathbb{R}^{n-k}$ be the subspace in $\mathbb{R}^n$ given by the equations $x'' = (x_{n-k+1},\ldots,x_n) = (0,\ldots,0)$. Let $x = (x', x'')$, where $x' = (x_1,\ldots, x_{n-k})$. Prove that $[u(x',0)]_{s-\frac{k}{2}} \leq C\|u\|_s$ for $s > \frac{k}{2}$, $\forall u \in C_0^\infty(\mathbb{R}^N)$, where $[u(x',0)]_{s-\frac{k}{2}}$ is the norm in $H_{s-\frac{k}{2}}(\mathbb{R}^{n-k})$.

20. Denote by $H_s^p(\mathbb{R}^n)$, $p>1$, $s \in R^1$, the space of tempered distributions such that

$$\|u\|_{s,p}^p = \int_{\mathbb{R}^n}(1+|\xi|)^{ps}|\tilde{u}(\xi)|^p d\xi < +\infty$$

(i.e., $H_s^2(\mathbb{R}^n)$ is the Sobolev space $H_s(\mathbb{R}^n)$). Let $q = \frac{p}{p-1}$. Prove that

a) $[u(x',0)]_{s-\frac{1}{q},p} \leq C\|u\|_{s,p}$ if $s > \frac{1}{q}$ for all $u \in C_0^\infty$, where $[u(x',0)]_{s-\frac{1}{q},p}$ is the norm in $H_{s-\frac{1}{q}}^p(\mathbb{R}^{n-1})$.

b) $\sup_{x\in\mathbb{R}^n}|u(x)| \leq C\|u\|_{s,p}$ if $sq>n$, $\forall u \in C_0^\infty$.

c) $(H_s^p)^* = H_{-s}^q$, where $(H_s^p)^*$ is the dual space to H_s^p with respect to the extension of the L_2 scalar product (φ, ψ).

21. **a)** Prove that $\frac{d^{2m} \ln |x_1|}{dx_1^{2m}}$ is the only even distribution in $\mathbb{R}^1$ that is homogeneous of degree $-2m$, $m \geq 1$.

b) Prove that $\delta^{(2m)}(x_1)$ is the only even distribution in $\mathbb{R}^1$ that is homogeneous of degree $-2m-1$, $m \geq 0$.

Chapter III

Applications of Distributions to Partial Differential Equations

16. Partial differential equations with constant coefficients

Let $P(-i\frac{\partial}{\partial x})$ be a partial differential operator with constant coefficients. A distribution $E \in \mathcal{D}'$ is called a fundamental solution to the equation $P(-i\frac{\partial}{\partial x})u = f$ if

$$P\left(-i\frac{\partial}{\partial x}\right)E = \delta. \tag{16.1}$$

If $|P(\xi)| \geq C > 0$ for all $\xi \in \mathbb{R}^n$, then $\frac{1}{P(\xi)}$ is a regular functional in $S'(\mathbb{R}^n)$ and $E = F^{-1}\frac{1}{P(\xi)} \in S'(\mathbb{R}^n)$ will be a fundamental solution. In the next section we construct a fundamental solution in $\mathcal{D}'(\mathbb{R}^n)$ if $P(\xi)$ is an arbitrary polynomial.

It is clear that the fundamental solution E is not unique, since for any $v \in \mathcal{D}'$ such that $P(-i\frac{\partial}{\partial x})v = 0$ in $\mathbb{R}^n$ we have $P(-i\frac{\partial}{\partial x})(E + v) = \delta$.

If f is a distribution with a compact support, then

$$u = E * f \tag{16.2}$$

gives a particular solution to

$$P\left(-i\frac{\partial}{\partial x}\right)u = f, \tag{16.3}$$

since $P(-i\frac{\partial}{\partial x})(E * f) = P(-i\frac{\partial}{\partial x}E) * f = \delta * f = f$. □

Let

$$P\left(-i\frac{\partial}{\partial x}, -i\frac{\partial}{\partial t}\right) u(x,t) = 0 \tag{16.4}$$

be a partial differential equation in the half-space $\mathbb{R}^{n+1}_+ = \{(x,t) : x \in \mathbb{R}^n, t > 0\}$. We assume that $u(x,t)$ is smooth in $t \in \mathbb{R}^1_+$ with values in $S'(\mathbb{R}^n)$. Then, performing the Fourier transform in x', we get an ordinary differential equation in t depending on $\xi \in \mathbb{R}^n$ as parameters:

$$P\left(\xi, -i\frac{\partial}{\partial t}\right) \tilde{u}(\xi,t) = 0, \quad t > 0. \tag{16.5}$$

We can find explicitly the solution of (16.5) and then, applying the inverse Fourier transform, we can find the solution of (16.4) subject to appropriate conditions at $t = 0$.

In the remaining part of this section we will illustrate this approach by considering some important equations of mathematical physics.

16.1. The heat equation.

Example 16.1. The equation

$$\frac{\partial u(x,t)}{\partial t} - a^2 \Delta u(x,t) = 0 \tag{16.6}$$

is called the heat equation. The solution $u(x,t)$ of (16.6) has the physical meaning of the temperature at (x,t), where $x \in \mathbb{R}^n$ is the space variable and $t \in (0,\infty)$ is the time variable. We consider (16.6) in the half-space $t > 0$ subject to the initial condition at $t = 0$:

$$u(x,0) = u_0(x), \tag{16.7}$$

i.e., $u_0(x)$ is the initial temperature.

We assume that $u(x,t) \in C(\overline{\mathbb{R}^1_+}, S'(\mathbb{R}^n))$, i.e., $u(x,t)$ is continuous in t, $t \geq 0$, with values in $S'(\mathbb{R}^n)$.

Taking the Fourier transform in x, we get

$$\frac{\partial \tilde{u}(\xi,t)}{\partial t} = -a^2|\xi|^2 \tilde{u}(\xi,t), \tag{16.8}$$

$$\tilde{u}(\xi,0) = \tilde{u}_0(\xi), \tag{16.9}$$

where $\tilde{u}(\xi,t) = F_x u(x,t)$. The solution of the initial value problem (Cauchy problem) (16.8), (16.9) is

$$\tilde{u}(\xi,t) = e^{-a^2 t|\xi|^2} \tilde{u}_0(\xi). \tag{16.10}$$

Denote $G(x,t) = F_\xi^{-1} e^{-a^2t|\xi|^2}$. We have

$$G(x,t) = \frac{1}{(2\pi)^n}\int_{\mathbb{R}^n} e^{-a^2|\xi|^2 t + ix\cdot\xi} d\xi$$
$$= \frac{1}{(2\pi)^n}\int_{\mathbb{R}^n} e^{-a^2t(\xi - \frac{ix}{2a^2t})\cdot(\xi - \frac{ix}{2a^2t})} \cdot e^{-\frac{|x|^2}{4a^2t}} d\xi.$$

Using the Cauchy integral theorem, we replace the integration over $\operatorname{Im}\xi = 0$ by the integration over $\operatorname{Im}\xi = \frac{ix}{2a^2t}$. We get

$$G(x,t) = \frac{1}{(2\pi)^n}\int_{\mathbb{R}^n} e^{-a^2t|\xi|^2} d\xi \cdot e^{-\frac{|x|^2}{4a^2t}}.$$

Changing the variables $\xi = \frac{\eta}{a\sqrt{t}}$ and taking into account that $\int_{-\infty}^{\infty} e^{-\eta_j^2} d\eta_j = \sqrt{\pi}, j = 1, \dots, n$, we finally get

$$G(x,t) = \frac{1}{(4\pi a^2 t)^{\frac{n}{2}}} e^{-\frac{|x|^2}{4a^2t}}, \qquad t > 0. \tag{16.11}$$

The function $G(x,t)$ is called the heat kernel. Note that $G(x,t) \to \delta$ as $t \to 0$ (cf. Example 4.4, c)). Using Proposition 12.1, we obtain from (16.10) that

$$u(x,t) = \frac{1}{(4\pi a^2 t)^{\frac{n}{2}}}\int_{\mathbb{R}^n} e^{-\frac{|x-y|^2}{4a^2t}} u_0(y) dy. \tag{16.12}$$

Note that it was assumed in Proposition 12.1 that $u_0(x)$ has a compact support. However, from (16.12) one can see that it is enough to assume that $u_0(x)$ is continuous and bounded in $\mathbb{R}^n$. Then (16.12) satisfies (16.6) for $t > 0$ and $u(x,t) \to u_0(x)$ as $t \to 0$ and x is fixed (cf. Example 4.4, a)). □

Consider equation (16.6), assuming that $u_0(x) \in L_2(\mathbb{R}^n)$. Then, by Plancherel's theorem, $\tilde{u}_0(\xi) \in L_2(\mathbb{R}^n)$ and (16.10) implies that

$$\|u(\cdot,t)\|_0 \le \|u_0\|_0, \quad t \ge 0. \tag{16.13}$$

It follows from the Lebesgue convergence theorem that $\tilde{u}(\xi,t)$ is continuous in $t \ge 0$ with values in $L_2(\mathbb{R}^n)$. Therefore, $u(x,t) \in C(\overline{\mathbb{R}^1_+}, L_2(\mathbb{R}^n))$. In particular,

$$\|u(\cdot,t) - u_0\|_0^2 = \frac{1}{(2\pi)^n}\int_{\mathbb{R}^n} |e^{-a^2t|\xi|^2} - 1|^2 |\tilde{u}_0(\xi)|^2 d\xi \to 0$$

as $t \to 0$. □

Let $E(x,t) = G(x,t)$ for $t \ge 0$ and $E(x,t) = 0$ for $t < 0$. Then $E(x,+0) - E(x,-0) = \delta(x)$. Therefore $(\frac{\partial}{\partial t} - a^2\Delta)E(x,t) = \delta(x)\delta(t)$ in

$\mathbb{R}^{n+1}$, i.e., $E(x,t)$ is a fundamental solution of (16.6) in $\mathbb{R}^{n+1}$. Let $f(x,t)$ be bounded and continuous in $\mathbb{R}^{n+1}$, and $f(x,t)=0$ for $t<0$. Then

$$u(x,t) = E * f = \int_0^t \int_{\mathbb{R}^n} E(x-y,t-\tau)f(y,\tau)dyd\tau$$

is the solution of

$$\frac{\partial u}{\partial t} - a^2\Delta u = f(x,t) \text{ in } \mathbb{R}^{n+1}. \tag{16.14}$$

Remark 16.1. The Cauchy problem (16.6), (16.7) has a unique solution in the class $C(\mathbb{R}^1, S'(\mathbb{R}^n))$ of $u(x,t)$ continuous in t, $t \geq 0$, with values in $S'(\mathbb{R}^n)$, since (16.10) is obviously a unique solution of (16.8), (16.9).

16.2. The Schrödinger equation.

Example 16.2. The Schrödinger equation

$$-ih\frac{\partial u(x,t)}{\partial t} = \frac{h^2}{2m}\Delta u(x,t), \quad x \in \mathbb{R}^n, \ t > 0, \tag{16.15}$$

describes a nonrelativistic electron in quantum mechanics. Here $h>0$ is the Planck constant and m is the mass of the electron. The quantity $|u(x,t)|^2$ has the physical meaning of the probability density, i.e., $|u(x,t)|^2dx$ is the probability that the electron is located in a neighborhood of $x \in \mathbb{R}^n$ at time t. As for the heat equation, we consider the initial value problem (the Cauchy problem) of finding the solution of (16.5) for $t>0$ satisfying the initial condition

$$u(x,0) = u_0(x). \tag{16.16}$$

We assume that $u(x,t) \in C(\overline{\mathbb{R}^1_+}, S'(\mathbb{R}^n))$. The Fourier transform in x gives

$$-ih\frac{\partial \tilde{u}(\xi,t)}{\partial t} = -\frac{h^2}{2m}|\xi|^2\tilde{u}(x,t), \quad t>0,$$
$$\tilde{u}(\xi,0) = \tilde{u}_0(\xi).$$

Therefore

$$\tilde{u}(\xi,t) = e^{-\frac{iht|\xi|^2}{2m}}\tilde{u}_0(\xi). \tag{16.17}$$

Note that $u(x,t) = F_\xi^{-1}\big(e^{-\frac{iht|\xi|^2}{2m}}\tilde{u}_0\big)$ is the unique solution of (16.15), (16.16) in $C(\overline{\mathbb{R}^1_+}, S'(\mathbb{R}^n))$ (cf. Remark 16.1).

Let $g(x,t) = F_\xi^{-1}e^{-\frac{iht|\xi|^2}{2m}}$. As in (16.11), we have

$$g(x,t) = \frac{m^{\frac{n}{2}}e^{-in\frac{\pi}{4}}}{(2\pi ht)^{\frac{n}{2}}}e^{\frac{im|x|^2}{2ht}}.$$

Then

$$u(x,t) = \int_{\mathbb{R}^n} g(x-y,t)u_0(y)dy.$$

Assuming that $u_0(x)$ is continuous and $\int_{\mathbb{R}^n}(1+|x|^2)|u_0(x)|dx < +\infty$, we conclude that $u(x,t)$ is a classical solution of (16.15) (i.e., $u(x,t)$ has a continuous derivative with respect to t and second order continuous derivatives with respect to x), and $u(x,t) \to u_0(x)$ as $t \to 0$.

Suppose $u_0(x) \in L_2(\mathbb{R}^n)$. Then $\tilde{u}_0(\xi) \in L_2(\mathbb{R}^n)$ (Plancherel's theorem). It follows from (16.17) that

$$\|u(\cdot,t)\|_0^2 = \|u_0\|_0^2, \quad \forall t. \tag{16.18}$$

Also we have that $u(x,t) \in C(\overline{\mathbb{R}^1_+}, L_2(\mathbb{R}^n))$, i.e., $u(x,t)$ is continuous in t, $t \geq 0$, with values in $L_2(\mathbb{R}^n)$ and

$$u(x,+0) = u_0(x).$$

16.3. The wave equation.

Example 16.3. The wave equation has the form

$$\frac{\partial^2 u(x,t)}{\partial t^2} - a^2 \Delta u(x,t) = 0, \quad t > 0, \ x \in \mathbb{R}^n. \tag{16.19}$$

It describes the propagation of waves (disturbances) in medium. Here a is the speed of the wave propagation. Consider the Cauchy problem for (16.19):

$$u(x,0) = \varphi(x), \qquad \frac{\partial u(x,0)}{\partial t} = \psi(x), \quad x \in \mathbb{R}^n, \tag{16.20}$$

i.e., we prescribe the initial form of the wave $\varphi(x)$ and the initial velocity $\psi(x)$.

Similarly to the case of the heat equation, we consider $u(x,t) \in C^1(\overline{\mathbb{R}^1_+}, S'(\mathbb{R}^n))$, i.e., $u(x,t)$ is continuously differentiable with respect to t, $t \geq 0$, with values in $S'(\mathbb{R}^n)$. Making the Fourier transform in x we get

$$\frac{\partial^2 \tilde{u}(\xi,t)}{\partial t^2} + a^2|\xi|^2 \tilde{u}(\xi,t) = 0, \tag{16.21}$$

$$\tilde{u}(\xi,0) = \tilde{\varphi}(\xi), \qquad \frac{\partial u(\xi,0)}{\partial t} = \tilde{\psi}(\xi). \tag{16.22}$$

Solving the Cauchy problem for the ordinary differential equation (16.21), (16.22), we obtain

$$\tilde{u}(\xi,t) = (\cos a|\xi|t)\tilde{\varphi}(\xi) + \frac{\sin a|\xi|t}{a|\xi|}\tilde{\psi}(\xi). \tag{16.23}$$

Consider the case where $\varphi(x) \in H_1(\mathbb{R}^n)$, $\psi(x) \in H_0(\mathbb{R}^n) = L_2(\mathbb{R}^n)$ (cf. §13). It follows from (16.23) that $\tilde{u}(\xi,t) \in L_2(\mathbb{R}^n)$ for all $t > 0$ and $\tilde{u}(\xi,t)$ depends continuously on t in $L_2(\mathbb{R}^n)$.

Consider

$$\varepsilon(\xi,t)=\left|\frac{\partial\tilde{u}(\xi,t)}{\partial t}\right|^2+a^2|\xi|^2|\tilde{u}(\xi,t)|^2.$$

Differentiating and using (16.21), we get

$$\begin{aligned}\frac{\partial\varepsilon}{\partial t}&=\frac{\partial^2\tilde{u}}{\partial t^2}\frac{\partial\overline{\tilde{u}}}{\partial t}+\frac{\partial\tilde{u}}{\partial t}\frac{\partial^2\overline{\tilde{u}}}{\partial t^2}+a^2|\xi|^2\left(\frac{\partial\tilde{u}}{\partial t}\overline{\tilde{u}}+\tilde{u}\frac{\partial\overline{\tilde{u}}}{\partial t}\right)\\&=\left(\frac{\partial^2\tilde{u}}{\partial t^2}+a^2|\xi|^2)\tilde{u}\right)\frac{\partial\overline{\tilde{u}}}{\partial t}+\frac{\partial\tilde{u}}{\partial t}\left(\frac{\partial^2\overline{\tilde{u}}}{\partial t^2}+a^2|\xi|^2\overline{\tilde{u}}\right)=0.\end{aligned}$$

Therefore

$$\left|\frac{\partial\tilde{u}(\xi,t)}{\partial t}\right|^2+a^2|\xi|^2|\tilde{u}(\xi,t)|^2=\left|\frac{\partial\tilde{u}(\xi,0)}{\partial t}\right|^2+a^2|\xi|^2|\tilde{u}(\xi,0)|^2. \tag{16.24}$$

Let

$$\mathcal{E}(t)=\frac{1}{2}\int_{\mathbb{R}^n}\left(\left|\frac{\partial u(x,t)}{\partial t}\right|^2+a^2\left|\frac{\partial u(x,t)}{\partial x}\right|^2\right)dx. \tag{16.25}$$

The integral $\mathcal{E}(t)$ has the physical meaning of energy at time t.

Integrating (16.24) and using Parseval's equality, we get that

$$\mathcal{E}(t)=\mathcal{E}(0),\quad \forall t\geq 0, \tag{16.26}$$

i.e., the energy $\mathcal{E}(t)$ is conserved.

Therefore, we conclude that for any $\varphi\in H_1(\mathbb{R}^n)$ and $\psi\in L_2(\mathbb{R}^n)$, there exists a unique solution $u(x,t)$ of the Cauchy problem (16.19), (16.20) such that $u(x,t)\in C^1(\overline{\mathbb{R}^1_+},L_2(\mathbb{R}^n))\cap C(\overline{\mathbb{R}^1_+},H_1(\mathbb{R}^n))$, i.e., $u(x,t)$ depends continuously on t with values in $H_1(\mathbb{R}^n)$ and $u(x,t)$ is a continuously differentiable function of t with values in $L_2(\mathbb{R}^n)$. The existence follows directly from (16.23) and the uniqueness holds in the more general class $C^1(\overline{\mathbb{R}^1_+},S'(\mathbb{R}^n))$ (cf. Remark 16.1). □

16.4. Fundamental solutions for the wave equations.

The fundamental solution for the Cauchy problem is $e(x,t)$ such that

$$\begin{gathered}\left(\frac{\partial^2}{\partial t^2}-a^2\Delta\right)e(x,t)=0\ \text{ for }\ t>0,\ x\in\mathbb{R}^n,\\ e(x,+0)=0,\qquad \frac{\partial e}{\partial t}(x,+0)=\delta(x).\end{gathered} \tag{16.27}$$

To compute $e(x,t)$, we make the Fourier transform in x and, solving the ordinary differential equation, we get

$$\tilde{e}(\xi,t)=\frac{\sin a|\xi|t}{a|\xi|}.$$

Therefore

$$e(x,t) = F_\xi^{-1} \frac{\sin a|\xi|t}{a|\xi|}. \tag{16.28}$$

Let

$$E(x,t) = \begin{cases} e(x,t) & \text{for } t > 0, \\ 0 & \text{for } t < 0. \end{cases} \tag{16.29}$$

Then $E(x,t)$ is a continuous function of t with values in $S'(\mathbb{R}^n)$ and $\frac{\partial E}{\partial t}$ has a jump at $t = 0$:

$$\frac{\partial E(x,+0)}{\partial t} - \frac{\partial E(x,-0)}{\partial t} = \delta(x).$$

Therefore

$$\left(\frac{\partial^2}{\partial t^2} - a^2\Delta\right) E(x,t) = \delta(x)\delta(t), \tag{16.30}$$

i.e., $E(x,t)$ is a fundamental solution.

Let

$$e_\varepsilon(x,t) = \frac{1}{(2\pi)^n} \int_{\mathbb{R}^n} e^{-\varepsilon|\xi|} \frac{\sin a|\xi|t}{a|\xi|} e^{ix\cdot\xi} d\xi.$$

Similarly to Example 11.2, b) we have $e_\varepsilon \to e$ in $S'(\mathbb{R}^n)$ as $\varepsilon \to 0$.

If $n = 3$, we can compute $e_\varepsilon(x,t)$ using spherical coordinates (r,θ,φ), where $r = |\xi|$, $\cos\theta = \frac{x\cdot\xi}{|x|\,|\xi|}$:

$$e_\varepsilon(x,t) = \frac{1}{(2\pi)^3} \int_0^\infty \int_0^\pi \int_0^{2\pi} e^{-\varepsilon r} \frac{\sin art}{ar} e^{i|x|r\cos\theta} r^2 \sin\theta \, dr d\theta d\varphi. \tag{16.31}$$

Computing the integral with respect to φ and θ, we get,

$$\int_0^{2\pi} \int_0^\pi e^{i|x|r\cos\theta} \sin\theta \, d\theta d\varphi = 2\pi \Big(\frac{-e^{i|x|r\cos\theta}}{i|x|r} \Big)\Big|_0^\pi = \frac{2\pi}{i|x|r} \big(e^{i|x|r} - e^{-i|x|r}\big).$$

Assuming that $t > 0$ we obtain

(16.32)

$$\begin{aligned} e_\varepsilon(x,t) &= \frac{1}{(2\pi)^2} \int_0^\infty e^{-\varepsilon r} \frac{\sin art}{a} \cdot \frac{2\sin|x|r}{|x|} dr \\ &= \frac{1}{4\pi^2 a|x|} \int_0^\infty e^{-\varepsilon r} \left(\cos(at - |x|)r - \cos(at + |x|)r\right) dr \\ &= \frac{1}{8\pi^2 a|x|} \left(\frac{1}{\varepsilon - i(at - |x|)} + \frac{1}{\varepsilon + i(at - |x|)} \right. \\ &\qquad\qquad \left. - \frac{1}{\varepsilon - i(at + |x|)} - \frac{1}{\varepsilon + i(at + |x|)} \right) \\ &= \frac{1}{4\pi^2 a|x|} \left(\frac{\varepsilon}{(at - |x|)^2 + \varepsilon^2} - \frac{\varepsilon}{(at + |x|)^2 + \varepsilon^2} \right). \end{aligned}$$

Hence, as in Example 4.4, b), we have $e_\varepsilon \to \frac{1}{4\pi a|x|}\delta(|x|-at)$ in S'. Therefore, for $n = 3$,

$$e(x,t) = \frac{1}{4\pi a|x|}\delta(|x|-at) = \frac{1}{4\pi a^2 t}\delta(|x|-at) = \frac{1}{2\pi a}\delta(|x|^2 - a^2t^2). \tag{16.33}$$

Let $f(x,t)$ be continuous and $f(x,t) = 0$ for $t < 0$. Then $u = E * f$ is the solution of

$$\left(\frac{\partial^2}{\partial t^2} - a^2\Delta\right)u = f(x,t). \tag{16.34}$$

Note that $u = 0$ for $t < 0$, since $E = 0$ for $t < 0$ and $f = 0$ when $t < 0$. For $t > 0$ we have

$$u(x,t) = \frac{1}{4\pi a^2}\int_{\mathbb{R}^3} \frac{f\left(y, t - \frac{|x-y|}{a}\right)}{|x-y|}dy. \tag{16.35}$$

Integral (16.35) is called a retarded potential. □

Now consider the case $n = 2$. Introducing polar coordinates (r, φ) and integrating with respect to r, we have:

$$\begin{aligned}(16.36)\quad & \frac{1}{(2\pi)^2}\int_0^\infty\int_0^{2\pi} e^{-\varepsilon r}\frac{\sin art}{ar}e^{i|x|r\cos\varphi}rdrd\varphi \\ &= \frac{1}{8\pi^2 ia}\int_0^{2\pi}\left(\frac{1}{\varepsilon - iat - i|x|\cos\varphi} - \frac{1}{\varepsilon + iat - i|x|\cos\varphi}\right)d\varphi \\ &= \frac{1}{8\pi^2 a|x|}\left[I\left(\frac{at + i\varepsilon}{|x|}\right) - I\left(\frac{-at + i\varepsilon}{|x|}\right)\right],\end{aligned}$$

where $I(z) = \int_0^{2\pi}\frac{d\varphi}{\cos\varphi + z}$. Note that $I(z)$ is an analytic function in the complex plane z with the cut along the real axis from -1 to $+1$. Computing $I(z)$ using residues we get $I(z) = \frac{2\pi}{\sqrt{z^2-1}}$, where the branch of the square root is such that $\sqrt{z^2-1} = z + O(\frac{1}{z})$ for large $|z|$. Therefore $I(x) = \frac{2\pi}{\sqrt{x^2-1}}$ for $x > 1$, $I(x) = \frac{-2\pi}{\sqrt{x^2-1}}$ for $x < -1$ and $I(x + i0) = \frac{-2\pi i}{\sqrt{1-x^2}}$ for $|x| < 1$. Here $\sqrt{x^2-1}$ for $|x| > 1$ and $\sqrt{1-x^2}$ for $|x| < 1$ are positive square roots. Therefore, taking the limit as $\varepsilon \to 0$, we get for $t > 0$:

$$e(x,t) = \frac{1}{8\pi^2 a|x|}\frac{4\pi}{\sqrt{\frac{(at)^2}{|x|^2} - 1}} = \frac{1}{2\pi a\sqrt{a^2t^2 - |x|^2}} \tag{16.37}$$

if $|x| < at$, and

$$e(x,t) = \frac{1}{8\pi^2 a|x|}\left(\frac{-2\pi i}{\sqrt{1 - \frac{a^2t^2}{|x|^2}}} + \frac{2\pi i}{\sqrt{1 - \frac{a^2t^2}{|x|^2}}}\right) = 0 \tag{16.38}$$

if $|x| > at$. □

Note that $\frac{\partial e(x,t)}{\partial t}$ satisfies (16.19) and the following initial conditions:

(16.39)
$$\frac{\partial e(x,0)}{\partial t} = \delta(x),$$
$$\frac{\partial^2 e(x,0)}{\partial t^2} = a^2 \Delta e(x,0) = 0.$$

Let $\psi(x)$ be continuous and let $\varphi(x)$ be continuously differentiable. The solution $u(x,t)$ of the Cauchy problem (16.19), (16.20) can be written in the following form:

(16.40)
$$u(x,t) = \frac{\partial e(x,t)}{\partial t} * \varphi(x) + e(x,t) * \psi(x)$$
$$= \frac{\partial}{\partial t} \frac{1}{4\pi a^2 t} \int_{|x-y|=at} \varphi(y) d\sigma + \frac{1}{4\pi a^2 t} \int_{|x-y|=at} \psi(y) d\sigma, \quad t > 0,$$

where $d\sigma$ is the area element of the sphere of radius at centered at $x \in \mathbb{R}^n$.

We can make some important conclusions from formula (16.40):

1) The value of $u(x,t)$ at a point (x,t) depends only on $\varphi(y)$ and $\psi(y)$ on the sphere $\{y : |x-y| = at\}$. The sphere $\{y : |x-y| = at\}$ is called the domain of dependence of the point (x,t).

2) Suppose $\operatorname{supp}\varphi \in B_R$ and $\operatorname{supp}\psi \in B_R$, where B_R is the ball $\{y : |y| < R\}$. Then $\operatorname{supp} u(x,t) \subset \mathcal{D}(t)$, where $\mathcal{D}(t)$ is the union of all spheres $\{x : |x-y| = at\}$ such that $y \in B_R$. The union $\bigcup_{t>0} \mathcal{D}(t)$ is called the domain of influence of B_R. Note that $\mathcal{D}(t)$ is a bounded domain in $\mathbb{R}^n$ for each t. This means that the disturbance originated in B_R propagates with a finite speed.

3) If t is large, $at > 2R$, then $\mathcal{D}(t)$ does not intersect the ball B_{t-2R}, i.e., $u(x,t) = 0$ in B_{t-2R}. This phenomenon is called the strict Huygens principle. □

Consider the case $n = 2$. Then the solution of the Cauchy problem (16.19), (16.20) has the form (cf. (16.37), (16.38)):
(16.41)
$$u(x,t) = \frac{\partial}{\partial t} \frac{1}{2\pi a} \int_{B_{at}(x)} \frac{\varphi(y)dy}{\sqrt{a^2t^2 - |x-y|^2}} + \frac{1}{2\pi a} \int_{B_{at}(x)} \frac{\psi(y)dy}{\sqrt{a^2t^2 - |x-y|^2}},$$

where $B_{at}(x)$ is the disk of radius at centered at $x \in \mathbb{R}^2$.

Note the difference between formulas (16.40) and (16.41):

In (16.40) the integration is over the sphere of radius at centered at x, but in (16.41) the integration is over the disk.

The finite propagation speed holds for the case $n = 2$ too (it will be shown in Chapter VI that it holds for general hyperbolic equations). However, the strict Huygens principle does not hold for $n = 2$.

16.5. The Laplace equation.

Example 16.4. The Laplace equation

$$\Delta u(x) = 0$$

describes stationary processes. It is obtained from the wave, heat, or Schrödinger equation when the solution $u(x)$ is independent of t. We find the fundamental solution of the Laplacian for $n \geq 3$:

$$-\Delta E = \delta. \tag{16.42}$$

Using the Fourier transforms we get

$$|\xi|^2 \tilde{E} = 1.$$

It follows from Problem 15.8 that

$$E = F^{-1}\frac{1}{|\xi|^2} = \frac{\Gamma\left(\frac{n-2}{2}\right)}{4\pi^{\frac{n}{2}}}|x|^{2-n}, \quad n \geq 3. \tag{16.43}$$

Therefore, for a bounded compactly supported $f(x)$, we have

$$u(x) = \frac{\Gamma\left(\frac{n-2}{2}\right)}{4\pi^{\frac{n}{2}}} \int_{\mathbb{R}^n} \frac{f(y)}{|x-y|^{n-2}} dy.$$

If $n = 3$, this integral is called the Newtonian potential. □

Now we study the Laplace equation in the half-space $\mathbb{R}^n_+ = \{(x', x_n) : x' \in \mathbb{R}^{n-1}, x_n > 0\}$ with the Dirichlet boundary condition:

$$\Delta u(x', x_n) = 0, \quad x \in \mathbb{R}^n_+, \tag{16.44}$$

$$u(x', 0) = h(x'), \tag{16.45}$$

where $x' = (x_1, \ldots, x_{n-1})$. Assuming that $u(x', x_n) \in C(\overline{\mathbb{R}^1_+}, L_2(\mathbb{R}^{n-1}))$ and making the Fourier transform in x', we get

$$\frac{\partial^2 \tilde{u}(\xi', x_n)}{\partial x_n^2} - |\xi'|^2 \tilde{u}(\xi', x_n) = 0, \quad x_n > 0, \tag{16.46}$$

$$\tilde{u}(\xi', 0) = \tilde{h}(\xi).$$

The general solution of (16.46) is

$$\tilde{u}(\xi, x_n) = C_1(\xi)e^{-x_n|\xi|} + C_2(\xi)e^{x_n|\xi|}.$$

Since we consider the solutions of (16.45) that belong to $L_2(\mathbb{R}^{n-1})$ in x', we put

$$C_2(\xi) = 0.$$

Therefore

$$\tilde{u}(\xi, t) = \tilde{h}(\xi)e^{-x_n|\xi|}$$

and

(16.47)
$$u(x,t) = F^{-1}\tilde{h}(\xi)e^{-x_n|\xi|}.$$

Suppose $h(x) \in L_2(\mathbb{R}^n)$. In view of (16.47), $u(x,t) \in C(\overline{\mathbb{R}^1_+}, L_2(\mathbb{R}^{n-1}))$ and

(16.48)
$$\|u(\cdot, x_n)\|_0 \leq C\|h\|_0, \quad \forall x_n \geq 0.$$

Note that the continuity in x_n in the L_2-norm follows, for example, from the Lebesgue convergence theorem. Now we prove the uniqueness of the solution of (16.44), (16.45) in the space $C(\overline{\mathbb{R}^1_+}, L_2(\mathbb{R}^{n-1}))$. Let $u(x', x_n) \in C(\overline{\mathbb{R}^1_+}, S'(\mathbb{R}^{n-1}))$, $\Delta u = 0$ for $x_n > 0$, and $u(x', 0) = 0$. Extending $u(x', x_n)$ as an odd function to $x_n < 0$, we get $\Delta u = 0$ in $\mathbb{R}^n$, $u \in S'(\mathbb{R}^n)$. By the Liouville theorem (cf. Example 10.3) we have that $u(x', x_n)$ is a polynomial $P(x', x_n)$ that is odd in x_n. Therefore, for any class of $u \in C(\overline{\mathbb{R}^1_+}, S'(\mathbb{R}^n))$ that does not contain such polynomials we have the uniqueness of (16.44), (16.45). In particular, the uniqueness holds for $u \in C(\overline{\mathbb{R}^1_+}, L_2(\mathbb{R}^n))$. □

To find an explicit formula for the solution of the Dirichlet problem we compute $F^{-1}_{\xi'} e^{-x_n|\xi'|}$ for $n \geq 3$. We shall show that

(16.49)
$$F^{-1}_{\xi'} e^{-x_n|\xi'|} = \frac{\Gamma\left(\frac{n-2}{2}\right)(n-2)}{2\pi^{\frac{n}{2}}} \frac{x_n}{(x_n^2 + |x'|^2)^{\frac{n}{2}}} = -2\frac{\partial}{\partial x_n}E(x, x_n),$$

where $E(x', x_n)$ is the same as in (16.43).

For the references we need the following well-known lemma of complex analysis:

Proposition 16.1 (The Jordan lemma). *Let $f(\zeta)$, $\zeta = \xi_1 + i\eta_1$, be an analytic function for $\eta_1 \geq 0$ except for a finite number of poles at $\zeta_1, \dots, \zeta_m$, $\operatorname{Im}\zeta_k \neq 0$, $1 \leq k \leq m$. Suppose $|f(\zeta)| = O(\frac{1}{|\zeta|^\alpha})$, $\alpha > 0$, for $\eta_1 \geq 0$. Then for $x_1 > 0$ we have*

(16.50)
$$\int_{-\infty}^{\infty} f(\xi_1)e^{ix_1\xi_1}d\xi_1 = \int_{C_N} f(\zeta)e^{ix_1\zeta}d\zeta = 2\pi i\sum_{k=1}^{m} \operatorname{res}_{\zeta_k}(f(\zeta)e^{ix_1\zeta}),$$

where C_N is the semicircle $|\zeta| = N$, $\eta_1 \geq 0$, enclosing all the poles of $f(\zeta)$. □

Applying the Jordan lemma, we get for $x_n > 0$:

(16.51)
$$\int_{-\infty}^{\infty} \frac{\xi_n e^{ix_n\xi_n}}{(\xi_n^2 + |\xi'|^2)}d\xi_n = \pi i e^{-x_n|\xi'|}.$$

Therefore

(16.52)
$$F^{-1}_{\xi'} e^{-x_n|\xi'|} = \frac{1}{(2\pi)^{n-1}\pi i}\int_{\mathbb{R}^n} \frac{\xi_n e^{ix\cdot\xi}}{\xi_n^2 + |\xi'|^2}d\xi' d\xi_n.$$

On the other hand, differentiating with respect to x_n for $x_n > 0$, we get

$$\frac{\partial E(x',x_n)}{\partial x_n} = F^{-1}\frac{i\xi_n}{\xi_n^2+|\xi'|^2}. \tag{16.53}$$

Comparing (16.52) and (16.53), we obtain for $x_n > 0$ that

$$-2\frac{\partial E(x',x_n)}{\partial x_n} = F_{\xi'}^{-1}e^{-x_n|\xi'|}.$$

Let $h(x')$ be continuous in $\mathbb{R}^n$ and $h(x') = O(\frac{1}{(1+|x'|)^\alpha})$, $\alpha > 0$. Then

$$\begin{aligned} u(x',x_n) &= -2\frac{\partial E}{\partial x_n} * (h(x')\delta(x_n)) \\ &= \frac{\Gamma\left(\frac{n-2}{2}\right)(n-2)}{2\pi^{\frac{n}{2}}}\int_{\mathbb{R}^{n-1}}\frac{x_n h(y')dy'}{(x_n^2+|x'-y'|^2)^{\frac{n}{2}}} \end{aligned} \tag{16.54}$$

is a solution of (16.44) in $\mathbb{R}_+^n, n \ge 3$, and $u(x',x_n) \to h(x')$ as $x_n \to 0$, $\forall x' \in \mathbb{R}^{n-1}$. □

Consider now the Neumann problem in $\mathbb{R}_+^n$, i.e.,

$$\Delta u(x',x_n) = 0, \quad x \in \mathbb{R}_+^n, \quad n \ge 3, \tag{16.55}$$

$$\frac{\partial u(x',0)}{\partial x_n} = g(x'), \quad x' \in \mathbb{R}^{n-1}. \tag{16.56}$$

Using the Fourier transform as in the case of the Dirichlet problem, we get

$$\tilde{u}(\xi',x_n) = -\frac{\tilde{g}(\xi')}{|\xi|}e^{-x_n|\xi'|}, \quad x_n > 0. \tag{16.57}$$

We compute $F_{\xi'}^{-1}\frac{e^{-x_n|\xi'|}}{|\xi'|}$ for $n \ge 3$. As in (16.51), we have, using the Jordan lemma:

$$\int_{-\infty}^{\infty}\frac{e^{ix_n\xi_n}}{\xi_n^2+|\xi'|^2}d\xi_n = \pi\frac{e^{-x_n|\xi'|}}{|\xi'|}, \quad x_n > 0.$$

Therefore, using (16.43), we obtain:

$$F_{\xi'}^{-1}\frac{e^{-x_n|\xi'|}}{|\xi'|} = 2F_\xi^{-1}\frac{1}{\xi_n^2+|\xi'|^2} = 2E(x',x_n), \quad x_n > 0, \tag{16.58}$$

where $E(x',x_n)$ is the same as in (16.43).

Let $g(x')$ be continuous and $g(x') = O(\frac{1}{|x|^{1+\alpha}})$, $\alpha > 0$. Then

$$\begin{aligned} u(x',x_n) &= -2E(x',x_n) * (g(x')\delta(x_n)) \\ &= -\frac{\Gamma\left(\frac{n-2}{2}\right)}{2\pi^{\frac{n}{2}}}\int_{\mathbb{R}^{n-1}}\frac{g(y')dy'}{(x_n^2+|x'-y'|^2)^{\frac{n-2}{2}}} \end{aligned} \tag{16.59}$$

is a solution of the Neumann problem (16.55), (16.56), and $u(x) \to 0$ as $|x| \to \infty$, $x_n \geq 0$. The uniqueness of such a solution of the Neumann problem can be proven similarly to the uniqueness of the Dirichlet problem.

Let $\Delta u = 0$ for $x_n > 0$ and $\frac{\partial u}{\partial x_n} = 0$ as $x_n = 0$. Extend $u(x', x_n)$ to $x_n < 0$ as an even function of x_n. Then $\Delta u = 0$ in $\mathbb{R}^n$, and by the Liouville theorem, $u(x, x_n)$ is an even polynomial in x_n. Therefore, $u = 0$ since $u \to 0$ as $|x| \to \infty$.

16.6. The reduced wave equation.

Example 16.5. The equation

$$(-\Delta - k^2)v(x) = 0, \quad x \in \mathbb{R}^n, \tag{16.60}$$

is called the reduced wave equation, or the Helmholtz equation. If we are looking for a solution of the wave equation in the form $u(x,t) = e^{ikt}v(x)$, then $v(x)$ will satisfy the reduced wave equation. We start with a simpler case of the operator $-\Delta + k^2$, where k is replaced by ik in (16.60). For an arbitrary $n \geq 2$ we define

$$E = F^{-1}\Big(\frac{1}{|\xi|^2 + k^2}\Big).$$

We have (cf. Example 11.2, a)):

$$E = \lim_{R\to\infty} E_R \quad \text{in} \quad S', \qquad \text{where} \quad E_R(x) = \frac{1}{(2\pi)^n}\int_{|\xi|\leq R} \frac{e^{ix\cdot\xi}}{|\xi|^2 + k^2} d\xi.$$

For $n = 3$, using the spherical coordinates $\xi = (r, \theta, \varphi)$, where $r = |\xi|$, $\cos\theta = \frac{x\cdot\xi}{|x|r}$, and integrating with respect to φ and θ (cf. (16.31)), we get:

$$\begin{aligned} E_R(x) &= \frac{1}{(2\pi)^3}\int_0^R\int_0^\pi\int_0^{2\pi} \frac{e^{ir|x|\cos\theta} r^2 \sin\theta}{r^2 + k^2} dr d\theta d\varphi \\ &= \frac{1}{(2\pi)^2}\int_0^R \frac{r^2}{r^2+k^2}\left(-\frac{1}{ir|x|}\right)\left(e^{-ir|x|} - e^{ir|x|}\right) dr \\ &= \frac{1}{4\pi^2|x|i}\int_{-R}^R \frac{re^{ir|x|}}{r^2+k^2} dr. \end{aligned} \tag{16.61}$$

Taking the limit as $R \to \infty$ and computing the integral using the residues, we get

$$E(x) = \frac{1}{4\pi|x|}e^{-k|x|}, \quad n = 3. \tag{16.62}$$

Now we construct fundamental solutions for $(-\Delta - k^2)$. Different regularizations of $\frac{1}{|\xi|^2 - k^2}$ (see Example 5.4) will give us different fundamental solutions. In particular, let

$$(16.63)\quad E_+ = F^{-1}\left(\frac{1}{|\xi|^2-(k+i0)^2}\right), \quad E_- = F^{-1}\left(\frac{1}{|\xi|^2-(k-i0)^2}\right).$$

$$E_0 = F^{-1}\left(\text{p.v.}\,\frac{1}{|\xi|^2-k^2}\right) = \frac{1}{2}E_+ + \frac{1}{2}E_-.$$

For $n=3$ we get, analogously to (16.62):

$$F^{-1}\left(\frac{1}{|\xi|^2-(k+i\varepsilon)^2}\right) = F^{-1}\left(\frac{1}{|\xi|^2+(\varepsilon-ik)^2}\right) = \frac{1}{4\pi|x|}e^{-(\varepsilon-ik)|x|},$$

and

$$F^{-1}\left(\frac{1}{|\xi|^2-(k-i\varepsilon)^2}\right) = F^{-1}\left(\frac{1}{|\xi|^2+(\varepsilon+ik)^2}\right) = \frac{1}{4\pi|x|}e^{-(\varepsilon+ik)|x|}.$$

Therefore, taking the limit as $\varepsilon \to +0$ we obtain
(16.64)

$$E_+(x) = \frac{1}{4\pi|x|}e^{ik|x|}, \quad E_-(x) = \frac{1}{4\pi|x|}e^{-ik|x|}, \quad E_0(x) = \frac{1}{4\pi|x|}\cos k|x|.$$

Remark 16.2. Let $f(x) \in C_0^\infty(\mathbb{R}^3)$. Then $u_+(x) = E_+ * f$, $u_- = E_- * f$, and $u_0 = E_0 * f$ are three distinct solutions of

$$(16.65)\qquad (-\Delta - k^2)u(x) = f(x),$$

decaying as $O(\frac{1}{|x|})$ for $|x| \to \infty$. In order to get a uniqueness class, one needs to impose additional conditions at infinity (the radiation conditions). This will be done in §19. □

Consider now the reduced wave equation (16.60) in the half-space $\mathbb{R}^n_+$ with the Dirichlet boundary condition

$$(16.66)\qquad u(x',0) = h(x'), \quad x' \in \mathbb{R}^{n-1},$$

or with the Neumann boundary condition

$$(16.67)\qquad \frac{\partial u(x',0)}{\partial x_n} = g(x'), \quad x' \in \mathbb{R}^{n-1}.$$

Suppose, for simplicity, that $h(x') \in C_0^\infty(\mathbb{R}^{n-1})$ and $g(x) \in C_0^\infty(\mathbb{R}^{n-1})$. First, we define the simple and the double layer potentials for the reduced wave equation.

Let $f(x') \in C_0^\infty(\mathbb{R}^{n-1})$. The single layer potential is the following integral:

$$(16.68)\qquad Sf = E_+ * (f(x')\delta(x_n)) = \int_{\mathbb{R}^{n-1}} E_+(x'-y',x_n)f(y')dy'.$$

Using the Fourier transform, we can rewrite (16.68) in the following form:

$$Sf = \frac{1}{(2\pi)^n} \int_{\mathbb{R}^n} \frac{e^{ix'\cdot\xi' + ix_n\xi_n} \tilde{f}(\xi') d\xi' d\xi_n}{\xi_n^2 + |\xi'|^2 - k^2 - i0}. \tag{16.69}$$

Replacing $k^2 + i0$ by $k^2 + i\varepsilon$, $\varepsilon > 0$, computing the integrals with respect to ξ_n using the Jordan lemma, and taking the limit as $\varepsilon \to +0$, we get (cf. (16.58)):

$$Sf = \frac{1}{(2\pi)^n} \int_{\mathbb{R}^{n-1}} \frac{\pi e^{ix'\cdot\xi' - |x_n|\sqrt{|\xi'|^2 - k^2 - i0}}}{\sqrt{|\xi'|^2 - k^2 - i0}} \tilde{f}(\xi') d\xi', \tag{16.70}$$

where $\sqrt{|\xi|' - k^2 - i\varepsilon}$ is the branch of the square root such that $\operatorname{Re}\sqrt{|\xi'|^2 - k^2 - i\varepsilon} > 0$ for $\varepsilon > 0$, and

$$\sqrt{|\xi'|^2 - k^2 - i0} = \lim_{\varepsilon \to 0} \sqrt{|\xi'|^2 - k^2 - i\varepsilon}.$$

It follows from (16.70) that Sf is continuous in x_n for $-\infty < x_n < +\infty$. Denote by $S_0 f$ the restriction of Sf to the plane $x_n = 0$, i.e.,

$$S_0 f = F_{\xi'}^{-1} \frac{\tilde{f}(\xi')}{2\sqrt{|\xi'|^2 - k^2 - i0}}, \tag{16.71}$$

where $F_{\xi'}^{-1}$ is the inverse Fourier transform in $\mathbb{R}^{n-1}$.

The double layer potential is the following integral:

$$Df = -\frac{\partial E_+}{\partial x_n} * (f(x')\delta(x_n)). \tag{16.72}$$

Differentiating (16.70) with respect to x_n for $x_n \neq 0$, we get

$$Df = \frac{1}{2} \operatorname{sgn} x_n F_{\xi'}^{-1}(e^{-|x_n|\sqrt{|\xi'|^2 - k^2 - i0}} \tilde{f}(\xi')), \quad x_n \neq 0, \tag{16.73}$$

where $\operatorname{sgn} x_n = +1$ for $x_n > 0$, and $\operatorname{sgn} x_n = -1$ for $x_n < 0$. Therefore

$$\lim_{x_n \to +0} Df = \frac{1}{2} f(x'), \qquad \lim_{x_n \to -0} Df = -\frac{1}{2} f(x'). \tag{16.74}$$

Note that $u(x', x_n) = 2Dh$ is the solution of the equation

$$(-\Delta - k^2)u = 0 \quad \text{in} \quad \mathbb{R}_+^n \tag{16.75}$$

that satisfy the Dirichlet boundary condition

$$u(x', +0) = \lim_{x_n \to +0} u(x', x_n) = h(x'). \tag{16.76}$$

Also $u = -2Sg$ is the solution of (16.75) with the Neumann boundary condition (16.67). Therefore we obtained solutions of the Dirichlet and the Neumann problems in $\mathbb{R}_+^n$ such that $u(x) \to 0$ as $|x| \to \infty$. However, the conditions on the decay of $u(x)$ are not sufficient to determine the uniqueness class. One needs also the radiation condition (cf. Problem 21.5).

Remark 16.3. Differentiating (16.72) for $x_n \neq 0$, we get
(16.77)
$$\frac{\partial}{\partial x_n} Df = -\frac{1}{2} F_{\xi'}^{-1}(\sqrt{|\xi'|^2 - k^2 - i0}\; e^{-|x_n|\sqrt{|\xi'|^2-k^2-i0}} \tilde{f}(\xi')), \quad x_n \neq 0.$$
Note that $\frac{\partial}{\partial x_n} Df$ has the same limits for $x_n \to +0$ and $x_n \to -0$, and
$$\lim_{x_n \to +0} \frac{\partial}{\partial x_n} Df = -\frac{1}{2} F_{\xi'}^{-1}(\sqrt{|\xi'|^2 - k^2 - i0}\tilde{f}(\xi')).$$
Let

(16.78)
$$\Lambda f = F_{\xi'}^{-1}(\sqrt{|\xi'|^2 - k^2 - i0}\tilde{f}(\xi')).$$
Note that $u = 2Df$ solves the Dirichlet problem (16.75), (16.76) and
$$\frac{\partial u(x', +0)}{\partial x_n} = \lim_{x_n \to +0} \frac{\partial}{\partial x_n}(2Df)$$
is the Neumann boundary data. Therefore the operator Λ maps the Dirichlet data $f(x')$ to the Neumann data $-\frac{\partial u(x',+0)}{\partial x_n}$ (cf. Remark 16.4). Operator Λ is called the Dirichlet-to-Neumann operator. We have

(16.79)
$$\Lambda S_0 f = S_0 \Lambda f = \frac{1}{2} f, \quad \forall f \in C_0^\infty(\mathbb{R}^{n-1}).$$

16.7. Faddeev's fundamental solutions for $(-\Delta - k^2)$.

Example 16.6. Let ν be a unit vector in $\mathbb{R}^n$ and $-k \leq \sigma \leq k$. Denote

(16.80)
$$E_{\nu,\sigma} = F^{-1} \frac{1}{|\xi|^2 - k^2 + i0(\xi_\nu - \sigma)}, \qquad \xi_\nu = \xi \cdot \nu,$$
where the distribution $\frac{1}{|\xi|^2 - k^2 + i0(\xi_\nu - \sigma)}$ is the limit of the regular functionals $\frac{1}{|\xi|^2 - k^2 + i\varepsilon(\xi_\nu - \sigma)}$, i.e.,

(16.81)
$$\frac{1}{|\xi|^2 - k^2 + i0(\xi_\nu - \sigma)}(\varphi) = \lim_{\varepsilon \to +0} \int_{\mathbb{R}^n} \frac{\varphi(\xi)}{|\xi|^2 - k^2 + i\varepsilon(\xi_\nu - \sigma)} d\xi.$$
It is obvious that

(16.82)
$$(-\Delta - k^2) E_{\nu,\sigma} = \delta,$$
since
$$\frac{1}{|\xi|^2 - k^2 + i0(\xi_\nu - \sigma)}(|\xi|^2 - k^2) = 1.$$
We have
$$\int_{\mathbb{R}^n} \frac{\varphi(\xi)}{|\xi|^2 - k^2 + i\varepsilon(\xi_\nu - \sigma)} d\xi$$
$$= \int_{\mathbb{R}^n} \frac{(|\xi|^2 - k^2)\varphi(\xi)}{(|\xi|^2 - k^2)^2 + \varepsilon^2(\xi_\nu - \sigma)^2} d\xi + \int_{\mathbb{R}^n} \frac{-i\varepsilon(\xi_\nu - \sigma))\varphi(\xi)}{(|\xi|^2 - k^2)^2 + \varepsilon^2(\xi_\nu - \sigma)^2} d\xi.$$

Therefore analogously to Example 5.4 we obtain, taking the limit in ε:

$$\int_{\mathbb{R}^n} \frac{\varphi(\xi)}{|\xi|^2 - k^2 + i0(\xi_\nu - \sigma)} d\xi = \text{p.v.} \int_{\mathbb{R}^n} \frac{\varphi(\xi)}{|\xi|^2 - k^2} d\xi - \frac{i\pi}{2k} \int_{|\xi|=k,\xi_\nu>\sigma} \varphi(\xi)ds + \frac{i\pi}{2k} \int_{|\xi|=k,\xi_\nu<\sigma} \varphi(\xi)ds, \tag{16.83}$$

where ds is the area element of the sphere $|\xi|^2 = k^2$. Therefore

$$\begin{aligned} E_{\nu,\sigma}(x) &= E_0(x) - \frac{i\pi}{2k(2\pi)^n} \int_{|\xi|=k,\xi_\nu>\sigma} e^{ix\cdot\xi} ds \\ &\quad + \frac{i\pi}{2k(2\pi)^n} \int_{|\xi|=k,\xi_\nu<\sigma} e^{ix\cdot\xi} ds \\ &= E_+(x) - \frac{2i\pi}{2k(2\pi)^n} \int_{|\xi|=k,\xi_\nu>\sigma} e^{ix\cdot\xi} ds. \end{aligned} \tag{16.84}$$

Faddeev's fundamental solutions are used in scattering theory (see Chapter V).

Remark 16.4. It is customary to define the Neumann problem as $\frac{\partial u}{\partial n}\big|_{\partial\Omega} = g$, where $n(x')$ is the outward normal (cf. §20). Therefore, the Neumann problem in $\mathbb{R}^n_+$ has the form $-\frac{\partial u}{\partial x_n}\big|_{x_n=0} = g(x')$. To have the definition of the double layer potential consistent with the definition of §20 we introduced minus in (16.72) (cf. §57), and we defined the D-to-N operator as a map of $u(x', +0)$ to $-\frac{\partial u(x',+0)}{\partial x_n}$.

17. Existence of a fundamental solution

Theorem 17.1. *Let $P(\xi)$ be an arbitrary nonzero polynomial of degree m. Then there exists $E \in \mathcal{D}'(\mathbb{R}^n)$ such that*

$$P\left(-i\frac{\partial}{\partial x}\right) E = \delta.$$

Proof: If $P(\xi)$ is such that $|P(\xi)| \geq C$ for all $\xi \in \mathbb{R}^n$, then, obviously, $E = F^{-1}(\frac{1}{P(\xi)})$ and $E(\varphi) = \tilde{E}(\check{\varphi})$, where

$$\check{\varphi}(\xi) = \frac{1}{(2\pi)^n} \int_{\mathbb{R}^n} \varphi(x) e^{ix\cdot\xi} dx = \frac{1}{(2\pi)^n} \tilde{\varphi}(-\xi)$$

is the inverse Fourier transform of φ. Therefore

$$E(\varphi) = \frac{1}{(2\pi)^n} \int_{\mathbb{R}^n} \frac{1}{P(\xi)} \tilde{\varphi}(-\xi) d\xi = \frac{1}{(2\pi)^n} \int_{\mathbb{R}^n} \frac{1}{P(-\xi)} \tilde{\varphi}(\xi) d\xi.$$

Without loss of generality we may assume that

$$P(\xi_1, \ldots, \xi_n) = \xi_n^m + \sum_{k=0}^{m-1} p_k(\xi')\xi_n^k,$$

where deg $p_k(\xi') \le m-k$. This can always be achieved by a linear change of variables $(\xi_1, \dots, \xi_n)$. We have

$$P(\xi_1, \dots, \xi_{n-1}, \xi_n + i\tau) = \prod_{k=1}^{m} (\xi_n + i\tau - \lambda_k(\xi')),$$

where $\lambda_k(\xi')$ are continuous in ξ', $\xi' = (\xi_1, \dots, \xi_{n-1})$. For fixed $\xi_0' \in \mathbb{R}^{n-1}$ one can find τ_0 such that $|\tau_0 - \operatorname{Im} \lambda_k(\xi_0')| > 1$ and $|\tau_0| \le m+1$.Then

$$|P(\xi_0', \xi_n + i\tau_0)| \ge \prod_{k=1}^{m} |\xi_n + i\tau_0 - \lambda_k(\xi_0')| > 1.$$

Since the $\lambda_k(\xi')$ are continuous, there exists a ball $B_{r_0}(\xi_0')$ such that $|P(\xi', \xi_n + i\tau_0)| > 1$ for $\xi' \in B_{r_0}(\xi_0')$. Thus we have an open cover of $\mathbb{R}^{n-1}$ by balls. Since $\mathbb{R}^{n-1}$ is locally compact, there exists a countable sequence of balls $B_{r_j}(\xi_j')$ that covers $\mathbb{R}^{n-1}$ and is such that any ball $\overline{B_N(0)}$ is covered by a finite number of $B_{r_j}(\xi_j')$.

Let $\{\chi_j(\xi')\}$ be a partition of unity subordinate to the cover $\{B_{r_j}(\xi_j')\}$, i.e., $\chi_j(\xi') \in C_0^\infty(\mathbb{R}^{n-1})$, $\operatorname{supp} \chi_j \subset B_{r_j}(\xi_j')$, $\chi_j(\xi') \ge 0$ and $\sum_{j=1}^\infty \chi_j(\xi') = 1$.

We define $E \in \mathcal{D}'$ by the following formula:

$$E(\varphi) = \sum_{j=1}^{\infty} \frac{1}{(2\pi)^n} \int_{\mathbb{R}^{n-1}} \chi_j(\xi') \int_{-\infty}^{\infty} \frac{\tilde{\varphi}(\xi', \xi_n + i\tau_j)}{P(-\xi', -(\xi_n + i\tau_j))} d\xi_n d\xi'.$$

Note that $E(\varphi)$ is a linear continuous functional over $\mathcal{D}(\mathbb{R}^n)$. Indeed,

$$|\tilde{\varphi}(\xi', \xi_n + i\tau_j)| \le C_N(1 + |\xi'| + |\xi_n + i\tau_j|)^{-N} e^{R|\tau_j|}$$

$$\le C_N(1 + |\xi'| + |\xi_n|)^{-N}, \quad \forall N,$$

since $|\tau_j| \le m+1$. Note that $\varphi_k \to \varphi$ in $\mathcal{D}$ implies

$$\sup_{|\tau| \le m+1} |\tilde{\varphi}_k(\xi', \xi_n + i\tau) - \tilde{\varphi}(\xi', \xi_n + i\tau)| \le C_N(1 + |\xi'| + |\xi_n|)^{-N}$$

for all k and $\tilde{\varphi}_k(\xi', \xi_n + i\tau) \to \tilde{\varphi}(\xi', \xi_n + i\tau)$ as $k \to \infty$. Therefore $E(\varphi_k) \to E(\varphi)$, i.e., $E \in \mathcal{D}'$.

Since $F(P(i\frac{\partial}{\partial x})\varphi) = P(-\xi)\tilde{\varphi}(\xi)$, we get

$$P\left(-i\frac{\partial}{\partial x}\right)E(\varphi) = E\left(P\left(i\frac{\partial}{\partial x}\right)\varphi\right)$$

$$= \sum_{j=1}^{\infty}\frac{1}{(2\pi)^n}\int_{\mathbb{R}^{n-1}}\chi_j(\xi')\int_{-\infty}^{\infty}\frac{P(-\xi',-(\xi_n+i\tau_j))\tilde{\varphi}(\xi',\xi_n+i\tau_j)}{P(-\xi',-(\xi_n+i\tau_j))}d\xi_n d\xi'$$

$$= \sum_{j=1}^{\infty}\frac{1}{(2\pi)^n}\int_{\mathbb{R}^{n-1}}\chi_j(\xi')\int_{-\infty}^{\infty}\tilde{\varphi}(\xi',\xi_n+i\tau_j)d\xi_n d\xi'$$

$$= \frac{1}{(2\pi)^n}\int_{\mathbb{R}^n}\tilde{\varphi}(\xi',\xi_n)d\xi' d\xi_n = \varphi(0) = \delta(\varphi).$$

We used that, by the Cauchy theorem,

$$\int_{-\infty}^{\infty}\tilde{\varphi}(\xi',\xi_n+i\tau_j)d\xi_n = \int_{-\infty}^{\infty}\tilde{\varphi}(\xi',\xi_n)d\xi_n.$$

Therefore $P(-i\frac{\partial}{\partial x})E = \delta$. □

Corollary 17.2. *Let $P(-i\frac{\partial}{\partial x})$ be an arbitrary differential operator with constant coefficients and f an arbitrary distribution with a compact support. Then the nonhomogeneous equation $P(-i\frac{\partial}{\partial x})u = f$ has a solution $u = E * f$ in $\mathcal{D}'(\mathbb{R}^n)$, where E is the fundamental solution constructed above.*

18. Hypoelliptic equations

We say that u is a distribution solution of $P(-i\frac{\partial}{\partial x})u = f$ in an open set Ω if $P(-i\frac{\partial}{\partial x})u(\varphi) = f(\varphi)$ for all $\varphi \in C_0^\infty(\Omega)$.

A partial differential operator $P(-i\frac{\partial}{\partial x})$ is called hypoelliptic if for any open Ω and any $f \in C^\infty(\Omega)$, each distribution solution u of $P(-i\frac{\partial}{\partial x})u = f$ in Ω is $C^\infty(\Omega)$, i.e., u is a regular functional in Ω corresponding to a $C^\infty(\Omega)$ function.

To show that $P(-i\frac{\partial}{\partial x})$ is hypoelliptic, it is enough to show that any distribution solution in Ω of the homogeneous equation $P(-i\frac{\partial}{\partial x})v = 0$ is $C^\infty(\Omega)$. Indeed, let $P(x, -i\frac{\partial}{\partial x}) = f$ in Ω, where $f \in C^\infty(\Omega)$. Take any $x_0 \in \Omega$. Consider $\chi(x) \in C_0^\infty(\mathbb{R}^n)$, $\chi(x) = 1$ for $|x| < 1$, $\chi(x) = 0$ for $|x| > 2$. Take $\delta > 0$ such that $B_{2\delta}(x_0) \subset \Omega$. Set $\chi_{x_0,\delta} = \chi(\frac{x-x_0}{\delta})$.

Let $E \in \mathcal{D}'(\mathbb{R}^n)$ be a fundamental solution, i.e., $P(-i\frac{\partial}{\partial x})E = \delta$. Since $\chi_{x_0,\delta}f \in C_0^\infty(\Omega)$, we have that $w = E * \chi_{x_0,\delta}f$ is a $C^\infty(\mathbb{R}^n)$ function (see §7). Note that $P(-i\frac{\partial}{\partial x})w = \chi_{x_0,\delta}f$. Therefore $P(-i\frac{\partial}{\partial x})(u - w) = (1 - \chi_{x_0,\delta}(x))f(x) = 0$ in $B_\delta(x_0)$. Thus if $u - w \in C^\infty$ in $B_\delta(x_0)$, then $u \in C^\infty$ in $B_\delta(x_0)$. Since x_0 is an arbitrary point of Ω, we have proven that $u \in C^\infty(\Omega)$. □

Note that if $P(-i\frac{\partial}{\partial x})$ is hypoelliptic, then any fundamental solution to $P(-i\frac{\partial}{\partial x})$ is a C^∞ function in $\mathbb{R}^n \setminus \{0\}$.

The converse is also true.

Theorem 18.1. *If there exists a fundamental solution E to $P(-i\frac{\partial}{\partial x})$ such that $E \in C^\infty(\mathbb{R}^n \setminus \{0\})$, then $P(-i\frac{\partial}{\partial x})$ is hypoelliptic.*

Proof: Let u be any distribution solution of $P(-i\frac{\partial}{\partial x})u = 0$ in Ω. We prove that $u \in C^\infty$ in a neighborhood of every point $x_0 \in \Omega$. Let $\delta > 0$ be such that $B_{12\delta}(x_0) \subset \Omega$. Consider the convolution $(\chi_{x_0,\delta}E) * (\chi_{x_0,5\delta}u)$. We have (cf. (7.4))

$$\text{(18.1)}\quad \left(P\Big(-i\frac{\partial}{\partial x}\Big)(\chi_{x_0,\delta}E)\right) * (\chi_{x_0,5\delta}u) = (\chi_{x_0,\delta}E) * \left(P\Big(-i\frac{\partial}{\partial x}\Big)(\chi_{x_0,5\delta}u)\right).$$

Note that

$$\text{(18.2)}\quad P\left(-i\frac{\partial}{\partial x}\right)(\chi_{x_0,\delta}E) = \chi_{x_0,\delta}P\left(-i\frac{\partial}{\partial x}\right)E + \sum_{|\alpha|=1}^{m} \frac{\partial^\alpha \chi_{x_0,\delta}}{\partial x^\alpha} Q_\alpha\left(\frac{\partial}{\partial x}\right)E = \delta + e(x),$$

where $e(x) \in C_0^\infty(\Omega)$ since $E \in C^\infty(\mathbb{R}^n\setminus\{0\})$ and $\frac{\partial^\alpha \chi_{x_0,\delta}}{\partial x^\alpha} = 0$ for $|x-x_0| < \delta$. Here $Q_\alpha(\frac{\partial}{\partial x})$ are differential operators of degree less than m.

Since $P(-i\frac{\partial}{\partial x})u = 0$ in Ω we get

$$\text{(18.3)}\quad P\left(-i\frac{\partial}{\partial x}\right)(\chi_{x_0,5\delta}u) = \chi_{x_0,5\delta}P\left(-i\frac{\partial}{\partial x}\right)u + \sum_{|\alpha|=1}^{m} \frac{\partial^\alpha \chi_{x_0,5\delta}}{\partial x^\alpha} Q_\alpha\left(\frac{\partial}{\partial x}\right)u = w,$$

where $\operatorname{supp} w$ is contained in the annulus domain $A = \{5\delta < |x-x_0| < 10\delta\}$. Therefore

$$\text{(18.4)}\qquad (\delta + e) * \chi_{x_0,5\delta}u = (\chi_{x_0,\delta}E) * w$$

or

$$\text{(18.5)}\qquad \chi_{x_0,5\delta}u = -e * \chi_{x_0,5\delta}u + (\chi_{x_0,\delta}E) * w.$$

Note that $\operatorname{supp}(\chi_{x_0,\delta}E) * w$ is contained in the annulus domain $A_1 = \{3\delta < |x - x_0| < 12\delta\}$. This follows from Proposition 7.1 since $\operatorname{supp}\chi_{x_0,\delta}E \subset B_{2\delta}(x_0)$ and $A + B_{2\delta}(x_0) \subset A_1$.

Therefore $\chi_{x_0,\delta}((\chi_{x_0,\delta}E) * w) = 0$ and we conclude that

$$\chi_{x_0,\delta}u = -\chi_{x_0,\delta}(e * \chi_{x_0,5\delta}u) \in C_0^\infty(\Omega),$$

since $e \in C_0^\infty(\Omega)$. □

18.1. Characterization of hypoelliptic polynomials.

Theorem 18.2. *Operator $P(-i\frac{\partial}{\partial x})$ is hypoelliptic if and only if there exists $\alpha > 0$ such that*

$$\left|\frac{\partial^k P}{\partial \xi^k}(\xi)\right| \leq C_1|P(\xi)|(1+|\xi|)^{-\alpha|k|} \tag{18.6}$$

for $|\xi| \geq C_2$ and all k.

Proof: We prove only the sufficiency of condition (18.6).

Let $\chi_1(\xi) = \chi(\frac{\xi}{C_2})$, where $\chi(\xi)$ is the same as in the proof of Theorem 18.1. We shall show that $E_1 = F^{-1}((1-\chi_1(\xi))\frac{1}{P(\xi)}) \in C^\infty(\mathbb{R}^n \setminus \{0\})$. We have, as in Example 11.2, b), that $E_{1\varepsilon} \to E_1$ in S', where

$$\begin{aligned} E_{1\varepsilon}(x) &= F^{-1}\left(\chi(\varepsilon\xi)(1-\chi_1(\xi))\frac{1}{P(\xi)}\right) \\ &= \frac{1}{(2\pi)^n}\int_{\mathbb{R}^n} \chi(\varepsilon\xi)(1-\chi_1(\xi))\frac{1}{P(\xi)}e^{ix\cdot\xi}d\xi. \end{aligned} \tag{18.7}$$

For any $x \neq 0$ and any N, we have

$$\left(\sum_{k=1}^n \frac{(-i)x_k}{|x|^2}\frac{\partial}{\partial\xi_k}\right)^N e^{ix\cdot\xi} = e^{ix\cdot\xi}. \tag{18.8}$$

Substituting the left hand side of (18.8) in (18.7) and integrating by parts, we get
(18.9)

$$E_{1\varepsilon}(x) = \frac{1}{(2\pi)^n}\int_{\mathbb{R}^n}\left[\left(\sum_{k=1}^n \frac{ix_k}{|x|^2}\frac{\partial}{\partial\xi_k}\right)^N \left(\chi(\varepsilon\xi)(1-\chi_1(\xi))\frac{1}{P(\xi)}\right)\right] e^{ix\cdot\xi}d\xi.$$

Since

$$\frac{\partial^p}{\partial\xi^p}\left(\chi(\varepsilon\xi)(1-\chi_1(\xi))\frac{1}{P(\xi)}\right) \to \frac{\partial^p}{\partial\xi^p}\left((1-\chi_1(\xi))\frac{1}{P(\xi)}\right)$$

in S' as $\varepsilon \to 0$, we have

$$E_1 = \frac{1}{(2\pi)^n}\int_{\mathbb{R}^n}\left[\left(\sum_{k=1}^n \frac{ix_k}{|x|^2}\frac{\partial}{\partial\xi_k}\right)^N \left((1-\chi_1(\xi))\frac{1}{P(\xi)}\right)\right] e^{ix\cdot\xi}d\xi. \tag{18.10}$$

Note that

$$\left| \left(\sum_{k=1}^{n} \frac{ix_k}{|x|^2} \frac{\partial}{\partial \xi_k} \right)^N \left((1 - \chi_1(\xi)) \frac{1}{P(\xi)} \right) \right| \leq C_N (1 + |\xi|)^{-N\alpha} |x|^{-N}. \tag{18.11}$$

Therefore $E_1(x)$ is continuous and has continuous derivatives up to order m when $x \neq 0$ and $N\alpha > n + 1 + m$. Note that

$$|E_1(x)| \leq \frac{C_1}{|x|^N}. \tag{18.12}$$

Since N is arbitrary, we see that $E_1 \in C^\infty(\mathbb{R}^n \setminus \{0\})$. Note that E_1 satisfies the equation

$$P\left(-i \frac{\partial}{\partial x} \right) E_1 = \delta - e_1(x),$$

where $e_1(x) = F^{-1}(\chi_1(\xi)) \in C^\infty(\mathbb{R}^n)$. The proof of Theorem 18.1 does not change when we substitute E_1 instead of E. Therefore, any distribution solution of $P(-i\frac{\partial}{\partial x})u = 0$ in Ω is C^∞, i.e., $P(\xi)$ is a hypoelliptic polynomial. □

18.2. Examples of hypoelliptic operators.

Example 18.1.

a) Elliptic polynomials. A polynomial $P(\xi)$ of degree m is called elliptic if $P_0(\xi) \neq 0$ for all $\xi \neq 0$, where $P_0(\xi)$ is the principal part of the polynomial $P(\xi)$, i.e., $P(\xi) = P_0(\xi) + P_1(\xi)$, where $P_0(\xi)$ is a homogeneous polynomial of degree m and the degree of P_1 is less than m. Since $P_0(\xi) \neq 0$ for $\xi \neq 0$, we have

$$C_1|\xi|^m \leq |P_0(\xi)| \leq C_2|\xi|^m.$$

Since $|P_1(\xi)| \leq C(1 + |\xi|)^{m-1}$, we conclude that $P(\xi)$ satisfies (18.6) with $\alpha = 1$, i.e., $P(\xi)$ is hypoelliptic. Note that Δ and $\Delta \pm k^2$ are elliptic.

b) The heat operator $\frac{\partial}{\partial t} - \Delta$ is hypoelliptic with $\alpha = \frac{1}{2}$, but the wave operator $\frac{\partial^2}{\partial t^2} - \Delta$ is not hypoelliptic. □

We say that $u(x_1, \dots, x_n)$ is real analytic in a domain $\Omega \subset \mathbb{R}^n$ if for each $x^{(0)} = (x_1^{(0)}, \dots, x_n^{(0)})$ in Ω there exists $\delta > 0$ such that

$$u(x_1, \dots, x_n) = \sum_{k_1 \geq 0, \dots, k_n \geq 0} a_{k_1, \dots, k_n} (x_1 - x_1^{(0)})^{k_1} \cdots (x_n - x_n^{(0)})^{k_n}$$

and the power series converges when $|x - x^{(0)}| < \delta$. Equivalently, $u(x_1, \dots, x_n)$ is a real analytic function in a neighborhood $U \subset \Omega$ of $x^{(0)}$ if $u(x_1, \dots, x_n)$ extends to a complex analytic function $u(z_1, \dots, z_n)$ in the domain $\{(z_1, \dots, z_n) : |z_k - x_k^{(0)}| < \delta, k = 1, \dots, n\}$.

Theorem 18.3. *If $P(\xi)$ is elliptic, then every distribution solution of $P(-i\frac{\partial}{\partial x})u = 0$ in Ω is real analytic.*

Proof: Since $P(\xi)$ is an elliptic polynomial, we already know that $u \in C^\infty(\Omega)$ and any fundamental solution to the operator $P(-i\frac{\partial}{\partial x})$ is a C^∞ function when $x \neq 0$. One can prove that there exists a fundamental solution $E(x)$ to the elliptic $P(-i\frac{\partial}{\partial x})$ such that $E(x)$ is real analytic when $x \neq 0$ (cf. Problem 21.14).

Note that Examples 16.4 and 16.5 show that fundamental solutions to $-\Delta$ and $-\Delta \pm k^2$ $(n = 3)$, are real analytic when $x \neq 0$. The same is true for any $n \geq 2$. Having a fundamental solution $E(x)$ that is real analytic for $x \neq 0$, we shall prove that each solution of $P(-i\frac{\partial}{\partial x})u = 0$ in Ω is real analytic. Let $\chi(x)$ be $C_0^\infty(\mathbb{R}^n)$, $\chi(x) = 1$ if $|x| < 1$, and $\chi(x) = 0$ if $|x| > 2$.

Let $y^{(0)}$ be an arbitrary point in Ω. Choose $\varepsilon > 0$ small enough that $\operatorname{supp}\chi(\frac{x-y^{(0)}}{\varepsilon}) \subset \Omega$. We assume that $x \in B_{\frac{\varepsilon}{2}}(y^{(0)})$, i.e., $|x - y^{(0)}| < \frac{\varepsilon}{2}$. Denote by δ_x the delta-function at x, i.e., $\delta_x(\varphi) = \varphi(x)$ for $\forall\varphi \in C_0^\infty(\mathbb{R}^n)$. We have $P(i\frac{\partial}{\partial y})E(x-y) = \delta_x$. Since $x \in B_{\frac{\varepsilon}{2}}(y^{(0)})$, $P(-i\frac{\partial}{\partial x})u = 0$ in Ω and since $\operatorname{supp}\frac{\partial^\alpha}{\partial y^\alpha}\chi(\frac{y-y^{(0)}}{\varepsilon}) \subset \overline{B}_{2\varepsilon}(y^{(0)}) \setminus B_\varepsilon(y^{(0)})$, we get (cf. 18.3)):

(18.13)

$$\begin{aligned} u(x) &= \delta_x\left(\chi\Big(\frac{y-y^{(0)}}{\varepsilon}\Big)u(y)\right) = P\Big(i\frac{\partial}{\partial y}\Big)E(x-y)\left(\chi\Big(\frac{y-y^{(0)}}{\varepsilon}\Big)u(y)\right) \\ &= E(x-y)\left(P\Big(-i\frac{\partial}{\partial y}\Big)\left(\chi\Big(\frac{y-y^{(0)}}{\varepsilon}\Big)u(y)\right)\right) \\ &= \int_{B_{2\varepsilon}(y^{(0)})\setminus B_\varepsilon(y^{(0)})} E(x-y)\sum_{|\alpha|=1}^{m}\frac{\partial^\alpha\chi(\frac{y-y^{(0)}}{\varepsilon})}{\partial y^\alpha}Q_\alpha\Big(\frac{\partial}{\partial y}\Big)u(y)dy. \end{aligned}$$

Since $|y-x| > \frac{\varepsilon}{2}$ if $|y-y^{(0)}| \geq \varepsilon$ and $|x-y^{(0)}| \leq \frac{\varepsilon}{2}$, and since $E(x-y)$ is real analytic for $|y-x| > \frac{\varepsilon}{2}$, we see that the integral (18.13) is real analytic for $x \in B_{\frac{\varepsilon}{2}}(y^{(0)})$. □

19. The radiation conditions

19.1. The Helmholtz equation in $\mathbb{R}^3$.

Consider the Helmholtz equation (the reduced wave equation) in $\mathbb{R}^3$,

(19.1) $$(-\Delta - k^2)u = f(x),$$

where

(19.2) $$|f(x)| \leq C(1+|x|)^{-3-\alpha}, \quad \alpha > 0.$$

Choosing the fundamental solution $E_+(x) = \frac{1}{4\pi|x|}e^{ik|x|}$, we get a particular solution

$$u_+(x) = \frac{1}{4\pi}\int_{\mathbb{R}^3} \frac{e^{ik|x-y|}}{|x-y|} f(y)dy. \tag{19.3}$$

Lemma 19.1. *Solution* (19.3) *can be represented in the form*

$$u_+(x) = \frac{1}{4\pi}\frac{e^{ik|x|}}{|x|}\tilde{f}(k\theta) + O\left(\frac{1}{|x|^{1+\alpha'}}\right), \tag{19.4}$$

where $\theta = \frac{x}{|x|}$, $|x| \to \infty$, $\alpha' = \min(\frac{\alpha}{2} - \varepsilon, 1)$, $\varepsilon > 0$ *is arbitrary, and* $\tilde{f}(k\theta)$ *is the Fourier transform of* $f(x)$.

Proof: We have

$$u_+(x) = u_1 + u_2,$$

where

$$u_1 = \frac{1}{4\pi}\int_{|y|<\frac{|x|}{2}} \frac{e^{ik|x-y|}}{|x-y|} f(y)dy$$

and $u_2 = u - u_1$. For $|y| < \frac{|x|}{2}$ we have

$$\begin{aligned}|x-y| &= \left(|x|^2 - 2x\cdot y + |y|^2\right)^{\frac{1}{2}} = |x|\left(1 - \frac{2\theta\cdot y}{|x|} + \frac{|y|^2}{|x|^2}\right)^{\frac{1}{2}} \\ &= |x|\left(1 - \frac{\theta\cdot y}{|x|} + \frac{|y|^2}{2|x|^2} + O\Big(\frac{|y|^2}{|x|^2}\Big)\right) = |x| - \theta\cdot y + O\Big(\frac{|y|^2}{|x|}\Big).\end{aligned}$$

Therefore

$$\frac{e^{ik|x-y|}}{|x-y|} = \frac{e^{ik|x| - ik\theta\cdot y + O(\frac{|y|^2}{|x|})}}{|x|\left(1 + O\Big(\frac{|y|}{|x|}\Big)\right)} = \frac{e^{ik|x| - ik\theta\cdot y + O(\frac{|y|^2}{|x|})}}{|x|}\left(1 + O\Big(\frac{|y|}{|x|}\Big)\right). \tag{19.5}$$

Note that $|e^{iO(\frac{|y|^2}{|x|})} - 1| \le C_\delta(\frac{|y|^2}{|x|})^\delta$ for any $0 \le \delta \le 1$.

Thus, we have

$$u_1 = \frac{1}{4\pi}\frac{e^{ik|x|}}{|x|}\tilde{f}(k\theta) + u_3, \tag{19.6}$$

where

$$\begin{aligned}|u_3| &\le C\int_{|y|<\frac{|x|}{2}}\left(\frac{|y|^{2\alpha_1}}{|x|^{1+\alpha_1}} + \frac{|y|}{|x|^2}\right)(1+|y|)^{-3-\alpha}dy \\ &\le \frac{C}{|x|^{1+\alpha_1}}\int_{|y|<\frac{|x|}{2}}(|y|^{2\alpha_1} + |y|^{\alpha_1})(1+|y|)^{-3-\alpha}dy \le \frac{C_1}{|x|^{1+\alpha_1}},\end{aligned} \tag{19.7}$$

with $2\alpha_1 < \alpha$, $\alpha_1 \le 1$.

Further,

$$|u_2| \le C \int_{|y|>\frac{|x|}{2}} \frac{dy}{|x-y||y|^{3+\alpha}}.$$

Making the change of variables $y = |x|t$, we obtain

$$|u_2| \le \frac{C}{|x|^{1+\alpha}} \int_{|t|>\frac{1}{2}} \frac{dt}{|\theta - t||t|^{3+\alpha}} \le \frac{C_1}{|x|^{1+\alpha}}. \tag{19.8}$$

Thus (19.4) is proven. □

Note that $\frac{i}{2\pi}\frac{\sin k|x|}{|x|} = E_+(x) - E_-(x)$ is a solution of the homogeneous equation $(-\Delta - k^2)u = 0$. Therefore u_+ is not a unique solution in the class of functions decaying as $O(\frac{1}{|x|})$.

19.2. Radiation conditions.

We say that $v(x)$ satisfies the radiation conditions in $\mathbb{R}^3$ if

$$v(x) = O\left(\frac{1}{|x|}\right) \quad \text{and} \quad \frac{\partial}{\partial r} v(x) - ik\, v(x) = o\left(\frac{1}{|x|}\right),$$

where $r = |x|$, $\frac{\partial}{\partial r} = \frac{x}{|x|} \cdot \frac{\partial}{\partial x}$.

Theorem 19.2. *Under the assumption* (19.2) *there is a unique solution of* (19.1) *satisfying the radiation conditions. This solution has the form* (19.4).

Proof: It was proven that (19.3) has the form (19.4). We shall show that (19.3) satisfies the radiation conditions. We have

$$\frac{\partial u_+}{\partial r} = \frac{1}{4\pi} \int_{\mathbb{R}^3} \frac{\frac{\partial}{\partial r}(e^{ik|x-y|})}{|x-y|} f(y)dy + \frac{1}{4\pi} \int_{\mathbb{R}^3} e^{ik|x-y|} \frac{\partial}{\partial r}\left(\frac{1}{|x-y|}\right) f(y)dy \stackrel{\text{def}}{=} v_1 + v_2.$$

For $|x| > 1$ we obtain

$$|v_2| \le C \int_{\mathbb{R}^3} \frac{dy}{|x-y|^2|y|^{3-\varepsilon}}.$$

Changing the variables $y = |x|t$, we get $|v_2| \le \frac{C}{|x|^{2-\varepsilon}}$.

Note that

$$\frac{\partial}{\partial r}(e^{ik|x-y|}) = ik\frac{x}{|x|} \cdot \frac{x-y}{|x-y|} e^{ik|x-y|},$$

and $\frac{x-y}{|x-y|} = \frac{x}{|x|} + O(\frac{|y|}{|x|})$ for $|y| < \frac{|x|}{2}$.

Therefore, if $|y| < \frac{|x|}{2}$, we obtain, as in (19.5), (19.6), (19.7),

$$(19.9)\quad \frac{\partial}{\partial r}(e^{ik|x-y|}) = ike^{ik|x|-ik\theta\cdot y}\left(1+O\Big(\frac{|y|}{|x|}\Big)\right)\left(1+O\Big(\frac{|y|^2}{|x|}\Big)^{\delta}\right)+O\Big(\frac{|y|}{|x|}\Big)$$

for any $0 \le \delta \le 1$.

Finally, as in the proof of Lemma 19.1, we get

$$\frac{\partial}{\partial r}u_+ - iku_+ = O\Big(\frac{1}{|x|^{1+\alpha_1}}\Big), \quad \forall \alpha_1 < \frac{\alpha}{2}, \quad \alpha_1 \le 1.$$

It remains to prove the uniqueness of the solution that satisfies the radiation conditions. Suppose

$$(19.10)\qquad (-\Delta - k^2)v = 0, \quad v = O\Big(\frac{1}{|x|}\Big), \quad \frac{\partial}{\partial r}v - ik\, v = o\Big(\frac{1}{|x|}\Big).$$

Let x_0 be any point and $B_R = \{x : |x| < R\}$, where $R > |x_0|$, and let $n(y)$ be the outward normal to Ω. Applying the Green formula

$$\int_\Omega [((\Delta + k^2)u)\, v - u(\Delta + k^2)v]\, dx = \int_{\partial\Omega}\left(\frac{\partial u}{\partial n}v - u\frac{\partial v}{\partial n}\right) ds$$

with $\Omega = B_R \setminus \overline{B_\varepsilon(x_0)}$ and $u = \frac{1}{4\pi}\frac{e^{ik|x-x_0|}}{|x-x_0|}$, and taking the limit as $\varepsilon \to 0$, we obtain

$$(19.11)\qquad \begin{aligned} -v(x_0) &= \int_{\partial B_R}\left[\left(\frac{\partial}{\partial r}\frac{e^{ik|x-x_0|}}{4\pi|x-x_0|}\right)v - \frac{e^{ik|x-x_0|}}{4\pi|x-x_0|}\frac{\partial v}{\partial r}\right] ds \\ &= \int_{\partial B_R}\left[\left(\frac{\partial}{\partial r}\left(\frac{e^{ik|x-x_0|}}{4\pi|x-x_0|}\right) - ik\frac{e^{ik|x-x_0|}}{4\pi|x-x_0|}\right)v \right. \\ &\qquad \left. - \frac{e^{ik|x-x_0|}}{4\pi|x-x_0|}\left(\frac{\partial v}{\partial r} - ikv\right)\right] ds, \end{aligned}$$

where ds is the area element of the sphere ∂B_R.

Analogously to (19.9), one can show that

$$(19.12)\qquad \frac{\partial}{\partial r}\frac{e^{ik|x-x_0|}}{|x-x_0|} - ik\frac{e^{ik|x-x_0|}}{|x-x_0|} = O\left(\frac{1}{|x|^2}\right).$$

It follows from (19.10), (19.11) and (19.12) that

$$v(x_0) = O\left(\frac{1}{R^2}\right)\cdot o\left(\frac{1}{R}\right)\cdot 4\pi R^2 - O\left(\frac{1}{R}\right)\cdot o\left(\frac{1}{R}\right)\cdot 4\pi R^2 = o(1).$$

Taking the limit as $R \to \infty$, we get $v(x_0) = 0$.

Note that the radiation conditions (19.10) are satisfied if we choose the fundamental solution $E_+(x)$. Another choice of fundamental solution, for example, $E_-(x)$, will require imposing different radiation conditions. □

19.3. The stationary phase lemma.

Now we generalize Theorem 19.2 to the case of any $n \geq 2$.

Lemma 19.3. *The fundamental solution* $E_+ = F^{-1}\frac{1}{|\xi|^2-k^2-i0}$ *has the following asymptotics as* $|x| \to \infty$*:*

$$E_+(x) = C_n \frac{e^{ik|x|}}{|x|^{\frac{n-1}{2}}} + O\left(\frac{1}{|x|^{\frac{n+1}{2}}}\right), \tag{19.13}$$

where

$$C_n = \frac{k^{\frac{n-3}{2}} e^{-i\frac{\pi}{4}(n-3)}}{2^{\frac{n+1}{2}} \pi^{\frac{n-1}{2}}}. \tag{19.14}$$

To prove (19.13) we use the stationary phase method.

Lemma 19.4 (The stationary phase lemma)**.** *Consider the integral*

$$I(\lambda) = \int_{\mathbb{R}^n} a(x) e^{i\lambda\varphi(x)} dx, \tag{19.15}$$

where $a(x) \in C_0^\infty(\mathbb{R}^n)$, $\varphi(x) \in C^\infty(\mathbb{R}^n)$, x_0 *is the only critical point of* $\varphi(x)$ *on* $\operatorname{supp} a(x)$*, i.e.,* $\varphi_x(x_0) = \nabla\varphi(x_0) = 0$, $\varphi_x(x) \neq 0$ *for* $x \neq x_0$*. Denote* $\varphi_{xx}(x_0) = [\frac{\partial^2\varphi(x_0)}{\partial x_j \partial x_k}]_{j,k=1}^n$*. We assume that* $\det\varphi_{xx}(x_0) \neq 0$*. Then as* $\lambda \to \infty$*,*

$$I(\lambda) = \left(\frac{2\pi}{\lambda}\right)^{\frac{n}{2}} \frac{e^{i\lambda\varphi(x_0) + i\frac{\pi}{4}\operatorname{sgn}\varphi_{xx}(x_0)}}{\sqrt{|\det\varphi_{xx}(x_0)|}} \left(a(x_0) + \sum_{j=1}^N a_j(x_0)\lambda^{-j}\right) + O\left(\frac{1}{\lambda^{\frac{n}{2}+N+1}}\right) \tag{19.16}$$

for some $a_j(x_0), 1 \leq j \leq N$*, where* $\operatorname{sgn}\varphi_{xx}(x_0)$ *is the difference between the number of positive and the number of negative eigenvalues of the matrix* $\varphi_{xx}(x_0)$*,* N *is arbitrary.*

Proof of Lemma 19.4. Let $\chi(x) \in C_0^\infty(\mathbb{R}^n)$ be such that $\varphi_x(x) \neq 0$ on $\operatorname{supp}\chi(x)$. Then, using the identity

$$e^{i\lambda\varphi} = \frac{\varphi_x \cdot \frac{\partial}{\partial x}}{i\lambda|\varphi_x|^2} e^{i\lambda\varphi} = \left(\frac{\varphi_x \cdot \frac{\partial}{\partial x}}{i\lambda|\varphi_x|^2}\right)^N e^{i\lambda\varphi(x)}, \ \forall N, \tag{19.17}$$

and integrating by parts N times, we get

$$\int_{\mathbb{R}^n} a(x)\chi(x) e^{i\lambda\varphi(x)} dx = \frac{1}{\lambda^N}\int_{\mathbb{R}^n}\left(i\frac{\partial}{\partial x}\cdot\frac{a\chi\varphi_x}{|\varphi_x|^2}\right)^N e^{i\lambda\varphi(x)} dx = O\left(\frac{1}{\lambda^N}\right). \tag{19.18}$$

Let x_0 be a critical point, i.e., $\varphi_x(x_0) = 0$. Then, using the Taylor formula, we obtain

$$\varphi(x) = \varphi(x_0) + \frac{1}{2}\sum_{j,k=1}^{n} \varphi_{x_j x_k}(x_0)(x_j - x_j^{(0)})(x_k - x_k^{(0)}) + O(|x-x_0|^3).$$

We make the change of variables $y = A(x - x_0)$, where A is an orthogonal matrix such that $A\varphi_{xx}(x_0)A^{-1}$ is a diagonal matrix. Then in y-coordinates we have

$$\varphi(x) = \varphi(x_0) + \frac{1}{2}\sum_{j=1}^{n} \alpha_j y_j^2 + r(y), \tag{19.19}$$

where $r(y) \in C^\infty, r(y) = O(|y|^3)$. Note that all $\alpha_j \neq 0, 1 \le j \le n$, since $\det \varphi_{xx}(x_0) = \prod_{j=1}^n \alpha_j \neq 0$.

Let $\beta(t) \in C_0^\infty(\mathbb{R}^1), \beta(t) = 1$ for $|t| < \delta$ and $\beta(t) = 0$ for $|t| > 2\delta$. Denote $\chi(y) = \prod_{j=1}^n \beta(y_j)$. We study the integral

$$I_1(\lambda) = \int_{\mathbb{R}^n} a(x_0 + A^{-1}y)\chi(y)e^{i\lambda\varphi(x_0) + \frac{i\lambda}{2}\sum_{j=1}^n \alpha_j y_j^2 + i\lambda r(y)}dy.$$

Expanding $a(x_0 + A^{-1}y)$ and $e^{i\lambda r(y)}$ in y by the Taylor formula up to the term of order N, we get

(19.20)

$$I_1(\lambda) = e^{i\lambda\varphi(x_0)}\int_{\mathbb{R}^n}\Big(\sum_{|k|=0}^{N}\sum_{0\le p\le\frac{|k|}{3}} b_{pk}\lambda^p y^k + r_1(\lambda, y)\Big)e^{\frac{i\lambda}{2}\sum_{j=1}^n \alpha_j y_j^2}\prod_{j=1}^n \beta(y_j)dy,$$

where $r_1(\lambda, y) = O\big((1+\lambda^{\frac{N}{3}})|y|^N\big)$.

The main term in (19.20) has the form

$$I_0(\lambda) = \int_{\mathbb{R}^n} a(x_0)e^{i\lambda\varphi(x_0)}\, e^{\frac{i\lambda}{2}\sum_{j=1}^n \alpha_j y_j^2}\prod_{j=1}^n \beta(y_j)dy.$$

We have

$$\int_{-\infty}^{\infty} e^{\frac{i\lambda}{2}\alpha_j y_j^2}\beta(y_j)dy_j = \int_{-\infty}^{\infty} e^{\frac{i\lambda}{2}\alpha_j y_j^2}dy_j + \int_{-\infty}^{\infty} e^{\frac{i\lambda}{2}\alpha_j y_j^2}(1-\beta(y_j))dy_j. \tag{19.21}$$

Since $\alpha_j y_j^2$ has no critical points on $\operatorname{supp}(1-\beta(y_j))$, we get, upon integrating by parts:

$$\int_{-\infty}^{\infty}(1-\beta(y_j))e^{\frac{i\lambda}{2}\alpha_j y_j^2}dy_j = O\left(\frac{1}{\lambda^N}\right), \quad \forall N. \tag{19.22}$$

Suppose $\alpha_j > 0$. Then by the Cauchy integral theorem,

$$\int_{-\infty}^{\infty} e^{\frac{i\lambda}{2}\alpha_j y_j^2}dy_j = 2\int_0^{\infty} e^{\frac{i\lambda}{2}\alpha_j y_j^2}dy_j = 2\int_{\Gamma_1} e^{\frac{i\lambda}{2}\alpha_j z^2}dz,$$

where Γ_1 is the ray $z = e^{i\frac{\pi}{4}}t, 0 \le t < +\infty$. Therefore

$$2\int_{\Gamma_1} e^{i\frac{\lambda}{2}\alpha_j z^2}dz = 2e^{i\frac{\pi}{4}}\int_0^\infty e^{-\frac{\lambda}{2}\alpha_j t^2}dt = \frac{2e^{i\frac{\pi}{4}}}{\sqrt{\frac{\lambda}{2}\alpha_j}}\int_0^\infty e^{-t^2}dt = \frac{\sqrt{2\pi}e^{i\frac{\pi}{4}}}{\sqrt{\lambda\alpha_j}}.$$

If $\alpha_j < 0$, then again by the Cauchy theorem,

$$2\int_0^\infty e^{i\lambda\alpha_j y_j^2}dy_j = 2\int_{\Gamma_2} e^{i\lambda\alpha_j z^2}dz,$$

where Γ_2 is the ray $z = e^{-i\frac{\pi}{4}}t, 0 \le t < \infty$. Therefore

$$2\int_{\Gamma_2} e^{i\lambda\alpha_j z^2}dz = 2e^{-i\frac{\pi}{4}}\int_0^\infty e^{\frac{\lambda}{2}\alpha_j t^2}dt = \frac{\sqrt{2\pi}e^{-i\frac{\pi}{4}}}{\sqrt{\lambda|\alpha_j|}}.$$

Thus

$$I_0(\lambda) = a(x_0)e^{i\lambda\varphi(x_0)}\left(\frac{2\pi}{\lambda}\right)^{\frac{n}{2}}\frac{e^{i\frac{\pi}{4}\operatorname{sgn}\varphi_{xx}(x_0)}}{|\det\varphi_{xx}(x_0)|^{\frac{1}{2}}} + O\left(\frac{1}{\lambda^N}\right), \quad \forall N, \tag{19.23}$$

since $|\det\varphi_{xx}| = \prod_{j=1}^n|\alpha_j|$ and $\operatorname{sgn}\varphi_{xx}$ is the number of positive α_j minus the number of negative α_j.

Consider an integral in $\mathbb{R}^n$ of the form

$$I_{q_1q_2}(y',\lambda) = \int_{-\infty}^\infty g(y',\lambda)\lambda^{q_1}y_n^{q_2}e^{i\frac{\lambda}{2}\alpha_n y_n^2}\beta(y_n)dy_n, \tag{19.24}$$

where $g(y',\lambda)$ does not depend on y_n and $q_2 \ge 1$, $y' = (y_n,\dots,y_{n-1})$. Since $y_n e^{i\frac{\lambda}{2}\alpha_n y_n^2} = \frac{\partial}{\partial y_n}\frac{1}{i\lambda\alpha_n}e^{i\frac{\lambda}{2}\alpha_n y_n^2}$, we get, integrating by parts:

$$I_{q_1q_2}(y',\lambda) = -\int_{-\infty}^\infty g(y',\lambda)\lambda^{q_1-1}\frac{(q_2-1)}{i\alpha_n}y_n^{q_2-2}e^{i\frac{\lambda}{2}\alpha_n y_n^2}\beta(y_n)dy_n + O\left(\frac{1}{\lambda^N}\right).$$

We have used that

$$\int_{-\infty}^\infty y_n^{q_2-1}\frac{\partial\beta}{\partial y_n}e^{i\frac{\lambda}{2}\alpha_n y_n^2}dy_n = O\left(\frac{1}{\lambda^N}\right), \tag{19.25}$$

since y_n^2 has no critical points on $\operatorname{supp}\frac{\partial\beta(y_n)}{\partial y_n}$.

Suppose q_2 is even. Then after $\frac{q_2}{2}$ integrations by parts we get

$$I_{q_1q_2}(y',\lambda) = C\int_{-\infty}^\infty g(y',\lambda)\lambda^{q_1-\frac{q_2}{2}}e^{i\frac{\lambda}{2}\alpha_n y_n^2}\beta(y_n)dy_n + O\left(\frac{1}{\lambda^N}\right), \tag{19.26}$$

where C is a constant. Computing the integral with respect to y_n as in (19.23), we get

$$I_{q_1q_2}(y',\lambda) = C_1 g(y',\lambda)\frac{e^{i\frac{\pi}{4}\operatorname{sgn}\alpha_n}}{\sqrt{\lambda|\alpha_n|}}\lambda^{q_1-\frac{q_2}{2}} + O\left(\frac{1}{\lambda^N}\right).$$

If q_2 is odd, then, after $\frac{q_2+1}{2}$ integrations by part, we get an integral of the form (19.25), and therefore $I_{q_1q_2} = O(\frac{1}{\lambda^N}), \forall N$, if q_2 is odd.

Analogously we integrate by parts with respect to $y_1, \dots, y_{n-1}$. Note that the contribution of $r_1(\lambda, y)$ will be $O(\lambda^{\frac{N}{3}-\frac{N}{2}}) = O(\lambda^{-\frac{N}{6}})$ after $\frac{N}{2}$ integrations by parts.

Combining the contribution of all terms in (19.20), we get (19.16). □

19.4. Radiation conditions for $n \geq 2$.

Proof of Lemma 19.3. Suppose $\chi(t) \in C_0^\infty(\mathbb{R}^1)$, $\chi(-t) = \chi(t)$, $\chi(t) = 1$ for $|t| < \frac{k}{4}$, and $\chi(t) = 0$ for $|t| \geq \frac{k}{2}$. We have (cf. Example 5.4):

$$E_+(x) = F^{-1}\left(\frac{1}{|\xi|^2 - (k+i0)^2}\right)$$
$$= \frac{1}{(2\pi)^n}\int_{\mathbb{R}^n} \frac{\chi(|\xi|-k)e^{ix\cdot\xi}}{|\xi|^2-(k+i0)^2}d\xi + F^{-1}\left(\frac{1-\chi(|\xi|-k)}{|\xi|^2-k^2}\right).$$

Introducing polar coordinates $\xi = r\omega$, we get

$$E_+(x) = \int_0^\infty \frac{\chi(r-k)}{r-k-i0}\psi(r,x)dr + F^{-1}\left(\frac{1-\chi}{|\xi|^2-k^2}\right), \tag{19.27}$$

where

$$\psi(r,x) = \frac{1}{(2\pi)^n}\frac{r^{n-1}}{r+k}\int_{|\omega|=1} e^{ix\cdot r\omega}d\omega. \tag{19.28}$$

Since $\frac{1-\chi}{|\xi|^2-k^2} \in C^\infty(\mathbb{R}^n)$ and $|\frac{\partial^p}{\partial\xi^p}(\frac{1-\chi}{|\xi|^2-k^2})| \leq C_p(1+|\xi|)^{-2-|p|}$, we see that

$$x^p F^{-1}\left(\frac{1-\chi}{|\xi|^2-k^2}\right) = F^{-1}\left(\left(i\frac{\partial}{\partial\xi}\right)^p \frac{1-\chi}{|\xi|^2-k^2}\right) \tag{19.29}$$

is bounded when $|p| > n-1$ (cf. estimates (18.7)–(18.12)). Therefore

$$F^{-1}\left(\frac{1-\chi}{|\xi|^2-k^2}\right) = O\left(\frac{1}{|x|^N}\right) \quad \text{for any} \quad N > n-1. \tag{19.30}$$

Let $\theta = \frac{x}{|x|}$ and assume that $|x| \to \infty$ and θ is fixed. The function $\varphi = \theta\cdot r\omega$ has exactly two critical points $\omega = \pm\theta$ on the sphere $|\omega| = 1$. We have:

$$\psi(r,x) = \sum_{j=1}^3 \psi_j(r,x), \tag{19.31}$$

where

$$\psi_j(r,x) = \frac{1}{(2\pi)^n}\frac{r^{n-1}}{r+k}\int_{|\omega|=1} \alpha_j(\omega)e^{i|x|\theta\cdot r\omega}d\omega, \tag{19.32}$$

$\alpha_j(\omega) \in C_0^\infty(S^{n-1})$, $\alpha_1(\omega) = 1$ for $|\omega-\theta| < \varepsilon$, $\alpha_1(\omega) = 0$ for $|\omega-\theta| > 2\varepsilon$, $\alpha_2(\omega) = 1$ for $|\omega+\theta| < \varepsilon$, $\alpha_2(\omega) = 0$ for $|\omega+\theta| > 2\varepsilon$, $\alpha_3 = 1-\alpha_1-\alpha_2$.

It follows from the computations of Example 14.4 that

$$|\psi_3(r,x)| \leq \frac{C_N}{|x|^N} \quad \text{for any} \quad N \tag{19.33}$$

and also

$$\left|\frac{\partial\psi_3(r,x)}{\partial r}\right| \leq \frac{C_{N1}}{|x|^N} \quad \text{for any} \quad N. \tag{19.34}$$

Since $\frac{\psi_3(r,x)-\psi_3(k,x)}{r-k} = \frac{\partial\psi_3(r',x)}{\partial r}$, where r' is between r and k, we get:

$$\begin{aligned} &\left|\int_0^\infty \frac{\chi(r-k)}{r-k-i0}\psi_3(r,x)dr\right| \\ &\leq \left|\int_0^\infty \frac{\chi(r-k)}{r-k-i0}\psi_3(k,x)dr\right| + \left|\int_0^\infty \chi(r-k)\frac{\partial\psi_3(r',x)}{\partial r}dr\right| \\ &\leq C|\psi_3(k,x)| + C\sup_{r'\in[\frac{k}{2},\frac{3k}{2}]}\left|\frac{\partial\psi_3(r',x)}{\partial r}\right| \leq \frac{C}{|x|^N},\ \forall N. \end{aligned} \tag{19.35}$$

It remains to study the contribution of ψ_1 and ψ_2. We have for $j = 1, 2$,

$$\begin{aligned} &\int_0^\infty \frac{\chi(r-k)}{r-k-i0}\psi_j(r,x)dr \\ &= \frac{1}{(2\pi)^n}\int_{|\omega|=1}\alpha_j(\omega)\left(\int_0^\infty \frac{\chi(r-k)}{r-k-i0}\frac{r^{n-1}}{r+k}e^{ir|x|\theta\cdot\omega}dr\right)d\omega. \end{aligned} \tag{19.36}$$

Note that $\theta\cdot\omega > 0$ on $\operatorname{supp}\alpha_1(\omega)$ and $\theta\cdot\omega < 0$ on $\operatorname{supp}\alpha_2(\omega)$.

Let $\sigma = |x|\theta\cdot\omega$. We have

$$\int_0^\infty \frac{\chi(r-k)}{r-k-i0}\frac{r^{n-1}}{r+k}e^{ir\sigma}dr = I_1 + I_2 + I_3, \tag{19.37}$$

where

$$I_1 = \int_0^\infty \frac{\chi(r-k)}{r-k-i0}\left(\frac{r^{n-1}}{r+k} - \frac{k^{n-1}}{2k}\right)e^{ir\sigma}dr,$$

$$I_2 = -\frac{k^{n-1}}{2k}\int_{-\infty}^\infty \frac{(1-\chi(r-k))}{r-k-i0}e^{ir\sigma}dr, \quad I_3 = \frac{k^{n-1}}{2k}\int_{-\infty}^\infty \frac{e^{ir\sigma}dr}{r-k-i0}.$$

Since I_1 is the Fourier transform of a C_0^∞ function, we have

$$|I_1| \leq \frac{C_N}{|\sigma|^N},\quad \forall N. \tag{19.38}$$

Integrating by parts in I_2, we also get (cf. (19.30))

$$|I_2| \leq \frac{C_N}{|\sigma|^N},\quad \forall N. \tag{19.39}$$

Finally (cf. Example 10.2):

$$I_3 = \frac{k^{n-1}}{2k}2\pi i\theta(\sigma)e^{ik\sigma}, \tag{19.40}$$

where $\theta(\sigma) = 1$ for $\sigma > 0$, and $\theta(\sigma) = 0$ for $\sigma < 0$, $\sigma = |x|\theta\cdot\omega$. Note that I_3 can also be computed using the Jordan lemma (cf. Proposition 16.1) for $k + i\varepsilon$ and then taking the limit as $\varepsilon \to +0$.

Therefore the contribution of $\psi_2(r,x)$ is $O(\frac{1}{|x|^N})$, $\forall N$, since $\sigma < 0$ on $\operatorname{supp}\alpha_2(\omega)$, and we have

$$E_+(x) = 2\pi i\psi_1(k,x) + O\left(\frac{1}{|x|^N}\right), \quad \forall N. \tag{19.41}$$

It remains to study the contribution of $\psi_1(k,x)$. We apply the stationary phase lemma to $\psi_1(k,x)$. The phase function here is $\varphi = k\omega\cdot\theta$. There is only one critical point $\omega = \theta$ on $\operatorname{supp}\alpha_1(\omega)$.

Make an orthogonal change of coordinates $x = Oy$ such that θ is the point $(0,\dots,1)$. Then the equation of the unit sphere in a neighborhood of θ will be

$$y_n = \sqrt{1-\sum_{k=1}^{n-1} y_k^2} = 1 - \frac{1}{2}\sum_{k=1}^{n-1} y_k^2 + O(|y'|^4), \quad y' = (y_1,\dots,y_{n-1}).$$

Since $\varphi = k\omega\cdot\theta = ky_n(y')$, we conclude that $\varphi_{y'y'} = -kI_{n-1}$ at $\omega = \theta$, where I_{n-1} is the identity matrix in $\mathbb{R}^{n-1}$.

Note that $\operatorname{sgn}\varphi_{y'y'} = -(n-1)$. Therefore Lemma 19.4 implies:

$$\psi_1(k,x) = \frac{1}{(2\pi)^n}\frac{k^{n-2}}{2}\left(\frac{2\pi}{|x|}\right)^{\frac{n-1}{2}} \frac{\alpha_1(\theta)e^{-i\frac{\pi}{4}(n-1)}}{k^{\frac{n-1}{2}}} e^{ik|x|} + O\left(\frac{1}{|x|^{\frac{n+1}{2}}}\right). \tag{19.42}$$

Thus (19.13) holds. □

Theorem 19.5. *Suppose $f(x) = O(\frac{1}{|x|^{n+\alpha}})$, $\alpha > 0$, $x\in\mathbb{R}^n$. Then $u = E_+ * f$ is a solution of*

$$(-\Delta - k^2)u = f \quad \text{in} \quad \mathbb{R}^n \tag{19.43}$$

such that

$$u = C_n \frac{e^{ik|x|}}{|x|^{\frac{n-1}{2}}}\tilde{f}(k\theta) + O\left(\frac{1}{|x|^{\frac{n-1}{2}+\alpha_1}}\right), \tag{19.44}$$

where $\theta = \frac{x}{|x|}$ is fixed, $|x|\to\infty$, C_n is the same as in (19.14), $\tilde{f}(k\theta)$ is the Fourier transform of $f(x)$ and $0 < \alpha_1 < \frac{\alpha}{2}$, $\alpha_1 \le 1$.

Moreover, $u = E_+ * f$ is the unique solution of (19.43) in the class of functions satisfying radiation conditions in $\mathbb{R}^n$:

$$u = O\left(\frac{1}{r^{\frac{n-1}{2}}}\right), \quad \frac{\partial u}{\partial r} - iku = o\left(\frac{1}{r^{\frac{n-1}{2}}}\right), \tag{19.45}$$

where $r = |x|$.

Proof: It follows from Lemma 19.3 that $E_+(x)$ has the form (19.13) if $|x| \geq 1$. We have, for $|x| \leq 1$:

$$|E_+(x)| \leq \frac{C}{|x|^{n-2}} \quad \text{if} \quad n \geq 3,$$

$$|E_+(x)| \leq C|\ln |x|| \quad \text{if} \quad n = 2,$$

because the main part of $E_+(x)$ for $|x| \leq 1$ is the same as the fundamental solution of the Laplacian $-\Delta$.

We proceed as in the proof of Lemma 19.1. We estimate $u_1 = \int_{|y|<\frac{|x|}{2}} E_+(x-y)f(y)dy$ using (19.13). Analogously to (19.5), (19.6), (19.7) we obtain

$$u_1 = C_n \frac{e^{ik|x|}}{|x|^{\frac{n-1}{2}}} \left(\tilde{f}(k\theta) + O\Big(\frac{1}{|x|^{\alpha_1}}\Big)\right), \quad \alpha_1 \leq 1, \ \alpha_1 < \frac{\alpha}{2}.$$

Then we split $u_2 = \int_{|y|>\frac{|x|}{2}} E_+(x-y)f(y)dy$ into two terms: u_{21} where $|y| > \frac{|x|}{2}$ and $|x-y| > 1$, and u_{22} where $|y| > \frac{|x|}{2}$ and $|x-y| < 1$. To estimate u_{21} we use that $|E_+(x-y)| \leq \frac{C}{|x-y|^{\frac{n-1}{2}}}$ and proceed as in (19.8). We get

$$|u_{21}| \leq \frac{C}{|x|^{\frac{n-1}{2}+\alpha}}.$$

To estimate u_{22} we get for $n \geq 3$,

$$|u_{22}| \leq \int_{|x-y|\leq 1} \frac{C\,dy}{|x-y|^{n-2}(1+|y|)^{n+\alpha}} \leq \frac{C_1}{|x|^{n+\alpha}}.$$

For $n = 2$ we have

$$|u_{22}| \leq \int_{|x-y|\leq 1} \frac{|\ln|x-y||dy}{(1+|y|)^{2+\alpha}} \leq \frac{C}{|x|^{2+\alpha}}.$$

Therefore (19.44) is proven. The proof of the uniqueness of the solution satisfying the radiation conditions (19.45) is the same as in the case of $n = 3$. □

19.5. The limiting amplitude principle.

Consider the nonhomogeneous wave equation with the right hand side of the form $f(x)e^{-ikt}$, $f(x) \in C_0^\infty(B_R)$:

$$\frac{\partial^2 u(x,t)}{\partial t^2} - \Delta u(x,t) = f(x)e^{-ikt}. \tag{19.46}$$

One can look for the solution of (19.46) in the form $u(x,t) = v(x)e^{-ikt}$. Then $v(x)$ satisfies the reduced wave equation

$$-\Delta v(x) - k^2 v(x,t) = f(x). \tag{19.47}$$

The question arises what fundamental solution one has to use to solve (19.47).

Consider the Cauchy problem for $t > 0, x \in \mathbb{R}^3$, with zero initial conditions:

$$\frac{\partial^2 w}{\partial t^2} - \Delta w = f(x)e^{-ikt}, \quad t > 0, \tag{19.48}$$

$$w(x,0) = w_t(x,0) = 0. \tag{19.49}$$

We extend e^{-ikt} by zero to $t < 0$. The unique solution of (19.48), (19.49) is given by the retarded potential (cf. (16.35))

$$w(x,t) = \frac{1}{4\pi}\int_{|x-y|<t} \frac{f(y)e^{-ik(t-|x-y|)}}{|x-y|}dy.$$

Note that $f(y) = 0$ when $|y| > R$. We have

$$w(x,t) = \frac{1}{4\pi}\int_{\mathbb{R}^3} \frac{e^{-ik(t-|x-y|)}}{|x-y|}f(y)dy = e^{-ikt}(E_+ * f)$$

for $t > |x| + R$, since the ball $\{y : |y| < R\}$ is contained in $\{y : |x-y| = t\}$ for $t > |x| + R$. Therefore

$$\lim_{t\to+\infty} e^{ikt}w(x,t) = v(x), \tag{19.50}$$

where $v(x) = E_+ * f$ is the outgoing solution of the reduced wave equation (19.47).

We have proven for $n = 3$ the limiting amplitude principle:

The $\lim_{t\to+\infty} e^{ikt}w(x,t)$ for the Cauchy problem (19.48), (19.49) singles out the solution $v(x) = E_+ * f$, i.e., the solution satisfying the radiation conditions (19.45) for $n = 3$. The same result holds for any $n \geq 2$ and for more general equations. Note that if we change k to $-k$ in (19.46), then $\lim_{t\to+\infty} e^{-ikt}w(x,t) = v_1$, where $v_1 = E_- * f$, i.e., the radiation conditions (19.45) will hold with k replaced by $-k$.

20. Single and double layer potentials

20.1. Limiting values of double layers potentials.

Let Ω be a bounded domain with a smooth boundary $\partial\Omega$. Let $f(x)$ be a continuous function on $\partial\Omega$ and let $E(x)$ be the fundamental solution for $-\Delta$, $E(x) = \frac{\Gamma(\frac{n-2}{2})}{4\pi^{\frac{n}{2}}}|x|^{2-n}$ for $n \geq 3$ and $E(x) = -\frac{1}{2\pi}\ln|x|$ for $n = 2$. The integral operator

$$(Sf)(x) = \int_{\partial\Omega} E(x-y')f(y')ds \tag{20.1}$$

is called a single layer potential, where ds is the area element of the surface $\partial\Omega$. Let $n(y')$ be the outward unit normal to Ω at $y' \in \partial\Omega$. The double layer potential is the following integral operator:

$$(Df)(x) = \int_{\partial\Omega} \frac{\partial E(x-y')}{\partial n(y')} f(y')ds. \tag{20.2}$$

Note that

$$\Delta Sf = 0 \quad \text{in} \quad \mathbb{R}^n \setminus \partial\Omega \tag{20.3}$$

and

$$\Delta Df = 0 \quad \text{in} \quad \mathbb{R}^n \setminus \partial\Omega. \tag{20.4}$$

Since the integral (20.1) converges absolutely on $\partial\Omega$, we conclude, using the Lebesgue convergence theorem, that $(Sf)(x)$ is a continuous function in $\mathbb{R}^n$. This result remains true if $f(y')$ is only in $L_2(\partial\Omega)$. All other results of this section can also be proven for a larger class of $f(y')$ using the technique of pseudodifferential operators (cf. §55).

Let $w(x)$ be a continuous function in $\overline{\Omega}$. Denote by $w^+(x')$, $x' \in \partial\Omega$, the values of $w(x)$ on $\partial\Omega$, i.e.,

$$w^+(x') = \lim_{\substack{x\to x' \\ x\in\Omega}} w(x).$$

Analogously, if $w(x)$ is continuous in $\mathbb{R}^n \setminus \Omega$, we denote by $w^-(x')$ the values of $w(x)$ on $\partial\Omega$:

$$w^-(x') = \lim_{\substack{x\to x' \\ x\in\mathbb{R}^n\setminus\overline{\Omega}}} w(x).$$

We have

$$\frac{\partial}{\partial n(y')} E(x-y') = -C\frac{(x-y')\cdot n(y')}{|x-y'|^n}, \tag{20.5}$$

where $C = \frac{\Gamma(\frac{n-2}{2})(2-n)}{4\pi^{\frac{n}{2}}}$ for $n \geq 3$ and $C = -\frac{1}{2\pi}$ for $n = 2$. Since $(x'-y')\cdot n(y') = O(|x'-y'|^2)$ if $x' \in \partial\Omega$, $y' \in \partial\Omega$, we get that $\frac{\partial}{\partial n(y')}E(x'-y') = O(\frac{1}{|x'-y'|^{n-2}})$. Therefore the integral

$$\int_{\partial\Omega} \frac{\partial}{\partial n(y')} E(x'-y')f(y')ds$$

converges absolutely and uniformly in $x' \in \partial\Omega$. We denote this integral by $(D_0 f)(x')$:

$$(D_0 f)(x') = \int_{\partial\Omega} \frac{\partial}{\partial n(y')} E(x'-y')f(y')ds, \quad x' \in \partial\Omega. \tag{20.6}$$

The function $(D_0 f)(x')$ is continuous on $\partial\Omega$.

Theorem 20.1. *Suppose f is continuous on $\partial\Omega$. Then the double layer potential Df is continuous in $\overline{\Omega}$ and in $\mathbb{R}^n \setminus \Omega$. Moreover,*

$$(Df)^+(x') = (D_0 f)(x') - \frac{1}{2} f(x'), \tag{20.7}$$

$$(Df)^-(x') = (D_0 f)(x') + \frac{1}{2} f(x'). \tag{20.8}$$

Proof: Denote by Ω_δ the neighborhood of $\partial\Omega$ in $\mathbb{R}^n$ consisting of points x such that their distance to $\partial\Omega$ is less than $\delta > 0$. Any point $x \in \Omega_\delta$ can be represented in the form $x = z' + z_n n(z')$, where $z' \in \partial\Omega$, $|z_n| < \delta$. Note that $n(y')$ is the outward normal. Therefore $z_n > 0$ outside of Ω and $z_n < 0$ inside Ω. We will use (z', z_n) as coordinates in Ω_δ. In the new coordinates we have

$$\begin{aligned}\frac{\partial E(x-y')}{\partial n(y')} &= -C\frac{(x-y')\cdot n(y')}{|x-y'|^n} \\ &= -C\frac{(z_n n(z') + z' - y')\cdot n(y')}{|z_n n(z') + z' - y'|^n} = E_0 + E_1,\end{aligned}$$

where

$$\begin{aligned}E_0 &= -\frac{C z_n}{|x-y'|^n}, \\ E_1 &= -C\frac{(z'-y')\cdot n(y') + z_n(n(z') - n(y'))\cdot n(y')}{|z_n n(z') + z' - y'|^n} = O\left(\frac{1}{|x-y'|^{n-2}}\right),\end{aligned}$$

since $(z'-y')\cdot n(y') = O(|z'-y'|^2)$, $n(z') - n(y') = O(|z'-y'|)$, and $|x-y'|^n$ is equivalent to $(z_n^2 + |z'-y'|^2)^{\frac{n}{2}}$.

Therefore $E_1 f$ is continuous in $\mathbb{R}^n$, and we can take the $\lim_{z_n \to 0} E_1 f$. This limit coincides with (20.6).

Let $\chi(y') \in C^\infty(\partial\Omega)$, $\chi(y') = 1$ for $|y'-z'| < \delta_1$, and $\chi = 0$ for $|y'-z'| > 2\delta_1$. We choose $\delta_1 > 0$ such that $|f(y') - f(z')| < \varepsilon$ on $\operatorname{supp}\chi$. Since $|y'-z'| > 0$ on $\operatorname{supp}(1-\chi)$, we have that $\lim_{z_n\to 0} E_0(1-\chi)f = 0$. On $\operatorname{supp}\chi$ we take the projection of $\partial\Omega$ on the tangent plane to $\partial\Omega$ at the point z'. Then $y' - z' \equiv (t, O(|t|^2))$, where $t = (t_1, \dots, t_{n-1}) \in \mathbb{R}^{n-1}$ are the coordinates of the projection of $y' - z'$ on the tangent plane. We have $(E_0\chi f)(z', z_n) = E_0\chi(f(y') - f(z')) + E_0\chi f(z')$. Making the change of variables $t' = \frac{t}{|z_n|}$ and passing to the limit as $|z_n| \to 0$, we get

$$\begin{aligned}(E_0\chi)f(z') &= -C\int_{\partial\Omega} \frac{z_n\chi\big(z' + (t, O(|t|^2))\big) f(z')(1+O(t))dt}{(|z_n + O(|t|^2)|^2 + |t|^2)^{\frac{n}{2}}} \\ &\to -C \operatorname{sgn} z_n f(z') \int_{\mathbb{R}^n} \frac{dt'}{(1+|t'|^2)^{\frac{n}{2}}} = \frac{1}{2}\operatorname{sgn} z_n f(z').\end{aligned}$$

Also we have

$$|E_0\chi(f(y') - f(z'))| \le C_1\varepsilon \int_{\mathbb{R}^{n-1}} \frac{|z_n|dt}{(|z_n|^2 + |t|^2)^{\frac{n}{2}}} \le C_2\varepsilon.$$

Since $\varepsilon > 0$ is arbitrary, we get (20.8) for $z_n > 0$ and (20.7) for $z_n < 0$. □

We denote by $n(x)$ a smooth extension of the field of normal vectors $n(x)$ from $\partial\Omega$ to Ω_δ, where Ω_δ is the same as in Theorem 20.1.

Theorem 20.2. *Let Df be a double layer potential, where f is twice continuously differentiable on $\partial\Omega$. Then $\frac{\partial}{\partial n(x)} Df$ is continuous in $\Omega_\delta \cap \overline{\Omega}$ and $\Omega_\delta \cap (\mathbb{R}^n \setminus \Omega)$, and*

$$\left(\frac{\partial}{\partial n(x)} Df\right)^+ (x') = \left(\frac{\partial}{\partial n(x)} Df\right)^- (x'), \quad x' \in \partial\Omega. \tag{20.9}$$

Proof: (See Courant and Hilbert [**CH**]) Let $w(y)$ be any twice continuously differentiable function in $\overline{\Omega}$ such that $w(y') = f(y')$ on $\partial\Omega$ and $\frac{\partial w(y')}{\partial n(y')} = 0$ on $\partial\Omega$. Let $x \in \mathbb{R}^n \setminus \overline{\Omega}$. Applying Green's formula in Ω, we get

$$\begin{aligned} &\int_{\partial\Omega} \left(\frac{\partial}{\partial n(y')} E(x-y')w(y') - E(x-y')\frac{\partial w(y')}{\partial n(y')}\right) ds \\ &\quad = \int_\Omega ((\Delta_y E(x-y))w(y) - E(x-y)\Delta w(y))dy. \end{aligned} \tag{20.10}$$

Taking into account that $\Delta_y E(x-y) = 0$, $w(y') = f(y')$, and $\frac{\partial w(y')}{\partial n(y')} = 0$, we have

$$(Df)(x) = -\int_\Omega E(x-y)\Delta w(y)dy.$$

Here the right hand side is continuously differentiable in $\mathbb{R}^n$. Therefore $(Df)(x)$ is continuously differentiable in $\mathbb{R}^n \setminus \Omega$ and

$$\left(\frac{\partial}{\partial n(x')} Df\right)^- (x') = -\int_\Omega \frac{\partial}{\partial n(x')} E(x'-y)\Delta w(y)dy, \quad x' \in \partial\Omega. \tag{20.11}$$

Consider $x \in \Omega$. Applying Green's formula to $\Omega \setminus B_\varepsilon(x)$ and taking the limit as $\varepsilon \to 0$, we obtain:

$$\int_{\partial\Omega} \frac{\partial}{\partial n(y')} E(x-y')f(y')ds + w(x) = -\int_\Omega E(x-y)\Delta w(y)dy. \tag{20.12}$$

Since the right hand side of (20.12) and $w(x)$ are continuously differentiable in $\overline{\Omega}$, we see that Df is continuously differentiable in $\overline{\Omega}$ and

$$\left(\frac{\partial}{\partial n(x')} Df\right)^+ (x') = -\frac{\partial w(x')}{\partial n(x')} - \int_\Omega \frac{\partial}{\partial n(x')} E(x'-y)\Delta w(y)dy. \tag{20.13}$$

Since $\frac{\partial w(x')}{\partial n(x')} = 0$, we get (20.9) from (20.11) and (20.13). □

20.2. Limiting values of normal derivatives of single layer potentials.

Now we study the behavior of the normal derivatives of single layer potentials.

Theorem 20.3. *Let Sf be a single layer potential and let $f(y')$ be continuous on $\partial\Omega$. Then $\frac{\partial}{\partial n(x)}Sf$ is continuous in $\overline{\Omega}\cap\Omega_\delta$ and in $(\mathbb{R}^n\setminus\Omega)\cap\Omega_\delta$. Moreover,*

$$\left(\frac{\partial}{\partial n(x')}Sf\right)^+(x') = \frac{1}{2}f(x') + \left(\frac{\partial}{\partial n(x')}Sf\right)_0(x'), \tag{20.14}$$

$$\left(\frac{\partial}{\partial n(x')}Sf\right)^-(x') = -\frac{1}{2}f(x') + \left(\frac{\partial}{\partial n(x')}Sf\right)_0(x'). \tag{20.15}$$

Here

$$\begin{aligned}\left(\frac{\partial}{\partial n(x')}Sf\right)_0(x') &= \int_{\partial\Omega}\frac{\partial}{\partial n(x')}E(x'-y')f(y')ds \\ &= \int_{\Omega}\frac{Cn(x')\cdot(x'-y')}{|x'-y'|^n}f(y')ds,\end{aligned} \tag{20.16}$$

where C is the same constant as in (20.5).

Proof: Note that $n(x')\cdot(x'-y') = O(|x'-y'|^2)$. Therefore, $(\frac{\partial}{\partial n(x')}Sf)_0(x')$ is a well-defined and continuous function on $\partial\Omega$. Let

$$Tf = \frac{\partial}{\partial n(x)}Sf + Df, \tag{20.17}$$

where Df is a double layer potential. We have

$$(Tf)(x) = \int_{\partial\Omega}\frac{C(n(x)-n(y'))\cdot(x-y')}{|x-y'|^n}f(y')ds$$

and

$$(Tf)(x') = \left(\frac{\partial}{\partial n(x')}Sf\right)_0(x') - (D_0f)(x'). \tag{20.18}$$

Since $n(x)-n(y') = O(|x-y'|)$, we see that Tf is a continuous in $\mathbb{R}^n$. Therefore $\frac{\partial}{\partial n(x)}Sf$ is continuous in $\overline{\Omega}$ and in $\mathbb{R}^n\setminus\Omega$, and (20.14), (20.15) follow from (20.17), (20.18) and (20.7), (20.8). □

Remark 20.1. In Chapter V we use single and double layer potentials in the case where the fundamental solution $E(x-y)$ is replaced by $E_+(x-y)$ (see (16.63)). Note that

$$\frac{\partial^p}{\partial x^p}(E_+(x-y)-E(x-y)) = O\left(\frac{1}{|x-y|^{n-3+|p|}}\right) \text{ for } n\geq 3,\ |p|\leq 2, \tag{20.19}$$

and

$$(20.20)\quad \frac{\partial^p}{\partial x^p}(E_+(x-y)-E(x-y)) = O(|x-y|^{1-|p|}) \text{ for } n=2,\ 1 \le |p| \le 2.$$

Therefore all results of §20 remain true in the case where $E(x-y)$ is replaced by $E_+(x-y)$.

21. Problems

1. Show that $E(x_1, x_2) = \frac{1}{4}$ for $|x_1| \le |x_2|$, $E(x_1, x_2) = 0$ otherwise, is a fundamental solution for $L = \frac{\partial^2}{\partial x_2^2} - \frac{\partial^2}{\partial x_1^2}$.

2. Find a fundamental solution for $\frac{\partial}{\partial t}$ in $\mathbb{R}^{n+1} = \mathbb{R}^n \times \mathbb{R}$.

3. Find the fundamental solution $E(x,t)$, equal to zero for $t < 0$, to the equation $\frac{\partial u(x,t)}{\partial t} + \sum_{k=1}^n \alpha_k \frac{\partial u(x,t)}{\partial x_k} + \alpha_0 u(x,t) = f(x,t)$, where $\alpha_0, \alpha_1, \ldots, \alpha_n$ are constants.

4. **a)** Using the Green formula

$$\int_{|x|>\varepsilon} (\Delta E\varphi - E\Delta\varphi)dx = -\int_{|x|=\varepsilon} \left(\frac{\partial E}{\partial r}\varphi - E\frac{\partial \varphi}{\partial r}\right) d\sigma,$$

where $\varphi \in C_0^\infty(\mathbb{R}^n)$, $r = |x|$, and $d\sigma$ is the area element of the sphere $|x| = \varepsilon$, and passing to the limit as $\varepsilon \to 0$, prove that $E = \frac{1}{4\pi|x|}$ is a fundamental solution for $-\Delta$ in $\mathbb{R}^3$.

b) Analogously prove that $E = -\frac{1}{2\pi}\ln|x|$ is a fundamental solution for $-\Delta$ in $\mathbb{R}^2$.

5. **a)** Show that $E = \frac{1}{8\pi}|x|^2 \ln|x|$ is a fundamental solution for Δ^2 in $\mathbb{R}^2$.

b) Using a fundamental solution for Δ, show that $E = \frac{1}{2\pi(x_1+ix_2)}$ is a fundamental solution for the Cauchy-Riemann operator $L = \frac{\partial}{\partial x_1} + i\frac{\partial}{\partial x_2}$.

6. **a)** Prove that $(-\Delta + k^2)u = f$ has a unique solution $u \in H_2(\mathbb{R}^n)$ for any $f \in L_2(\mathbb{R}^n)$.

b) Consider $\Delta u = f$ in $\mathbb{R}^n$, $n \ge 3$, where $f(x) = O(\frac{1}{(1+|x|)^{n+\varepsilon}})$. Prove that there exists a solution $u = O(\frac{1}{|x|^{n-2}})$.

c) Show that if $\Delta u = 0$ in $\mathbb{R}^n$, and $|u| \to 0$ as $|x| \to \infty$, then $u = 0$.

7. Find the asymptotics of the Faddeev fundamental solution (see Example 16.6) using the stationary phase method.

8. Consider single and double layer potentials with $E(x-y)$ replaced by $E_{\nu,\sigma}(x-y)$, where $E_{\nu,\sigma}(x-y)$ are the Faddeev fundamental solutions (see (16.80)). Prove that all the results of §20 remain true.

9. Let $u \in \mathcal{D}'(\mathbb{R}^{n+1})$ be a distribution solution of the wave equation

$$\frac{\partial^2 u}{\partial t^2} - \sum_{k=1}^{n} \frac{\partial^2 u}{\partial x_k^2} = 0.$$

Prove that for each $x \in \mathbb{R}^n$, u is a distribution in t depending smoothly on x. (Hint: Study the wave front set of u and use Corollary 14.4).

10. **a)** Prove that the Dirichlet problem

$$\Delta u(x_1, x_2) = 0, \ x \in \mathbb{R}^2_+, \ u(x_1, 0) = h(x_1),$$

has a unique solution bounded in $\overline{\mathbb{R}^2_+}$, assuming that $h(x_1)$ is continuous and bounded.

b) Prove that the Neumann problem

$$\Delta u(x_1, x_2) = 0, \ x \in \mathbb{R}^2_+, \ \frac{\partial u(x_1, 0)}{\partial x_2} = g(x_1),$$

has a unique solution such that $u(x) \to 0$ as $|x| \to \infty$, $x \in \mathbb{R}^2_+$, assuming that $g(x_1)$ is continuous, has a compact support and $\int_{-\infty}^{\infty} g(x_1)dx_1 = 0$.

11. Show that $e(x,t) = \frac{1}{4\pi a|x|}\delta(|x| - at)$ is the fundamental solution of the Cauchy problem for the wave equation in $\mathbb{R}^3$ without using the Fourier transform. (Hint: Introduce the spherical coordinates and look for $e(x,t)$ depending on $r = |x|$ and t only. Show that $e_1(r,t) = re(r,t)$ satisfies the one-dimensional wave equation $\frac{\partial^2 e_1(r,t)}{\partial t^2} - \frac{a^2 \partial^2 e_1(r,t)}{\partial r^2} = 0$.)

12. Derive the fundamental solution $e_2(x_1, x_2, t)$ of the Cauchy problem for the wave equation in two dimensions (cf. (16.37), (16.38)) from the formula for the three-dimensional case by the method of descent. Namely, consider $e_2(x_1, x_2, t)$ as a solution of the three-dimensional wave equation that does not depend on x_3. It has the initial conditions $e_2(x_1, x_2, 0) = 0$ and $e_{2t}(x_1, x_2, 0) = \delta(x_1)\delta(x_2)$. Apply formula (16.40) with $\varphi(x_1, x_2, x_3) = 0$ and $\psi(x_1, x_2, x_3) = \delta(x_1)\delta(x_2)$. Finally, compute the integral with respect to x_3.

13. Consider the Laplace equation in $\mathbb{R}^n_+$, $n \geq 2$, with the Robin boundary condition

$$\frac{\partial u(x', 0)}{\partial x_n} - bu(x', 0) = f(x'), \quad b > 0.$$

Prove that for any $f \in L_2(\mathbb{R}^{n-1})$ there exists a unique solution $u(x', x_n)$, continuous and bounded with respect to x_n, with values in $H_1(\mathbb{R}^{n-1})$.

14. Suppose $P(-i\frac{\partial}{\partial x})$ is an elliptic operator with constant coefficients. Prove that there exists a fundamental solution $E(x)$ of $P(-i\frac{\partial}{\partial x})$ that is real analytic if $x \neq 0$. (Hint: Since $|P(\xi)| \geq C_0|\xi|^m$ when $|\xi| > R$, it is enough to prove that

$$E_1(x) = \frac{1}{(2\pi)^n} \int_{|\xi|\geq R} \frac{e^{ix\cdot\xi}d\xi}{P(\xi)(1+|\xi|^2)^r}, \quad 2r + m > n,$$

is real analytic if $x \neq 0$. It follows from the ellipticity of $P(\xi)$ that $|\eta| \geq C_1|\xi|$ if $P(\xi + i\eta) = 0$ and $|\xi| > R$. Therefore, by the Cauchy theorem, one can deform the plane of integration $\eta = 0$ to the surface $\eta = \eta(\xi)$, where $x \cdot \eta(\xi) > \varepsilon|\xi|$, z is small. Then $E_1(x)$ can be extended analytically to $z = x + iy$, where $|y| < \frac{\varepsilon}{2}$ since $-x \cdot \eta(\xi) - y \cdot \xi < 0$).

15. Let $\Delta u + k^2 u = 0$ in $\mathbb{R}^n_+$ and let $u(x)$ satisfy the radiation conditions in $\mathbb{R}^n_+$. Prove that either $u(x', 0) = 0$ or $\frac{\partial u(x',0)}{\partial x_n} = 0$ implies that $u = 0$ in $\mathbb{R}^n_+$.

Chapter IV

Second Order Elliptic Equations in Bounded Domains

Introduction to Chapter IV

This chapter deals with the Dirichlet and the Neumann problem for second order elliptic equations in a bounded domain. We start in §22 with the background on the Sobolev spaces in a bounded domain. In §23 we prove the existence and the uniqueness for the Dirichlet problem in $H_1(\Omega)$ using methods of functional analysis. The regularity of solutions of the Dirichlet problem in a closed domain is proven in §24. In §25 we consider a more restrictive but important variational approach to elliptic boundary value problems. In particular, we define a weak solution of the homogeneous Neumann problem in $H_1(\Omega)$ and prove the uniqueness and the existence of this weak solution. Then we prove a regularity result for the Neumann problem.

The choice of topics in the last two sections (§26 and §27) is less common. In §27 we consider elliptic equations in $L_2(\Omega)$ with smooth right hand sides (in particular, homogeneous equations), and we prove that there exist restrictions (traces) to $\partial\Omega$ of $u(x)$ and its normal derivatives $\frac{\partial^k u}{\partial\nu^k}$ of any order, $k \geq 0$, and that $\frac{\partial^k u}{\partial\nu^k}|_{\partial\Omega} \in H_{-\frac{1}{2}-k}(\partial\Omega)$, $k \geq 0$ (cf. \$7). This result is a variant of the partial hypoellipticity property of elliptic equations. As an application, we consider the distribution solution in $H_1(\Omega)$ of the nonhomogeneous Neumann problem (see also Lions-Magenes [**LM**]). In the last §27 we consider variational inequalities. These are the problems of minimization of quadratic functionals on convex sets. The general theory is quite simple, but it allows us to solve some nontrivial boundary value problems with mixed boundary conditions (see Lions [**L**] and Duvault-Lions [**DL**]).

22. Sobolev spaces in domains with smooth boundaries

22.1. The spaces $\overset{\circ}{H}_s(\Omega)$ and $H_s(\Omega)$.

The Sobolev space $H_s(\mathbb{R}^n), s \in \mathbb{R}$, consists of all distributions $f \in S'(\mathbb{R}^n)$ such that their Fourier transforms $\tilde{f}(\xi)$ are Lebesgue measurable functions with the norm (cf. §13):

$$\|f\|_s^2 = \int_{\mathbb{R}^n} |\tilde{f}(\xi)|^2(1+|\xi|)^{2s}d\xi < +\infty. \tag{22.1}$$

Let Ω be a bounded domain with a smooth boundary $\partial\Omega$. The space $\overset{\circ}{H}_s(\Omega)$ is defined as a subspace of $H_s(\mathbb{R}^n)$ consisting of all $f \in H_s(\mathbb{R}^n)$ with supports in $\overline{\Omega}$, i.e., $f(\varphi) = 0$ for all $\varphi \in C_0^\infty(\mathbb{R}^n \setminus \overline{\Omega})$ (cf. §6).

Lemma 22.1. *$\overset{\circ}{H}_s(\Omega)$ is a closed subspace of $H_s(\mathbb{R}^n)$.*

Proof: Suppose $f_n \in \overset{\circ}{H}_s(\Omega)$ and $f_n \to f$ in $H_s(\mathbb{R}^n)$. We have $f_n(\varphi) \to f(\varphi)$ as $n \to \infty$ because $|f_n(\varphi) - f(\varphi)| \le (2\pi)^{-n}\|f_n - f\|_s\|\varphi\|_{-s}$ (cf. (13.2)). Since $f_n(\varphi) = 0$ for $\varphi \in C_0^\infty(\mathbb{R}^n \setminus \overline{\Omega})$, we conclude that $f(\varphi) = 0, \forall\varphi \in C_0^\infty(\mathbb{R}^n \setminus \overline{\Omega})$. Therefore $f \in \overset{\circ}{H}_s(\Omega)$. □

Now we introduce the space $H_s(\Omega)$. Let $g \in H_s(\mathbb{R}^n)$. Denote by pg the restriction of the distribution $g \in H_s(\mathbb{R}^n)$ to the open set Ω. Recall that a distribution g_1 in Ω is a linear continuous functional $g_1(\varphi)$ on $\mathcal{D}(\Omega)$ (cf. §2). Suppose we started with a distribution f in Ω, and suppose that there exists an extension $lf \in H_s(\mathbb{R}^n)$ of f to $\mathbb{R}^n$. Then, obviously, $plf = f$. Note that if $l_1 f$ and $l_2 f$ are two extensions of f, then $p(l_1 f - l_2 f) = 0$ in Ω, i.e., $l_1 f - l_2 f \in \overset{\circ}{H}_s(\mathbb{R}^n \setminus \overline{\Omega})$. We define the space $H_s(\Omega)$ as the space of distributions f in Ω having extensions $lf \in H_s(\mathbb{R}^n)$. The norm in $H_s(\Omega)$ is defined as the infimum of the norms of all extensions $lf \in H_s(\mathbb{R}^n)$:

$$\|f\|_{s,\Omega} = \inf_{lf} \|lf\|_s. \tag{22.2}$$

If we fix one extension $l_0 f$ of f, then for any other extension lf we have $lf = l_0 f + f_-$, where $f_- \in \overset{\circ}{H}_s(\mathbb{R}^n \setminus \overline{\Omega})$. Therefore

$$\|f\|_{s,\Omega} = \inf_{f_- \in \overset{\circ}{H}_s(\mathbb{R}^n\setminus\overline{\Omega})} \|l_0 f + f_-\|_s. \tag{22.3}$$

Hence $H_s(\Omega) = pH_s(\mathbb{R}^n)$ and the norm in $H_s(\Omega)$ is the norm of the quotient space $H_s(\mathbb{R}^n)/\overset{\circ}{H}_s(\mathbb{R}^n \setminus \overline{\Omega})$. Recall that if B is a Banach space and B_0 is a closed subspace of B, then X belongs to the quotient space B/B_0 if $X = x + B_0$, where $x \in B$ and $\|X\|_{B/B_0} = \inf_{y \in B_0} \|x + y\|_B$.

Note that $H_0(\Omega) = L_2(\Omega)$ and $\|f\|_{0,\Omega}^2 = \int_\Omega |f(x)|^2 dx$, since the infimum in (22.2) is attained by $lf = 0$ in $\mathbb{R}^n \setminus \overline{\Omega}$.

22.2. Equivalent norm in $H_m(\Omega)$.

Let $s = m \geq 0$ be an integer. Denote by $\|f\|'_{m,\Omega}$ the following norm:

$$(\|f\|'_m)^2 = \sum_{|k|=0}^{m} \int_\Omega \left| \frac{\partial^k f(x)}{\partial x^k} \right|^2 dx. \tag{22.4}$$

Theorem 22.2. *The norms $\|f\|_{m,\Omega}$ and $\|f\|'_{m,\Omega}$ are equivalent (cf. (13.7)), i.e.,*

$$C_1 \|f\|'_{m,\Omega} \leq \|f\|_{m,\Omega} \leq C_2 \|f\|'_{m,\Omega}, \quad \forall f \in H_m(\Omega). \tag{22.5}$$

Proof: If $f \in H_m(\Omega)$, then for any extension $lf \in H_m(\mathbb{R}^n)$ we have (cf. (13.7)) $\|lf\|_m \geq C_1 \|lf\|'_m \geq C_1 \|f\|'_{m,\Omega}$. Therefore $\|f\|_{m,\Omega} = \inf \|lf\|_m \geq C_1 \|f\|'_{m,\Omega}$. It remains to prove that $\|f\|_{m,\Omega} \leq C_2 \|f\|'_{m,\Omega}$. It follows from Theorem 13.2 and the definition of the norm (22.2) that $C^\infty(\overline{\Omega})$ is dense in $H_s(\Omega)$. Therefore it is enough to prove the inequalities (22.5) for smooth functions.

Consider first the case where $\Omega = \mathbb{R}^n_+ = \{(x', x_n) : x_n > 0\}$.

Lemma 22.3. *The norms $\|f\|_{m,\mathbb{R}^n_+}$ and $\|f\|'_{m,\mathbb{R}^n_+}$ are equivalent.*

Proof: Let $f \in C^\infty(\overline{\mathbb{R}^n_+})$, $f = 0$ for large $|x|$. Denote by $L_N f$ the following extension of $f(x', x_n)$ for $x_n < 0$:

$$L_N f = \begin{cases} f(x', x_n) & \text{for } x_n > 0, \\ \sum_{p=1}^N \lambda_p f(x', -p x_n) & \text{for } x_n < 0, \end{cases} \tag{22.6}$$

where

$$\sum_{p=1}^N \lambda_p (-p)^r = 1, \quad 0 \leq r \leq N-1, \ N \geq m+1. \tag{22.7}$$

Conditions (22.7) imply that $L_N f \in C^{N-1}(\mathbb{R}^n)$. We have:

$$(22.8)\quad (\|L_N f\|'_{m,\mathbb{R}^n})^2 = \sum_{|k|=0}^{m} \int_{\mathbb{R}^n_+} \left| \frac{\partial^k f(x)}{\partial x^k} \right|^2 dx + \sum_{|k|=0}^{m} \int_{\mathbb{R}^n_-} \left| \sum_{p=1}^{N} \lambda_p \frac{\partial^k}{\partial x^k} f(x', -px_n) \right|^2 dx \le C(\|f\|_{m,\mathbb{R}^n_+})^2.$$

Therefore $\|f\|^2_{m,\mathbb{R}^n_+} \le \|L_N f\|^2_{m,\mathbb{R}^n_+} \le C(\|f\|'_{m,\mathbb{R}^n_+})^2$. □

Now consider the case of an arbitrary bounded domain Ω with a smooth boundary $\partial\Omega$. The boundary $\partial\Omega$ is called smooth if for any $x_0 \in \partial\Omega$ there exist a neighborhood $U_0 \subset \mathbb{R}^n$ and a change of coordinates $y = s(x)$ in U_0 such that the equation of $\partial\Omega \cap U_0$ is $y_n = 0$ and $\Omega \cap U_0$ is contained in $y_n > 0$. We call $y = s(x)$ local coordinates in U_0.

Let $\{\alpha_j(x), U_j\}, j = 1, \dots, N$, be a partition of unity in $\overline{\Omega}$, i.e., $\alpha_j(x) \in C_0^\infty(U_j)$ and $\sum_{j=1}^N \alpha_j(x) = 1$ in $\overline{\Omega}$. If $\operatorname{supp} \alpha_j \subset \Omega$, then we see that the norms $\|\alpha_j f\|_m$ and $\|\alpha_j f\|'_m$ are equivalent (cf. (13.7)). If $U_j \cap \partial\Omega \ne \emptyset$, we introduce local coordinates $y = s_j(x)$ in U_j such that $\overline{\Omega} \cap U_j$ is contained in the half-space $y_n \ge 0$. Applying Lemma 22.3 and using the invariance of the Sobolev spaces under the change of variables (cf. Theorem 13.8), we get

$$(22.9)\qquad C_1\|\alpha_j f\|'_{m,\Omega} \le \|\alpha_j f\|_{m,\Omega} \le C_2\|\alpha_j f\|'_{m,\Omega}, \quad 1 \le j \le N.$$

It is clear that (22.9) implies (22.5). □

22.3. The density of C_0^∞ in $\overset{\circ}{H}_s(\Omega)$.

Theorem 22.4. *The $C_0^\infty(\Omega)$ functions are dense in $\overset{\circ}{H}_s(\Omega)$, i.e., for any $f \in \overset{\circ}{H}_s(\Omega)$ there exists a sequence $\varphi_n \in C_0^\infty(\Omega)$ such that $\|f - \varphi_n\|_s \to 0$ as $n \to \infty$.*

Again, we first consider the case $\Omega = \mathbb{R}^n_+$.

Lemma 22.5. *$C_0^\infty(\mathbb{R}^n_+)$ is dense in $\overset{\circ}{H}_s(\mathbb{R}^n_+)$.*

Proof: Let $f \in \overset{\circ}{H}_s(\mathbb{R}^n_+)$ be arbitrary. In the case where $f(x', x_n)$ is a regular functional, let $f_h(x', x_n) = f(x', x_n - h)$, where $h > 0$. Since $f(x', x_n) = 0$ for $x_n < 0$, we obtain that $f_h = 0$ for $x_n < h$. Note that we can define f_h for any distribution (cf. (3.8)). The Fourier transform of f_h is $e^{i\xi_n h}\tilde{f}(\xi', \xi_n)$. Therefore $f_h \in H_s(\mathbb{R}^n)$, $f_h = 0$ for $x_n < h$. Moreover, $\|f - f_h\|_s^2 = \int_{\mathbb{R}^n} (1 + |\xi|)^{2s} |1 - e^{-i\xi_n h}|^2 |\tilde{f}(\xi)|^2 d\xi \to 0$ as $h \to 0$. Applying the proof of Theorem 13.2 to f_h instead of f, we get $f_{h\delta} = (f_h * \beta_\delta)\beta(\delta_1 x) \in C_0^\infty(\mathbb{R}_+^n)$ provided $0 < \delta < h$ and $\beta_\delta = 0$ for $|x| > \delta$. Therefore, as in the proof of Theorem 13.2, we see that $C_0^\infty(\mathbb{R}_+^n)$ is dense in $\overset{\circ}{H}_s(\mathbb{R}_+^n)$. □

To prove Theorem 22.4 for an arbitrary smooth domain, we use the partition of unity as in the proof of Theorem 22.2, applying Theorem 13.2 if $\operatorname{supp} \alpha_j f \cap \partial\Omega = \emptyset$ and applying Lemma 22.5 in local coordinates in case $\operatorname{supp} \alpha_j f \cap \partial\Omega \neq \emptyset$. □

22.4. Restrictions to $\partial\Omega$.

Consider now the restrictions of $f \in H_s(\mathbb{R}^n)$ to $\partial\Omega$. The restrictions of f to $\partial\Omega$ are also called the traces of f on $\partial\Omega$. It follows from Theorem 13.6 that if $s > \frac{1}{2}$, then for any $f \in H_s(\mathbb{R}^n)$ the trace $p'f$ of f on the hyperplane $x_n = 0$ exists and

$$[p'f]_{s-\frac{1}{2}} \leq C\|f\|_s, \tag{22.10}$$

where $[p'f]_{s-\frac{1}{2}}$ is the norm in $H_{s-\frac{1}{2}}(\mathbb{R}^{n-1})$. In particular, if $s = m > 0$ is an integer, then the traces $\frac{\partial^k f(x',0)}{\partial x_n^k}$ exist for $0 \leq k \leq m-1$ and

$$\left[\frac{\partial^k f(x',0)}{\partial x_n^k}\right]_{m-k-\frac{1}{2}} \leq C\|f\|_m, \quad 0 \leq k \leq m-1. \tag{22.11}$$

We discuss the relation between $\overset{\circ}{H}_m(\mathbb{R}_+^n)$ and $H_m(\mathbb{R}_+^n)$.

Suppose $f(x', x_n) \in \overset{\circ}{H}_m(\mathbb{R}_+^n)$. Then $f(x', x_n) = 0$ for $x_n < 0$. It follows from Theorem 13.6 that $\frac{\partial^k f(x',x_n)}{\partial x_n^k}$ are continuous functions of x_n with values in $H_{m-k-\frac{1}{2}}(\mathbb{R}^{n-1})$. Therefore $\frac{\partial^k f(x',0)}{\partial x_n^k} = 0$ for $0 \leq k \leq m-1$.

Consider $f \in H_m(\mathbb{R}_+^n)$. Then $\frac{\partial^k f(x',0)}{\partial x_n^k}$ are defined and belong to the space $H_{s-k-\frac{1}{2}}(\mathbb{R}^{n-1})$. Suppose $\frac{\partial^k f(x',0)}{\partial x_n^k} = 0$, $0 \leq k \leq m-1$. Denote by $f_0(x', x_0)$ the extension of $f(x', x_n)$ by zero to $x_n < 0$. Then f_0 will have m square integrable derivatives in $\mathbb{R}^n$, and therefore $f_0 \in \overset{\circ}{H}_m(\mathbb{R}^n)$. □

We wish to extend these results to the case of domains Ω with smooth boundaries $\partial\Omega$. First, we define the Sobolev spaces $H_s(\partial\Omega)$.

Let f be a smooth function in $\mathbb{R}^n$. Denote by $p'f$ the trace of f on $\partial\Omega$. Let $\{\alpha_j(x), U_j\}$, $1 \leq j \leq N$, be the partition of the unity as in the proof of Theorem 22.2. We have

$$p'f = \sum_j{}' p'\alpha_j f, \tag{22.12}$$

where summation is over j such that $\operatorname{supp}\alpha_j \cap \partial\Omega \neq \emptyset$. Note that $p'\alpha_j f$ in local coordinates is a function on $\mathbb{R}^{n-1}$. Therefore, we can define $[p'\alpha_j f]_s$ in local coordinates. Then we define $[p'f]_{s,\partial\Omega}$ as the sum:

$$[p'f]^2_{s,\partial\Omega} = \sum_j{}' [p'\alpha_j f]^2_s. \tag{22.13}$$

The space $H_s(\partial\Omega)$ is a closure of $C^\infty(\partial\Omega)$ in the norm (22.13). If we replace $\{\alpha_j, U_j\}$ by another partition of unity and use other local coordinates in U_j, we get a norm that is equivalent to (22.13). We denote by $[v]_{s,\partial\Omega}, v \in H_s(\partial\Omega)$, any one of the equivalent norms in $H_s(\partial\Omega)$. Therefore, the following generalization of Theorem 13.6 holds.

Theorem 22.6. *Let $s > \frac{1}{2}$. Then for any $f \in H_s(\mathbb{R}^n)$ there exists a trace $p'f \in H_{s-\frac{1}{2}}(\partial\Omega)$ and*

$$[p'f]_{s-\frac{1}{2},\partial\Omega} \leq C\|f\|_s.$$

Moreover, if $f \in \overset{\circ}{H}_m(\Omega)$, where $m > 0$ is an integer, then $\frac{\partial^k f}{\partial\nu^k} = 0$ on $\partial\Omega, 0 \leq k \leq m-1$, where $\frac{\partial}{\partial\nu}$ is the normal derivative to $\partial\Omega$. Vice versa, if $f \in H_m(\Omega)$ and $\frac{\partial^k f}{\partial\nu^k} = 0,\ 0 \leq k \leq m-1$, on $\partial\Omega$ then $f_0 \in \overset{\circ}{H}_m(\Omega)$, where f_0 is the extension of f by zero outside Ω. □

22.5. Duality of Sobolev spaces in Ω.

Finally, we discuss the duality for the spaces $\overset{\circ}{H}_s(\Omega)$ and $H_s(\Omega)$. Recall that if B is a Banach space then the Banach space B^* of all linear continuous functionals on B is called the space dual to B. It was shown in §13 (see (13.24)) that the L_2 scalar product $(\varphi, \psi) = \int_{\mathbb{R}^n} \varphi(x)\overline{\psi(x)}dx$ can be extended for any $\varphi \in H_s(\mathbb{R}^n)$ and $\psi \in H_{-s}(\mathbb{R}^n)$, s is arbitrary.

Lemma 22.7. *For any $\varphi_+ \in \overset{\circ}{H}_s(\Omega)$ we have:*

$$(\varphi_+, \psi_-) = 0 \tag{22.14}$$

for all $\psi_- \in \overset{\circ}{H}_{-s}(\mathbb{R}^n \setminus \overline{\Omega})$.

Proof: It follows from Theorem 22.4 that for any $\psi_- \in \overset{\circ}{H}_{-s}(\mathbb{R}^n \setminus \overline{\Omega})$ there exists a sequence $\psi_n \in C_0^\infty(\mathbb{R}^n \setminus \overline{\Omega})$ such that $\|\psi_- - \psi_n\|_{-s} \to 0$. We have $(\varphi_+, \psi_n) = \varphi_+(\overline{\psi_n}) = 0$ since $\operatorname{supp}\varphi_+ \subset \overline{\Omega}$. Passing to the limit as $n \to \infty$, we get (22.14). □

Let $f \in H_s(\Omega)$ and let $lf \in H_s(\mathbb{R}^n)$ be an arbitrary extension of f. For any $\varphi_+ \in \overset{\circ}{H}_{-s}(\Omega)$ the pairing

$$(lf, \varphi_+) \tag{22.15}$$

is independent of the choice of extension lf: if $l_1 f$ is another extension, then $lf - l_1 f = \psi_- \in \overset{\circ}{H}_{-s}(\mathbb{R}^n \setminus \overline{\Omega})$ and, therefore, $(l_1 f, \varphi_+) = (lf, \varphi_+)$ by Lemma 22.7. It follows from (13.24) that

$$|(lf, \varphi_+)| \le (2\pi)^{-n} \|lf\|_s \|\varphi_+\|_{-s}, \tag{22.16}$$

and since the left hand side in (22.16) does not depend on the extension, we have

$$|(lf, \varphi_+)| \le (2\pi)^{-n} \|f\|_{s,\Omega} \|\varphi_+\|_{-s}. \tag{22.17}$$

Therefore, any $\varphi_+ \in \overset{\circ}{H}_{-s}(\Omega)$ defines a linear continuous functional on $H_s(\Omega)$. Any linear continuous functional on $H_s(\Omega)$ can be represented as a linear continuous functional $\Phi(lf)$ on $H_s(\mathbb{R}^n)$ such that $\Phi(l_1 f) = \Phi(l_2 f)$ if $pl_1 f = pl_2 f$. By Lemma 13.7 there exists $\varphi \in H_{-s}(\mathbb{R}^n)$ such that $\Phi(lf) = (lf, \varphi)$. Since (lf, φ) is independent of the extension lf, we obtain, as in Lemma 22.7, that $\varphi \in \overset{\circ}{H}_s(\Omega)$. Hence any linear continuous functional on $H_s(\Omega)$ has the form (lf, φ_+), where $\varphi_+ \in \overset{\circ}{H}_{-s}(\Omega)$.

Therefore we proved the following theorem:

Theorem 22.8. *We have* $(H_s(\Omega))^* = \overset{\circ}{H}_{-s}(\Omega)$ *and, similarly,* $(\overset{\circ}{H}_s(\Omega))^* = H_{-s}(\Omega)$. □

Remark 22.1. Let $s = m + \gamma$, where $m \ge 0$ is an integer and $0 < \gamma < 1$. Then, as in Theorem 13.5, we get that $\|f\|_{m+\gamma,\Omega}$ is equivalent to the following norm:

$$\sum_{|k|=m} \int_\Omega \int_\Omega \frac{\left|\frac{\partial^k f(x)}{\partial x^k} - \frac{\partial^k f(y)}{\partial y^k}\right|^2}{|x-y|^{m+2\gamma}} dx dy + \int_\Omega |f(x)|^2 dx.$$

23. Dirichlet problem for second order elliptic PDEs

Consider the Dirichlet problem:

$$L(x, D)u = f(x), \quad x \in \Omega, \tag{23.1}$$

$$u|_{\partial\Omega} = 0, \tag{23.2}$$

where Ω is a bounded domain in $\mathbb{R}^n$ with a smooth boundary $\partial\Omega$, and $L(x, D)$ is a second order elliptic operator, i.e.,

$$L(x, D)u = -\sum_{j,k=1}^n a_{jk}(x) \frac{\partial^2 u}{\partial x_j \partial x_k} + \sum_{j=1}^n b_j(x) \frac{\partial u}{\partial x_j} + c(x)u(x). \tag{23.3}$$

We assume that the coefficients in (23.3) are $C^\infty(\overline{\Omega})$ functions, the $a_{jk}(x) = a_{kj}(x)$ are real-valued, and the ellipticity condition holds:

$$\sum_{j,k=1}^{n} a_{jk}(x)\xi_j\xi_k \geq C_0 \sum_{j=1}^{n} \xi_j^2 \tag{23.4}$$

for all $x \in \overline{\Omega}$, $(\xi_1, \dots, \xi_n) \in \mathbb{R}^n$.

Consider the bilinear form

$$(L\varphi, \psi) = \int_\Omega L\varphi\overline{\psi(x)}dx,$$

where $\varphi, \psi \in C_0^\infty(\Omega)$. Integrating by parts twice, we get:

$$(L\varphi, \psi) = (\varphi, L^*\psi), \tag{23.5}$$

where

$$\begin{aligned} L^*\psi &= -\sum_{j,k=1}^{n} \frac{\partial^2}{\partial x_j \partial x_k}(a_{jk}(x)\psi) - \sum_{j=1}^{n} \frac{\partial}{\partial x_j}(\overline{b_j(x)}\psi) + \overline{c(x)}\psi(x) \\ &= -\sum_{j,k=1}^{n} a_{jk}(x)\frac{\partial^2\psi}{\partial x_j \partial x_k} + \sum_{j=1}^{n} b_j^{(1)}(x)\frac{\partial\psi}{\partial x_j} + c_1(x)\psi. \end{aligned} \tag{23.6}$$

The operator L^* is called formally adjoint to L. Note that L^* is also an elliptic operator with the same principal part as L. In particular, if L has a selfadjoint form:

$$L\varphi = -\sum_{j,k=1}^{n} \frac{\partial}{\partial x_j}\left(a_{jk}(x)\frac{\partial\varphi}{\partial x_k}\right) + q(x)\varphi(x), \tag{23.7}$$

where $q(x)$ is real-valued, then $L^* = L$.

23.1. The main inequality.

Lemma 23.1 (The Poincaré inequality). *Let Ω be bounded and $\varphi \in C_0^\infty(\Omega)$. Then there exists a constant α_0 depending on Ω and such that*

$$\int_\Omega \sum_{j=1}^{n} \left|\frac{\partial\varphi}{\partial x_j}\right|^2 dx \geq \alpha_0 \int_\Omega |\varphi(x)|^2 dx, \quad \forall\varphi \in C_0^\infty(\Omega). \tag{23.8}$$

Proof: Since Ω is bounded, there exists $a > 0$ such that Ω is contained in a strip $\{(x', x_n) : |x_n| \leq a, x' \in \mathbb{R}^{n-1}\}$. We have

$$\varphi(x', x_n) = \int_{-a}^{x_n} \frac{\partial\varphi(x', t)}{\partial x_n}dt,$$

since $\varphi(x', -a) = 0$. By the Cauchy-Schwartz inequality, we get

(23.9)
$$|\varphi(x', x_n)|^2 \leq \left(\int_{-a}^{x_n} 1^2 dt\right) \int_{-a}^{x_n} \left|\frac{\partial \varphi(x', t)}{\partial x_n}\right|^2 dt \leq 2a \int_{-a}^{a} \left|\frac{\partial \varphi(x', t)}{\partial x_n}\right|^2 dt.$$

Integrating (23.9) over the strip, we get (23.8) with $\alpha_0 = (4a^2)^{-1}$. □

Consider the operator (23.7) with $q(x) \geq 0$. Integrating by parts, we have, using (23.4):

$$(23.10)\quad (L\varphi, \varphi) = \int_\Omega \left(\sum_{j,k=1}^n a_{jk}(x) \frac{\partial \varphi}{\partial x_j} \overline{\frac{\partial \varphi}{\partial x_k}} + q(x)|\varphi(x)|^2\right) dx$$
$$\geq C_0 \int_\Omega \sum_{j=1}^n \left|\frac{\partial \varphi}{\partial x_j}\right|^2 dx.$$

Using the Poincaré inequality, we get

$$(L\varphi, \varphi) \geq \frac{C_0}{2} \int_\Omega \sum_{j=1}^n \left|\frac{\partial \varphi}{\partial x_k}\right|^2 dx + \frac{C_0}{2}\alpha_0 \int_\Omega |\varphi(x)|^2 dx \geq C_1 \|\varphi\|_1^2.$$

Consider now the operator of the form (23.3). Assuming that $\operatorname{Re} c(x) \geq \lambda_0$, where λ_0 is sufficiently large, and using (23.4) and the inequality of the form

$$\left|\int_\Omega b(x) \frac{\partial \varphi}{\partial x_k} \overline{\varphi(x)} dx\right| \leq \frac{\varepsilon^2}{2} \int_\Omega \left|\frac{\partial \varphi}{\partial x_k}\right|^2 dx + \frac{C}{2\varepsilon^2} \int_\Omega |\varphi(x)|^2 dx,$$

we conclude that

$$(23.11)\qquad \operatorname{Re}(L\varphi, \varphi) \geq C_1 \|\varphi\|_1^2$$

and

$$(23.12)\qquad \operatorname{Re}(L^*\varphi, \varphi) \geq C_2 \|\varphi\|_1^2$$

for all $\varphi \in C_0^\infty(\Omega)$.

In this section we assume that the inequalities (23.11) and (23.12) hold. It will be convenient to extend the coefficients of (23.3) from $\overline{\Omega}$ to $\mathbb{R}^n$ as $C^\infty(\mathbb{R}^n)$ functions such that $a_{jk}(x) = \delta_{jk},\ b_j = 0, c = 0$ for large $|x|$.

Lemma 23.2. *The operator $L(x, D)$ is a bounded operator from $\overset{\circ}{H}_s(\Omega)$ to $H_{s-2}(\Omega)$.*

Proof: The operators $\frac{\partial^k}{\partial x^k}$ map $H_s(\mathbb{R}^n)$ to $H_{s-|k|}(\mathbb{R}^n)$ since

$$\left\|\frac{\partial^k u}{\partial x^k}\right\|_t^2 = \int_{\mathbb{R}^n} (1 + |\xi|)^{2t} |\xi^k \tilde{u}|^2 d\xi \leq C\|u\|_{t+|k|}^2$$

for all $t \in \mathbb{R}$. Also, the operator of multiplication by $a(x) \in C_0^\infty(\mathbb{R}^n)$ is bounded in $H_s(\mathbb{R}^n) : \|au\|_s \le C\|u\|_s$ (see Theorem 13.3). Therefore

$$\|Lu\|_{s-2,\Omega} \le \|Lu\|_{s-2} \le C\|u\|_{s,\Omega}, \quad \forall s \in \mathbb{R}, \tag{23.13}$$

i.e., $L(x,D)$ is a bounded operator from $\overset{\circ}{H}_s(\Omega)$ to $H_{s-2}(\Omega)$. □

Analogously, we have that L^* is a bounded operator from $\overset{\circ}{H}_s(\Omega)$ to $H_{s-2}(\Omega)$, $\forall s$. Using Theorem 22.4, Lemma 23.2, and Theorem 22.8, we see that (23.11) and (23.12) hold for any $\varphi \in \overset{\circ}{H}_1(\Omega)$. Note that $L(x,D)$ is also bounded from $H_s(\Omega)$ to $H_{s-2}(\Omega)$: Take any extension $lu \in H_s(\mathbb{R}^n)$. Since $L(x,D)lu$ is an extension of $L(x,D)u$, we have $\|Lu\|_{s-2,\Omega} \le \|L(lu)\|_{s-2} \le C\|lu\|_s$. Since lu is arbitrary, we get $\|Lu\|_{s-2,\Omega} \le C\inf_l \|lu\|_s = C\|u\|_{s,\Omega}$.

23.2. Uniqueness and existence theorem in $\overset{\circ}{H}_1(\Omega)$.

Theorem 23.3. *Suppose* (23.11) *and* (23.12) *hold. Then for any* $f \in H_{-1}(\Omega)$ *there exists a unique* $u \in \overset{\circ}{H}_1(\Omega)$ *such that* $Lu = f$.

Since $u \in \overset{\circ}{H}_1(\Omega)$ implies that (23.2) holds, we conclude that $u \in \overset{\circ}{H}_1(\Omega)$ solves the Dirichlet problem (23.1), (23.2). Note that $u(x) \in \overset{\circ}{H}_1(\Omega)$ is a distribution solution of the equation $L(x,D)u = f$ in Ω, i.e., $(Lu,\varphi) = (lf,\varphi)$, for all $\varphi \in C_0^\infty(\Omega)$.

Proof of Theorem 23.3: It follows from (23.11) that

$$\|u\|_1^2 \le C\,\mathrm{Re}(Lu,u) \le C\|Lu\|_{-1,\Omega}\|u\|_1.$$

Therefore

$$\|u\|_1 \le C\|Lu\|_{-1,\Omega}, \quad \forall u \in \overset{\circ}{H}_1(\Omega). \tag{23.14}$$

Analogously

$$\|v\|_1 \le C\|L^*v\|_{-1,\Omega}, \quad \forall v \in \overset{\circ}{H}_1(\Omega). \tag{23.15}$$

The uniqueness of the solution of the Dirichlet problem (23.1), (23.2) in $\overset{\circ}{H}_1(\Omega)$ follows immediately from (23.14).

Now we prove the existence. Denote by R the subspace (not closed) of $H_{-1}(\Omega)$ that consists of all $g \in H_{-1}(\Omega)$ having the form: $g = L^*\varphi, \quad \varphi \in C_0^\infty(\Omega)$. Consider a linear functional Φ on $R \subset H_{-1}(\Omega)$ of the form

$$\Phi(L^*\varphi) = (lf,\varphi), \quad \forall \varphi \in C_0^\infty(\Omega), \tag{23.16}$$

where $f \in H_{-1}(\Omega)$ is the right hand side of (23.1) and $lf \in H_{-1}(\mathbb{R}^n)$ is an arbitrary extension of f (cf. (22.15)).

It follows from (23.15) and (22.16) that

$$|\Phi(L^*\varphi)| \le (2\pi)^{-n}\|f\|_{-1,\Omega}\|\varphi\|_1 \le C\|f\|_{-1,\Omega}\|L^*\varphi\|_{-1,\Omega}. \tag{23.17}$$

Therefore Φ is a bounded linear functional on $R \subset H_{-1}(\Omega)$. By the Hahn-Banach theorem, $\Phi(g)$ can be extended to a bounded linear functional on $H_{-1}(\Omega)$, and the norm of the functional Φ is less than or equal to $C\|f\|_{-1,\Omega}$. By Theorem 22.8, any bounded linear functional on $H_{-1}(\Omega)$ has the form (u, lg) for some $u \in \overset{\circ}{H}_1(\Omega)$. Here $g \in H_{-1}(\Omega)$ is arbitrary and $lg \in H_{-1}(\mathbb{R}^n)$ is an arbitrary extension of g. Therefore,

$$(u, L^*\varphi) = (lf, \varphi), \quad \forall\varphi \in C_0^\infty(\Omega). \tag{23.18}$$

Since $C_0^\infty(\Omega)$ is dense in $\overset{\circ}{H}_1(\Omega)$, we get from (23.5):

$$(Lu, \varphi) = (u, L^*\varphi), \quad \forall\varphi \in C_0^\infty(\Omega). \tag{23.19}$$

Therefore, (23.18) and (23.19) imply that

$$(Lu, \varphi) = (lf, \varphi), \quad \forall\varphi \in C_0^\infty(\Omega), \tag{23.20}$$

i.e., $u \in \overset{\circ}{H}_1(\Omega)$ is a distribution solution of (23.1). Note that (cf. (23.14))

$$\|u\|_1 \le C\|f\|_{-1,\Omega}. \tag{23.21}$$

□

23.3. Nonhomogeneous Dirichlet problem.

Consider now the nonhomogenous Dirichlet problem

$$Lu = f \quad \text{in} \quad \Omega, \qquad u|_{\partial\Omega} = g. \tag{23.22}$$

Theorem 23.4. *Assume that inequalities* (23.11) *and* (23.12) *hold. For any* $f \in H_{-1}(\Omega)$ *and* $g \in H_{\frac{1}{2}}(\partial\Omega)$ *there exists a unique* $u \in H_1(\Omega)$ *such that* (23.22) *holds and*

$$\|u\|_{1,\Omega} \le C(\|f\|_{-1,\Omega} + [g]_{\frac{1}{2},\partial\Omega}), \tag{23.23}$$

where $[g]_{\frac{1}{2},\partial\Omega}$ *is the norm in* $H_{\frac{1}{2}}(\partial\Omega)$.

Proof: We shall reduce the proof of Theorem 23.4 to the case of the zero Dirichlet condition. It was proven in Example 13.3 that for any $g \in H_{s-\frac{1}{2}}(\mathbb{R}^{n-1})$, there exists $v \in H_s(\mathbb{R}^n)$, $s \in \mathbb{R}$, such that $v|_{x_n=0} = g$ and

$$\|v\|_s = C\ [g]_{s-\frac{1}{2}}.$$

Using Example 13.3 and a partition of unity, one can prove that for any $g \in H_{s-\frac{1}{2}}(\partial\Omega)$, there exists $v \in H_s(\mathbb{R}^n)$ such that $v|_{\partial\Omega} = g$ and

$$\|v\|_s \le C\ [g]_{s-\frac{1}{2},\partial\Omega}. \tag{23.24}$$

We look for the solution $u(x)$ of the Dirichlet problem (23.22) in the form

$$u = v + w \quad \text{in} \quad \Omega, \tag{23.25}$$

where $w \in \overset{\circ}{H}_1(\Omega)$. Then w satisfies the equation

$$Lw = Lu - Lv = f - Lv, \ x \in \Omega. \tag{23.26}$$

Note that $Lv \in H_{-1}(\mathbb{R}^n)$ since $g \in H_{\frac{1}{2}}(\partial\Omega)$. By Theorem 23.3, there exists a unique $w \in \overset{\circ}{H}_1(\Omega)$ that satisfies (23.26) and

$$\|w\|_1 \leq C\|f - pLv\|_{-1,\Omega} \leq C_1\|f\|_{-1,\Omega} + C_2\|v\|_{1,\Omega}. \tag{23.27}$$

Therefore $u = v + w \in H_1(\Omega)$ and solves (23.22). We have, using (23.24) and (23.27), that

$$\|u\|_{1,\Omega} \leq \|v\|_{1,\Omega} + \|w\|_1 \leq C(\|f\|_{-1,\Omega} + [g]_{\frac{1}{2},\partial\Omega}). \tag{23.28}$$

□

24. Regularity of solutions for elliptic equations

In this section we prove that if the right hand sides of (23.22) are smooth, then the solution $u(x)$ is also smooth in $\overline{\Omega}$. In particular, we show that a classical solution of (23.22) (i.e., a solution with continuous partial derivatives up to the second order) exists.

Theorem 24.1. *Suppose $f \in H_{-1+m}(\Omega)$ and $g \in H_{\frac{1}{2}+m}(\partial\Omega)$, where $m \geq 0$ is an integer. Let $u \in H_1(\Omega)$ be the solution of (23.22). Then $u \in H_{1+m}(\Omega)$.*

Proof: Analogously to (23.24), there exists $v \in H_{1+m}(\Omega)$ such that $v|_{\partial\Omega} = g$ and

$$\|v\|_{1+m,\Omega} \leq C \ [g]_{\frac{1}{2}+m,\partial\Omega}. \tag{24.1}$$

Replacing u by $v + w$ as in the proof of Theorem 23.4, we get

$$Lw = f_1, \tag{24.2}$$

where $f_1 = f - pLv \in H_{-1+m}(\Omega)$, $w \in \overset{\circ}{H}_1(\Omega)$ and

$$\|w\|_1 \leq C\|f_1\|_{-1,\Omega}. \tag{24.3}$$

It remains to prove that $w \in H_{1+m}(\Omega)$.

Let $\{\alpha_j(x), U_j\}$, $1 \leq j \leq N$, be a partition of unity in $\overline{\Omega}$. It is enough to prove that $\alpha_j w \in H_{1+m}(\Omega)$ for all $j, 1 \leq j \leq N$. We consider separately two cases:

a) $\operatorname{supp}\chi \cap \partial\Omega = \emptyset$ (interior regularity);

b) $\operatorname{supp}\chi \cap \partial\Omega \neq \emptyset$ (boundary regularity).

Here $\chi(x)$ is any of the $\alpha_j(x)$, $1 \leq j \leq N$.

24.1. Interior regularity.

Assume that $f_1 \in H_0(\Omega)$, i.e., $m = 1$. Let $\chi(x) \in C_0^\infty(\Omega)$. Multiplying $Lw = f_1$ by $\chi(x)$, we get

$$L\chi w = \chi f_1 + L_1 w,$$

where $L_1 w = L\chi w - \chi L w$ is a first order differential operator. Since $w \in \overset{\circ}{H}_1(\Omega)$, we have that $\|L_1 w\|_0 \leq C\|w\|_1$, i.e., $L_1 w \in H_0(\Omega)$. Therefore

$$Lw_1 = f_2, \tag{24.4}$$

where $f_2 = \chi f_1 + L_1 w \in H_0(\Omega)$, $w_1 = \chi w \in \overset{\circ}{H}_1(\Omega)$ and $\operatorname{supp} w_1 \subset \Omega$, $\operatorname{supp} f_2 \subset \Omega$. Let $w_{1h_k} = w_1(x + \vec{h}_k)$, where $\vec{h}_k = (0, \ldots, h_k, 0, \ldots, 0)$, and let

$$D_{h_k} w_1 = \frac{w_{1h_k}(x) - w_1(x)}{h_k}. \tag{24.5}$$

Lemma 24.2. *Let $u \in H_s(\mathbb{R}^n)$. Then $u(x) \in H_{s+1}(\mathbb{R}^n)$ iff $\|D_{h_k} u\|_s \leq C$ for all small h_k, $1 \leq k \leq n$. Here C is independent of h_k.*

Proof: The Fourier transform of $D_{h_k} u$ has the form $\tilde{u}(\xi)\frac{e^{ih_k\xi_k}-1}{h_k}$. Note that $\left|\frac{e^{ih_k\xi_k}-1}{h_k}\right| \leq C|\xi_k|$. If $u \in H_{s+1}(\mathbb{R}^n)$, then

$$\|D_{h_k} u\|_s^2 \leq C \int_{\mathbb{R}^n} (1 + |\xi|)^{2s} |\xi_k \tilde{u}|^2 d\xi \leq C\|u\|_{s+1}^2, \ 1 \leq k \leq n.$$

Vice versa, let $\|D_{hk} u\|_s^2 = \int_{\mathbb{R}^n} (1 + |\xi|)^{2s} |\tilde{u}(\xi)|^2 \left|\frac{e^{ih_k\xi_k}-1}{h_k}\right|^2 d\xi \leq C$ for all h_k, $1 \leq k \leq n$. Recall the Fatou lemma:

Lemma 24.3 (Fatou). *If $g_n(\xi) \geq 0$, $g_n(\xi) \to g(\xi)$ almost everywhere as $n \to \infty$ and if $\int_{\mathbb{R}^n} g_n(\xi) d\xi \leq C$, then $\int_{\mathbb{R}^n} g(\xi) d\xi \leq C$.*

Note that $\left|\frac{e^{ih_k\xi_k}-1}{h_k}\right|^2 \to |\xi_k|^2$ as $h_k \to 0$. Using the Fatou lemma, we get

$$\int_{\mathbb{R}^n} (1 + |\xi|)^{2s} |\xi_k|^2 |\tilde{u}(\xi)|^2 d\xi \leq C, \quad k = 1, \ldots, n.$$

Therefore $u \in H_{s+1}(\mathbb{R}^n)$. □

Consider the case $m = 1$. Apply operator D_{h_k} to (24.4). Note that the supports of $D_{h_k} w_1$ and $D_{h_k} f_2$ are contained in Ω if h_k is small. Note also that

$$D_{h_k}\left(a(x)\frac{\partial^k w}{\partial x^k}\right) = a\left(\frac{\partial^k}{\partial x^k} D_{h_k} w\right) + (D_{h_k} a(x))\frac{\partial^k}{\partial x^k} w_{h_k}(x).$$

Therefore

$$D_{h_k}(Lw_1) = L(D_{h_k} w_1) + L_{h_k} w_{1h_k}, \tag{24.6}$$

where L_{h_k} is the second order differential operator with each coefficient $a(x)$ replaced by $D_{h_k}a$. It follows from (24.4) and (24.6) that

$$L(D_{h_k}w_1) = D_{h_k}f_2 - L_{h_k}w_{1h_k}. \tag{24.7}$$

Since $D_{h_k}w_1 \in \overset{\circ}{H}_1(\Omega)$ for small h_k, we have, using (23.21), that

$$\|D_{h_k}w_1\|_1 \le C\|D_{h_k}f_2\|_{-1} + C\|L_{h_k}w_{1h_k}\|_{-1}. \tag{24.8}$$

Since $f_2 \in H_0(\mathbb{R}^n)$, we obtain, using Lemma 24.2, that

$$\|D_{h_k}f_2\|_{-1} \le C\|f_2\|_0. \tag{24.9}$$

Note that (the mean value theorem)

$$D_{h_k}a(x) = \int_0^1 \frac{\partial a}{\partial x_k}(x + t\vec{h}_k)dt. \tag{24.10}$$

It follows from (24.10) that $D_{h_k}a$ and its derivatives are uniformly bounded. Therefore

$$\|L_{h_k}w_{1h_k}\|_{-1} \le C\|w_{1h_k}\|_1 = C\|w_1\|_1. \tag{24.11}$$

Combining (24.8), (24.9), (24.11), we get:

$$\|D_{h_k}w_1\|_1 \le C\|f_2\|_0 + C\|w_1\|_1.$$

By Lemma 24.2, $w_1 = \chi(x)w(x) \in \overset{\circ}{H}_2(\Omega)$ and

$$\|\chi w\|_2 \le C(\|f_2\|_0 + \|w_1\|_1) \le C_1(\|f_1\|_{0,\Omega} + \|w\|_1) \le C_2\|f_1\|_{0,\Omega}. \tag{24.12}$$

24.2. Boundary regularity.

Now we consider the regularity of $w(x)$ near the boundary $\partial\Omega$, where $w(x)$ is the solution of (24.2). Suppose $\chi(x) \in C_0^\infty(U_0)$, where $U_0 \cap \partial\Omega \ne \emptyset$ is the same as in the proof of Theorem 22.2, i.e., there exists a change of variables $y = s(x)$ such that the equation of $\partial\Omega \cap U_0$ is $y_n = 0$ and $\overline{\Omega} \cap U_0$ is contained in the half-space $y_n \ge 0$. Multiplying (24.2) by χ and rewriting (24.2) in y-coordinates, we get

$$\hat{L}(\hat{\chi}\hat{w}) = \hat{\chi}\hat{f}_1 - \hat{L}_1\hat{\chi}_1\hat{w}, \quad y_n > 0, \tag{24.13}$$

where $\chi_1 \in C_0^\infty(U_0)$, $\chi_1 = 1$ in a neighborhood of $\operatorname{supp}\chi$, $\hat{w}, \hat{\chi}, \hat{\chi}_1, \hat{L}, \hat{f}_1$ are w, χ, χ_1, L, f_1 in y-coordinates, and $\operatorname{ord}\hat{L}_1 = 1$.

Consider the case $m = 1$. Use finite differences as in the interior case, but only in the tangential variables $y_1, \dots, y_{n-1}$, i.e., $\hat{w}_{h_k}(y) = \hat{w}(y + \vec{h}_k)$, where $k = 1, \dots, n-1$. Then exactly as in the interior case, we get

$$\left\|\frac{\partial}{\partial y_k}(\hat{\chi}\hat{w})\right\|_{1,\mathbb{R}^n_+} \le C\left(\|\hat{\chi}f_1\|_{0,\mathbb{R}^n_+} + \|\hat{\chi}_1\hat{w}\|_{1,\mathbb{R}^n_+}\right), \quad k = 1, \dots, n-1. \tag{24.14}$$

It remains to estimate $\frac{\partial}{\partial y_n}(\hat{\chi}\hat{w})$. It follows from (24.13) that

$$\hat{a}_{nn}(y)\frac{\partial^2}{\partial y_n^2}(\hat{\chi}\hat{w}) = \hat{L}_2\frac{\partial}{\partial y_n}(\hat{\chi}\hat{w}) + \hat{L}_3(\hat{\chi}\hat{w}) + \hat{\chi}\hat{f}_1 - \hat{L}_1\hat{\chi}_1\hat{w}, \tag{24.15}$$

where $\hat{a}_{nn} \neq 0$, $y_n > 0$, $\hat{L}_2$ is a first order differential operator in $\frac{\partial}{\partial y_1}, \dots, \frac{\partial}{\partial y_{n-1}}$ and $\hat{L}_3$ is a second order differential operator in $\frac{\partial}{\partial y_1}, \dots, \frac{\partial}{\partial y_{n-1}}$. Therefore (24.15) implies:

$$\left\|\frac{\partial^2}{\partial y_n^2}(\hat{\chi}\hat{w})\right\|_{0,\mathbb{R}^n_+} \leq C\sum_{j=1}^{n-1}\left\|\frac{\partial}{\partial y_k}(\hat{\chi}\hat{w})\right\|_{1,\mathbb{R}^n_+} + C\|w\|_1 + C\|f_1\|_{0,\Omega}. \tag{24.16}$$

Combining (24.14) and (24.16) and using the invariance of the Sobolev spaces under the change of variables, we get

$$\|\chi w\|_{2,\Omega} \leq C\|f_1\|_{0,\Omega} + C\|w\|_1. \tag{24.17}$$ □

Since we proved that $\alpha_j w \in H_2(\Omega)$ for all $j = 1, \dots, N$, we have

$$\|w\|_{2,\Omega} \leq C\|f_1\|_{0,\Omega} + C\|w\|_1 \leq C\|f_1\|_{0,\Omega}. \tag{24.18}$$

Now let $m = 2$ and repeat the arguments for the interior and the boundary steps starting with the estimate (24.18) instead of (24.3). We get

$$\|w\|_{3,\Omega} \leq C\|f_1\|_{1,\Omega} + C\|w\|_{2,\Omega}.$$

Continuing this process, we obtain

$$\|w\|_{1+m,\Omega} \leq C\|f_1\|_{-1+m,\Omega} \leq C(\|f\|_{-1+m,\Omega} + [g]_{\frac{1}{2}+m,\partial\Omega}) \tag{24.19}$$

for arbitrary m. In particular, if $f \in C^\infty(\overline{\Omega})$ and $g \in C^\infty(\partial\Omega)$, we see that $u \in C^\infty(\overline{\Omega})$, where $u(x)$ is the solution of (23.22).

Remark 24.1. It follows from the proof of Theorem 24.1 that it is not necessary to assume the uniqueness of the solution to have the regularity result. It is enough to replace the estimate (24.3) by a weaker estimate

$$\|w\|_1 \leq C\|f_1\|_{-1,\Omega} + C\|w\|_{0,\Omega}. \tag{24.20}$$

Such an estimate follows from the ellipticity condition without any restrictions on the lower order terms.

25. Variational approach. The Neumann problem

In this section we consider the selfadjoint elliptic equation of the form (23.7) with $q(x) \geq 0$. We also assume that $u(x)$ and $f(x)$ are real-valued.

Let $a(u, v)$ be the following bilinear form:

$$a(u,v) = \int_\Omega \left(\sum_{j,k=1}^n a_{jk}(x)\frac{\partial u}{\partial x_j}\frac{\partial v}{\partial x_k} + q(x)uv\right) dx,$$

where $q(x) \geq 0$. We have (cf. Lemma 23.1):

$$a(v,v) \geq C\|v\|^2_{1,\Omega}, \quad \forall v \in \overset{\circ}{H}_1(\Omega). \tag{25.1}$$

Let $f(u)$ be a linear continuous functional on $\overset{\circ}{H}_1(\Omega)$. Note that $f(v)$ has the form:

$$f(v) = \int_\Omega f(x)v(x)dx$$

if $f(x) \in L_2(\Omega)$, and

$$f(v) = (lf, v) \tag{25.2}$$

if $f \in H_{-1}(\Omega)$, where lf is an arbitrary extension of f (cf. (22.15)).

Consider the following continuous functional on $\overset{\circ}{H}_1(\Omega)$:

$$\Phi(u) = \frac{1}{2}a(u,u) - f(u). \tag{25.3}$$

Theorem 25.1. *Assume that* (25.1) *holds. For every* $f \in H_{-1}(\Omega)$ *there exists a unique* $u_0 \in \overset{\circ}{H}_1(\Omega)$ *such that*

$$\Phi(u_0) = \min_{u \in \overset{\circ}{H}_1(\Omega)} \Phi(u). \tag{25.4}$$

The proof of Theorem 25.1 will be given in §27.

We shall derive the equation for u_0. Let $v \in \overset{\circ}{H}_1(\Omega)$ be arbitrary. We have

$$\Phi(u_0 + \varepsilon v) = \Phi(u_0) + \varepsilon[a(u_0, v) - f(v)] + \frac{\varepsilon^2}{2}a(v,v). \tag{25.5}$$

Since (25.5) has a minimum at $\varepsilon = 0$, we get:

$$a(u_0, v) - f(v) = 0 \quad \text{for all} \quad v \in \overset{\circ}{H}_1(\Omega). \tag{25.6}$$

The function $u_0 \in \overset{\circ}{H}_1(\Omega)$ satisfying (25.6) is called a weak solution of the Dirichlet problem (23.1), (23.2). Assuming $v \in C_0^\infty(\Omega)$ in (25.6) and using the definition of the derivative of a distribution, we see that (25.6) is equivalent to $(Lu_0 - lf, v) = 0$, $\forall v \in C_0^\infty(\Omega)$, i.e., $Lu = f$ in Ω in the distribution sense. Therefore the weak solution (25.6) is equivalent to the solution found in §23. □

25.1. Weak solution of the Neumann problem.

Consider $\Phi(u)$ on $H_1(\Omega)$. Now $f(u)$ is a linear continuous functional on $H_1(\Omega)$, i.e., $f(u) = (f, lu)$, where $f \in \overset{\circ}{H}_{-1}(\Omega)$, and lu is an arbitrary extension of u. Assume that $q(x) \geq 0$ and $q(x) > 0$ on a set of positive measure in Ω. Then

$$a(v,v) \geq C\|v\|_1^2, \quad \forall v \in H_1(\Omega).$$

As in the case of Theorem 25.1, one can prove (see §27) that there exists a unique $w_0 \in H_1(\Omega)$ such that

$$\Phi(w_0) = \min_{u \in H_1(\Omega)} \Phi(u). \tag{25.7}$$

Analogously to (25.6) one obtains that w_0 satisfies the following equation:

$$a(w_0, v) - f(v) - 0 \quad \text{for all} \quad v \in H_1(\Omega). \tag{25.8}$$

If $v \in C_0^\infty(\Omega)$, we have, integrating by parts as in (25.6), that

$$Lw_0 - f = 0 \quad \text{in} \quad \Omega, \tag{25.9}$$

i.e., w_0 is a distribution solution of (22.1). Substituting $v = w_0$ in (25.8) and using that

$$a(w_0, w_0) \geq C\|w_0\|_{1,\Omega}^2, \qquad |f(w_0)| = |(f, lw_0)| \leq \|f\|_{-1}\|w_0\|_{1,\Omega},$$

we get

$$\|w_0\|_{1,\Omega} \leq C\|f\|_{-1}, \quad \forall f \in \overset{\circ}{H}_{-1}(\Omega). \tag{25.10}$$

We now analyze what kind of boundary value problem does w_0 solve. If $w_0 \in H_2(\Omega)$ and $v \in C^\infty(\overline{\Omega})$, we see, integrating by parts in (25.8) and using the Green formula, that

$$\int_\Omega (Lw_0 - f)v(x)dx + \int_{\partial\Omega} \frac{\partial w_0}{\partial \nu_L} v(x)ds = 0, \tag{25.11}$$

where $\frac{\partial w_0}{\partial \nu_L} = \sum_{j,k=1}^n a_{jk}(x) \frac{\partial w_0}{\partial x_j} \nu_k(x)$ is the conormal derivative and $(\nu_1, \dots, \nu_n)$ is the outward unit normal to $\partial\Omega$. It follows from (25.9) and (25.11) that

$$\int_{\partial\Omega} \frac{\partial w_0}{\partial \nu_L} v(x)ds = 0. \tag{25.12}$$

Since $v(x)$ is an arbitrary smooth function on $\partial\Omega$, we conclude that

$$\left.\frac{\partial w_0}{\partial \nu_L}\right|_{\partial\Omega} = 0, \tag{25.13}$$

i.e., w_0 is the solution of $Lw_0 = f$ in Ω with the zero Neumann boundary condition. In case $w_0 \in H_1(\Omega)$ and satisfies (25.8), we call w_0 a weak solution of the Neumann problem (25.9), (25.13).

We have proven the following theorem.

Theorem 25.2. *Suppose* $a(v,v) \geq C\|v\|^2_{1,\Omega}$, *for all* $v \in H_1(\Omega)$ *and that* $f \in \overset{\circ}{H}_{-1}(\Omega)$. *Then there exists a unique weak solution of the Neumann problem* (25.9), (25.13), *i.e., there exists* $w_0 \in H_1(\Omega)$ *such that* (25.8) *holds for all* $v \in H_1(\Omega)$. *Moreover, the estimate* (25.10) *holds.* □

25.2. Regularity of weak solution of the Neumann problem.

We now study the regularity of the solution w_0 of the Neumann problem.

Theorem 25.3. *Let* $w_0 \in H_1(\Omega)$ *be the solution of* (25.8), *and let* $f \in H_{-1+m}(\Omega)$, $m \geq 1$. *Then* $w_0 \in H_{1+m}(\Omega)$, $\frac{\partial w_0}{\partial \nu_L}\big|_{\partial\Omega} = 0$, *and*

$$\|w_0\|_{1+m,\Omega} \leq C\|f\|_{-1+m,\Omega}. \tag{25.14}$$

Proof: Since $Lw_0 - f = 0$ in Ω (cf. (25.9)), the proof of the interior regularity for the Neumann problem is exactly the same as for the Dirichlet problem. The proof of the boundary regularity is slightly different.

Let $\chi(x), U_0$ be the same as in (24.13). Take χv instead of v in (25.8) and introduce local coordinates in U_0 as in (24.13). Commuting χ with $\frac{\partial}{\partial x_j}, 1 \leq j \leq n$, we can represent (25.8) in y-coordinates in the following form:

$$\begin{aligned}(25.15)\quad &\int_{y_n>0} \sum_{j,k=1}^{n} \hat{a}_{jk}(y)\frac{\partial}{\partial y_j}(\hat{\chi}\hat{w}_0)\frac{\partial}{\partial y_k}\hat{v}(y)dy \\ &= \int_{y_n>0}\left(\sum_{j=1}^{n} b_j^{(1)}(y)\frac{\partial \hat{w}_0}{\partial y_j}\hat{v} + \sum_{j=1}^{n} b_j^{(2)}(y)\hat{w}_0\frac{\partial \hat{v}}{\partial y_j} - \hat{q}\hat{\chi}\hat{w}_0\hat{v} + \hat{\chi}\hat{f}\hat{v}\right)dy,\end{aligned}$$

where $b_j^{(1)}, b_j^{(2)}$ are equal to zero for $|y| > C$, $y_n \geq 0$, and $\hat{w}_0, \hat{v}, \hat{\chi}, \hat{f}, \hat{q}$ are $w_0, v, \chi(x), f, q$ in y-coordinates. As in §24, let $D_{h_k}\hat{v} = \frac{\hat{v}(y+\vec{h}_k)-\hat{v}(y)}{h_k}, 1 \leq k \leq n-1$. Substitute $-D_{-h_k}D_{h_k}(\hat{\chi}\hat{w}_0)$ in (25.15) instead of $\hat{v}(y)$. Note that the change of variables $y_k - h_k = y_k'$ gives

$$\int_{y_n>0} \hat{u}D_{-h_k}\hat{v}dy = -\int_{y_n>0}(D_{h_k}\hat{u})\hat{v}dy. \tag{25.16}$$

Using (25.16), we can rewrite (25.15) in the form:

$$\begin{aligned}(25.17)\quad &\int_{y_n>0} \sum_{j,k=1}^{n} \hat{a}_{jk}(y)\frac{\partial}{\partial y_j}(D_{h_k}(\hat{\chi}\hat{w}_0))\frac{\partial}{\partial y_k}(D_{h_k}(\hat{\chi}\hat{w}_0))dy \\ &= b_{h_k}(D_{h_k}(\hat{\chi}w_0), w_0) - (\hat{\chi}\hat{f}, D_{-h_k}D_{h_k}\hat{\chi}\hat{w}_0),\end{aligned}$$

where

$$(25.18)\quad \begin{aligned} &|b_{h_k}(D_{h_k}(\hat{\chi}\hat{w}_0),\hat{w}_0)| \\ &\quad\le C\|D_{h_k}(\hat{\chi}\hat{w}_0)\|_{1,\mathbb{R}^n_+}\left(\|\hat{\chi}_1\hat{w}_0\|_{1,\mathbb{R}^n_+}+\|D_{h_k}(\hat{\chi}_1\hat{w}_0)\|_{0,\mathbb{R}^n_+}\right) \\ &\qquad\le \varepsilon\|D_{h_k}(\hat{\chi}\hat{w}_0)\|^2_{1,\mathbb{R}^n_+}+C_\varepsilon\|\hat{\chi}_1\hat{w}_0\|^2_{1,\mathbb{R}^n_+}, \end{aligned}$$

$\hat{\chi}_1(y)\in C_0^\infty(U_0)$, $\hat{\chi}_1=1$ in a neighborhood of $\operatorname{supp}\hat{\chi}$, ε is small.

Since

$$\|D_{h_k}(\hat{\chi}\hat{w}_0)\|^2_{1,\mathbb{R}^n_+}\le C\int_{y_n>0}\sum_{j,k=1}^n \hat{a}_{jk}\frac{\partial}{\partial y_j}(D_{h_k}(\hat{\chi}w_0))\frac{\partial}{\partial y_k}(D_{h_k}(\hat{\chi}\hat{w}_0))dy + C\|D_{h_k}(\hat{\chi}\hat{w}_0)\|^2_{0,\mathbb{R}^n_+},$$

and since

$$|(\hat{\chi}\hat{f},D_{-h_k}D_{h_k}(\hat{\chi}\hat{w}_0))|\le C\|\hat{\chi}\hat{f}\|_{0,\mathbb{R}^n_+}\|D_{h_k}(\hat{\chi}\hat{w}_0)\|_{1,\mathbb{R}^n_+} \le \varepsilon\|D_{h_k}(\hat{\chi}\hat{w}_0)\|^2_{1,\mathbb{R}^n_+}+C_\varepsilon^{(1)}\|\hat{\chi}\hat{f}\|_{0,\mathbb{R}^n_+}$$

we get:

(25.19)

$$\|D_{h_k}(\hat{\chi}\hat{w}_0)\|^2_{1,\mathbb{R}^n_+}\le 2\varepsilon\|D_{h_k}(\hat{\chi}\hat{w}_0)\|^2_{1,\mathbb{R}^n_+}+C_\varepsilon^{(2)}\left(\|\hat{\chi}\hat{f}\|_0^2+\|\hat{\chi}_1\hat{w}_0\|^2_{1,\mathbb{R}^n_+}\right).$$

Therefore, since ε is small, Lemma 24.2 implies that $\frac{\partial}{\partial y_k}\hat{\chi}\hat{w}_0\in H_1(\mathbb{R}^n_+)$, $k=1,\dots,n-1$. The estimate of $\frac{\partial}{\partial y_n}\hat{\chi}\hat{w}_0$ is the same as in the Dirichlet case. Hence, we get as in §24:

$$(25.20)\qquad \|w_0\|_{2,\Omega}\le C\|f\|_{0,\Omega}.$$

It follows from (25.11) and (25.12) that (25.13) holds. Therefore we have proven that for any $f\in L_2(\Omega)$ there exists a unique solution $w_0(x)\in H_2(\Omega)$ satisfying (25.9), (25.13) and that the estimate (25.20) holds. If $f\in H_m(\Omega)$, $m\ge 1$, then starting from (25.20) and repeating the proof of regularity in the Dirichlet case we see that $w_0(x)\in H_{m+2}(\Omega)$ and $\|w_0\|_{m+2,\Omega}\le C\|f\|_{m,\Omega}$. □

Note that the proof of the regularity does not require any restrictions on the lower order terms of L (cf. Remark 24.1).

26. Boundary value problems with distribution boundary data

26.1. Partial hypoellipticity property of elliptic equations.

We have had only a weak formulation (25.8) of the homogeneous Neumann problem since the trace $\frac{\partial u}{\partial \nu}\big|_{\partial\Omega}$ may not exist for an arbitrary $u\in H_1(\Omega)$.

In this section we show that if u is a distribution solution of a homogeneous elliptic equation $Lu = 0$ in Ω, then the traces of u and its normal derivatives are well defined. This property is called the partial hypoellipticity of elliptic equations (cf. Proposition 7.6). The same is true for the solutions of nonhomogeneous equations $Lu = f$ if f is smooth. The following theorem holds:

Theorem 26.1. *Let $u(x) \in L_2(\Omega)$ be a distribution solution of $Lu = 0$ in Ω. Then there exist traces $u|_{\partial\Omega} = h_0$ and $\frac{\partial^k u}{\partial \nu_L^k}|_{\partial\Omega} = h_k, k \geq 1$, where $h_k \in H_{-\frac{1}{2}-k}(\partial\Omega)$, $k \geq 0$, and*

$$\left[\frac{\partial^k u}{\partial \nu_L^k}\right]_{-\frac{1}{2}-k,\partial\Omega} \leq C\|u\|_{0,\Omega}. \tag{26.1}$$

Proof: It is enough to prove that the trace of χu on $\partial\Omega$ is well defined, where $\chi(x) \in C_0^\infty(U_0)$, $\operatorname{supp}\chi \cap \partial\Omega \neq \emptyset$, and U_0 and χ are the same as in the proof of the boundary regularity (see §24).

Rewriting the equation $0 = \chi L u = L\chi u + L_1 u$ in y-coordinates, we get (cf. (24.13)):

$$\hat{L}(\hat{\chi}\hat{u}) + \hat{L}_1(\hat{\chi}_1\hat{u}) = 0, \quad y_n > 0, \tag{26.2}$$

where $\hat{u}, \hat{L}, \hat{\chi}$ are $u(x), L, \chi(x)$ in y-coordinates and $\hat{L}_1$ is a first order differential operator in $\frac{\partial}{\partial y_k}, 1 \leq k \leq n$, $\hat{\chi}_1 \in C_0^\infty(U_0)$, and $\hat{\chi}_1 = 1$ in a neighborhood of $\operatorname{supp}\hat{\chi}$. Analogously to (24.15), we can rewrite (26.2) in the following form for $y_n \geq 0$:

$$\frac{\partial^2}{\partial y_n^2} u_1(y', y_n) = \frac{\partial}{\partial y_n}(\hat{L}_2\hat{u}_2) + \hat{L}_3\hat{u}_2, \tag{26.3}$$

where $u_1 = \hat{\chi}\hat{u}$, $L_k, k = 2, 3$, are differential operators in $\frac{\partial}{\partial y_1}, \dots, \frac{\partial}{\partial y_{n-1}}$ of order $k-1$, and $\hat{u}_2 = \hat{\chi}_1\hat{u}$. Note that $\hat{u}_2(y', y_n) \in L_2(\mathbb{R}_+^n)$.

Let $v_1 = \hat{L}_2\hat{u}_2$ and $v_2 = L_3\hat{u}_2$. In (26.3) make the Fourier transform in $y_1, \dots, y_{n-1}$. We obtain

$$\frac{\partial^2}{\partial y_n^2}\tilde{u}_1(\xi', y_n) = \frac{\partial}{\partial y_n}\tilde{v}_1(\xi', y_n) + \tilde{v}_2(\xi', y_n), \quad y_n > 0. \tag{26.4}$$

Note that

$$\int_0^\infty \int_{\mathbb{R}^{n-1}} |\tilde{v}_r(\xi', y_n)|^2 (1 + |\xi'|)^{-2r} d\xi' dy_n \leq \|\hat{u}_2\|_{0,\mathbb{R}_+^n}, \quad r = 1, 2. \tag{26.5}$$

Denote by A_p, $p = 1, 2$, the following operator:

$$A_p\hat{u} = -\int_{y_n}^\infty (y_n - t)^p e^{(y_n - t)\Lambda(\xi')}\hat{u}(\xi', t)dt, \quad p = 0, 1, \ y_n > 0, \tag{26.6}$$

where $\Lambda(\xi') = 1 + |\xi'|$.

Integrating by parts with respect to t, we get

$$A_1\frac{\partial\hat{u}}{\partial y_n}=\Lambda A_1\hat{u}+A_0\hat{u},\tag{26.7}$$
$$A_1\frac{\partial^2\hat{u}}{\partial y_n^2}=\Lambda^2A_1\hat{u}+2\Lambda A_0\hat{u}+\hat{u}(\xi',y_n).$$

Therefore, applying A_1 to both sides of (26.4), we get

$$\tilde{u}_1(\xi',y_n)=-\Lambda^2A_1\tilde{u}_1-2\Lambda A_0\tilde{u}_1+\Lambda A_1\tilde{v}_1+A_0\tilde{v}_1+A_1\tilde{v}_2.\tag{26.8}$$

Since each term in the right hand side of (26.8) contains an integration with respect to y_n, it is a continuous function of y_n with values in a Sobolev space of distributions in $y'=(y_1,\dots,y_{n-1})$. Therefore, $\lim_{y_n\to+0}\tilde{u}(\xi',y_n)=\tilde{u}(\xi',0)$ exists. Now we estimate $\tilde{u}_1(\xi',0)$. By the Cauchy-Schwartz inequality,

$$\left|\int_0^\infty te^{-t\Lambda}\tilde{u}_1(\xi',t)dt\right|^2\le\left(\int_0^\infty t^2e^{-2t\Lambda}dt\right)\left(\int_0^\infty|\tilde{u}_1(\xi',t)|^2dt\right)\tag{26.9}$$
$$\le C\Lambda^{-3}\int_0^\infty|\tilde{u}_1(\xi',t)|^2dt.$$

Analogously,

$$\left|\int_0^\infty e^{-t\Lambda}\tilde{u}_1(\xi',t)dt\right|^2\le C\Lambda^{-1}\int_0^\infty|\tilde{u}_1(\xi',t)|^2dt.\tag{26.10}$$

Multiplying (26.8) by $\Lambda^{-\frac{1}{2}}$ and taking the limit as $y_n\to+0$, we get

$$|\Lambda^{-\frac{1}{2}}\tilde{u}_1(\xi',0)|^2\tag{26.11}$$
$$\le|-\Lambda^{\frac{3}{2}}A_{10}\tilde{u}_1-2\Lambda^{\frac{1}{2}}A_{00}\tilde{u}_1+\Lambda^{\frac{1}{2}}A_{10}\tilde{v}_1+\Lambda^{-\frac{1}{2}}A_{00}\tilde{v}_1+\Lambda^{-\frac{1}{2}}A_{10}\tilde{v}_2|^2,$$

where A_{p0} is A_p for $y_n=0, p=0,1$.

Integrate (26.11) with respect to ξ' and apply the estimates of the form (26.9), (26.10). Using (26.5) and $u_1=\hat{\chi}\hat{u}=\hat{\chi}\hat{u}_2$, we get:

$$[\hat{\chi}\hat{u}]_{-\frac{1}{2},\mathbb{R}^{n-1}}\le C\|\hat{u}_2\|^2_{0,\mathbb{R}^n_+}.\tag{26.12}$$

Differentiating (26.8) with respect to y_n and using (26.12), we obtain

$$\left[\frac{\partial\hat{u}_1}{\partial y_n}\right]_{-\frac{3}{2},\mathbb{R}^{n-1}}\le C\|\hat{u}_2\|_{0,\mathbb{R}^n_+}.\tag{26.13}$$

After m differentiations, we see that $\frac{\partial^m\hat{u}_1}{\partial y_n^m}$ has a limit as $y_n\to+0$ and

$$\left[\frac{\partial^m\hat{u}_1}{\partial y_n^m}\right]_{-\frac{1}{2}-m,\mathbb{R}^{n-1}}\le C\|\hat{u}_2\|_{0,\mathbb{R}^n_+}.\tag{26.14}$$
□

Let $u(x)\in L_2(\Omega)$ be a solution of a nonhomogeneous equation $Lu=f$ in Ω, where $f\in H_{-1+m}(\Omega), m\ge1$. Let $u_0\in H_{1+m}(\Omega)$ be a particular solution. Then $\frac{\partial^k u_0}{\partial \nu_L^k}\big|_{\partial\Omega} \in H_{m-k+\frac{1}{2}}(\partial\Omega)$ exists for $0 \leq k \leq m$. Therefore, we can apply Theorem 26.1 to $u - u_0$ and get that $\frac{\partial^k u}{\partial \nu_L^k}\Big|_{\partial\Omega} \in H_{-\frac{1}{2}-k}(\partial\Omega)$ for $0 \leq k \leq m$.

Analogously, we have the following theorem:

Theorem 26.2. *Let $Lu = f$ in Ω, $f \in H_{-1+m}(\Omega)$, $u \in H_1(\Omega)$. Then $\frac{\partial^k u}{\partial \nu_L^k}\big|_{\partial\Omega} \in H_{\frac{1}{2}-k}(\partial\Omega)$, $0 \leq k \leq m$.* □

26.2. Applications to nonhomogeneous Dirichlet and Neumann problems.

Let $Lu = f$ in Ω, $u \in L_2(\Omega)$, $f \in L_2(\Omega)$, and let $\varphi \in H_2(\Omega)$. It follows from Theorem 26.1 that $u|_{\partial\Omega} \in H_{-\frac{1}{2}}(\partial\Omega)$, $\frac{\partial u}{\partial \nu_L}\big|_{\partial\Omega} \in H_{-\frac{3}{2}}(\partial\Omega)$. The following Green formula holds:

$$(f, \varphi)_\Omega - (u, L\varphi)_\Omega = \left[\frac{\partial u}{\partial \nu_L}, \varphi\right]_{\partial\Omega} - \left[u, \frac{\partial \varphi}{\partial \nu_L}\right]_{\partial\Omega}, \tag{26.15}$$

where $(\ ,\ )_\Omega$ is the L_2 scalar product in Ω, and $[f, g]_{\partial\Omega}$ is the extension of the L_2 scalar product on $\partial\Omega$ to $f \in H_s(\partial\Omega), g \in H_{-s}(\partial\Omega)$ (cf. §13).

To justify (26.15), consider domains $\Omega_\varepsilon \subset \Omega$ that approach Ω as $\varepsilon \to 0$. By the interior regularity, $u \in H_2(\Omega_\varepsilon)$ (cf. §24). Applying formula (26.15) to Ω_ε and taking the limit as $\varepsilon \to 0$, we get (26.15). Note that the limits of the boundary integrals exist by Theorem 26.1.

Theorem 26.3. *For each $h \in H_{-\frac{1}{2}}(\partial\Omega)$, there exists $u \in L_2(\Omega)$ such that $Lu = 0$ in Ω, $u|_{\partial\Omega} = h$, and*

$$\|u\|_{0,\Omega} \leq C[h]_{-\frac{1}{2},\partial\Omega}. \tag{26.16}$$

We assume that (25.1) *holds.*

Proof: Assuming that the solution $u \in L_2(\Omega)$ exists, we get from (26.15), with $f = 0, \varphi|_{\partial\Omega} = 0$, that

$$(u, L\varphi)_\Omega = \left[u, \frac{\partial \varphi}{\partial \nu_L}\right]_{\partial\Omega}. \tag{26.17}$$

Now we prove that such $u(x)$ exists. As in §23, denote by R the subspace (not closed) of $L_2(\Omega)$ consisting of functions $g(x) = L\varphi$, where $\varphi \in H_2(\Omega)$, $\varphi|_{\partial\Omega} = 0$. Let $F(g)$ be a linear functional on R given by the formula

$$F(L\varphi) = \left[h, \frac{\partial \varphi}{\partial \nu_L}\right]_{\partial\Omega}.$$

It follows from Theorem 22.6 that

$$\left[\frac{\partial \varphi}{\partial \nu_L}\right]_{\frac{1}{2},\partial\Omega} \leq C\|\varphi\|_{2,\Omega}. \tag{26.18}$$

We also have (see (24.19)) that

$$\|\varphi\|_{2,\Omega} \leq C\|L\varphi\|_{0,\Omega}. \tag{26.19}$$

Therefore, using (26.18), (26.19), we have

$$|F(L\varphi)| = \left|\left[h, \frac{\partial\varphi}{\partial\nu_L}\right]_{\partial\Omega}\right| \leq C[h]_{-\frac{1}{2},\partial\Omega}\left[\frac{\partial\varphi}{\partial\nu_L}\right]_{\frac{1}{2},\partial\Omega} \leq C[h]_{-\frac{1}{2},\partial\Omega}\|L\varphi\|_{0,\Omega},$$

i.e., F is bounded on R. Therefore, by the Hahn-Banach and the Riesz theorems, $F(L\varphi) = (u, L\varphi)$ for some $u \in L_2(\Omega)$ and $\|u\|_{0,\Omega} \leq C[h]_{-\frac{1}{2},\partial\Omega}$.

The uniqueness follows from (26.17) since for any $g \in L_2(\Omega)$ there exists $\varphi \in H_2(\Omega)$ such that $L\varphi = g$, $\varphi\big|_{\partial\Omega} = 0$. □

An analogous theorem holds for the Neumann problem.

Theorem 26.4. *Let $a(v, v) \geq C\|v\|_{1,\Omega}^2$, $\forall v \in H_1(\Omega)$. For each $h \in H_{-\frac{1}{2},\partial\Omega}$, there exists $u \in H_1(\Omega)$ such that $Lu = 0$ in Ω, $\frac{\partial u}{\partial\nu_L}\big|_{\partial\Omega} = h$, and*

$$\|u\|_{1,\Omega} \leq C[h]_{-\frac{1}{2},\partial\Omega}. \tag{26.20}$$

Proof: Let $\varphi \in H_2(\Omega)$ and $\frac{\partial\varphi}{\partial\nu_L}\big|_{\partial\Omega} = 0$. Extend $L\varphi$ by zero to $\mathbb{R}^n \setminus \overline{\Omega}$.

Let $\Phi(L\varphi)$ be a linear functional on $R \subset \overset{\circ}{H}_{-1}(\Omega)$ such that (cf. (26.15))

$$\Phi(L\varphi) = -[h, \varphi]_{\partial\Omega}.$$

We have: $[\varphi]_{\frac{1}{2},\partial\Omega} \leq C\|\varphi\|_{1,\Omega}$ (Theorem 22.6) and $\|\varphi\|_{1,\Omega} \leq C\|L\varphi\|_{-1}$ since $\frac{\partial\varphi}{\partial\nu_L}\big|_{\partial\Omega} = 0$ (see (25.10)). Therefore

$$|\Phi(L\varphi)| \leq C[h]_{-\frac{1}{2},\partial\Omega}[\varphi]_{\frac{1}{2},\partial\Omega} \leq C[h]_{-\frac{1}{2},\partial\Omega}\|L\varphi\|_{-1}, \tag{26.21}$$

i.e., Φ can be extended as a linear continuous functional to $\overset{\circ}{H}_{-1}(\Omega)$. By Theorem 22.8, there exists $u \in H_{1,\Omega}$ such that

$$(lu, L\varphi) = -[h, \varphi]_{\partial\Omega}, \tag{26.22}$$

where lu is an arbitrary extension of u. Note that $(lu, L\varphi) = (u, L\varphi)_\Omega$ since $L\varphi = 0$ outside of Ω. For any $\varphi \in C_0^\infty(\Omega)$, we get from (26.22) that $(lu, L\varphi) = 0$, i.e., $Lu = 0$ in Ω. It follows from Theorem 26.2 that the trace $\frac{\partial u}{\partial\nu_L}\big|_{\partial\Omega}$ exists and belongs to $H_{-\frac{1}{2}}(\partial\Omega)$. Therefore, (26.15) implies that

$$(u, L\varphi)_\Omega = -\left[\frac{\partial u}{\partial\nu_L}, \varphi\right]_{\partial\Omega} \tag{26.23}$$

for any $\varphi \in H_2(\Omega)$, $\frac{\partial\varphi}{\partial\nu_L}\big|_{\partial\Omega} = 0$. Since we can choose $\varphi|_{\partial\Omega} \in H_{\frac{3}{2}}(\partial\Omega)$ arbitrarily, we get from (26.22) and (26.23) that $\frac{\partial u}{\partial\nu_L}\big|_{\partial\Omega} = h$.

The uniqueness follows from (26.23), since for any $g \in H_0(\Omega)$ one can find $\varphi \in H_2(\Omega)$ such that $L\varphi = g$, $\frac{\partial\varphi}{\partial\nu_L}\big|_{\partial\Omega} = 0$ (cf. Theorem 25.3). □

27. Variational inequalities

27.1. Minimization of a quadratic functional on a convex set.

Let B be a Banach space. The set $K \in B$ is called convex if $u_1 \in K$ and $u_2 \in K$ imply that $tu_1 + (1-t)u_2 \in K$ for all $t \in [0,1]$. A continuous functional $\Phi(u)$ on a closed convex set K is called convex if

$$\Phi(tv + (1-t)u) \leq t\Phi(v) + (1-t)\Phi(u) \tag{27.1}$$

for all $u, v \in K$ and $t \in [0,1]$. If $\Phi(tv + (1-t)u) < t\Phi(v) + (1-t)\Phi(u)$ for $t \in (0,1)$, then $\Phi(u)$ is called strictly convex. In particular, for $t = \frac{1}{2}$ we have

$$\Phi(u) + \Phi(v) - 2\Phi\left(\frac{u+v}{2}\right) > 0. \tag{27.2}$$

We call $\Phi(u)$ strongly convex if

$$\Phi(u) + \Phi(v) - 2\Phi\left(\frac{u+v}{2}\right) \geq C\|u-v\|_B^p, \tag{27.3}$$

where $\|u\|_B$ is the norm in B, $p > 0$.

Our main class of functionals will be

$$\Phi(u) = \frac{1}{2}a(u,u) - f(u), \tag{27.4}$$

where $B = H_1(\Omega)$, $a(u,u)$ is the quadratic form

$$a(u,u) = \int_\Omega \left[\sum_{j,k=1}^n a_{jk}(x)\frac{\partial u}{\partial x_j}\frac{\partial u}{\partial x_k} + q(x)u^2\right] dx,$$

and $f(u)$ is a linear continuous functional on $H_1(\Omega)$. We assume that $a(v,v) \geq C\|v\|_{1,\Omega}^2$, $\forall v \in H_1(\Omega)$.

We consider three choices of the convex set K:

a) $K = H_1(\Omega)$, **b)** $K = \overset{\circ}{H}_1(\Omega)$, **c)** $K = \{v \in H_1(\Omega),\ v \geq 0 \text{ on } \partial\Omega\}$.

Since $a(u,v)$ is a bilinear form, the following "parallelogram" identity holds:

$$a\left(\frac{u+v}{2}, \frac{u+v}{2}\right) + a\left(\frac{u-v}{2}, \frac{u-v}{2}\right) = \frac{1}{2}a(u,u) + \frac{1}{2}a(v,v).$$

Therefore

$$\Phi(u) + \Phi(v) - 2\Phi\left(\frac{u+v}{2}\right) = a\left(\frac{u-v}{2}, \frac{u-v}{2}\right) \geq C\|u-v\|_1^2, \tag{27.5}$$

i.e., $\Phi(u)$ is a strongly convex functional on $H_1(\Omega)$. Note that the functional (27.4) is bounded below since $a(v,v) \geq C\|v\|_{1,\Omega}^2$ and $|f(v)| \leq \|f\|_{-1}\|v\|_{1,\Omega} \leq \frac{\varepsilon^2}{2}\|v\|_{1,\Omega} + \frac{1}{2\varepsilon^2}\|f\|_{-1}^2$.

Theorem 27.1. *Let Φ be a strongly convex functional on a closed convex set K and let Φ be bounded below. Then there exists a unique $u_0 \in K$ such that*

$$\Phi(u_0) = \min_{u\in K} \Phi(u).$$

Proof: Let $\mu = \inf_{u\in K} \Phi(u)$. There exists a sequence $u_n \in K$ such that $\Phi(u_n) \to \mu$ as $n \to \infty$. Note that $\frac{u_m+u_n}{2} \in K$ since K is convex and $\Phi(\frac{u_m+u_n}{2}) \geq \mu$. Applying (27.3), we get:

$$\begin{aligned} (27.6)\quad C\|u_m - u_n\|_B^p &\leq \Phi(u_m) + \Phi(u_n) - 2\Phi(\frac{u_m+u_n}{2}) \\ &\leq (\Phi(u_m) - \mu) + (\Phi(u_n) - \mu). \end{aligned}$$

Since $\Phi(u_n)-\mu \to 0$, we see that $\{u_n\}$ is a Cauchy sequence in B. Therefore, there exists $u_0 \in K$ such that $\|u_n - u_0\|_B \to 0$. Passing to the limit in (27.6) as $m \to \infty$, we get a useful estimate

$$(27.7)\qquad C\|u_n - u_0\|_B^p \leq \Phi(u_n) - \mu.$$

Since $\Phi(u)$ is continuous, we have $\Phi(u_0) = \lim_{n\to\infty} \Phi(u_n) = \mu$. The uniqueness of u_0 follows from (27.2): If u_1 and u_2 are two points of minimum, i.e., $\Phi(u_1) = \Phi(u_2) = \mu$, $u_1 \neq u_2$, then $\Phi(\frac{u_1+u_2}{2}) < \frac{1}{2}(\Phi(u_1) + \Phi(u_2)) < \mu$, i.e., there is a contradiction. □

Applying Theorem 27.1 to $\Phi(v) = \frac{1}{2}a(v,v) - f(v)$ and $K = \overset{\circ}{H}_1(\Omega)$ or $H_1(\Omega)$, we justify the claims (25.4) and (25.7) (see Theorems 25.1 and 25.2).

27.2. Characterization of the minimum point.

Now we give a characterization of the point of minimum.

When $\Phi(u) = \min_{v\in K} \Phi(v)$, we have

$$(27.8)\qquad \Phi(u + t(v-u)) - \Phi(u) \geq 0$$

for every $v \in K$ and every $t \in [0,1]$.

When $\Phi(v)$ has the form (27.4), we get

$$(27.9)\qquad \Phi(u + t(v-u)) - \Phi(u) = t[a(u, v-u) - f(v-u)] + O(t^2).$$

Dividing (25.9) by t and taking the limit as $t \to 0$, we obtain:

$$(27.10)\qquad a(u, v-u) \geq f(v-u), \quad \forall v \in K.$$

Inequality (27.10) is called the variational inequality. If u is an interior point of K, we can take $v = u + \tau w \in K$, $\forall w \in K$, where $|\tau|$ is small enough. Therefore, $a(u,\tau w) - f(\tau w) \geq 0$ implies that

$$(27.11)\qquad a(u,w) - f(w) = 0,$$

since τ can be positive or negative. The equation (25.11) coincides with (25.6) or (25.8) if K is equal to $\overset{\circ}{H}_1(\Omega)$ or $H_1(\Omega)$, respectively.

Consider in more detail the case where $K = \{v \in H_1(\Omega), v|_{\partial\Omega} \geq 0\}$. Choosing $v = 0$ and $v = 2u$ in (27.10), we get $-a(u,u) \geq -f(u)$ and $a(u,u) \geq f(u)$. Therefore

$$a(u,u) = f(u). \tag{27.12}$$

Adding (27.12) to (27.10), we get

$$a(u,v) \geq f(v), \quad \forall v \in K, \tag{27.13}$$

i.e., (27.12), (27.13) are consequences of (27.10). Subtracting (27.12) from (27.13), we obtain (27.10), i.e., (27.10) is equivalent to (27.12), (27.13).

Let $\varphi \in C_0^\infty(\Omega)$ be arbitrary. Since $\varphi|_{\partial\Omega} = 0$, we have that $\pm\varphi \in K$. Substituting $v = \pm\varphi$ in (27.13), we get:

$$a(u,\varphi) - f(\varphi) = 0, \ \forall \varphi \in C_0^\infty(\Omega). \tag{27.14}$$

Equation (27.14) is the same as (25.6). Therefore, we have that

$$Lu = f \quad \text{in} \quad \Omega. \tag{27.15}$$

Assume that $f \in L_2(\Omega)$. Then Theorem 26.2 implies that $u|_{\partial\Omega} \in H_{\frac{1}{2}}(\partial\Omega)$, $\frac{\partial u}{\partial \nu_L}|_{\partial\Omega} \in H_{-\frac{1}{2}}(\partial\Omega)$. Similarly to (26.15), we have the following Green formula:

(27.16)

$$a(u,v) - f(v) = \left[\frac{\partial u}{\partial \nu_L}, v\right]_{\partial\Omega} + (Lu - f, v) = \left[\frac{\partial u}{\partial \nu_L}, v\right]_{\partial\Omega} \geq 0, \quad \forall v \in K,$$

where we took into account (27.15). Replacing v by u in (27.16) and using (27.12), we get:

$$\left[\frac{\partial u}{\partial \nu_L}, u\right] = 0. \tag{27.17}$$

Suppose $\frac{\partial u}{\partial \nu_L} \in L_2(\partial\Omega)$. Then (27.16) and (27.17) have the form:

$$\int_{\partial\Omega} \frac{\partial u}{\partial \nu_L} v ds \geq 0, \quad \forall v \geq 0, \quad v \in H_{\frac{1}{2}}(\partial\Omega), \tag{27.18}$$

$$\int_{\partial\Omega} \frac{\partial u}{\partial \nu_L} u ds = 0. \tag{27.19}$$

It follows from (27.18) that $\frac{\partial u}{\partial \nu_L} \geq 0$ on $\partial\Omega$ because $v \geq 0$ is arbitrary. Since $u \geq 0$ and $\frac{\partial u}{\partial \nu_L} \geq 0$ on $\partial\Omega$, we get from (27.19) that

$$\frac{\partial u}{\partial \nu_L} u(x) = 0 \quad \text{a.e. on} \quad \Omega. \tag{27.20}$$

Therefore, the variational inequality (27.10) is equivalent to $Lu = f$ in Ω and

$$(27.21) \qquad u|_{\partial\Omega} \geq 0, \quad \frac{\partial u}{\partial \nu_L}\Big|_{\partial\Omega} \geq 0, \quad \frac{\partial u}{\partial \nu_L} u = 0 \quad \text{on } \partial\Omega,$$

if we additionally assume that $\frac{\partial u}{\partial \nu_L} \in L_2(\partial\Omega)$.

Let $\Gamma_1 \subset \partial\Omega$ be the set where $u|_{\partial\Omega} > 0$ and let $\Gamma_2 \subset \partial\Omega$ be the set where $\frac{\partial u}{\partial \nu_L}\Big|_{\partial\Omega} > 0$. It follows from (27.20) that $\frac{\partial u}{\partial \nu_L} = 0$ on Γ_1 and $u = 0$ on Γ_2. Note that the sets Γ_1 and Γ_2 are unknown in advance. □

We shall give a physical example of the variational inequality of the form (27.21):

Example 27.1. Let $w(x)$ be the temperature in $\overline{\Omega}$. Consider a functional $\Phi(w)$ of the form (27.4) on the convex set $K_1 = \{w \in H_1(\Omega),\ w \leq T \text{ on } \partial\Omega\}$, where T is a given constant. Replacing $u = T - w$, we get a functional $\Phi_1(u) = \Phi(T - u)$ on the convex set

$$K_1 = \{u \in H_1(\Omega), u \geq 0 \text{ on } \partial\Omega\}.$$

Therefore, if $w(x)$ is the solution of the variational inequality, then $w(x)$ satisfies

$$(27.22) \qquad w|_{\partial\Omega} \leq T, \quad \frac{\partial w}{\partial \nu_L}\Big|_{\partial\Omega} \leq 0, \quad (T - w)\frac{\partial w}{\partial \nu_L} = 0 \text{ on } \partial\Omega,$$

provided $\frac{\partial w}{\partial \nu_L} \in L_2(\partial\Omega)$.

The physical interpretation of (27.22) is the following: on the set Γ_1, where $w|_{\partial\Omega} < T$, we have $\frac{\partial w}{\partial \nu_L}\Big|_{\partial\Omega} = 0$, i.e., the portion Γ_1 of $\partial\Omega$ is insulated. The insulation breaks down, i.e., $\frac{\partial w}{\partial \nu_L}\Big|_{\Omega} < 0$, only on the set where $w|_{\partial\Omega} = T$.

28. Problems

1. Prove the existence of the weak solution of the Neumann problem for the operator L of the form (23.3), i.e., L is not formally selfadjoint.

2. Prove (cf. Theorem 26.2) that for any $h \in H_{-\frac{3}{2}}(\partial\Omega)$, there exists $u \in L_2(\Omega)$ such that $(-\Delta + 1)u = 0$ in Ω, $\frac{\partial u}{\partial \nu}\Big|_{\partial\Omega} = h$.

3. Study the solvability in $H_1(\Omega)$ of the weak solution of the Neumann problem

$$\Delta u = f \text{ in } \Omega, \quad \frac{\partial u}{\partial \nu}\Big|_{\partial\Omega} = 0.$$

4. Define the weak solution of the boundary value problem

$$\Delta u = f \text{ in } \Omega, \quad \frac{\partial u}{\partial \nu} + a(x)u = 0 \text{ on } \partial\Omega,$$

where $a(x) > 0$, $a(x) \in C^\infty(\partial\Omega)$ and ν is the outward normal. Prove the existence and the uniqueness of this solution in $H_1(\Omega)$.

Hint: Consider the functional

$$\Phi(v) = \frac{1}{2}a(v,v) - (f,v) + \frac{1}{2}\int_{\partial\Omega} a(x)v^2(x)ds.$$

5. Let $F(v)$ be a functional in $H_1(\Omega)$ of the form (27.4). Find the minimum of $F(v)$ in $H_1(\Omega)$ subject to the constraint $g(v) = C$, where $g(v)$ is a linear continuous functional on $H_1(\Omega)$.

6. Let $F(v)$ be the same as in (27.4) and let

$$A = \{v \in H_1(\Omega),\ v \geq h(x) \text{ a.e. in } \Omega\},$$

where $h(x)$ is a continuous function in $\overline{\Omega}$. Find $u(x) \in H_1(\Omega)$ such that $F(u) = \min_{v\in A} F(v)$. What is the variational inequality that $u(x)$ satisfies?

7. Let $F(v)$ be the same as in Problem 6, and let

$$A_1 = \Big\{v \in H_1(\Omega),\ \sum_{j=1}^{n}\Big(\frac{\partial u}{\partial x_j}\Big)^2 \leq 1 \text{ a.e. in } \Omega\Big\}.$$

Prove that A_1 is a closed convex set in $H_1(\Omega)$. Find the variational inequality for $u \in H_1(\Omega)$, where $F(u) = \min_{v\in A_1} F(v)$.

8. **a)** Show that $F^{-1}\frac{\xi_j}{|\xi|^2+1} = iK_j(x)$ satisfy $|K_j(x)| \leq C_\varepsilon\frac{e^{-(1-\varepsilon)|x|}}{|x|^{n-1}}$, $\forall\varepsilon > 0$, $F^{-1}\frac{1}{1+|\xi|^2} = K_0(x)$ satisfies $|K_0(x)| \leq C\frac{e^{-|x|}}{|x|^{n-2}}$ for $n > 2$, and $|K_0(x)| \leq C_\varepsilon\frac{e^{-|x|}}{|x|^\varepsilon}$, $\forall\varepsilon > 0$, for $n = 2$.

b) Using the identity $1 = \sum_{k=1}^n \xi_k\frac{\xi_k}{|\xi|^2+1} + \frac{1}{|\xi|^2+1}$, prove that

$$u(x) = \sum_{j=1}^{n}\int_{\mathbb{R}^n} K_j(x-y)\frac{\partial u(y)}{\partial y_j}dy + \int_{\mathbb{R}^n} K_0(x-y)u(y)dy, \tag{28.1}$$

$\forall u \in C_0^\infty(\mathbb{R}^n)$.

9. Let $u(x) = K * f = \int_{\mathbb{R}^n} K(x-y)f(y)dy$. Prove that:

a) $\max_{x\in\mathbb{R}^n}|u(x)| \leq \|K\|_{L_q}\|f\|_{L_p}$, $\forall f \in C_0^\infty(\mathbb{R}^n)$, where $\|f\|_{L_p} = \left(\int_{\mathbb{R}^n}|f(x)|^p dx\right)^{\frac{1}{p}}$, $p > 1$, $\frac{1}{q} + \frac{1}{p} = 1$.

b) $\|u\|_{L_q} \leq C\|K\|_{L_r}\|f\|_{L_p}$, $\forall f \in C_0^\infty(\mathbb{R}^n)$, where $\frac{1}{p} + \frac{1}{r} = 1 + \frac{1}{q}$.

10. Denote by $W^{1,p}(\mathbb{R}^n)$ the space of L_p functions such that their distribution derivatives (weak derivatives) also belong to $L_p(\mathbb{R}^n)$. Let $\|u\|_{1,p} = \left(\sum_{k=1}^n \left\|\frac{\partial u}{\partial x_k}\right\|_{L_p}^p + \|u\|_{L_p}^p\right)^{\frac{1}{p}}$ be the norm in $W^{1,p}(\mathbb{R}^n)$.

Using Problems 8 and 9, prove Sobolev's inequality

$$\|u\|_{L_q} \le C\|u\|_{1,p},$$

where $p < n$, $\frac{1}{q} > \frac{1}{p} - \frac{1}{n}$.

11. Use Problems 8 and 9 to prove that

$$\max_{x\in\mathbb{R}^n} |u(x)| \le C\|u\|_{1,p}, \quad \forall u \in C_0^\infty(\mathbb{R}^n)$$

when $p > n$.

12. The maximum principle (cf. Evans [**Ev**]). Let

$$Lu = -\sum_{j,k=1}^{n} a_{jk} u_{x_j} u_{x_k} + \sum_{j=1}^{n} b_j(x) u_{x_j} = 0$$

in Ω, where Ω is a bounded domain, and $u(x)$ is continuous in $\overline{\Omega}$. Prove that

(28.2) $$\max_{x\in\overline{\Omega}} u(x) = \max_{x\in\partial\Omega} u(x).$$

Hints:

a) Show that if x_0 is a point of maximum inside Ω, then

$$-\sum_{j,k=1}^{n} a_{jk}(x_0) w_{x_j x_k}(x_0) \ge 0.$$

b) Show that if $Lw < 0$ in Ω, then $w(x)$ has no points of maximum inside Ω, i.e.,

(28.3) $$\max_{x\in\overline{\Omega}} w(x) = \max_{x\in\partial\Omega} w(x).$$

c) Let $Lu = 0$. Prove that $L(u + \varepsilon e^{\lambda x_1}) < 0$, where $\lambda > 0$ is large and $\varepsilon > 0$ is arbitrary.

d) Use (28.2) for $w = u(x) + \varepsilon e^{\lambda x_1}$ and pass to the limit as $\varepsilon \to 0$.

Chapter V

Scattering Theory

Introduction to Chapter V

In this chapter we study the nonhomogeneous Schrödinger equation in $\mathbb{R}^n$,

$$(-\Delta + q(x) - k^2)u(x) = f(x),$$

with applications to scattering theory. The main technical tool is the Agmon's estimate ([**A2**]) described in §29. This is an estimate of the resolvent of the nonperturbed operator $-\Delta$. We consider two classes of potentials $q(x)$: $q(x) = O(\frac{1}{(1+|x|)^{n+\alpha}})$, $\alpha > 0$, and $q(x) = O(\frac{1}{(1+|x|)^{1+\varepsilon}})$. The first class is more restrictive, but the asymptotics of solutions are much nicer, and the proofs are simpler. The crucial step in proving the solvability of the nonhomegeneous Schrödinger equation is the uniqueness theorem (see Agmon [**A2**] and Hörmander [**H1**]). The uniqueness result is related to the absence of the point spectrum of the operator $-\Delta + q(x)$ for any $k^2 > 0$. The limiting absorption principle is proven in §32. The scattering problem is discussed in §33 for both classes of potentials separately.

We are mostly preoccupied with the existence of distorted plane waves and their asymptotics as $|x| \to \infty$. We are not developing the general scattering theory since it requires the knowledge of the theory of unbounded self-adjoint operators, the notion of absolutely continuous spectrum, and other facts of functional analysis, which would further increase the size of this chapter (see Hörmander [**H1**] and Agmon [**A2**] for more results). Instead, we are focusing on the inverse scattering problems, i.e., on the recovery of the potential from the scattering amplitude $a(\theta, \omega, k)$. It is done in §33, where we assume that $a(\theta, \omega, k)$ is given for all $\theta \in S^{n-1}$, $\omega \in S^{n-1}$, and $k > 0$. This is an overdetermined problem, since $q(x)$ is a function of n variables,

$n \geq 2$, and $a(\theta, \omega, k)$ depends on $n-1+n-1+1 = 2n-1$. In §36 we consider less overdetermined inverse problems such as the inverse obstacle problem where we recover the obstacle from $a(\theta, \omega, k)$ given for a fixed k, all $\theta \in S^{n-1}$, and countably many ω_j, $j = 1, \ldots$ (see Colton-Kress [**CK**] for references and additional results on the scattering by obstacles). In §37 we consider the inverse scattering problem in the case of exponentially decaying potential, assuming that $a(\theta, \omega, k)$ is given for k fixed and all $\theta \in S^{n-1}$, $\omega \in S^{n-1}$. The fact that k is fixed and the potential has a noncompact support makes the problem quite difficult, and the analytic continuation of the scattering amplitude is the key to solving this inverse problem. In §§34–35 we consider the inverse boundary value problem (see Sylverter-Uhlmann [**SU**], Novikov [**No1**]), and we study the relation between the inverse boundary value problem and the inverse scattering problem (see Nachman [**Na1**] and Novikov [**No1**]). Finally, in §38 we state the results on the inverse backscattering problem of recovering $q(x)$ from $a(\theta, -\theta, k)$. This is not an overdetermined problem; however, the results of §38 are only local.

29. Agmon's estimates

Denote by $V_{m,s}(\mathbb{R}^n)$ the weighted Sobolev space with the norm

$$\|u\|_{m,s}^2 = \sum_{|k|=0}^{m} \int_{\mathbb{R}^n} (1+|x|)^{2s} \left| \frac{\partial^k u}{\partial x^k} \right|^2 dx. \tag{29.1}$$

Theorem 29.1. *Let E_+ be the operator*

$$E_+ h = \int_{\mathbb{R}^n} E_+(x-y) h(y) dy = \frac{1}{(2\pi)^n} \int_{\mathbb{R}^n} \frac{e^{ix\cdot\xi}\tilde{h}(\xi)d\xi}{|\xi|^2 - k^2 - i0}.$$

Then for any $\varepsilon > 0$,

$$\|E_+ h\|_{p,-\frac{1}{2}-\varepsilon} \leq C_\varepsilon k^{p-1} \|h\|_{0,\frac{1}{2}+\varepsilon}, \quad 0 \leq p \leq 2, \tag{29.2}$$

where C_ε is independent of k for $k \geq \varepsilon_0 > 0$.

We prove Theorem 29.1 as a particular case of the following more general theorem:

Theorem 29.2. *Denote by $E_{\nu,\sigma,\tau}h$ the operator*

$$E_{\nu,\sigma,\tau} h = \frac{1}{(2\pi)^n} \int_{\mathbb{R}^n} \frac{e^{ix\cdot\xi}\tilde{h}(\xi)d\xi}{|\xi|^2 - k^2 - \tau^2 + 2i\tau(\xi_\nu - \sigma)}, \tag{29.3}$$

where $|\nu| = 1$, $\xi_\nu = \xi \cdot \nu$, $|\sigma| \leq k$, $0 \leq \tau \leq \tau_0$, and $E_{\nu,\sigma,0}h$ means the limit of $E_{\nu,\sigma,\tau}h$ as $\tau \to 0$.

Then for any $\varepsilon > 0$,

$$\|E_{\nu,\sigma,\tau} h\|_{p,-\frac{1}{2}-\varepsilon} \leq C_\varepsilon k^{p-1} \|h\|_{0,\frac{1}{2}+\varepsilon}, \tag{29.4}$$

where C_ε *is independent of* τ, ν, σ, k, $k \geq \varepsilon_0$, $0 \leq \tau \leq \tau_0$, τ_0 *is small, and* $0 \leq p \leq 2$.

Note that $E_{\nu,\sigma,0}$ is Faddeev's fundamental solution, $E_{\nu,\sigma,0} = E_+$ for $\sigma > k$, and $E_{\nu,\sigma,0} = E_-$ for $\sigma < -k$. In the case $\sigma \neq 0$ we need only $\tau \in [0, \tau_0]$, where $\tau_0 > 0$ is small. The operator $E_{\nu,0,\tau}$, $\tau > 0$, $\sigma = 0$, will be used in §§28 and 30 in the case where k is fixed, but τ is large.

Theorem 29.3. *For any* $\varepsilon > 0$ *the following estimate holds:*

$$\|E_{\nu,0,\tau}h\|_{p,-\frac{1}{2}-\varepsilon} \leq C_\varepsilon(k^2+\tau^2)^{\frac{p-1}{2}}\|h\|_{0,-\frac{1}{2}-\varepsilon}, \tag{29.5}$$

where $k \geq 0$, $\tau \geq 0$, $k^2+\tau^2 \geq \varepsilon_0^2$, C_ε *does not depend on* k, τ, ν, $0 \leq p \leq 2$.

Proof of Theorem 29.2. For any $h \in C_0^\infty(\mathbb{R}^n)$, $g \in C_0^\infty(\mathbb{R}^n)$, and $\tau > 0$ we have:

$$(E_{\nu,\sigma,\tau}h, g) = \frac{1}{(2\pi)^n}\int_{\mathbb{R}^n} \frac{\tilde{h}(\xi)\overline{\tilde{g}(\xi)}d\xi}{|\xi|^2 - k^2 - \tau^2 + 2i\tau(\xi_\nu - \sigma)}. \tag{29.6}$$

Let $\chi_0(t) \in C_0^\infty(\mathbb{R}^1)$, $\chi_0(t) \geq 0$, $\chi_0(-t) = \chi_0(t)$, $\chi_0(t) = 1$ for $|t| < \frac{1}{2}$, $\chi_0(t) = 0$ for $|t| > 1$, and let $\chi(|\xi|) = \chi_0\Big(\frac{r-\sqrt{k^2+\tau^2}}{\frac{1}{2}\sqrt{k^2+\tau^2}}\Big)$. Then

$$(E_{\nu,\sigma,\tau}h, g) = I_1 + I_2 + I_3, \tag{29.7}$$

where

(29.8)

$$I_1(\tau) = \frac{1}{(2\pi)^n}\int_0^\infty\int_{|\omega|=1} \frac{\chi^2(r)\big(\tilde{h}(r\omega)\overline{\tilde{g}(r\omega)} - \tilde{h}(\sqrt{k^2+\tau^2}\omega)\overline{\tilde{g}(\sqrt{k^2+\tau^2}\omega)}\big)}{r^2 - k^2 - \tau^2 + 2i\tau(\xi_\nu - \sigma)} \times r^{n-1}drd\omega,$$

$\xi = r\omega$, $r = |\xi|$, $\omega = \frac{\xi}{|\xi|}$, $\xi_\nu = r\omega_\nu$, $\omega_\nu = \omega \cdot \nu$, and

(29.9)

$$I_2(\tau) = \frac{1}{(2\pi)^n}\int_0^\infty\int_{|\omega|=1} \frac{\chi^2(r)\tilde{h}(\sqrt{k^2+\tau^2}\omega)\overline{\tilde{g}(\sqrt{k^2+\tau^2}\omega)}r^{n-1}drd\omega}{r^2 - k^2 - \tau^2 + 2i\tau(r\omega_\nu - \sigma)},$$

$$I_3(\tau) = \frac{1}{(2\pi)^n}\int_0^\infty\int_{|\omega|=1} \frac{(1-\chi^2(|\xi|))\tilde{h}(\xi)\overline{\tilde{g}(\xi)}d\xi}{r^2 - k^2 - \tau^2 + 2i\tau(\xi_\nu - \sigma)}. \tag{29.10}$$

We have

$$\begin{aligned} |I_3(\tau)| &\leq \frac{C}{\sqrt{\tau^2+k^2}}\int_{\mathbb{R}^n} |\tilde{h}(\xi)||\tilde{g}(\xi)|d\xi \\ &\leq \frac{C}{\sqrt{\tau^2+k^2}}\|\tilde{h}\|_0\|\tilde{g}\|_0 \leq \frac{C}{\sqrt{\tau^2+k^2}}\|h\|_{0,\frac{1}{2}+\varepsilon}\|g\|_{0,\frac{1}{2}+\varepsilon}, \end{aligned} \tag{29.11}$$

Let

$$a(r,\omega_\nu,\sigma,\tau)=\frac{r^{n-1}}{r^2-k^2-\tau^2+2i\tau(r\omega_\nu-\sigma)}. \tag{29.12}$$

Then

$$I_2(\tau)=\frac{1}{(2\pi)^n}\int_{|\omega|=1}\Big[\int_0^\infty a(r,\omega_\nu,\sigma,\tau)\chi^2(r)dr\Big] \times \tilde{h}(\sqrt{k^2+\tau^2}\omega)\overline{\tilde{g}(\sqrt{k^2+\tau^2}\omega)}d\omega. \tag{29.13}$$

We prove that

$$\Big|\int_0^\infty a(r,\omega_\nu,\sigma,\tau)\chi^2(r)dr\Big|\le C(k^2+\tau^2)^{\frac{n-2}{2}}, \tag{29.14}$$

where C is independent of τ,σ and ω_ν, $0\le\tau\le\tau_0$, and τ_0 is small.

Let $p(r,\omega_\nu,\sigma,\tau)=r^2-k^2-\tau^2+2i\tau(r\omega_\nu-\sigma)$ and let $r_0=\sqrt{k^2+\tau^2}$. We have

$$p=(r-r_0+i\tau\alpha)(r+r_0-i\tau\alpha)+2i\tau(r-r_0)\omega_\nu-\tau^2\alpha^2, \tag{29.15}$$

where $\alpha=\omega_\nu-\frac{\sigma}{r_0}$.

For $\frac{\sigma}{r_0}-\omega_\nu<0$, denote by Γ_+ the contour in the complex plane consisting of the positive real axis (without the interval $|r-r_0|<\varepsilon$) and the upper semicircle $|r-r_0|=\varepsilon$.

For $\frac{\sigma}{r_0}-\omega_\nu>0$, denote by Γ_- the contour in the complex plane consisting of the positive real axis (without the interval $|r-r_0|<\varepsilon$) and the lower semicircle $|r-r_0|=\varepsilon$.

We assume that $\varepsilon>0$ is much smaller than ε_0 and $\chi(r)=1$ for $|r-r_0|\le\varepsilon$. We show that $|p|\ge C\varepsilon r_0$ on Γ_+ for $r_0\omega_\nu-\sigma>0$ and $|p|\ge C\varepsilon r_0$ on Γ_- for $r_0\omega_\nu-\sigma<0$, where C is independent of $\tau>0,\omega_\nu,\sigma,r$, and $0<\tau\le\tau_0$.

If r is real and $|r-r_0|\ge\varepsilon$, we have

$$|p|\ge|\operatorname{Re}p|=|r-r_0||2r_0+r-r_0|\ge\varepsilon(2r_0-\varepsilon)\ge\varepsilon r_0.$$

If $r=r_0+\varepsilon e^{i\varphi}$, $0<\varphi<\pi$, and $\alpha=\omega_\nu-\frac{\sigma}{r_0}>0$, we have $|r-r_0+i\tau\alpha|\ge\varepsilon$ for all $\tau\ge 0$. Therefore, $|p|\ge\varepsilon|2r_0-\varepsilon-2\tau|-\alpha^2\tau^2\ge\varepsilon r_0$ since $0<\tau\le\tau_0$, τ_0 is small, and $|\alpha|\le 2$.

Hence $|p|\ge\varepsilon r_0$ on Γ_+ when $\alpha=\omega_\nu-\frac{\sigma}{r_0}>0$. If $\alpha=\omega_\nu-\frac{\sigma}{r_0}<0$, then analogously $|p|\ge\varepsilon r_0$ on Γ_-. Note that $a(r,\omega_\nu,\nu,\tau)$ has no poles inside the upper semicircle $|r-r_0|=\varepsilon$ for $\tau>0$ and $\alpha>0$, and it has no poles inside the lower semicircle for $\tau>0,\alpha<0$.

Therefore, using the Cauchy theorem, we can replace the integral over the positive real axis in (29.13) by Γ_+ if $\frac{\sigma}{r_0}-\omega_\nu<0$, and by Γ_- if $\frac{\sigma}{r_0}-\omega_\nu>0$.

Since the denominator in (29.12) is bounded below by $C(\tau^2+k^2)^{\frac{1}{2}}$ on either Γ_+ or Γ_-, we get (29.14). Hence

$$|I_2(\tau)| \le C(k^2+\tau^2)^{\frac{n-2}{2}} \int_{|\omega|=1} |\tilde{h}(\sqrt{k^2+\tau^2}\omega)||\tilde{g}(\sqrt{k^2+\tau^2}\omega)|d\omega. \tag{29.16}$$

It follows from (29.16), Remark 13.1 (cf. (13.27)), and the Cauchy-Schwartz inequality that

$$|I_2| \le \frac{C}{\sqrt{k^2+\tau^2}}\|\tilde{h}\|_s\|\tilde{g}\|_s, \quad s = \frac{1}{2}+\varepsilon. \tag{29.17}$$

Now we estimate $I_1(\tau)$. We have:

$$\begin{aligned}(29.18)\quad &|I_1(\tau)| \\ &\le \frac{1}{(2\pi)^n}\int_0^\infty\int_{|\omega|-1} \frac{\chi^2(r)\left|\left(\tilde{h}(r\omega)\overline{\tilde{g}(r\omega)} - \tilde{h}(\sqrt{k^2+\tau^2}\omega)\overline{\tilde{g}(\sqrt{k^2+\tau^2}\omega)}\right)\right|}{\sqrt{k^2+\tau^2}\left|r-\sqrt{k^2+\tau^2}\right|} \\ &\qquad\qquad \times r^{n-1}drd\omega.\end{aligned}$$

It follows from (13.28) that

$$\begin{aligned}(29.19)\quad &\int_{|\omega|=1} r^{n-1}\chi(r)|\tilde{f}(r\omega)-\tilde{f}(\sqrt{k^2+\tau^2}\omega)|^2 d\omega \\ &\qquad\qquad \le C|r-\sqrt{k^2+\tau^2}|^{2\varepsilon}\|\tilde{f}\|_s^2,\end{aligned}$$

where $s > \frac{1}{2}$, $0<\varepsilon\le 1$, $\varepsilon < s-\frac{1}{2}$. Also we have (see (13.27)):

$$r^{n-1}\chi(r)\int_{|\omega|=1}|\tilde{f}(r\omega)|^2 d\omega \le C\|\tilde{f}\|_s^2, \quad s>\frac{1}{2}. \tag{29.20}$$

Since

$$\begin{aligned}|(\tilde{h}(r\omega)\overline{\tilde{g}(r\omega)} &- \tilde{h}(\sqrt{k^2+\tau^2}\omega)\overline{\tilde{g}(\sqrt{k^2+\tau^2}\omega)}| \\ &\le |\tilde{h}(r\omega)-\tilde{h}(\sqrt{k^2+\tau^2}\omega)||\tilde{g}(r\omega)| \\ &\qquad + |\tilde{h}(\sqrt{k^2+\tau^2}\omega)||\tilde{g}(r\omega)-\tilde{g}(\sqrt{k^2+\tau^2}\omega)|,\end{aligned}$$

we get, using the Cauchy-Schwartz inequality:

$$(29.21)\quad \sqrt{k^2+\tau^2}|I_1| \le C\int_0^\infty \int_{|\omega|=1} r^{n-1}\chi^2(r) \times \frac{|(\tilde{h}(r\omega)\overline{\tilde{g}(r\omega)} - \tilde{h}(\sqrt{k^2+\tau^2}\omega)\overline{\tilde{g}(\sqrt{k^2+\tau^2}\omega)}|}{|r-\sqrt{k^2+\tau^2}|} dr d\omega$$

$$\le C\int_0^\infty \frac{r^{n-1}\chi^2(r)}{|r-\sqrt{k^2+\tau^2}|^{1-\varepsilon}}\left[\left(\int_{|\omega|=1}\frac{|\tilde{h}(r\omega)-\tilde{h}(\sqrt{k^2+\tau^2}\omega)|^2}{|r-\sqrt{k^2+\tau^2}|^{2\varepsilon}}d\omega\right)^{\frac{1}{2}} \times \left(\int_{|\omega|=1}|\tilde{g}(r\omega)|^2 d\omega\right)^{\frac{1}{2}}\right.$$

$$\left. + \left(\int_{|\omega|=1}\frac{|\tilde{g}(r\omega)-\tilde{g}(\sqrt{k^2+\tau^2}\omega)|^2}{|r-\sqrt{k^2+\omega^2}|^{2\varepsilon}}d\omega\right)^{\frac{1}{2}}\left(\int_{|\omega|=1}|\tilde{h}(\sqrt{k^2+\tau^2}\omega)|^2 d\omega\right)^{\frac{1}{2}}\right] dr.$$

Since

$$\int_0^\infty \frac{\chi(r)dr}{|r-\sqrt{k^2+\tau^2}|^{1-\varepsilon}} \le C,$$

we get, using (29.19) and (29.20), that

$$(29.22)\qquad |I_1| \le \frac{C}{\sqrt{k^2+\tau^2}}\|\tilde{h}\|_s\|\tilde{g}\|_s, \quad s > \frac{1}{2}.$$

Therefore

$$(29.23)\qquad |(E_{\nu,\sigma,\tau}h, g)| \le \frac{C}{\sqrt{k^2+\tau^2}}\|h\|_{0,s}\|g\|_{0,s},$$

where $s > \frac{1}{2}$. Since $V_{0,-s}$ is dual to $V_{0,s}$, and since $C_0^\infty(\mathbb{R}^n)$ is dense in $V_{0,s}$, we get

$$(29.24)\qquad \|E_{\nu,\sigma,\tau}h\|_{0,-s} \le \frac{C}{\sqrt{k^2+\tau^2}}\|h\|_{0,s}, \quad s > \frac{1}{2},$$

for every $h \in V_{0,s}$.

For any $1 \le |p| \le 2$ we have

$$\frac{\partial^p}{\partial x^p}E_{\nu,\sigma,\tau}h = \frac{1}{(2\pi)^n}\int_{\mathbb{R}^n}\frac{(i\xi)^p e^{ix\cdot\xi}\tilde{h}(\xi)d\xi}{|\xi|^2-k^2-\tau^2+i\tau(\xi_\nu-\sigma)}.$$

Therefore

$$(29.25)\qquad \left(\frac{\partial^p}{\partial x^p}E_{\nu,\sigma,\tau}h, g\right) = \frac{1}{(2\pi)^n}\int_{\mathbb{R}^n}\frac{(i\xi)^p\tilde{h}(\xi)\overline{\tilde{g}(\xi)}d\xi}{|\xi|^2-k^2-\tau^2+i\tau(\xi_\nu-\sigma)}.$$

Repeating the previous estimates with $(|\xi|^2-k^2-\tau^2+2i\tau(\xi_\nu-\sigma))^{-1}$ replaced by $(i\xi)^p(|\xi|^2-k^2-\tau^2+2i\tau(\xi_\nu-\sigma))^{-1}$, we get the estimate (29.3) with $p=1$ and $p=2$.

Now we consider the case $\tau = 0$. Taking the limit as $\tau \to 0$, we get

(29.26)
$$I_1(0) = \frac{1}{(2\pi)^n} \int_0^\infty \int_{|\omega|=1} \frac{\chi^2(r)(\tilde{h}(r\omega)\overline{\tilde{g}(r\omega)} - \tilde{h}(k\omega)\overline{\tilde{g}(k\omega)})r^{n-1}drd\omega}{r^2 - k^2},$$

(29.27)
$$I_3(0) = \frac{1}{(2\pi)^n} \int_{\mathbb{R}^n} \frac{(1 - \chi^2(|\xi|))\tilde{h}(\xi)\overline{\tilde{g}(\xi)}d\xi}{|\xi|^2 - k^2},$$

where $\chi(r) = \chi_0(r - k)$.

Therefore, the estimates (29.11), (29.18), and (29.22) hold for $\tau = 0$. When we change the integration in (29.13) from the positive real axis to Γ_+ or Γ_-, we can take the limit as $\tau \to 0$ and get the estimate (29.17) for $\tau = 0$. Therefore, (29.24) holds for $\tau = 0$. □

Note that $E_{\nu,\sigma,0}$ is the Faddeev's fundamental solution (cf. Example 16.6).

Proof of Theorem 29.3. The proof of Theorem 29.3 is almost a repetition of the proof of Theorem 29.2. The estimates of $I_1(\tau)$ and $I_2(\tau)$ are exactly the same (cf. (29.11) and (29.22)). The only change is in the proof of the estimate (29.14). We have

$$p(r, \omega_\nu, \tau) = r^2 - k^2 - \tau^2 + 2i\tau r\omega_\nu.$$

If $|\tau\omega_\nu| > \varepsilon$ and $r \in \operatorname{supp} \chi(r)$, we have

$$|p| \geq 2r|r\omega_\nu| \geq C\varepsilon r_0, \quad r_0 = \sqrt{k^2 + \tau^2}.$$

If $|\tau\omega_\nu| < \varepsilon$, we decompose as before,

$$p = (r - r_0 + i\tau\omega_\nu)(r + r_0 - i\tau\omega_\nu) - \tau^2\omega_\nu^2 + 2i(r - r_0)\tau\omega_\nu.$$

Therefore, on the upper semicircle $|r - r_0| = \varepsilon$ we have for $\omega_\nu > 0$ and $|\tau\omega_\nu| < \varepsilon$:

$$|p| \geq \varepsilon|2r_0 - 2\varepsilon| - \varepsilon^2 - 2\varepsilon^2 \geq \varepsilon r_0.$$

Analogously, we estimate $|p|$ on the lower semicircle for $\omega_\nu < 0, |\tau\omega_\nu| < \varepsilon$. Thus, the estimate (29.14) holds for all $\tau > 0, k \geq 0, \sqrt{k^2 + \tau^2} > \varepsilon_0$. Note that (29.14) implies (29.17), and this completes the proof of Theorem 29.3.

Remark 29.1. We indicate the simplifications in the proof of Theorem 29.1 in comparison with the proofs of Theorems 29.2 and 29.3.

Let $\chi(r)$ be the same as in (29.8) for $\tau = 0$, and let $p = r^2 - k^2 - 2ik\tau$, where $\tau > 0, \tau \to 0$. Estimates for $I_1(\tau)$ and $I_3(\tau)$ are exactly the same as before, and we can pass to the limit as $\tau \to 0$. In the integral of the form (29.13) with $r^2 - k^2 - \tau^2 + 2i\tau(r\omega_\nu - \sigma)$ replaced by $p = r^2 - k^2 - 2ik\tau$ we change the integration from the real axis to Γ_-. Then $|p| \geq \varepsilon k$ on Γ_- without the lower semicircle $r - k = \varepsilon e^{i\varphi}$, $\pi \leq \varphi \leq 2\pi$. On the semicircle we have $|p| = |(r-k-i\tau)(r+k+i\tau)-\tau^2| \geq \varepsilon|2k-2\varepsilon|-\varepsilon^2 \geq \varepsilon k$ for $0<\tau<\varepsilon$. Therefore we can pass to the limit as $\tau \to 0$. □

30. Nonhomogeneous Schrödinger equation

Consider the nonhomogeneous Schrödinger equation

$$(-\Delta - k^2 + q(x))v = f(x), \tag{30.1}$$

where $q(x)$ is real-valued. We consider separately two classes of potentials:

a) $|q(x)| = O\Big(\frac{1}{(1+|x|)^{\frac{n+1}{2}+\alpha+\varepsilon}}\Big)$, $\alpha > 0$;

b) a more general class $|q(x)| = O\Big(\frac{1}{(1+|x|)^{1+\varepsilon}}\Big)$, $\varepsilon > 0$.

An equation of the form (30.1) arises with $f(x) = 0$ when we consider the solutions $w(x,t)$ of the nonstationary Schrödinger equation

$$-i\frac{\partial\psi(x,t)}{\partial t} - \Delta\psi + q(x)\psi(x,t) = 0 \tag{30.2}$$

of the form $\psi = e^{-ik^2t}v(x)$.

30.1. The case of $q(x) = O\big(\frac{1}{(1+|x|)^{\frac{n+1}{2}+\alpha+\varepsilon}}\big)$.

Denote by $L_{\infty,\alpha}$ the Banach space of Lebesgue measurable functions $f(x)$ with the norm $\|f\|_{\infty,\alpha} = \sup_{x\in\mathbb{R}^n}(1+|x|)^\alpha|f(x)|$.

We assume in (30.1) that $q(x) \in L_{\infty,\frac{n+1}{2}+\alpha+\varepsilon}$, $\varepsilon > 0$, $q(x)$ is real-valued, and $f(x) \in L_{\infty,n+\alpha}$, $\alpha > 0$.

We say that v is an outgoing solution of (30.1) if $v = E_+g$, where $g \in L_{\infty,n+\alpha}$.

Theorem 30.1. *Suppose $q(x) \in L_{\infty,\frac{n+1}{2}+\alpha+\varepsilon}$. Then for any $f \in L_{\infty,n+\alpha}$ there exists a unique outgoing solution $v \in L_{\infty,\frac{n-1}{2}}$ of (30.1).*

Proof: Substituting $v = E_+g$ into (30.1), we obtain the following integral equation for $g(x)$:

$$g(x) + qE_+g = f. \tag{30.3}$$

Set $Tg = qE_+g$. We show that T is a compact operator in $L_{\infty,n+\alpha}$. Indeed, it follows from Theorem 19.5 that

$$(1+|x|)^{\frac{n-1}{2}}|E_+g| \leq C\|g\|_{\infty,n+\alpha}. \tag{30.4}$$

Moreover

$$(1+|x|)^{\frac{n-1}{2}}\left|\frac{\partial}{\partial x}(E_+g)\right| \leq C\|g\|_{\infty,n+\alpha}, \tag{30.5}$$

and E_+g and $\frac{\partial}{\partial x}(E_+g)$ are continuous functions.

Let

$$T_N g = q(x)\chi\left(\frac{x}{N}\right) E_+ g, \tag{30.6}$$

where $\chi(x) \in C_0^\infty(\mathbb{R}^n)$, $\chi(x) = 1$ for $|x| < \frac{1}{2}$, and $\chi(x) = 0$ for $|x| > 1$.

If g_m is a bounded sequence in $L_{\infty,n+\alpha}$, then $\chi(\frac{x}{N})E_+g_m$ is bounded and equicontinuous in $C(\overline{B_N})$, where $C(\overline{B_N})$ is the space of continuous functions in the closed ball $\overline{B_N}$ (cf. (30.4), (30.5)).

By the Arzelà-Ascoli theorem, $\chi(\frac{x}{N})E_+g$ is a compact operator from $L_{\infty,n+\alpha}$ to $C(\overline{B_N})$ and, therefore, $T_N g = q(x)\chi(\frac{x}{N})E_+g$ is a compact operator in $L_{\infty,n+\alpha}$. Since $\|(T - T_N)g\|_{\infty,n+\alpha} \le \frac{C}{(1+N)^\varepsilon}\|g\|_{\infty,n+\alpha}$, we see that T is also a compact operator in $L_{\infty,n+\alpha}$ as the limit of compact operators in the operator norm.

By the Fredholm alternative, the equation (30.2) is uniquely solvable in $L_{\infty,n+\alpha}$ if and only if the homogeneous equation

$$h + q(x)E_+h = 0 \tag{30.7}$$

has only a trivial solution in $L_{\infty,n+\alpha}$.

We prove the uniqueness in the next section. □

30.2. Asymptotic behavior of outgoing solutions (the case of $q(x) = O(\frac{1}{(1+|x|)^{\frac{n+1}{2}+\alpha+\varepsilon}}), \alpha > 0$).

Theorem 30.2. *Let $v(x) \in L_{\infty,\frac{n-1}{2}}$ be the outgoing solution of (30.1) obtained in Theorem 30.1. Then*

$$v(x) = C_n \frac{e^{ik|x|}\tilde{g}(k\theta)}{|x|^{\frac{n-1}{2}}} + O\left(\frac{1}{(1+|x|)^{\frac{n-1}{2}+\alpha_1}}\right),$$

where C_n is the same as in (19.14), $0 < \alpha_1 < \frac{\alpha}{2}$, $\alpha_1 \le 1$, $g(x)$ is the solution of (30.3), $\theta = \frac{x}{|x|}$ is fixed, and $|x| \to \infty$.

The proof of Theorem 30.2 follows from the representation $u = E_+g$, $g \in L_{\infty,n+\alpha}$ of the outgoing solution and from Theorem 19.5. □

30.3. The case of $q(x) = O(\frac{1}{(1+|x|)^{1+\varepsilon}})$.

Now we study equation (30.1) for a larger class of potentials $q(x)$ and right hand sides $f(x)$.

Theorem 30.3. *Suppose $q(x) = O(\frac{1}{|x|^{1+\varepsilon}})$, $\varepsilon > 0$. Then for any $f \in V_{0,\frac{1}{2}+\varepsilon_1}$, there exists a unique outgoing solution $v \in V_{2,-\frac{1}{2}-\varepsilon_1}$ of (30.1), where $0 < \varepsilon_1 < \frac{\varepsilon}{2}$ is arbitrary.*

Proof: Here we call a solution v outgoing if $v = E_+g$, where $g \in V_{0,\frac{1}{2}+\varepsilon_1}$. Substituting into (30.1), we obtain an equation (30.2) for g. By Theorem 29.1, $Tg = qE_+g$ is bounded from $V_{0,\frac{1}{2}+\varepsilon_1}(\mathbb{R}^n)$ to $V_{2,\frac{1}{2}+\varepsilon_1}(\mathbb{R}^n)$, $0 < \varepsilon_1 < \frac{\varepsilon}{2}$, since $q(x) = O(\frac{1}{|x|^{1+\varepsilon}})$. Here $V_{m,s}(\mathbb{R}^n)$ is the weighted Sobolev space with the norm (29.1).

We show that Tg is a compact operator in $V_{0,\frac{1}{2}+\varepsilon_1}$.

Denote by Λ^r the following operator:

$$p\Lambda^r f = F^{-1}(1 + |\xi|^2)^{\frac{r}{2}} \tilde{f}(\xi).$$

It is clear that Λ^r maps $H_s(\mathbb{R}^n)$ to $H_{s-r}(\mathbb{R}^n)$ for any $s \in \mathbb{R}^1$. Let $\lambda^r(x) = (1 + |x|^2)^{\frac{r}{2}}$. We have

$$Tg = q(x)\lambda^{\frac{1}{2}+\varepsilon-\varepsilon_1} T_0 \Lambda^2 \lambda^{-\frac{1}{2}-\varepsilon_1} E_+g,$$

where

$$T_0 f = \lambda^{-(\varepsilon-2\varepsilon_1)} \Lambda^{-2} f.$$

Note that $\Lambda^2 = (-\Delta + 1)$. Therefore, applying $(-\Delta + 1)$ to $\lambda^{-\frac{1}{2}-\varepsilon_1} E_+g$ and using Theorem 29.1, we get that $\Lambda^2 \lambda^{-\frac{1}{2}-\varepsilon_1} E_+$ is bounded from $V_{0,\frac{1}{2}+\varepsilon_1}$ to $L_2(\mathbb{R}^n)$. Also $q(x)\lambda^{\frac{1}{2}+\varepsilon-\varepsilon_1} = O(|x|^{-\frac{1}{2}-\varepsilon_1})$ is bounded from $L_2(\mathbb{R}^n)$ to $V_{0,\frac{1}{2}+\varepsilon}(\mathbb{R}^n)$.

We prove that T_0 is a compact operator in $L_2(\mathbb{R}^n)$. Then T will be a compact operator in $V_{0,\frac{1}{2}+\varepsilon_1}$, since the product of a compact operator and bounded operators is a compact operator. Let

$$T_N f = \lambda^{-(\varepsilon-2\varepsilon_1)} \chi\Big(\frac{x}{N}\Big) \Lambda^{-2} \chi\Big(\frac{D}{N}\Big) f, \tag{30.8}$$

where $\chi(x) \in C_0^\infty(\mathbb{R}^n)$ is the same as in (30.6), $\chi(\frac{D}{N})f = F^{-1}\chi(\frac{\xi}{N})\tilde{f}(\xi)$. We have

$$(\widetilde{T_N f})(\xi) = \int_{\mathbb{R}^n} \tilde{a}(\xi - \eta)(1 + |\eta|^2)^{-1} \chi\Big(\frac{\eta}{N}\Big) \tilde{f}(\eta) d\eta, \tag{30.9}$$

where

$$\tilde{a}(\xi) = \frac{1}{(2\pi)^n} \int_{\mathbb{R}^n} \chi\Big(\frac{x}{N}\Big)(1 + |x|^2)^{-\frac{\varepsilon}{2}+\varepsilon_1} e^{-ix\cdot\xi} d\xi.$$

Since $|\tilde{a}(\xi)| \leq C_M(1 + |\xi|)^{-M}$ for any M, the operator (30.9) is a Hilbert-Schmidt operator, i.e.

$$\int_{\mathbb{R}^n} \int_{\mathbb{R}^n} \Big| \tilde{a}(\xi - \eta)(1 + |\eta|^2)^{-1} \chi\Big(\frac{\eta}{N}\Big) \Big|^2 d\xi d\eta < +\infty.$$

Since the Hilbert-Schmidt operators are compact in $L_2(\mathbb{R}^n)$, we see that the operator (30.9) and, consequently, T_N are compact operators in $L_2(\mathbb{R}^n)$. Since $|||T_0 - T_N||| \to 0$ as $N \to \infty$, we obtain that T_0 is compact. Here $|||T_0 - T_N|||$ is the operator norm of $T_0 - T_N$.

Therefore, by the Fredholm alternative, the unique solvability of (30.2) in $V_{0,\frac{1}{2}+\varepsilon_1}$ will follow from the fact that the homogeneous equation (30.7) has only a trivial solution in $V_{0,\frac{1}{2}+\varepsilon_1}$ provided $q(x) = O(\frac{1}{|x|^{1+\varepsilon}})$. We prove this in the next section.

31. The uniqueness of outgoing solutions

Theorem 31.1. *Let $h \in V_{0,\frac{1}{2}+\varepsilon_1}(\mathbb{R}^n)$ be a solution of the homogeneous equation*

$$h + q(x)E_+h = 0, \tag{31.1}$$

where $q(x) = O(\frac{1}{|x|^{1+\varepsilon}})$, $0 < \varepsilon_1 < \frac{\varepsilon}{2}$, $\overline{q(x)} = q(x)$. Then $h = 0$.

Note that if $v = E_+h$, then $h = (-\Delta - k^2)v$ and (31.1) becomes

$$(-\Delta - k^2 + q(x))v = 0, \tag{31.2}$$

i.e., v is an outgoing solution of the homogeneous equation (31.2).

Proof of Theorem 31.1. By Theorem 29.1, $v = E_+h \in V_{0,-\frac{1}{2}-\varepsilon_1}$. Taking the L_2 scalar product of (31.1) with $v = E_+h$, we get

$$(E_+h, h) + (E_+h, qE_+h) = 0. \tag{31.3}$$

Note that $(qE_+h, E_+h) = \int_{\mathbb{R}^n} q(x)|v(x)|^2 dx < +\infty$ and

$$(E_+h, h) = \frac{1}{(2\pi)^n} \int_{\mathbb{R}^n} \frac{|\tilde{h}(\xi)|^2 d\xi}{|\xi|^2 - k^2 - i0}.$$

By Example 5.4 we have:

$$\begin{aligned}(E_+h, h) &= \lim_{\varepsilon\to 0} \frac{1}{(2\pi)^n} \int_{\mathbb{R}^n} \frac{|\tilde{h}(\xi)|^2 d\xi}{|\xi|^2 - k^2 - i\varepsilon} \\ &= \lim_{\varepsilon\to 0} \frac{1}{(2\pi)^n} \int_0^\infty \frac{r^{n-1}}{r^2 - k^2 - i\varepsilon} \left(\int_{|\omega|=1} |\tilde{h}(r\omega)|^2 d\omega \right) dr \\ &= \text{p.v.} \frac{1}{(2\pi)^n} \int_0^\infty \frac{r^{n-1}}{r^2 - k^2} \left(\int_{|\omega|=1} |\tilde{h}(r\omega)|^2 d\omega \right) dr \\ &\qquad + \frac{i\pi k^{n-1}}{2k(2\pi)^n} \int_{|\omega|=1} |\tilde{h}(k\omega)|^2 d\omega,\end{aligned} \tag{31.4}$$

where

$$\text{p.v.}\int_0^\infty \frac{r^{n-1}}{r^2-k^2}\left(\int_{|\omega|=1}|\tilde{h}(r\omega)|^2d\omega\right)dr = \int_{|r-k|<\frac{k}{2}}\frac{\psi(r)-\psi(k)}{r-k}dr$$
$$+\int_{|r-k|>\frac{k}{2}}\frac{\psi(r)}{r-k}dr,\ \psi(r)=\frac{r^{n-1}}{r+k}\int_{|\omega|=1}|h(r\omega)|^2d\omega.$$

Since $h \in V_{0,\frac{1}{2}+\varepsilon_1}$, we see that $\int_{|\omega|=1}|\tilde{h}(r\omega)|^2d\omega$ is a Hölder function of r (see Theorem 13.6 and Remark 13.1). Therefore, the principal value integral exists.

Taking the imaginary part of (31.3), we obtain:

$$\int_{|\omega|=1}|\tilde{h}(k\omega)|^2d\omega = 0. \tag{31.5}$$

Therefore, $\tilde{h}(k\omega)=0$ almost everywhere.

Lemma 31.2. *Suppose $\tilde{h}(\xi)\in H_s(\mathbb{R}^n)$, $s>\frac{1}{2}$, and $\tilde{h}(k\omega)=0$. Then $\tilde{v}(\xi)=\frac{\tilde{h}(\xi)}{|\xi|^2-k^2}\in H_{s-1}(\mathbb{R}^n)$ and*

$$\|\tilde{v}\|_{s-1}\le C\|\tilde{h}\|_s. \tag{31.6}$$

Proof: Making a partition of unity and a change of variables, we can reduce the proof of Lemma 31.2 to the case where $\tilde{v}(\xi',\xi_n)=\frac{\tilde{h}(\xi',\xi_n)}{\xi_n}$, $\tilde{h}(\xi',0)=0$, and $\tilde{h}(\xi',\xi_n)$ has a compact support in $\mathbb{R}^n$. If $\tilde{h}(\xi)\in H_s(\mathbb{R}^n)$, $s>\frac{1}{2}$, and $\tilde{h}(\xi',0)=0$, then there exists a sequence $\tilde{\psi}_n(\xi)\in C^\infty(\mathbb{R}^n)\cap H_s(\mathbb{R}^n)$ such that $\tilde{\psi}_n(\xi',0)=0$ and $\|\tilde{\psi}_n-\tilde{h}\|_s\to 0$.

Indeed, by Theorem 13.2, there exists $\tilde{\varphi}_n(\xi)\in C_0^\infty(\mathbb{R}^n)$ such that $\tilde{\varphi}_n\to\tilde{h}$ in $H_s(\mathbb{R}^n)$. By Example 13.3, there exists $\tilde{\varphi}_{n1}(\xi)\in C^\infty(\mathbb{R}^n)\cap H_s(\mathbb{R}^n)$ such that $\tilde{\varphi}_{n1}(\xi',0)=\tilde{\varphi}_n(\xi',0)$ and $\|\tilde{\varphi}_{n1}\|_s^2\le C[\tilde{\varphi}_n(\xi',0)]_{s-\frac{1}{2}}$. Note that $[\tilde{\varphi}_n(\xi',0)]_{s-\frac{1}{2}}\to 0$, since $[\tilde{h}(\xi',0)-\tilde{\varphi}_n(\xi',0)]_{s-\frac{1}{2}}\le C\|\tilde{h}-\tilde{\varphi}_n\|_s$ and $\tilde{h}(\xi',0)=0$. Therefore, $\tilde{\psi}_n(\xi)=\tilde{\varphi}_n(\xi)-\tilde{\varphi}_{n1}(\xi)\to\tilde{h}(\xi)$ in $H_s(\mathbb{R}^n)$ and $\tilde{\psi}_n(\xi',0)=0$.

Thus, it is enough to prove (31.6) for $\tilde{h}(\xi)\in C^\infty(\mathbb{R}^n)\cap H_s(\mathbb{R}^n)$. We have

$$\tilde{h}(\xi',\xi_n)=\tilde{h}(\xi',0)+\int_0^{\xi_n}\frac{\partial\tilde{h}(\xi',\eta_n)}{\partial\eta_n}d\eta_n. \tag{31.7}$$

Take $\eta_n=\tau\xi_n$. Then $\tilde{v}(\xi',\xi_n)=\frac{\tilde{h}(\xi',\xi_n)}{\xi_n}=\int_0^1\frac{\partial}{\partial\eta_n}\tilde{h}(\xi',\tau\xi_n)d\tau$.

Let $v(x',x_n)$ be the inverse Fourier transform of $\tilde{v}(\xi',\xi_n)$. Note that

$$v(x',x_n)=\frac{1}{(2\pi)^n}\int_0^1\int_{\mathbb{R}^n}\frac{\partial\tilde{h}(\xi',\tau\xi_n)}{\partial\eta_n}e^{ix'\cdot\xi'+ix_n\xi_n}d\xi'd\xi_n\,d\tau. \tag{31.8}$$

Making again the change of variables $\eta_n = \tau\xi_n$ and integrating by parts with respect to η_n, we obtain

$$v(x', x_n) = \int_0^1 \frac{(-ix_n)}{\tau^2} h\Big(x', \frac{x_n}{\tau}\Big) d\tau. \tag{31.9}$$

Therefore, (31.8) and (31.9) imply that
(31.10)

$$\|\tilde{v}\|_{s-1} \le \int_0^1 \Big(\int_{\mathbb{R}^n} \Big|\frac{x_n}{\tau^2}\Big|^2 \Big|h\Big(x', \frac{x_n}{\tau}\Big)\Big|^2 (1 + |x'| + |x_n|)^{2(s-1)} dx' dx_n\Big)^{\frac{1}{2}} d\tau.$$

In (31.10) we have used that the norm of the integral (with respect to τ) is less than or equal to the integral of the norm.

Make the change of variables $x_n = \tau y_n$. Then
(31.11)

$$\|\tilde{v}\|_{s-1} \le \int_0^1 \Big(\int_{\mathbb{R}^n} |y_n|^2 |h(x', y_n)|^2 (1 + |x'| + \tau|y_n|)^{2(s-1)} \tau^{-1} dx' dy_n\Big)^{\frac{1}{2}} d\tau.$$

When $s \ge 1$, we have $(1 + |x'| + \tau|y_n|)^{2(s-1)} \le (1 + |x'| + |y_n|)^{2(s-1)}$ and, therefore, $\|\tilde{v}\|_{s-1} \le C\|\tilde{h}\|_s$ since $\int_0^1 \tau^{-\frac{1}{2}} d\tau < +\infty$.

When $\frac{1}{2} < s < 1$, we get

$$|y_n|^2 (1 + |x'| + \tau|y_n|)^{2(s-1)} \le \frac{|y_n|^2}{(\tau|y_n|)^{2(1-s)}} = \frac{|y_n|^{2s}}{\tau^{(2-2s)}}.$$

Therefore, $\|\tilde{v}\|_{s-1} \le C\|\tilde{h}\|_s$ since $\int_0^1 \frac{d\tau}{\tau^{\frac{3}{2}-s}} < +\infty$. □

Lemma 31.3. *Let* $v = E_+ h$, $h \in V_{0,\frac{1}{2}+\varepsilon_1}$, *is an outgoing solution of the homogeneous equation* (31.2). *Then for all* N_1,

$$\int_{\mathbb{R}^n} (1 + |x|)^{2N_1} |v(x)|^2 dx < +\infty. \tag{31.12}$$

Proof: It follows from (31.2) that

$$h(x) = -q(x)v(x),$$

where $\tilde{v}(\xi) = \frac{\tilde{h}(\xi)}{|\xi|^2 - k^2}$, $\tilde{h}(k\omega) = 0$. Since $\tilde{h}(\xi) \in H_{s_0}(\mathbb{R}^n)$, where $s_0 = \frac{1}{2} + \varepsilon_1$, we get by Lemma 31.2 that $\tilde{v} \in H_{s_0 - 1}(\mathbb{R}^n)$. Since $q(x) = O(\frac{1}{|x|^{1+\varepsilon}})$, we have $h = -qv \in V_{0,1+\varepsilon+s_0-1} = V_{0,s_0+\varepsilon}$. We see that $\tilde{h} \in H_{s_0+\varepsilon}(\mathbb{R}^n)$, provided $\tilde{h} \in H_{s_0}(\mathbb{R}^n)$. Repeating this argument, we conclude that $\tilde{h} \in H_{s_0+N\varepsilon}(\mathbb{R}^n)$ for all N. Therefore, $\tilde{v} \in H_{N_1}(\mathbb{R}^n)$ for any N_1, i.e., $v \in V_{0,N_1}(\mathbb{R}^n)$. □

The last step in the proof of Theorem 31.1 is the following lemma:

Lemma 31.4. *Suppose* $(-\Delta - k^2 + q(x))v = 0$, $q(x) = O(\frac{1}{|x|^{1+\varepsilon}})$, *and* $v(x)$ *satisfies* (31.12) *for every* N_1. *Then* $v = 0$.

Proof: We prove for an arbitrary N the following estimate:

$$4k^2 N \int_{\mathbb{R}^n} |x|^N |v(x)|^2 dx \le \int_{\mathbb{R}^n} |x|^{N+2} |(\Delta + k^2)v|^2 dx, \tag{31.13}$$

provided $v(x)$ satisfies (31.12) for any N_1.

The weighted estimates of the form (31.13) are called Carleman's estimates.

Introducing polar coordinates, we have

$$(\Delta + k^2)v = \left(\frac{\partial^2}{\partial r^2} + \frac{n-1}{r}\frac{\partial}{\partial r} + \frac{1}{r^2}\Delta_\omega + k^2\right) v(r\omega),$$

where $r = |x|$, $\omega = \frac{x}{|x|}$, and Δ_ω is the Laplace-Beltrami operator on the unit sphere S^{n-1}. When $r = e^t$, we get

$$(\Delta + k^2)v = e^{-2t}\left(\frac{\partial}{\partial t^2} + (n-2)\frac{\partial}{\partial t} + \Delta_\omega + k^2 e^{2t}\right) v(e^t\omega).$$

Therefore

$$\begin{aligned} &\int_{\mathbb{R}^n} |x|^{2N+2} |(\Delta + k^2)v|^2 \\ &= \int_{-\infty}^{\infty} \int_{|\omega|=1} e^{(2N+2)t} \left| \left(\frac{\partial^2}{\partial t^2} + (n-2)\frac{\partial}{\partial t} + \Delta_\omega + k^2 e^{2t}\right) v\right|^2 e^{-4t} e^{nt} dt d\omega. \end{aligned} \tag{31.14}$$

Set $w(t,\omega) = e^{(N+\frac{n}{2}-1)t} v(e^t\omega)$, i.e., $v = e^{-(N+\frac{n}{2}-1)t} w$. Then

$$\begin{aligned} &\left(\frac{\partial^2}{\partial t^2} + (n-2)\frac{\partial}{\partial t} + \Delta_\omega + k^2 e^{2t}\right) v \\ &\quad = e^{-(N+\frac{n}{2}-1)t}\left(\frac{\partial^2}{\partial t^2} + (n-2)\frac{\partial}{\partial t} + \Delta_\omega + k^2 e^{2t}\right. \\ &\quad \left. + \left(N + \frac{n}{2} - 1\right)^2 - 2\left(N + \frac{n}{2} - 1\right)\frac{\partial}{\partial t} - (n-2)\left(N + \frac{n}{2} - 1\right)\right) w \\ &\quad = e^{-(N+\frac{n}{2}-1)t}(L_1 + L_2)w, \end{aligned} \tag{31.15}$$

where

$$L_1 = \frac{\partial^2}{\partial t^2} + \left(N + \frac{n}{2} - 1\right)\left(N - \frac{n}{2} + 1\right) + \Delta_\omega + k^2 e^{2t}, \tag{31.16}$$

$$L_2 = -2N\frac{\partial}{\partial t}. \tag{31.17}$$

It follows from (31.14)–(31.17) that

$$\int_{\mathbb{R}^n} |x|^{2N+2}|(\Delta + k^2)v|^2 dx = \int_{-\infty}^{\infty}\int_{|\omega|=1} |L_1 w + L_2 w|^2 dt d\omega$$

$$= \int_{-\infty}^{\infty}\int_{|\omega|=1} (|L_1 w|^2 + |L_2 w|^2 + (L_1 w)\overline{L_2 w} + (L_2 w)\overline{L_1 w}) dt d\omega$$

$$= I_1 + I_2 + I_3 + I_4.$$

Integrating I_3 and I_4 by parts, we get

$$I_3 = -\int_{-\infty}^{\infty}\int_{|\omega|=1} (L_2 L_1 w)\bar{w} dt d\omega,$$

$$I_4 = \int_{\infty}^{\infty}\int_{|\omega|=1} (L_1 L_2 w)\bar{w} dt d\omega.$$

Note that $L_2 L_1 w = L_1 L_2 w - 4Nk^2 e^{2t} w$.

Therefore, since $I_1 \geq 0, I_2 \geq 0$, we get

$$\int_{\mathbb{R}^n} |x|^{2N+2}|(\Delta + k^2)v|^2 dx \geq 4Nk^2 \int_{-\infty}^{\infty} e^{2t}|w|^2 dt d\omega$$

$$= 4Nk^2 \int_{-\infty}^{\infty} e^{(2N+n)t}|v|^2 dt d\omega = 4Nk^2 \int_{\mathbb{R}^n} |x|^{2N}|v(x)|^2 dx. \quad \square$$

We are now ready to finish the proof of Lemma 31.3. We have

$$(-\Delta - k^2)v = -q(x)v,$$

where v satisfies (31.12) for any N_1 and $|q(x)| \leq C(1+|x|)^{-1-\varepsilon}$. Applying estimate (31.13), we get

$$4k^2 N \int_{\mathbb{R}^n} |x|^{2N}|v|^2 dx \leq \int_{\mathbb{R}^n} |q(x)|^2|v|^2|x|^{2N+2} dx \leq C\int_{\mathbb{R}^n} |v|^2|x|^{2N} dx.$$

Choosing N such that $4k^2 N > C$, we conclude that $v = 0$. Therefore, the integral equation (31.1) has only a trivial solution in $V_{0,\frac{1}{2}+\varepsilon_1}(\mathbb{R}^n)$, and this completes the proof of Theorem 31.1 and, consequently, of Theorem 30.3. Theorem 31.1 obviously implies that (30.6) has only a trivial solution in $L_{\infty,n+\alpha}$, and so Theorem 30.1 holds as well.

31.1. Absence of discrete spectrum for $k^2 > 0$.

Similarly to the proof of Theorem 31.1 we can prove the following theorem:

Theorem 31.5. *Let $q(x) = O(\frac{1}{1+|x|^{1+\varepsilon}})$. If $v \in L_2(\mathbb{R}^n)$ is a solution of the homogeneous Schrödinger equation (31.2), then $v = 0$.*

This result means that no $k^2 > 0$ is an eigenvalue for the operator $-\Delta + q(x)$.

Proof of Theorem 31.5. Denote $h = (-\Delta - k^2)v$. Then $h = -q(x)v(x)$ and, therefore, $h \in V_{0,1+\varepsilon}(\mathbb{R}^n)$ and $\tilde{v}(\xi) = \frac{\tilde{h}(\xi)}{|\xi|^2-k^2}$. Since $\tilde{v}(\xi) \in L_2(\mathbb{R}^n)$ and $\tilde{h}(r\omega)$ is continuous in r with values in $V_{\frac{1}{2}+\varepsilon}(S^{n-1})$ (cf. Remark 13.1), we have that $\tilde{h}(k\omega) = 0$. Applying Lemma 31.2, we get, as in Lemma 31.3, that $v(x)$ satisfies estimates (31.12). Finally, Lemma 31.4 implies that $v = 0$. □

31.2. Existence of outgoing fundamental solution (the case of $q(x) = O(\frac{1}{(1+|x|)^{\frac{n+1}{2}+\delta}})$).

Theorem 31.6. *Let $q(x) \in L_{\infty,\frac{n+1}{2}+\delta}$, $\delta > 0$. There exists a fundamental solution $G(x,y)$ in $\mathbb{R}^n$:*

$$(31.18) \qquad (-\Delta - k^2 + q(x))G(x,y) = \delta(x-y),$$

satisfying the radiation conditions

$$(31.19) \qquad G(x,y) = O\left(\frac{1}{r^{\frac{n-1}{2}}}\right), \quad \frac{\partial G(x,y)}{\partial r} - ikG(x,y) = o\left(\frac{1}{r^{\frac{n-1}{2}}}\right),$$

where $r = |x-y|$, y is fixed, and $|x| \to \infty$.

Proof: We look for $G(x,y)$ in the form $G(x,y) = E_+(x-y) + w(x,y)$. Then we have

$$(31.20) \qquad (-\Delta - k^2 + q(x))w(x,y) = -q(x)E_+(x-y).$$

Note that $q(x)E_+(x-y) = O(\frac{1}{|x|^{n+1+\delta}})$ as $|x| \to \infty$. However, $q(x)E_+(x-y) \notin L_{\infty,n+1+\delta}$ because of the singularity at $x = y$.

Let $w(x,y) = -E_+q(x)E_+(x-y) + w_1(x,y)$. Then $w_1(x,y)$ satisfies the equation

$$(31.21) \qquad (-\Delta - k^2 + q(x))w_1 = q(x)E_+qE_+(x-y),$$

where $qE_+qE_+(x-y) \in L_{\infty,n+\delta}$ in x. Therefore, by Theorem 30.1 there exists an outgoing solution $w_1(x,y)$ of equation (31.21). Then $G(x,y) = E_+(x-y) - E_+qE_+(x-y) + w_1(x,y)$ will be an outgoing fundamental solution.

Note that $G(x,y) - E_+(x-y)$ is a continuous function in x and y. □

Theorem 31.7. *Let $v(x)$ be the solution in $\mathbb{R}^n$ of*

$$(-\Delta - k^2 + q(x))v(x) = 0,$$

where $q \in L_{\infty,\frac{n+1}{2}+\delta}$. If $v(x)$ satisfies the radiation conditions (19.45), then $v = 0$.

The proof of Theorem 31.7 is exactly the same as the proof of the uniqueness for the Helmholtz equation $(-\Delta - k^2)u = 0$. One needs only to replace E_+ by the outgoing fundamental solution $G(x, y)$.

Remark 31.1. Let $q(x) = O(\frac{1}{(1+|x|)^{1+\varepsilon}})$. It follows from Theorem 30.3 that for any $f \in V_{0,\frac{1}{2}+\varepsilon_1}$, $0 < \varepsilon_1 < \frac{\varepsilon}{2}$, equation (30.1) has a unique outgoing solution $v \in V_{2,-\frac{1}{2}-\varepsilon_1}$. Denote by G the operator $v = Gf$, i.e.,

$$(-\Delta - k^2 + q(x))Gf = f, \quad \forall f \in V_{0,\frac{1}{2}+\varepsilon_1}. \tag{31.22}$$

For $q = O(\frac{1}{|x|^{\frac{n+1}{2}+\delta}})$, the operator G is an integral operator with the kernel $G(x, y)$ (cf. (31.18)).

32. The limiting absorption principle

Consider the equation:

$$(-\Delta + q(x) - k^2 - i\varepsilon)u_\varepsilon = f, \tag{32.1}$$

where $\varepsilon > 0$, $q(x) = O(\frac{1}{(1+|x|)^{1+\delta}})$ is real-valued, $f \in H_{0,s}(\mathbb{R}^n)$, and $0 < s - \frac{1}{2} < \frac{\delta}{2}$.

Theorem 32.1. *Equation* (32.1) *has a unique solution* $u_\varepsilon \in H_2(\mathbb{R}^n)$ *for any* $f \in L_2(\mathbb{R}^n)$. *If* $f \in V_{0,s}(\mathbb{R}^n)$, *then there exists* $u_0 \in V_{2,-s}$ *such that*

$$\|u_\varepsilon - u_0\|_{2,-s} \to 0 \quad \text{as} \quad \varepsilon \to 0.$$

The function u_0 *satisfies*

$$(-\Delta + q(x) - k^2)u_0 = f \tag{32.2}$$

and is an outgoing solution of (32.2), *i.e.,* $u_0 = E_+g_0$ *for some* $g_0 \in V_{0,s}(\mathbb{R}^n)$.

Proof: We look for u_ε in the form

$$u_\varepsilon = E_{+,\varepsilon}g_\varepsilon = \frac{1}{(2\pi)^n}\int_{\mathbb{R}^n} \frac{e^{ix\cdot\xi}\tilde{g}_\varepsilon(\xi)}{|\xi|^2 - k^2 - i\varepsilon}d\xi, \tag{32.3}$$

where $g_\varepsilon \in L_2(\mathbb{R}^n)$. If $g_\varepsilon \in L_2$, we have that $u_\varepsilon \in H_2(\mathbb{R}^n)$, since $\tilde{u}_\varepsilon(\xi) = \frac{\tilde{g}_\varepsilon(\xi)}{|\xi|^2-k^2-i\varepsilon}$ and $\int_{\mathbb{R}^n}(1 + |\xi|)^4|\tilde{u}_\varepsilon(\xi)|^2 d\xi \le C_\varepsilon \int_{\mathbb{R}^n} |\tilde{g}_\varepsilon(\xi)|^2 d\xi$.

Substituting $u_\varepsilon = E_{+,\varepsilon}g_\varepsilon$ in (32.1), we obtain an integral equation

$$g_\varepsilon + q(x)E_{+,\varepsilon}g_\varepsilon = f. \tag{32.4}$$

The operator $E_{+,\varepsilon}$ is bounded from $L_2(\mathbb{R}^n)$ to $H_2(\mathbb{R}^n)$. Since $q(x) \to 0$ as $|x| \to \infty$, we see, as in §30, that $qE_{+,\varepsilon}$ is a compact operator in $L_2(\mathbb{R}^n)$.

Consider the homogeneous equation

$$h_\varepsilon + q(x)E_{+,\varepsilon}h_\varepsilon = 0. \tag{32.5}$$

If $h_\varepsilon \in L_2(\mathbb{R}^n)$ is a solution of (32.5), then $v_\varepsilon = E_\varepsilon h_\varepsilon$ belongs to $H_2(\mathbb{R}^n)$ and is a solution of

$$(32.6) \qquad (-\Delta - k^2 - i\varepsilon)v_\varepsilon + q(x)v_\varepsilon = 0.$$

We show that $v_\varepsilon = 0$. Indeed, taking the L_2 scalar product of (32.6) with v_ε, we obtain

$$(32.7) \qquad ((-\Delta - k^2 - i\varepsilon)v_\varepsilon, v_\varepsilon) + (q(x)v_\varepsilon, v_\varepsilon) = 0.$$

We have that

$$(q(x)v_\varepsilon, v_\varepsilon) = \int_{\mathbb{R}} q(x)|v_\varepsilon(x)|^2 dx$$

is real-valued since $q(x) = \overline{q(x)}$ and

$$((-\Delta - k^2 - i\varepsilon)v_\varepsilon, v_\varepsilon) = \frac{1}{(2\pi)^n}\int_{R^n} (|\xi|^2 - k^2 - i\varepsilon)|\tilde{v}(\xi)|^2 d\xi.$$

Therefore, taking the imaginary part of (32.7), we get $-\varepsilon \int_{\mathbb{R}^n} |v_\varepsilon|^2 dx = 0$, i.e., $v_\varepsilon = 0$. Therefore $h_\varepsilon = (\Delta - k^2 - i\varepsilon)v_\varepsilon = 0$. Since the homogeneous equation (32.5) has only a trivial solution, the Fredholm alternative implies that (32.4) has a unique solution in $L_2(\mathbb{R}^n)$ for any $f \in L_2(\mathbb{R}^n)$. Therefore, (32.1) has a unique solution in $H_2(\mathbb{R}^n)$ for any $f \in L_2(\mathbb{R}^n)$.

Suppose now that $f \in V_{0,s}(\mathbb{R}^n)$, $s > \frac{1}{2}$.

Lemma 32.2. *The operator $E_{+,\varepsilon}$ converges to E_+ as $\varepsilon \to 0$, in the norm of bounded operators from $V_{0,s}$ to $V_{0,-s}$.*

Proof: For $h, g \in V_{0,s}(\mathbb{R}^n)$, we have, as in the proof of Theorem 29.2:

$$\begin{aligned}(E_{+,\varepsilon}h, g) &= \frac{1}{(2\pi)^n}\int_{\mathbb{R}^n} \frac{\tilde{h}(\xi)\overline{\tilde{g}(\xi)}d\xi}{|\xi|^2 - k^2 - i\varepsilon} \\ &= \frac{1}{(2\pi)^n}\int_{\mathbb{R}^n} \frac{\tilde{h}(\xi)\tilde{g}(\xi) - \tilde{h}(k\omega)\overline{\tilde{g}(k\omega)}}{|\xi|^2 - k^2 - i\varepsilon}\chi^2(|\xi|)d\xi \\ + \frac{1}{(2\pi)^n}\int_{\mathbb{R}^n} &\frac{\chi^2(|\xi|)\tilde{h}(k\omega)\overline{\tilde{g}(k\omega)}}{|\xi|^2 - k^2 - i\varepsilon}d\xi + \frac{1}{(2\pi)^n}\int_{\mathbb{R}^n} \frac{(1 - \chi^2(|\xi|))\tilde{h}(\xi)\overline{\tilde{g}(\xi)}}{|\xi|^2 - k^2 - i\varepsilon}d\xi \\ &\overset{\text{def}}{=} I_1(\varepsilon) + I_2(\varepsilon) + I_3(\varepsilon),\end{aligned}$$

where $\chi(|\xi|)$ is the same as in (29.8) with $\tau = 0$.

If $\varepsilon = 0$, we have

$$(E_+h, g) = \sum_{m=1}^{3} I_m(0),$$

where $I_1(0)$ and $I_3(0)$ are obtained by letting $\varepsilon = 0$ in $I_1(\varepsilon)$ and $I_3(\varepsilon)$, and

$$I_2(0) = \lim_{\varepsilon\to 0} I_2(\varepsilon) = \lim_{\varepsilon\to 0} a(\varepsilon)\int_{|\omega|=1} \tilde{h}(k\omega)\overline{\tilde{g}(k\omega)}d\omega,$$

where

$$a(\varepsilon) = \frac{1}{(2\pi)^n} \int_0^\infty \frac{\chi^2(r) r^{n-1}}{r^2 - k^2 - i\varepsilon} dr. \tag{32.8}$$

As in the proof of Theorem 29.2 and in Remark 29.1 we have, using the Cauchy theorem:

$$a(\varepsilon) = \frac{1}{(2\pi)^n} \int_{\Gamma_-} \frac{\chi^2(r) r^{n-1} dr}{r^2 - k^2 - i\varepsilon}, \tag{32.9}$$

where Γ_- is a contour that coincides with the positive real axis when $|r-k| > \delta$ and with the lower semicircle $|r - k| = \delta$; δ is small. Therefore, $a(\varepsilon)$ is continuous for $\varepsilon \geq 0$ and $|a(\varepsilon) - a(0)| < C\varepsilon$.

We have

$$|I_3(\varepsilon) - I_3(0)| \leq C\varepsilon \|h\|_0 \|g\|_0 \leq C\varepsilon \|h\|_{0,s} \|g\|_{0,s}, \tag{32.10}$$

$$\begin{aligned} |I_2(\varepsilon) - I_2(0)| \leq |a(\varepsilon) - a(0)| \int_{|\omega|=1} |\tilde{h}(k\omega)| |\tilde{g}(k\omega)| d\omega \\ \leq C\varepsilon \|h\|_{0,s} \|g\|_{0,s}. \end{aligned} \tag{32.11}$$

Finally,

$$\begin{aligned} \left| \frac{1}{|\xi|^2 - k^2 - i\varepsilon} - \frac{1}{|\xi|^2 - k^2} \right| &= \frac{\varepsilon}{||\xi|^2 - k^2 - i\varepsilon| |\xi|^2 - k^2|} \\ &\leq \frac{C\varepsilon^\alpha}{||\xi|^2 - k^2|^{1+\alpha}} \leq \frac{C\varepsilon^\alpha}{||\xi| - k|^{1+\alpha}} \end{aligned}$$

for any $0 < \alpha < 1$, where $|\xi| \in \operatorname{supp} \chi(|\xi|)$. Therefore,

$$|I_1(\varepsilon) - I_1(0)| \leq C\varepsilon^\alpha \int_0^\infty \int_{|\omega|=1} \frac{\chi^2(r) |\tilde{h}(r\omega)\overline{\tilde{g}(r\omega)} - \tilde{h}(k\omega)\overline{\tilde{g}(k\omega)}| r^{n-1}}{|r-k|^{1+\alpha}} dr d\omega.$$

Estimating as in (29.21) with $2\alpha < s - \frac{1}{2}$, $2\alpha \leq 1$, we get:

$$|I_1(\varepsilon) - I_1(0)| \leq C\varepsilon^\alpha \int_0^\infty \frac{\chi^2(r) r^{n-1}}{|r-k|^{1-\alpha}} dr \|\tilde{h}\|_s \|\tilde{g}\|_s. \tag{32.12}$$

Therefore

$$|((E_{+,\varepsilon} - E_+)h, g)| \leq C(\varepsilon) \|h\|_{0,s} \|g\|_{0,s}, \tag{32.13}$$

where $C(\varepsilon) \to 0$ as $\varepsilon \to 0$.

Estimate (32.13) implies that

$$|||E_{+,\varepsilon} - E_+||| \leq C(\varepsilon) \to 0 \quad \text{as} \quad \varepsilon \to 0, \tag{32.14}$$

where $|||L|||$ is the norm of an operator from $V_{0,s}$ to $V_{0,-s}$. Since $q(x) = O(\frac{1}{(1+|x|)^{1+\delta}})$ and since we assume that $0 < s - \frac{1}{2} < \delta$, we conclude that

$q(E_{+,\varepsilon}-E_+)$ is a bounded operator from $V_{0,s}$ to $V_{0,s}$ and $|||q(E_{+,\varepsilon}-E_+)||| \to 0$ as $\varepsilon \to 0$.

Denote by $g_0 \in V_{0,s}$ the unique solution of

$$g_0 + q(x)E_+ g_0 = f. \tag{32.15}$$

We shall show that $\|g_\varepsilon - g_0\|_{0,s} \to 0$ as $\varepsilon \to 0$, where g_ε is the solution of (32.4). Subtracting (32.15) from (32.4), we get

$$(g_\varepsilon - g_0) + qE_+(g_\varepsilon - g_0) = -q(E_{+,\varepsilon} - E_+)g_\varepsilon.$$

Since $|||q(E_{+,\varepsilon}-E_+)||| \to 0$ as $\varepsilon \to 0$, we conclude that the norm of $(I+qE_+)^{-1}q(E_{+,\varepsilon}-E_+)$ is less than $C_1(\varepsilon)$, where $C_1(\varepsilon) \to 0$ as $\varepsilon \to 0$. We get:

$$\|g_\varepsilon - g_0\|_{0,s} \le C_1(\varepsilon)\|g_\varepsilon\|_{0,s} \le C_1(\varepsilon)\|g_\varepsilon - g_0\|_{0,s} + C_1(\varepsilon)\|g_0\|_{0,s}.$$

Therefore

$$\|g_\varepsilon - g_0\|_{0,s} \le \frac{C_1(\varepsilon)}{1-C_1(\varepsilon)}\|g_0\|_s \to 0 \quad \text{as} \quad \varepsilon \to 0.$$

Since $g_\varepsilon \to g_0$ in $V_{0,s}$, we have that $u_\varepsilon \to u_0 = E_+ g_0$ in $V_{2,-s}$. Indeed,

$$u_\varepsilon - u_0 = E_{+,\varepsilon}g_\varepsilon - E_+ g_0 = (E_{+,\varepsilon} - E_+)g_\varepsilon + E_+(g_\varepsilon - g_0) \to 0 \quad \text{in} \quad V_{2,-s},$$

since $\|g_\varepsilon\|_{0,s} \le C$, $\|g_\varepsilon - g_0\|_{0,s} \to 0$, and the norm of $E_{+,\varepsilon} - E_+$ as of an operator from $V_{0,s}$ to $V_{2,-s}$ tends to 0 as $\varepsilon \to 0$.

Note that $u_0 = E_0 g_0$ is an outgoing solution of (32.2). □

The way to choose a solution of (32.2) by considering the perturbed equation (32.1) and taking the limit as $\varepsilon \to 0$ is called the limiting absorption principle.

33. The scattering problem

33.1. The scattering problem (the case of $q(x) = O(\frac{1}{(1+|x|)^{n+\alpha}})$).

The Schrödinger equation

$$(-\Delta + q(x))u = k^2 u \tag{33.1}$$

describes an electron with potential $q(x)$. When $q(x) \equiv 0$, i.e., the electron is "free", the solutions of (33.1) are plane waves $e^{ik\omega\cdot x}$, where the ω are unit vectors. When $q(x) \neq 0$, the solution of the form $w = e^{ik\omega\cdot x} + v(x, k\omega)$ is called a distorted plane wave. Here v satisfies the nonhomogeneous equation

$$(-\Delta + q(x) - k^2)v = -q(x)e^{ik\omega\cdot x}. \tag{33.2}$$

We assume that $q(x) = O(\frac{1}{|x|^{n+\alpha}})$, and we look for the outgoing solution $v = E_+ h$ of (33.2). By Theorem 30.1, there exists a unique outgoing solution of

(33.2), where $h(x,k\omega)=O(\frac{1}{|x|^{n+\alpha}})$. Therefore, we have proven the existence of the distorted plane waves $w(x,k\omega,k)$ for any $\omega\in S^{n-1}$.

By Theorem 19.5, $v(x,k\omega)$ has the following asymptotics as $|x|\to\infty$ and $\theta=\frac{x}{|x|}$ is fixed:

$$v(x,k\omega,k)=C_n\frac{e^{ik|x|}}{|x|^{\frac{n-1}{2}}}\tilde{h}(k\theta,k\omega)+O\left(\frac{1}{|x|^{\frac{n-1}{2}+\alpha'}}\right), \tag{33.3}$$

where

$$\tilde{h}(k\theta,k\omega)=\int_{\mathbb{R}^n}h(x,k\omega)e^{-ik\theta\cdot x}dx,\ 0<\alpha'<\frac{\alpha}{2},\ \alpha'\le 1, \tag{33.4}$$

$$C_n=\frac{k^{\frac{n-3}{2}}e^{-i\frac{\pi}{4}(n-3)}}{2^{\frac{n+1}{2}}\pi^{\frac{n-1}{2}}}.$$

The function $a(\theta,\omega,k)=C_n\tilde{h}(k\theta,k\omega)$ is called the scattering amplitude.

Therefore, the following theorem holds:

Theorem 33.1. *For any $\omega\in S^{n-1}$ the distorted plane wave $u(x,k\omega,k)$ exists and has the following form as $|x|\to\infty$:*

$$w(x,k\omega)=e^{ik\omega\cdot x}+a(\theta,\omega,k)\frac{e^{ik|x|}}{|x|^{\frac{n-1}{2}}}+O\left(\frac{1}{|x|^{\frac{n-1}{2}+\alpha'}}\right),\ \theta=\frac{x}{|x|}. \quad \square \tag{33.5}$$

Theorem 33.2 (The case of $q(x)=O(\frac{1}{(1+|x|)^{n+\alpha}})$). *Let $G(x,y)$ be an outgoing fundamental solution, and let $w(x,k\omega)$ be a distorted plane wave. Then, as $|x|\to\infty$, $\theta=\frac{x}{|x|}$, and the y are fixed, we have:*

$$G(x,y)=C_n\frac{e^{ik|x|}w(y,-k\theta)}{|x|^{\frac{n-1}{2}}}+O\left(\frac{1}{|x|^{\frac{n-1}{2}+\alpha_1}}\right). \tag{33.6}$$

Proof: It follows from (33.2) that $v(x,k\omega)=-Gqe^{-ik\omega\cdot y}$, where G is the integral operator with the kernel $G(x,y)$. Therefore

$$w(x,k\omega)=e^{ik\omega\cdot x}-\int_{\mathbb{R}^n}G(x,y)q(y)e^{ik\omega\cdot y}dy. \tag{33.7}$$

We have two representations of the outgoing solution of the nonhomogeneous equation

$$(-\Delta-k^2+q(x))u=f,\ \ f\in C_0^\infty(\mathbb{R}^n):$$

1) $u=Gf$;

2) $u=E_+h$, where $h(x)$ satisfies $h+qE_+h=f$.

It follows from Theorem 30.2 that

$$u=Gf=E_+h=C_n\frac{e^{ik|x|}}{|x|^{\frac{n-1}{2}}}\tilde{h}(k\theta)+O\left(\frac{1}{|x|^{\frac{n-1}{2}+\alpha_1}}\right), \tag{33.8}$$

where

$$\tilde{h}(k\theta) = \int_{\mathbb{R}^n} h(x)e^{-ikx\cdot\theta}dx. \tag{33.9}$$

We have $h = f - qE_+h = f - qGf$, since $E_+h = Gf$. Therefore

$$\tilde{h}(k\theta) = \int_{\mathbb{R}^n} f(y)e^{-ik\theta\cdot y}dy - \int_{\mathbb{R}^n}\int_{\mathbb{R}^n} e^{-ik\theta\cdot x}q(x)G(x,y)f(y)dydx.$$

Note that $G(x, y) = G(y, x)$. Therefore

$$\tilde{h}(k\theta) = \int_{\mathbb{R}^n} f(y)\left[e^{-ik\theta\cdot y} - \int_{\mathbb{R}^n} G(y,x)q(x)e^{-ik\theta\cdot x}dx\right]dy. \tag{33.10}$$

Substituting (33.7) into (33.10) we get

$$\tilde{h}(k\theta) = \int_{\mathbb{R}^n} f(y)w(y, -k\theta)dy. \tag{33.11}$$

Therefore (33.8) is equivalent to (33.6), since $f(x) \in C_0^\infty$ is arbitrary.

33.2. Inverse scattering problem (the case of $q(x) = O(\frac{1}{(1+|x|^{n+\alpha}})$).

The inverse scattering problem consists of the recovery of the potential $q(x)$ from the scattering amplitude $a(\theta, \omega, k)$.

Note that the inverse scattering problem is overdetermined for $n \geq 2$, since $q(x)$ is a function of n variables and $a(\theta, \omega, k)$ is a function of $n - 1 + n - 1 + 1 = 2n - 1$ variables.

Theorem 33.3. *Suppose $q(x) = O(\frac{1}{|x|^{n+\alpha}})$. Knowing $a(\theta, \omega, k)$ for all $|\theta| = 1,\ |\omega| = 1,\ k > 0$, we can recover $q(x)$.*

Proof: We substitute $v = E_+h$ into (33.2) and get

$$h + q(x)E_+h = -q(x)e^{ik\omega\cdot x}. \tag{33.12}$$

Consider equation (33.12) in $V_{0,\frac{1}{2}+\alpha}$, $0 < \alpha \leq 1$. Then $E_+h \in V_{0,-\frac{1}{2}-\alpha}$ and $Th = qE_+h \in V_{0,n-\frac{1}{2}} \subset V_{0,\frac{1}{2}+\alpha}$. By Theorem 29.1,

$$\|qE_+h\|_{0,\frac{1}{2}+\alpha} \leq \frac{C}{k}\|h\|_{0,\frac{1}{2}+\alpha}, \tag{33.13}$$

where $k \geq 1$ and C is independent of k.

Therefore, the operator $I + T$ is invertible in $V_{0,\frac{1}{2}+\alpha}$ for large k, and the norm of $(I + T)^{-1}$ is uniformly bounded for large k. Since $\|qe^{ik\omega\cdot x}\|_{0,\frac{1}{2}+\alpha} = \|q\|_{0,\frac{1}{2}+\alpha}$, we obtain that $\|h\|_{0,\frac{1}{2}+\alpha}$ is bounded as $k \to \infty$ uniformly in ω.

Take the Fourier transform of (33.12):

$$\tilde{h}(\xi, k\omega) + F(qE_+h) = -\tilde{q}(\xi - k\omega), \tag{33.14}$$

where F is the operator of the Fourier transform. We have, using the Cauchy-Schwartz inequality and the estimate (29.2),

$$\begin{aligned}(33.15)\quad \int_{\mathbb{R}^n} |q(x)||E_+h|dx &\le C\int_{\mathbb{R}^n}(1+|x|)^{-n-\alpha}|E_+h|dx \\ &\le C\left(\int_{\mathbb{R}^n}(1+|x|)^{-2n-2\alpha+1+2\alpha}dx\right)^{\frac{1}{2}}\|E_+h\|_{0,-\frac{1}{2}-\alpha}\le\frac{C}{k}.\end{aligned}$$

Therefore

$$(33.16)\qquad |F(qE_+h)|\le\int_{\mathbb{R}^n}|qE_+h|dx\le\frac{C}{k}.$$

Let η be any vector orthogonal to ω : $\eta\cdot\omega=0$. Choose $\theta_k=\frac{k\omega+\eta}{|k\omega+\eta|}$, $\xi_k=k\theta_k$. Note that $\tilde{h}(\xi_k,k\omega)$ is known, since for any $|\theta_k|=1$, $|\omega|=1$, and $k>0$, $a(\theta_k,\omega,k)$ is known. Note that $|k\omega+\eta|^{-1}=(k^2+|\eta|^2)^{-\frac{1}{2}}=k^{-1}(1+O(\frac{|\eta|^2}{k^2}))$. We have

$$(33.17)\quad \tilde{q}(\xi_k-k\omega)=\tilde{q}\Big(\Big(1+O\Big(\frac{|\eta|^2}{k^2}\Big)\Big)(k\omega+\eta)-k\omega\Big)\to\tilde{q}(\eta)\text{ as }\ k\to\infty.$$

Therefore, substituting $\xi_k=k\theta_k$ into (33.14) and taking into account (33.16) and (33.17), we get

$$\tilde{q}(\eta)=\lim_{k\to\infty}\tilde{h}(\xi_k,k\omega).$$

Since ω is an arbitrary unit vector, we can recover $\tilde{q}(\eta)$ for any $\eta\in\mathbb{R}^n$. Therefore, $q(x)$ can be recovered from the scattering amplitude.

33.3. The scattering problem (the case of $q(x)=O(\frac{1}{(1+|x|)^{1+\varepsilon}})$).

The goal of this subsection is to extend the notions of distorted plane wave and scattering amplitude to the case of potentials $q(x)=O(\frac{1}{(1+|x|)^{1+\varepsilon}})$. This will be achieved by representing the distorted plane wave and the scattering amplitude as Schwartz's kernels of appropriate operators.

Lemma 33.4. *Let $\varphi(\omega)\in L_2(S^{n-1})$ and let $\hat{\varphi}(x)=\int_{S^{n-1}}e^{ix\cdot\omega}\varphi(\omega)d\omega$, i.e., $\hat{\varphi}(-x)$ is the Fourier transform of $\varphi(\omega)\delta(|\xi|-1)$, $\omega=\frac{\xi}{|\xi|}$. Then $\hat{\varphi}(x)\in V_{0,-\frac{1}{2}-\delta}(\mathbb{R}^n)$, $\forall\delta>0$, and*

$$(33.18)\qquad \|\hat{\varphi}\|_{0,-\frac{1}{2}-\delta}\le C_\delta[\varphi]_0,$$

where $[\varphi]_0^2=\int_{S^{n-1}}|\varphi(\omega)|^2d\omega$.

Proof: Consider

$$(33.19)\qquad g(\eta')\delta(\eta_n-\alpha),$$

where $\eta' \in \mathbb{R}^{n-1}$, a is a constant, and $g(\eta') \in L_2(\mathbb{R}^{n-1})$. Then $F_\eta g(\eta')\delta(\eta_n - a) = e^{-iay_n}(F_{\eta'}g)(y')$. Therefore

$$(33.20)\quad \|g(\eta')\delta(\eta_n - a)\|^2_{-\frac{1}{2}-\delta} \le C_1 \int_{\mathbb{R}^n} \frac{|(F_{\eta'}g)(y')|^2 dy_n dy'}{(1+|y'|+|y_n|)^{1+2\delta}}$$
$$\le C_1 \int_{\mathbb{R}^n} \frac{|(F_{\eta'}g)(y')|^2 dy_n dy'}{(1+|y_n|)^{1+2\delta}} \le C \int_{\mathbb{R}^{n-1}} |g(\eta')|^2 d\eta',$$

where $\|u\|_{-\frac{1}{2}-\delta}$ is the norm in the Sobolev space $H_{-\frac{1}{2}-\delta}(\mathbb{R}^n)$.

Take a partition of unity $\{\psi_j, U_j\}$ in a neighborhood of S^{n-1} in $\mathbb{R}^n$. Making the changes of variables that transform $\psi_j(\xi)\varphi(\omega)\delta(|\xi|-1)$ to the form (33.19) and using the invariance of the Sobolev spaces under the changes of variables (cf. Theorem 13.8), we get that $\varphi(\omega)\delta(|\xi|-1) \in H_{-\frac{1}{2}-\delta}(\mathbb{R}^n)$. Therefore, $\hat{\varphi}(x) \in V_{0,-\frac{1}{2}-\delta}(\mathbb{R}^n)$, and the estimate (33.18) holds. □

33.4. Generalized distorted plane waves.

Now we extend the notion of distorted plane wave to the case of potentials $q(x) = O(\frac{1}{(1+|x|)^{1+\varepsilon}})$. Note that

$$(-\Delta - k^2)\hat{\varphi}(kx) = 0 \quad \text{for any} \quad \varphi(\omega) \in L_2(S^{n-1}).$$

Therefore,

$$(-\Delta - k^2 + q(x))\hat{\varphi}(kx) = q(x)\hat{\varphi}(kx).$$

Since $\hat{\varphi}(kx) \in V_{0,-\frac{1}{2}-\delta}(\mathbb{R}^n)$ for all δ and $q(x) = O(\frac{1}{(1+|x|)^{1+\varepsilon}})$, we have $q(x)\hat{\varphi}(kx) \in V_{0,\frac{1}{2}+\varepsilon_1}(\mathbb{R}^n)$, $\varepsilon_1 = \frac{\varepsilon}{2}$. Define $v = -G(q\hat{\varphi}(kx))$, where G is the same as in Remark 31.1. Then

$$(33.21)\qquad W_k\varphi = \hat{\varphi}(kx) - G(q\hat{\varphi}(kx))$$

is a bounded operator from $L_2(S^{n-1})$ to $V_{0,-\frac{1}{2}-\varepsilon_1}(\mathbb{R}^n)$ and

$$(33.22)\qquad (-\Delta - k^2 + q(x))W_k\varphi = 0, \ \forall \varphi \in L_2(S^{n-1}).$$

Therefore, we can define the distorted plane wave as the Schwartz kernel (cf. §44 for the definition of Schwartz's kernels) of the operator W_k. In the case where $q(x) = O(\frac{1}{(1+|x|)^{n+\alpha}})$, $\alpha > 0$, the kernel of W_k is a continuous function of $x \in \mathbb{R}^n$ and $\omega \in S^{n-1}$, and it coincides with (33.7). If $q(x) = O(\frac{1}{(1+|x|)^{1+\varepsilon}})$, the Schwartz kernel is a distribution in $\mathcal{D}'(\mathbb{R}^n \times S^{n-1})$. □

33.5. Generalized scattering amplitude.

Now we generalize the notion of scattering amplitude to the case of $q(x) = O(\frac{1}{(1+|x|)^{1+\varepsilon}})$.

Lemma 33.5. *Let $\psi(\theta) \in C^\infty(S^{n-1})$, $h(x) \in V_{0,\frac{1}{2}+\varepsilon_1}(\mathbb{R}^n)$, and $u = E_+h$. Then*

$$\int_{S^{n-1}} \psi(\theta)u(x)d\theta = C_n \frac{e^{ik|x|}}{|x|^{\frac{n-1}{2}}} \int_{S^{n-1}} \psi(\omega)\tilde{h}(k\omega)d\omega + o\left(\frac{1}{|x|^{\frac{n-1}{2}}}\right), \tag{33.23}$$

where $\theta = \frac{x}{|x|}$, $|x| \to \infty$.

Proof: Since $u(x) \in V_{2,-\frac{1}{2}-\varepsilon_1}$ the restriction of $u(x)$ to the sphere S^{n-1} is well defined (cf. Remark 13.1). We have $u = u_1 + u_2 + u_3$, where

$$u_1 - \frac{1}{(2\pi)^n} \int_{\mathbb{R}^n} \frac{e^{ix\cdot\xi}(1 - \chi(|\xi|))\tilde{h}(\xi)d\xi}{|\xi|^2 - k^2},$$

$$u_2 = \frac{1}{(2\pi)^n} \int_{\mathbb{R}^n} \frac{e^{ix\cdot\xi}\chi(|\xi|)(\tilde{h}(|\xi|\omega) - \tilde{h}(k\omega))d\xi}{|\xi|^2 - k^2},$$

$$u_3 = \frac{1}{(2\pi)^n} \int_{\mathbb{R}^n} \frac{e^{ix\cdot\xi}\chi(|\xi|)\tilde{h}(k\omega)d\xi}{|\xi|^2 - k^2 - i0},$$

and $\chi(|\xi|)$ is the same as in (29.8) for $\tau = 0$.

Let $I_j(|x|) = \int_{S^{n-1}} \psi(\theta)u_j(|x|\theta)d\theta$, $1 \le j \le 3$.

First, we consider

$$\begin{aligned} I_3(|x|) &= \int_{S^{n-1}} \psi(\theta)u_3(|x|\theta)d\theta \\ &= \int_{S^{n-1}} \left(\frac{1}{(2\pi)^n} \int_0^\infty \int_{S^{n-1}} \frac{\chi(r)e^{i|x|r\theta\cdot\omega}}{r^2 - k^2 - i0} \psi(\theta)r^{n-1}drd\theta \right) \tilde{h}(k\omega)d\omega. \end{aligned}$$

Apply the stationary phase lemma (Lemma 19.4) to the integral with respect to (r, θ) as $|x| \to \infty$. The relevant critical point is $\theta = \omega$. Therefore, we get, as in the proof of Lemma 19.3:

$$\begin{aligned} I_3(|x|) &= \int_{S^{n-1}} \psi(\theta)u_3(|x|\theta)d\theta \\ &= C_n \frac{e^{ik|x|}}{|x|^{\frac{n-1}{2}}} \int_{S^{n-1}} \psi(\omega)\tilde{h}(k\omega)d\omega + O\left(\frac{1}{|x|^{\frac{n+1}{2}}}\right), \end{aligned} \tag{33.24}$$

where C_n is the same as in (33.4).

To estimate $I_1(|x|)$, we apply the stationary phase lemma to the integral $\int_{S^{n-1}} \psi(\theta)e^{i|x|\theta\cdot|\xi|\omega}d\theta$. Taking into account that there are two critical points

$\theta = \pm\omega$, we get

$$(33.25)\quad \int_{S^{n-1}} \psi(\theta)e^{i|x|\theta\cdot|\xi|\omega}d\theta = (2\pi)^{\frac{n-1}{2}}e^{i\frac{\pi}{4}(n-1)}\frac{e^{-i|x||\xi|}}{(|x||\xi|)^{\frac{n-1}{2}}}\psi(-\omega) + (2\pi)^{\frac{n-1}{2}}e^{-i\frac{\pi}{4}(n-1)}\frac{e^{i|x||\xi|}}{(|x||\xi|)^{\frac{n-1}{2}}}\psi(\omega) + O\left(\frac{1}{(|x||\xi|)^{\frac{n+1}{2}}}\right).$$

Therefore, $I_1(|x|)$ has the form $I_1(|x|) = I_1^+(|x|)+I_1^-(|x|)+O(\frac{1}{|x|^{\frac{n+1}{2}}})$, where

$$(33.26)\quad I_1^\pm(|x|) = \frac{C_n^\pm}{|x|^{\frac{n-1}{2}}}\int_0^\infty e^{\pm i|x|r}\frac{r^{\frac{n-1}{2}}(1-\chi(r))}{r^2-k^2}b_\pm(r)dr.$$

Here $r = |\xi|$, $b_\pm(r) = \int_{S^{n-1}}\psi(\pm\omega)\tilde{h}(r\omega)d\omega$, and $C_n^\pm$ are some constants.

It follows from (13.27) that

$$|b_\pm(r)| \le C\left(\int_{S^{n-1}}|\tilde{h}(r\omega)|^2d\omega\right)^{\frac{1}{2}} \le \frac{C}{r^{\frac{n-1}{2}}}\|h\|_{0,\frac{1}{2}+\varepsilon_1}.$$

Therefore, $\frac{1-\chi(r)}{r^2-k^2}r^{\frac{n-1}{2}}b_\pm(r)$ is absolutely integrable on $[0,+\infty)$, and hence its Fourier transform tends to zero as $|x|\to\infty$. Therefore,

$$(33.27)\quad I_1^\pm(|x|) = o\left(\frac{1}{|x|^{\frac{n-1}{2}}}\right).$$

Now consider $I_2(|x|)$. Again, using (33.25), we get $I_2(|x|) = I_2^+(|x|) + I_2^-(|x|) + O(\frac{1}{|x|^{\frac{n+1}{2}}})$, where

$$I_2^\pm(|x|) = \frac{C_n^\pm}{|x|^{\frac{n-1}{2}}}\int_0^\infty e^{i|x|r}\frac{r^{\frac{n-1}{2}}\chi(r)}{r^2-k^2}(b_\pm(r)-b_\pm(k))dr,$$

and $b_\pm(r)$ are the same as in (33.26).

It follows from (13.28) that

$$|b_\pm(r)-b_\pm(k)| \le C\left(\int_{S^{n-1}}|\tilde{h}(r\omega)-\tilde{h}(k\omega)|^2d\omega\right)^{\frac{1}{2}} \le C\frac{|r-k|^\delta}{k^{\frac{n-1}{2}}}\|h\|_{0,\frac{1}{2}+\varepsilon_1},$$

where $0<\delta<\varepsilon_1, \delta\le 1$. Therefore, $\frac{r^{\frac{n-1}{2}}\chi(r)}{r^2-k^2}(b_\pm(r)-b_\pm(k))$ is an absolutely integrable function on $[0,+\infty)$, and hence (cf. (33.27))

$$(33.28)\quad I_2^\pm = o\left(\frac{1}{|x|^{\frac{n-1}{2}}}\right).$$

Now Lemma 33.5 follows from (33.24), (33.27), and (33.28). □

Corollary 33.6. *Let $f \in V_{0,\frac{1}{2}+\varepsilon_1}$ and $u = Gf = E_+h$, where $h = f - qGf$. Then for all $\psi \in C^\infty(S^{n-1})$,*

(33.29)

$$\int_{S^{n-1}} \psi(\theta)(Gf)(|x|\theta)d\theta = C_n \frac{e^{ik|x|}}{|x|^{\frac{n-1}{2}}} \int_{\mathbb{R}^n} W_k(\psi(-\,\cdot))(y)f(y)dy + o\left(\frac{1}{|x|^{\frac{n-1}{2}}}\right),$$

where W_k is the same as in (33.21).

Proof: We have

$$\begin{aligned}\int_{S^{n-1}} \psi(\theta)\tilde{h}(k\theta)d\theta &= \int_{\mathbb{R}^n}\int_{S^{n-1}} \psi(\theta)e^{-ik\theta\cdot y}(f(y) - q(y)Gf)dyd\theta \\ &= \int_{\mathbb{R}^n} \hat{\psi}(-ky)(f(y) - q(y)Gf)dy \\ &= \int_{\mathbb{R}^n} \hat{\psi}(-ky)f(y)dy - \int_{\mathbb{R}^n} G(q\hat{\psi}(-ky))f(x)dx,\end{aligned}$$

since G is a symmetric operator. Hence (33.23) and (33.21) imply (33.29). Note that (33.23) generalizes (33.6) for the case $q(x) = O\left(\frac{1}{1+|x|)^{1+\varepsilon}}\right)$. □

Let $\varphi(\theta) \in L_2(S^{n-1})$ and $\psi(\theta) \in C^\infty(S^{n-1})$. Consider the asymptotics as $|x| \to \infty$ of the integral

$$\int_{S^{n-1}} \psi(\theta)(W_k\varphi)(x)d\theta, \quad x = |x|\theta,$$

where $W_k\varphi = \hat{\varphi}(kx) - (Gq\hat{\varphi}(k\cdot))(x)$.

Applying the stationary phase lemma to the integral

$$\int_{S^{n-1}} \psi(\theta)\hat{\varphi}(kx)d\theta = \int_{S^{n-1}} \left(\int_{S^{n-1}} \psi(\theta)e^{ik|x|\theta\cdot\omega}d\theta\right)\varphi(\omega)d\omega,$$

we get, as in (33.25):

$$\begin{aligned}\int_{S^{n-1}} \psi(\theta)\hat{\varphi}(k|x|\theta)d\theta &= i^{n-1}C_n^{(1)}\frac{e^{-ik|x|}}{|x|^{\frac{n-1}{2}}}\int_{S^n} \psi(-\omega)\varphi(\omega)d\omega \\ &+ C_n^{(1)}\frac{e^{ik|x|}}{|x|^{\frac{n-1}{2}}}\int_{S^{n-1}} \psi(\omega)\varphi(\omega)d\omega + O\left(\frac{1}{|x|^{\frac{n+1}{2}}}\right),\end{aligned} \tag{33.30}$$

where $C_n^{(1)} = \left(\frac{2\pi}{k}\right)^{\frac{n-1}{2}} e^{-i\frac{\pi}{4}(n-1)}$.

We have $G(q\hat{\varphi}(kx)) = E_+a(\varphi)$, where $a + qE_+a = q\hat{\varphi}(kx)$. Therefore,

$$a(\varphi)(x) = q\hat{\varphi}(kx) - qE_+a = q\hat{\varphi} - qG(q\hat{\varphi}) \in V_{0,\frac{1}{2}+\varepsilon_1}(\mathbb{R}^n).$$

Moreover,

$$\|a(\varphi)\|_{0,\frac{1}{2}+\varepsilon_1} \le C[\varphi]_0.$$

Denote by $\tilde{a}(\varphi)(\xi)$ the Fourier transform of $a(\varphi)(x)$. Using (33.30) and (33.23) with $\tilde{h}(\xi)$ replaced by $\tilde{a}(\varphi)(\xi)$, we obtain

$$\text{(33.31)} \quad \int_{S^{n-1}} \psi(\theta)(W_k\varphi)(x)dx = i^{n-1}C_n^{(1)}\frac{e^{-ik|x|}}{|x|^{\frac{n-1}{2}}}\int_{S^{n-1}}\psi(-\omega)\varphi(\omega)d\omega$$
$$+ C_n^{(1)}\frac{e^{ik|x|}}{|x|^{\frac{n-1}{2}}}\int_{S^{n-1}}\psi(\omega)(S_k\varphi)(\omega)d\omega + o\left(\frac{1}{|x|^{\frac{n-1}{2}}}\right),$$

where

$$(S_k\varphi)(\theta) = \varphi(\theta) - \frac{ik^{n-2}}{2(2\pi)^{n-1}}\tilde{a}(\varphi)(k\theta).$$

One can consider (33.31) as a generalization of (33.5). Note that $\tilde{a}(\varphi)(\xi) \in H_{\frac{1}{2}+\varepsilon_1}(\mathbb{R}^n)$, and therefore the restriction $\tilde{a}(\varphi)(k\theta) \in H_{\varepsilon_1}(S^{n-1}) \subset L_2(S^{n-1})$. Hence $\tilde{a}(\varphi)(k\theta)$ is a bounded operator in $L_2(S^{n-1})$. Operator S_k coincides with the scattering matrix (see Hörmander [**H1**]). Note that if $q(x) = O(\frac{1}{(1+|x|)^{n+\alpha}})$, then $\tilde{a}(\varphi)(k\theta)$ is an integral operator with continuous kernel $C_n^{-1}a(\theta,\omega,k)$, where $a(\theta,\omega,k)$ is the scattering amplitude (cf. (33.5)), and C_n is the same as in (33.4). In the case where $q(x) = O(\frac{1}{(1+|x|)^{1+\varepsilon}})$ we define the scattering amplitude as the Schwartz kernel of the operator $\frac{ik^{n-2}}{2(2\pi)^{n-1}}\tilde{a}(\varphi)(k\theta) = (S_k - I)\varphi$ multiplied by $C_n^{(1)} = \left(\frac{2\pi}{k}\right)^{\frac{n-1}{2}} e^{-i\frac{\pi}{4}(n-1)}$ (cf. (33.30)).

34. Inverse boundary value problem

Let Ω be a bounded domain in $\mathbb{R}^n$ with a smooth boundary $\partial\Omega$. Consider the Dirichlet problem in Ω,

$$\text{(34.1)} \qquad (-\Delta + q(x) - k^2)u = 0, \quad x \in \Omega,$$

$$\text{(34.2)} \qquad u|_{\partial\Omega} = h(x'), \quad x' \in \partial\Omega,$$

where $q(x)$ is bounded in Ω, and $h(x') \in H_{\frac{3}{2}}(\partial\Omega)$. Suppose k is fixed and the boundary value problem (34.1), (34.2) has a unique solution $u(x) \in H_2(\Omega)$. Denote by Λh the following operator on $\partial\Omega$:

$$\Lambda h = \left.\frac{\partial u}{\partial n}\right|_{\partial\Omega},$$

where n is the outward unit normal vector to $\partial\Omega$, and $u(x)$ is the solution of (34.1), (34.2)). Operator Λ is called the Dirichlet-to-Neumann operator.

The inverse boundary value problem consists in the recovery of $q(x)$ knowing the Dirichlet-to-Neumann operator Λ.

Let $q'(x) = q(x) - k^2$.

Theorem 34.1. *Suppose $n \geq 3$. Consider $L_j = -\Delta + q_j'(x)$, $j = 1,2$, where $q_j'(x)$ are bounded complex-valued functions in Ω. Assume that the Dirichlet problems (34.1), (34.2) have a unique solution for $L_j, j = 1,2$. Let Λ_j, $j = 1,2$, be the corresponding Dirichlet-to-Neumann operators on $\partial\Omega$. If $\Lambda_1 = \Lambda_2$, then $q_1(x) = q_2(x)$ a.e.*

Proof: Let u_j be solutions of $L_j u_j = 0$ in Ω, $j = 1,2$ and let v_j, $j = 1,2$, be solutions of $L_j^* v_j = 0$, where $L_j^* = -\Delta + \overline{q_j'(x)}$. Applying Green's formula

$$\int_\Omega (-\Delta u)\overline{v}dx - \int_\Omega u\overline{(-\Delta v)}dx = -\int_{\partial\Omega} \frac{\partial u}{\partial n}\overline{v}ds + \int_{\partial\Omega} u\frac{\partial \overline{v}}{\partial n}ds, \tag{34.3}$$

we get

$$0 = (L_j u_j, v_j) - (u_j, L_j^* v_j) = -\left[\frac{\partial u_j}{\partial n}, v_j\right] + \left[u_j, \frac{\partial v_j}{\partial n}\right], \tag{34.4}$$

where $[f,g] = \int_{\partial\Omega} f\overline{g}ds$ is the L_2 scalar product on $\partial\Omega$. We assume that $u_j \in H_2(\Omega)$, $v_j \in H_2(\Omega)$, $j = 1,2$. Denote $f_j = u_j|_{\partial\Omega}$, $g_j = v_j|_{\partial\Omega}$, $j = 1,2$. Then $\Lambda_j f_j = \frac{\partial u_j}{\partial n}|_{\partial\Omega}$, $j = 1,2$. It follows from (34.4) that

$$[\Lambda_j f_j, g_j] = \left[f_j, \frac{\partial v_j}{\partial n}\right],$$

i.e., the Dirichlet-to-Neumann operator for L_j^* is the adjoint of the Dirichlet-to-Neumann operator for $L_j : \Lambda_j^* g_j = \frac{\partial v_j}{\partial n}|_{\partial\Omega}$. We have

$$\begin{aligned} 0 &= (L_1 u_1, v_2) - (u_1, L_2^* v_2) \\ &= \int_\Omega \left(\frac{\partial u_1}{\partial x}\cdot\overline{\frac{\partial v_2}{\partial x}} + q_1' u_1 \overline{v_2}\right)dx - [\Lambda_1 f_1, g_2] \\ &\quad - \int_\Omega \left(\frac{\partial u_1}{\partial x}\cdot\overline{\frac{\partial v_2}{\partial x}} + q_2' u_1 \overline{v_2}\right)dx + [f_1, \Lambda_2^* g_2] \\ &= \int_\Omega (q_1 - q_2')u_1\overline{v_2}dx - [(\Lambda_1 - \Lambda_2)f_1, g_2]. \end{aligned} \tag{34.5}$$

If $\Lambda_1 = \Lambda_2$, then

$$\int_\Omega (q_1(x) - q_2(x))u_1\overline{v_2}dx = 0 \tag{34.6}$$

for all u_1, v_2 in $H_2(\Omega)$ such that $L_1u_1 = 0$, $L_2^* v_2 = 0$ in Ω. Let μ, ν be orthogonal unit vectors in $\mathbb{R}^n : \mu\cdot\nu = 0$, $|\mu| = |\nu| = 1$. Let $l \in \mathbb{R}^n$ be an arbitrary vector orthogonal to μ and ν. Denote

$$\zeta_1 = \frac{l}{2} + \sqrt{\tau^2 - \frac{|l|^2}{4}}\mu + i\tau\nu, \quad \zeta_2 = -\frac{l}{2} + \sqrt{\tau^2 - \frac{|l|^2}{4}}\mu - i\tau\nu,$$

where $\tau > \frac{|l|^2}{4}$. Note that $\zeta_1\cdot\zeta_1 = 0$, $\zeta_2\cdot\zeta_2 = 0$, $\zeta_1 - \overline{\zeta_2} = l$. We look for the solutions of $L_1 u_1 = 0$ and $L_2^* v_2 = 0$ in the form $u_1 = e^{ix\cdot\zeta_1}u_1'$ and

$v_2 = e^{ix\cdot\zeta_2}v_2'$, respectively. Then u_1', v_2' in Ω satisfy:

$$\left[\left(-i\frac{\partial}{\partial x}+\zeta_1\right)\cdot\left(-i\frac{\partial}{\partial x}+\zeta_1\right)+q_1'\right]u_1'=0, \tag{34.7}$$

$$\left[\left(-i\frac{\partial}{\partial x}+\zeta_2\right)\cdot\left(-i\frac{\partial}{\partial x}+\zeta_2\right)+\overline{q_2'}\right]v_2'=0. \tag{34.8}$$

We shall show that there exist solutions u_1', v_2' such that

$$u_1' = 1 + r_1, \quad v_2' = 1 + r_2,$$

where $\|r_j\|_{L_2(\Omega)} = O(\frac{1}{\tau})$ as $\tau \to \infty$, $j = 1, 2$.

Extend q_j' by zero outside of Ω, $j = 1, 2$. We are looking for a solution of (34.7) in $\mathbb{R}^n$ in the form

$$u_1' = 1 + Eh_1,$$

where

$$Eh_1 = \frac{1}{(2\pi)^n}\int_{\mathbb{R}^n}\frac{e^{ix\cdot\eta}\tilde{h}_1(\eta)d\eta}{(\eta+\zeta_1)\cdot(\eta+\zeta_1)}.$$

Then h_1 satisfies

$$h_1 + q_1'Eh_1 = -q_1'. \tag{34.9}$$

Making the change of variables $\eta + \operatorname{Re}\zeta_1 = \eta_1'$, we get that

$$E = e^{ix\cdot\operatorname{Re}\zeta_1}E_{\nu,0,\tau}e^{-ix\cdot\operatorname{Re}\zeta_1},$$

where $E_{\nu,0,\tau}$ is the same as in (29.5) with $k = 0$. Therefore, Agmon's estimate (29.5) gives

$$\|Eh_1\|_{0,-\frac{1}{2}-\varepsilon} \leq \frac{C}{\tau}\|h_1\|_{0,\frac{1}{2}+\varepsilon}, \quad \varepsilon > 0,\ \tau \geq 1. \tag{34.10}$$

When τ is large, the equation (34.9) has a solution $h_1 \in V_{0,\frac{1}{2}+\varepsilon}(\mathbb{R}^n)$ such that $\|h_1\|_{0,\frac{1}{2}+\varepsilon} \leq C$, where C is independent of τ. Since $r_1 = u_1' - 1 = Eh_1$, we have

$$\|r_1\|_{L_2(\Omega)} \leq C\|r_1\|_{0.-\frac{1}{2}-\varepsilon} \leq \frac{C}{\tau}. \tag{34.11}$$

Analogously, we can find $v_2' = 1 + r_2$, where $\|r_2\|_{L_2(\Omega)} \leq \frac{C}{\tau}$. Substituting $u_1 = e^{ix\cdot\zeta_1}(1 + r_1)$, $v_2 = e^{ix\cdot\zeta_2}(1 + r_2)$ in (34.6) and taking the limit as $\tau \to \infty$, we obtain

$$\int_\Omega (q_1(x) - q_2(x))e^{ix\cdot l}dx = 0. \tag{34.12}$$

Since l is arbitrary, we see that $q_1(x) - q_2(x) = 0$ a.e.

34.1. Electrical impedance tomography.

Example 34.1. Consider the conductivity equation in a smooth domain $\Omega \subset \mathbb{R}^n$, $n \geq 3$:

$$\nabla\gamma(x)\nabla u(x) = 0, \quad x \in \Omega, \tag{34.13}$$

where $\gamma(x) \in C^\infty(\overline{\Omega})$ is the conductivity, $\gamma(x) > 0$ on $\overline{\Omega}$, and $u(x)$ is the voltage. Suppose that on $\partial\Omega$ we can measure the voltage

$$u|_{\partial\Omega} = f \tag{34.14}$$

and the current flux

$$\gamma\frac{\partial u}{\partial n}\Big|_{\partial\Omega} = g, \tag{34.15}$$

where $\frac{\partial}{\partial n}$ is the normal derivative.

The electrical impedance problem consists in determining the conductivity $\gamma(x)$ in Ω by measuring the voltage and the current's flux on the boundary $\partial\Omega$. Note that the Dirichlet problem (34.13), (34.14) has a unique solution $u \in H_2(\Omega)$ provided $f \in H_{\frac{3}{2}}(\Omega)$. Therefore, the Dirichlet-to-Neumann operator $\Lambda_\gamma f = \gamma\frac{\partial u}{\partial n}\big|_{\partial\Omega}$ is well defined and maps $H_{\frac{3}{2}}(\Omega)$ to $H_{\frac{1}{2}}(\partial\Omega)$. Let $v = \gamma^{\frac{1}{2}}u$. Then $v(x)$ satisfies the equation

$$-\Delta v(x) + q(x)v(x) = 0, \quad x \in \Omega, \tag{34.16}$$

where $q(x) = \frac{\Delta\gamma^{\frac{1}{2}}}{\gamma^{\frac{1}{2}}}$.

Let Λ_q be the Dirichlet-to-Neumann operator corresponding to (34.16), i.e.,

$$\Lambda_q h = \frac{\partial v}{\partial n}\Big|_{\partial\Omega}, \tag{34.17}$$

where $v|_{\partial\Omega} = h$. Note that $h = \gamma^{\frac{1}{2}}u|_{\partial\Omega} = \gamma^{\frac{1}{2}}f$. We have

$$\begin{aligned}\Lambda_q h = \frac{\partial v}{\partial n}\Big|_{\partial\Omega} = \frac{\partial}{\partial n}(\gamma^{\frac{1}{2}}u)\Big|_{\partial\Omega} &= \gamma^{\frac{1}{2}}\frac{\partial u}{\partial n}\Big|_{\partial\Omega} + \frac{1}{2}\gamma^{-\frac{1}{2}}\frac{\partial\gamma}{\partial n}u\Big|_{\partial\Omega} \\ &= \gamma^{-\frac{1}{2}}\Lambda_\gamma f + \frac{1}{2}\gamma^{-\frac{1}{2}}\frac{\partial\gamma}{\partial n}f.\end{aligned} \tag{34.18}$$

Hence, knowing $\gamma|_{\partial\Omega}$, $\frac{\partial\gamma}{\partial n}\Big|_{\partial\Omega}$, and Λ_γ, we can determine Λ_q.

Therefore, we reduced the inverse boundary value problem for the conductivity equation to the inverse boundary value problem for the equation (34.16). By Theorem 34.1, Λ_q uniquely determines $q(x) = \frac{\Delta\gamma^{\frac{1}{2}}}{\gamma^{\frac{1}{2}}}$. Knowing $q(x)$, we can uniquely determine $\gamma^{\frac{1}{2}}$ as the unique solution of the elliptic equation $\Delta\gamma^{\frac{1}{2}} - q\gamma^{\frac{1}{2}} = 0$ given the Cauchy data $\gamma|_{\partial\Omega}$ and $\frac{\partial\gamma}{\partial n}|_{\partial\Omega}$. The only question remaining is how to determine $\gamma|_{\partial\Omega}$ and $\frac{\partial\gamma}{\partial n}|_{\partial\Omega}$. It can be done in various ways (see, for example, §57, where a more general class of equations is considered).

35. Equivalence of inverse BVP and inverse scattering

Let $q(x) \in L_\infty(\Omega)$, where Ω is a smooth bounded domain in $\mathbb{R}^n$, $n \geq 2$, such that $\mathbb{R}^n \setminus \Omega$ is connected. Extending $q(x)$ by zero outside of Ω we can consider the inverse scattering problem in $\mathbb{R}^n$ for the equation

$$(-\Delta + q(x) - k^2)u = 0, \quad x \in \mathbb{R}^n, \tag{35.1}$$

i.e., the problem of recovering $q(x)$ from the scattering amplitude $a(\theta, \omega, k)$ for a fixed k and all $(\theta, \omega) \in S^{n-1} \times S^{n-1}$. We can also consider the inverse boundary value problem for the equation (35.1) in Ω, i.e., the recovery of $q(x)$ from the Dirichlet-to-Neumann operator Λ on $\partial\Omega$. In this section, we prove that knowing Λ on $\partial\Omega$, one can recover $a(\theta, \omega, k)$ for all $(\theta, \omega) \in S^{n-1} \times S^{n-1}$, and vice versa.

Theorem 35.1. *Suppose $q_j(x) \in L_\infty(\Omega)$, $q_j(x) = \overline{q_j(x)}$, $j = 1, 2$. Suppose that the Dirichlet problem in Ω is uniquely solvable, $j = 1, 2$. Let Λ_j be the Dirichlet-to-Neumann operators corresponding to q_j, $j = 1, 2$, and let $a_j(\theta, \omega, k)$ be the scattering amplitudes corresponding to $q_j(x)$, $j = 1, 2$. Then $\Lambda_1 = \Lambda_2$ iff $a_1(\theta, \omega, k) = a_2(\theta, \omega, k)$ for all $(\theta, \omega) \in S^{n-1} \times S^{n-1}$, where k is fixed.*

Proof:

a) Suppose we have two potentials $q_1(x)$ and $q_2(x)$ such that the corresponding Dirichlet-to-Neumann operators are equal, i.e., $\Lambda_1 = \Lambda_2$. Let $u_j = e^{ik\omega\cdot x} + v_j$ be the distorted plane waves corresponding to $q_j(x)$, $j = 1, 2$, i.e., $(\Delta + k^2 + q_j(x))u_j = 0$ in $\mathbb{R}^n$.

Denote by u_3 the following function:

$$u_3 = u_1 \quad \text{in} \quad \mathbb{R}^n \setminus \overline{\Omega}, \tag{35.2}$$

$$(-\Delta + q_2(x) - k^2)u_3 = 0 \quad \text{in} \quad \Omega, \tag{35.3}$$

$$u_3\big|_{\partial\Omega} = u_1\big|_{\partial\Omega}. \tag{35.4}$$

We assume that k^2 is not a Dirichlet eigenvalue for $-\Delta + q_1$ and $-\Delta + q_2$. Thus, such a solution u_3 exists and is unique. Note that u_3 is the solution of $(-\Delta - k^2 + q_2(x))u = 0$ in $\mathbb{R}^n$. Indeed, u_3 is continuous in $\mathbb{R}^n$ because of (35.2), (35.4).

The limit of $\frac{\partial u_3}{\partial n}$ as $x \in \mathbb{R}^n \setminus \Omega$ approaches $\partial\Omega$ is $\frac{\partial u_1}{\partial n}|_{\partial\Omega}$. The limit of $\frac{\partial u_3}{\partial n}$ as $x \in \Omega$ approaches $\partial\Omega$ is $\Lambda_2(u_1\big|_{\partial\Omega})$. Since $\Lambda_2 = \Lambda_1$, we see that

$\Lambda_2(u_1|_{\partial\Omega}) = \Lambda_1(u_1|_{\partial\Omega}) = \frac{\partial u_1}{\partial n}|_{\partial\Omega}$. Therefore, $\frac{\partial u_3}{\partial n}$ is continuous near $\partial\Omega$ and hence $(-\Delta + q_2(x) - k^2)u_3 = 0$ in $\mathbb{R}^n$.

Since $u_3 = u_1$ in $\mathbb{R}^n \setminus \Omega$ and $u_1 - e^{ik\omega\cdot x}$ satisfies the radiation conditions, we conclude that $u_3 - e^{ik\omega\cdot x}$ also satisfies the radiation conditions. Therefore, $v = u_3 - u_2$ is the solution of $(\Delta + q_2(x) + k^2)v = 0$ in $\mathbb{R}^n$ satisfying the radiation condition. The uniqueness theorem (see Theorem 31.7) implies that $v = u_3 - u_2 = 0$. Finally, since $u_1 = u_2$ in $\mathbb{R}^n \setminus \Omega$, the scattering amplitudes $a_1(\theta, \omega, k)$ and $a_2(\theta, \omega, k)$ are equal for all $|\theta| = 1$, $|\omega| = 1$.

b) Now suppose that $a_1(\theta, \omega, k) = a_2(\theta, \omega, k)$ for k fixed and all $(\theta, \omega) \in S^{n-1} \times S^{n-1}$. We prove that $\Lambda_1 = \Lambda_2$ on $\partial\Omega$. Let $u_j(x, k\omega)$, $j = 1, 2$, be the distorted plane waves corresponding to $q_j(x)$, $j = 1, 2$, respectively. Since $a_1 = a_2$, we have

$$u = u_1 - u_2 = O\left(\frac{1}{|x|^{\frac{n+1}{2}}}\right),$$

i.e., $u \in L_2(\mathbb{R}^n)$. Note that $u(x, k\omega)$ satisfies the following differential equation:

$$(-\Delta - k^2)u = f, \tag{35.5}$$

where $f = q_2u_2 - q_1u_1$ has a compact support.

We will need the following lemma:

Lemma 35.2 (Rellich's lemma). *Let $u(x) \in L_2(\mathbb{R}^n)$ be a solution of* (35.5), *where $f \in L_2(\mathbb{R}^n)$ and has a compact support. Then $u(x) = 0$ for $|x| > R$, where R is large.*

Proof: Making the Fourier transform in (34.5), we get

$$\tilde{u}(\xi) = \frac{\tilde{f}(\xi)}{\sum_{j=1}^n \xi_j^2 - k^2}. \tag{35.6}$$

Assuming that $f(x) = 0$ for $|x| > R$, we see that $\tilde{f}(\xi) = \int_{|x|\in R} f(x)e^{-ix\xi}dx$ extends to $\mathbb{C}^n$ as an entire analytic function of $\zeta = \xi + i\eta$, and

$$|\tilde{f}(\xi + i\eta)| \le Ce^{R|\eta|}, \ \forall\eta \in \mathbb{R}^n. \tag{35.7}$$

Since $\tilde{u}(\xi) \in L_2$, we must have $\tilde{f}(\xi) = 0$ on $\sum_{j=1}^n \xi_j^2 - k^2 = 0$. Therefore, $\frac{\tilde{f}(\xi)}{\sum_{j=1}^n \xi_j^2 - k^2}$ is real analytic in $\mathbb{R}^n$ since $\frac{\partial S(\xi)}{\partial \xi} \neq 0$ for $S(\xi) = \sum_{j=1}^n \xi_j^2 - k^2 = 0$, $\xi \in \mathbb{R}^n$. Note that $\sum_{j=1}^n \zeta_j^2 - k^2 = 0$ is an irreducible polynomial. It is a basic fact of the theory of analytic functions in $\mathbb{C}^n, n > 1$, that if $\frac{\tilde{f}(\zeta)}{\zeta^2 - k^2}$ is analytic at some point ζ_0, $\zeta_0^2 - k^2 = 0$, then $\frac{f(\zeta)}{\zeta^2 - k^2}$ is analytic everywhere in $\mathbb{C}^n$, i.e., $\tilde{u}(\zeta) = \frac{\tilde{f}(\zeta)}{\zeta^2 - k^2}$ is also an entire function in $\mathbb{R}^n$. It follows from (34.7) that

(35.8) $$|\tilde{u}(\xi + i\eta)| \le Ce^{R|\eta|}.$$

Therefore, by Paley-Wiener theorem, $u(x) = 0$ for $|x| > R$.

Remark 35.1. One can also prove Rellich's lemma using Carleman's estimate (31.13)): replacing x by Rx we can assume that $f(x) = 0$ for $|x| > 1$. Then $\int_{\mathbb{R}^n} |x|^{N+2}|f(x)|^2 dx \le C$ for all N. Since N is arbitrary the estimate (31.13) implies that $u(x) = 0$ for $|x| > 1$. □

Since $f(x) = q_2 u_2 - q_1 u_1 = 0$ outside of Ω, we have

(35.9) $$(-\Delta - k^2)u = 0 \quad \text{in} \quad \mathbb{R}^n \setminus \overline{\Omega}.$$

Note that any solution of (35.9) is real analytic in $x = (x_1, \dots, x_n) \in \mathbb{R}^n \setminus \overline{\Omega}$ (see Theorem 18.3). Since $u = 0$ for $|x| > R$, and since $\mathbb{R}^n \setminus \overline{\Omega}$ is connected, we see that $u(x) = 0$ in $\mathbb{R}^n \setminus \overline{\Omega}$, i.e.,

(35.10) $$u_1(x, k\omega) = u_2(x, k\omega) \quad \text{in} \quad \mathbb{R}^n \setminus \overline{\Omega}.$$

By continuity, (35.10) holds on $\partial\Omega$ too. Differentiating (35.10), we get

(35.11) $$\frac{\partial u_1(x, k\omega)}{\partial n} = \frac{\partial u_2(x, k\omega)}{\partial n} \quad \text{on} \quad \partial\Omega.$$

Therefore, we have proven that $\Lambda_1 = \Lambda_2$ on the subspace H_0 of restrictions of the distorted plane waves to $\partial\Omega$. We show that H_0 is dense in $L_2(\partial\Omega)$.

Lemma 35.3. *Suppose $f \in L_2(\partial\Omega)$ satisfies*

(35.12) $$\int_{\partial\Omega} f(x)u(x, -k\omega)ds = 0$$

for all $\omega \in S^{n-1}$, where $u(x, -k\omega) \in H_0$. Assume that the Dirichlet problem in Ω has a unique solution. Then $f = 0$.

Proof: Let $G(x, y)$ be the outgoing fundamental solution of $-\Delta + q(x) - k^2$ constructed in §31:

$$(-\Delta + q(x) - k^2)G(x, y) = \delta(x - y).$$

Denote by Sf the single layer potential corresponding to $G(x, y)$:

(35.13) $$Sf = \int_{\partial\Omega} G(x, y')f(y')ds;$$

i.e., $Sf = G(f\delta_{\partial\Omega})$, where G is the operator with the kernel $G(x, y)$ and $\delta_{\partial\Omega}$ is the delta-function of $\partial\Omega$, i.e.,

(35.14) $$f\delta_{\partial\Omega}(\varphi) = \int_{\partial\Omega} f(x')\varphi(x')ds, \quad \forall \varphi \in C_0^\infty(\mathbb{R}^n).$$

Note that

(35.15) $$(-\Delta + q - k^2)Sf = f(x)\delta_{\partial\Omega},$$

and Sf is a continuous function in $\mathbb{R}^n$. Take the limit of Sf as $|x| \to \infty$ and $\theta = \frac{x}{|x|}$ is fixed. By Theorem 33.2 (see (33.6)) we have

$$Sf = C_n \frac{e^{ik|x|}}{|x|^{\frac{n-1}{2}}} \int_{\partial\Omega} w(y', -k\theta) f(y') ds + O\left(\frac{1}{|x|^{\frac{n+1}{2}}}\right), \tag{35.16}$$

where $w(y, -k\theta)$ is a distorted plane wave. It follows from (35.12) that $Sf = O(\frac{1}{|x|^{\frac{n+1}{2}}})$.

Rellich's lemma (Lemma 35.2) and the analyticity of Sf imply that $Sf = 0$ in $\mathbb{R}^n \setminus \overline{\Omega}$. By continuity, $Sf = 0$ on $\partial\Omega$. Therefore, Sf is a solution of $(-\Delta + q - k^2)u = 0$ in Ω with zero Dirichlet data. Then by assumption, $Sf = 0$ in Ω. Therefore, $Sf = 0$ in $\mathbb{R}^n$. Applying $(-\Delta + q - k^2)$ to Sf, we get $f = 0$. □

Let h be an arbitrary function in $H_{\frac{3}{2}}(\partial\Omega)$. By Lemma 35.3 there exists a sequence of distorted plane waves $u_1^{(p)} = u_2^{(p)} \to h$ in $L_2(\partial\Omega)$ as $p \to \infty$. Take any $\varphi \in C^\infty(\partial\Omega)$. Then $[\Lambda_j u_j^{(p)}, \varphi] = [u_j^{(p)}, \Lambda_j^* \varphi] \to [h, \Lambda_j^* \varphi] = [\Lambda_j h, \varphi]$ as $p \to \infty$, $j = 1, 2$. Since $\Lambda_1 u_1^{(p)} = \Lambda_2 u_2^{(p)}$ for all p (see (35.11)), we obtain that $[\Lambda_1 h, \varphi] = [\Lambda_2 h, \varphi]$, $\forall \varphi \in C_0^\infty(\partial\Omega)$. Therefore, $\Lambda_1 h = \Lambda_2 h$ for all $h \in H_{\frac{3}{2}}(\partial\Omega)$. □

36. Scattering by obstacles

Let Ω be a smooth bounded domain such that the complement $\mathbb{R}^n \setminus \Omega$ is connected. Let $u = e^{ik\omega\cdot x} + v$ be a distorted plane wave solution of

$$(-\Delta - k^2)u = 0 \quad \text{in} \quad \mathbb{R}^n \setminus \overline{\Omega}, \tag{36.1}$$

such that $v(x, k\omega)$ satisfies the radiation conditions

$$v = O\left(\frac{1}{|x|^{\frac{n-1}{2}}}\right), \quad \frac{\partial v}{\partial r} - ikv = O\left(\frac{1}{|x|^{\frac{n+1}{2}}}\right) \tag{36.2}$$

as $|x| \to \infty$, $r = |x|$, and $u(x, k\omega)$ satisfies the zero Dirichlet boundary condition

$$u|_{\partial\Omega} = (e^{ik\omega\cdot x} + v)|_{\partial\Omega} = 0. \tag{36.3}$$

The problem of finding $u(x, k\omega)$ satisfying (36.1), (36.2), (36.3) is called the obstacle problem.

One can also consider the problem of finding $u(x, k\omega)$ satisfying (36.1), (36.2), and the Neumann boundary condition:

$$\left.\frac{\partial u}{\partial n}\right|_{\partial\Omega} = 0. \tag{36.4}$$

We prove the existence and uniqueness of solutions for the problems (36.1), (36.2), (36.3) and (36.1), (36.2), (36.4).

Lemma 36.1. *Let $w(x)$ be a solution of* (36.1) *in $\mathbb{R}^n\setminus\overline{\Omega}$ that is smooth in $\mathbb{R}^n\setminus\Omega$ and satisfies the radiation conditions* (36.2). *Then*

$$w = b(\theta)\frac{e^{ik|x|}}{|x|^{\frac{n-1}{2}}} + O\left(\frac{1}{|x|^{\frac{n+1}{2}}}\right) \tag{36.5}$$

as $|x|\to\infty$, where $\theta = \frac{x}{|x|}$ is fixed.

Proof: Let x_0 be an arbitrary point in $\mathbb{R}^n\setminus\overline{\Omega}$ and let $B_R\supset\overline{\Omega}$, $|x_0|<R$. Applying Green's formula to $B_R\setminus(\overline{\Omega}\cup B_\varepsilon(x_0))$ and taking the limit as $\varepsilon\to 0$, we get, as in (19.11):

$$\begin{aligned}-w(x_0) = &\int_{\partial B_R}\left(\left(\frac{\partial}{\partial r}E_+(x-x_0)\right)w(x) - E_+(x-x_0)\frac{\partial w(x)}{\partial r}\right)ds\\ &-\int_{\partial\Omega}\left(\left(\frac{\partial}{\partial n}E_+(x-x_0)\right)w(x) - E_+(x-x_0)\frac{\partial w(x)}{\partial n}\right)ds,\end{aligned} \tag{36.6}$$

where n is the outward normal to $\partial\Omega$, $r=|x|$. Since $E_+(x-x_0)$ and $w(x)$ satisfy the radiation condition, we see (cf. (19.11), (19.12)) that the integral over ∂B_R tends to zero as $R\to\infty$. Therefore

$$w(x_0) = \int_{\partial\Omega}\left(\frac{\partial E_+(x-x_0)}{\partial n}w(x) - E_+(x-x_0)\frac{\partial w(x)}{\partial n}\right)ds. \tag{36.7}$$

Taking the limit as $|x_0|\to\infty$ for $\theta=\frac{x_0}{|x_0|}$ fixed, we get, as in the proofs of Theorems 19.2 and 19.5, that (36.5) holds. □

Theorem 36.2. *Suppose $u_j = e^{ik\omega\cdot x}+v_j$, $j=1,2$, are two solutions of the problem* (36.1), (36.2), (36.3). *Then $u_1\equiv u_2$.*

Proof: Denote $w=u_1-u_2=v_1-v_2$. Then $w(x)$ satisfies (36.1), the radiation condition (36.2), and $w|_{\partial\Omega}=0$. Let $B_R\supset\overline{\Omega}$. We have, applying Green's formula,

$$\begin{aligned}0 &= \int_{B_R\setminus\Omega}(-\Delta-k^2)w\overline{w}dx\\ &= -\int_{\partial B_R}\frac{\partial w}{\partial r}\overline{w}ds + \int_{\partial\Omega}\frac{\partial w}{\partial n}\overline{w}ds + \int_{B_R\setminus\Omega}(|\nabla w|^2 - k^2|w|^2)dx.\end{aligned} \tag{36.8}$$

Taking the imaginary part of (36.8), we obtain

$$\operatorname{Im}\int_{\partial B_R}\frac{\partial w}{\partial r}\overline{w}ds = \operatorname{Im}\int_{\partial\Omega}\frac{\partial w}{\partial n}\overline{w}ds.$$

Using the assymptotics (36.5) and taking the limit as $R \to \infty$, we get:

$$\operatorname{Im} \int_{\partial B_R} \frac{\partial w}{\partial r} \overline{w} ds = \operatorname{Im} \int_{\partial B_R} \overline{\left(\frac{b(\theta) ik e^{ik|x|}}{|x|^{\frac{n-1}{2}}} + O\left(\frac{1}{|x|^{\frac{n+1}{2}}} \right) \right)} \times \left(b(\theta) \frac{e^{ik|x|}}{|x|^{\frac{n-1}{2}}} + O\left(\frac{1}{|x|^{\frac{n+1}{2}}} \right) \right) R^{n-1} d\theta$$

$$\to \int_{|\theta|=1} k|b(\theta)|^2 d\theta,$$

where $d\theta$ is the surface area of the unit sphere S^{n-1}. Therefore

$$\operatorname{Im} \int_{\partial\Omega} \frac{\partial w}{\partial n} \overline{w} ds = \int_{|\theta|=1} k|b(\theta)|^2 d\theta. \tag{36.9}$$

Since $w|_{\partial\Omega} = 0$, we get that $b(\theta) = 0$. Therefore, $w(x) = O(\frac{1}{|x|^{\frac{n+1}{2}}})$ as $|x| \to \infty$. Then by Rellich's lemma (Lemma 35.2), we conclude that $w = 0$ in $\mathbb{R}^n \setminus \Omega$.

Remark 36.1. The same uniqueness result holds if the Dirichlet condition (36.3) is replaced by the Neumann condition (36.4), since $\frac{\partial w}{\partial n}|_{\partial\Omega} = 0$ implies (see (36.9)) that $b(\theta) = 0$.

Theorem 36.3. *There exists a unique solution of the problem* (36.1), (36.2), (36.3). *This solution has the following asymptotics:*

$$u(x, k\omega) = e^{ik\omega\cdot x} + \frac{a(\theta, \omega, k) e^{ik|x|}}{|x|^{\frac{n-1}{2}}} + O\left(\frac{1}{|x|^{\frac{n+1}{2}}} \right), \tag{36.10}$$

where $\theta = \frac{x}{|x|}$, $|x| \to \infty$. *The function* $a(\theta, \omega, k)$ *is called the scattering amplitude.*

Proof: Consider first the case where the Neumann problem

$$(-\Delta - k^2)v = 0 \quad \text{in} \quad \Omega, \tag{36.11}$$

$$\left(\frac{\partial v}{\partial n} \right)^+ \Big|_{\partial\Omega} = 0 \tag{36.12}$$

has only the zero solution. Here $(\frac{\partial v(x')}{\partial n(x')})^+$ is the limit of $\frac{\partial v}{\partial n}$ as $x \to x' \in \partial\Omega$, $x \in \Omega$. We look for the solution of the problem (36.1), (36.2), (36.3) in the form $u = e^{ik\omega\cdot x} + Dh$, where Dh is the double layer potential:

$$Dh = \int_{\partial\Omega} \frac{\partial}{\partial n(y')} E_+(x - y') h(y') ds, \quad x \in \mathbb{R}^n \setminus \overline{\Omega}, \tag{36.13}$$

and $\frac{\partial}{\partial n(y')}$ is the outward normal derivative at a point $y' \in \partial\Omega$. Then $v = Dh$ satisfies (36.1) and the radiation conditions (36.2). We choose $h \in C^2(\partial\Omega)$

such that (36.3) is satisfied. It follows from Theorem 20.1 and Remark 20.1 that

$$v^-(x') = \frac{1}{2}h + D_0 h = -e^{ik\omega\cdot x}|_{\partial\Omega}, \tag{36.14}$$

where $D_0 h$ is the same as in (20.6), only with $E(x'-y')$ replaced by $E_+(x'-y')$, and $v^-(x') = \lim v(x)$ as $x \to x' \in \partial\Omega$, $x \in \mathbb{R}^n \setminus \overline{\Omega}$. The operator D_0 is compact in $C^m(\partial\Omega)$ for every $m \geq 0$. By the Fredholm alternative, equation (36.14) will have a solution if the homogeneous equation $\frac{1}{2}h_0 + D_0 h_0 = 0$ has only a trivial solution in $C^2(\partial\Omega)$.

Set $v_0 = Dh_0$. Then $(-\Delta - k^2)v_0 = 0$ in $\mathbb{R}^n \setminus \overline{\Omega}$, v_0 satisfies (36.2), and $v_0^- = \frac{1}{2}h_0 + D_0 h_0 = 0$ on $\partial\Omega$. By Theorem 36.2, $Dh_0 = 0$ in $\mathbb{R}^n \setminus \overline{\Omega}$. Therefore $(\frac{\partial Dh_0}{\partial n})^- = 0$. Consider $v_0 = Dh_0$ in Ω. We have $(-\Delta - k^2)v_0 = 0$ in Ω. By Theorem 20.2, $(\frac{\partial Dh_0}{\partial n})^+ = (\frac{\partial Dh_0}{\partial n})^- = 0$. By assumption, the Neumann problem (36.11), (36.12) has only a trivial solution. Therefore, $v_0 = Dh_0 = 0$ in Ω.

Finally, using Theorem 20.1, we get:

$$h_0 = (Dh_0)^- - (Dh_0)^+ = 0. \tag{36.15}$$

Therefore, there exists $h \in C^2(\partial\Omega)$ such that (36.14) holds. □

Now we drop the assumption that the problem (36.11), (36.12) has only a trivial solution. We can always choose $q(x) \in C_0^\infty(\Omega)$ such that the Neumann problem

$$(-\Delta + q - k^2)v = 0 \quad \text{in} \quad \Omega, \tag{36.16}$$

$$\left(\frac{\partial v}{\partial n}\right)^+ = 0 \quad \text{on} \quad \partial\Omega, \tag{36.17}$$

has only a trivial solution. Extend q by zero outside of Ω. It was proven in §31 that there exists a fundamental solution $G(x,y)$ in $\mathbb{R}^n$ such that

$$(-\Delta + q - k^2)G(x,y) = \delta(x-y) \tag{36.18}$$

and $G(x,y)$ satisfies the radiation condition in x for each y. We look for the solution of (36.1), (36.2), (36.3) in the form $u = e^{ik\omega\cdot x} + v$, where

$$v(x) = \int_{\partial\Omega} \frac{\partial}{\partial n(y')} G(x,y')h(y')ds. \tag{36.19}$$

Then

$$(-\Delta - k^2)v = 0 \quad \text{in} \quad \mathbb{R}^n \setminus \overline{\Omega},$$
$$(-\Delta - k^2 + q)v = 0 \quad \text{in} \quad \Omega,$$

and $v(x)$ satisfies the radiation conditions. Since $G(x,y) - E(x-y)$ satisfies the same estimates as $E_+(x-y) - E(x-y)$ (cf. (20.19), (20.20)), we conclude that all the results of §20 remain true if $E(x-y)$ is replaced by $G(x,y)$ (cf. Remark 20.1). Therefore, we can repeat the proof of Theorem 36.3

with (36.13) replaced by (36.19) to remove the restriction that the problem (36.16), (36.17) has only a trivial solution.

36.1. The case of the Neumann conditions.

The existence and uniqueness result holds also for the case of the Neumann boundary conditions (36.4). Let $q(x) \in C_0^\infty(\Omega)$ be such that the Dirichlet problem

$$(-\Delta + q - k^2)v = 0 \quad \text{in} \quad \Omega, \tag{36.20}$$

$$v^+(x') = 0 \quad \text{on} \quad \Omega, \tag{36.21}$$

has only a trivial solution. Let $G(x,y)$ be the fundamental solution as in (36.18). Note that we could leave $E_+(x-y)$ instead of $G(x,y)$ if (36.20), (36.21) has only a trivial solution when $q = 0$. We look for the solution of (36.1), (36.2), (36.4) in the form $u = e^{ik\omega\cdot x} + v$, where

$$v(x) = \int_{\partial\Omega} G(x,y')h(y')ds. \tag{36.22}$$

Then $v(x)$ satisfies (36.1), (36.2), and we find $h(y')$ such that (cf. (20.15)):
(36.23)

$$\left(\frac{\partial v}{\partial n(x')}\right)^- = -\frac{1}{2}h(x') + \int_{\partial\Omega} \frac{\partial}{\partial n(x')} G(x',y')h(y')ds = -\frac{\partial e^{ik\omega\cdot x'}}{\partial n(x')}\Big|_{\partial\Omega}.$$

By the Fredholm alternative, (36.23) has a solution in $C^2(\partial\Omega)$ if the homogeneous equation $-\frac{1}{2}h_0 + G_1 h_0 = 0$ has only a trivial solution, where $G_1 h$ is the integral operator in (36.23). Let $v_0 = \int_{\partial\Omega} G(x,y')h_0(y')ds$. Then v_0 satisfies (36.1), (36.2) and $(\frac{\partial v_0}{\partial n(x')})^- = 0$. By Remark 36.1 we have $v_0 = 0$ in $\mathbb{R}^n \setminus \overline{\Omega}$. Hence $v_0^- = 0$ on $\partial\Omega$. Since the single layer potential is continuous in $\mathbb{R}^n$, we get $v_0^+ = v_0^- = 0$. Consequently, v_0 is a solution of the homogeneous Dirichlet problem (36.20), (36.21). Hence, by the choice of $q(x)$, we have $v_0 = 0$ in Ω. Then $(\frac{\partial v}{\partial n})^+ = 0$ on $\partial\Omega$. By Theorem 20.3 (cf. Remark 20.1) we have $h_0 = (\frac{\partial v_0}{\partial n})^+ - (\frac{\partial v_0}{\partial n})^- = 0$. Therefore, equation (36.23) has a unique solution in $C^2(\partial\Omega)$. □

36.2. Inverse obstacle problem.

Now we study the inverse obstacle problem, i.e., the recovery of Ω from the scattering amplitude. We consider only the case of Dirichlet boundary conditions.

Theorem 36.4. *Let*

$$u_j(x,k\omega) = e^{ik\omega\cdot x} + v_j = e^{ik\omega\cdot x} + a_j(\theta,\omega,k)\frac{e^{ik|x|}}{|x|^{\frac{n-1}{2}}} + O\left(\frac{1}{|x|^{\frac{n+1}{2}}}\right)$$

be the distorted plane wave solutions of $(-\Delta - k^2)u_j = 0$ *in* $\mathbb{R}^n \setminus \overline{\Omega_j}$, *satisfying the Dirichlet boundary conditions* $u_j|_{\partial\Omega_j} = 0,\ j = 1, 2$. *Assume that* $\mathbb{R}^n \setminus \Omega_j$ *is connected. Suppose* $a_1(\theta, \omega, k) = a_2(\theta, \omega, k)$ *for fixed* $k > 0$, *for all* $\theta \in S^{n-1}$, *and for infinitely many directions* $\omega_p,\ p = 1, 2, \ldots$. *Then* $\Omega_1 = \Omega_2$.

Proof: For fixed k and ω_p we have

$$w = u_1 - u_2 = v_1 - v_2 = O\left(\frac{1}{|x|^{\frac{n+1}{2}}}\right),$$

and $(-\Delta - k^2)w = 0$ in G, where G is the unbounded component of $\mathbb{R}^n \setminus (\overline{\Omega_1} \cup \overline{\Omega_2})$. Then, by the Rellich lemma (Lemma 35.2), we have $w = 0$ in G. Denote $\Omega_3 = \mathbb{R}^n \setminus \overline{G}$. By continuity, $w = 0$ on $\partial\Omega_3$. Therefore $u_1 = u_2$ on $\partial\Omega_3$. If $\Omega_1 \neq \Omega_2$, then either $\Omega_3 \setminus \overline{\Omega_1}$ or $\Omega_3 \setminus \overline{\Omega_2}$ is nonempty. Suppose, for the definiteness, that $\Omega_4 = \Omega_3 \setminus \overline{\Omega_1}$ is nonempty. Suppose also that Ω_4 is connected (otherwise, take a connected component of Ω_4). Since $\Omega_4 \subset \mathbb{R}^n \setminus \overline{\Omega_1}$, we see that u_1 is defined in Ω_4 and

$$(-\Delta - k^2)u_1 = 0 \quad \text{in} \quad \Omega_4. \tag{36.24}$$

Since the boundary of Ω_4 is contained in $\partial\Omega_1 \cup \partial\Omega_2$ and $u_1 = u_2 = 0$ on $\partial\Omega_2 \cap \partial\Omega_4$, we get that

$$u_1 = 0 \quad \text{on} \quad \partial\Omega_4. \tag{36.25}$$

Therefore, $u_1(x, k\omega_p), p = 1, 2, \ldots$, are the solutions of the homogeneous Dirichlet problem (36.24), (36.25).

We show that $u_1(x, k\omega_p),\ p = 1, 2, \ldots$, are linearly independent. If

$$\sum_{p=1}^{N} d_p u_1(x, k\omega_p) = 0 \tag{36.26}$$

in Ω_4, then, by the analyticity of $u_1(x, k\omega_p)$ in $\mathbb{R}^n \setminus \overline{\Omega_1}$ (see Theorem 18.3), (36.26) holds for all $x \in \mathbb{R}^n \setminus \overline{\Omega_1}$. Multiplying (36.26) by $\frac{1}{R^{n-1}} e^{-ik\omega_m \cdot x}$, integrating over the sphere ∂B_R, using (36.10), and taking the limit as $R \to \infty$, we get

$$Cd_m = 0,$$

where C is a constant, $C \neq 0$.

Therefore, $d_m = 0$ for any $m = 1, 2, \ldots$. However, the space of solutions $u(x, k\omega)$ satisfying (36.24), (36.25) is finite-dimensional, since the set

$$\{u : \|u\|_{L_2(\Omega_4)} = 1 \text{ and } u(x) \text{ satisfies (36.24), (36.25)}\}$$

is compact.

This contradiction proves that $\Omega_1 = \Omega_2$. □

37. Inverse scattering at a fixed energy

As was noted in §33, the recovery of $q(x)$ from the scattering amplitude known for all $|\theta| = 1$, $|\omega| = 1$, and $k > 0$ is an overdetermined problem if $n \geq 2$. A less overdetermined inverse problem is the recovery of $q(x)$ when the scattering amplitude $a(\theta, \omega, k)$ is known for all $|\theta| = 1$, $|\omega| = 1$, and a fixed $k = k_0$. Such inverse problems arise in geophysics.

Theorem 37.1. *Consider the reduced acoustic equation*

$$c^2(x)\Delta u + k^2 u = 0, \ x \in \mathbb{R}^n, \ n \geq 3, \tag{37.1}$$

where $0 < c_1 \leq c^2(x) \leq c_2$, $c(x) - 1 = O(e^{-\delta|x|})$, $\delta > 0$.

Note that (37.1) *can be rewritten in the form*

$$-\Delta u - k^2 u + q(x)u = 0, \tag{37.2}$$

where $q(x) = k^2(1 - c^{-2}(x)) = O(e^{-\delta|x|})$.

Knowing the scattering amplitude $a(\theta, \omega, k)$ *for all* $|\theta| = 1$, $|\omega| = 1$, *and fixed* $k > 0$, *one can recover* $q(x)$ *and, therefore,* $c(x)$.

Theorem 37.1 was proven by Novikov [**No2**]. The proof presented here follows Eskin-Ralston [**ER1**] and contains the procedure for the recovery of $q(x)$ from $a(\theta, \omega, k)$. The proof consists of several steps.

37.1. Relation between the scattering amplitude and the Faddeev's scattering amplitudes.

As in §33, the distorted plane wave is the solution of the form $u(x) = e^{ikx\cdot\omega} + v(x, k\omega)$, where $v(x, k\omega)$ is the outgoing solution to

$$(-\Delta - k^2 + q(x))v = -q(x)e^{ik\omega\cdot x}. \tag{37.3}$$

We have $v = E_+ h$, where $h(x, k\omega)$ is the solution of the integral equation

$$h(x, k\omega) + q(x)E_+ h = -q(x)e^{ik\omega\cdot x}. \tag{37.4}$$

Note that (see (33.3))

$$a(\theta, \omega, k) = C_n \tilde{h}(k\theta, k\omega), \tag{37.5}$$

where $\tilde{h}(\xi, x\omega)$ is the Fourier transform of $h(x, k\omega)$ and C_n is the same as in (33.4).

Let

$$E_{\nu,\sigma} = F^{-1}\frac{1}{|\xi|^2 - k^2 + i0(\xi_\nu - \sigma)}$$

be Faddeev's fundamental solution (see Example 16.6), where $|\nu| = 1$, $\xi_\nu = \xi \cdot \nu$. Consider the solution of (37.3) of the form $v_{\nu,\sigma} = E_{\nu,\sigma} h_{\nu,\sigma}$. Then

$h_{\nu,\sigma}(x,k\omega)$ satisfies an integral equation similar to (37.4):

$$(37.6)\quad h_{\nu,\sigma}(x,k\omega)+q(x)\frac{1}{(2\pi)^n}\int_{\mathbb{R}^n}\frac{\tilde{h}_{\nu,\sigma}(\eta,k\omega)e^{ix\cdot\eta}d\eta}{|\eta|^2-k^2+i0(\eta_\nu-\sigma)}=-q(x)e^{ix\cdot k\omega}.$$

One can find the asymptotics of $v_{\nu,\sigma}$ as $|x|\to\infty$ applying the stationary phase method to $E_{\nu,\sigma}h_{\nu,\sigma}$, but we will not need this.

A function $a_{\nu,\sigma}(\theta,\omega.k)=C_n\tilde{h}_{\nu,\sigma}(k\theta,k\omega)$ is called Faddeev's scattering amplitude. Since $E_{\nu,\sigma}$ is bounded from $V_{0,s}(\mathbb{R}^n)$ to $V_{2,-s}(\mathbb{R}^n)$ for any $s>\frac{1}{2}$ (see Theorem 29.2), we get, as in §30, that (37.6) is a second kind Fredholm equation in $V_{0,s}(\mathbb{R}^n)$, provided k and ω are fixed.

The following lemma gives a relation between $a(\theta,\omega,k)$ and Faddeev's scattering amplitude:

Lemma 37.2. *The following integral equation holds:*

$$(37.7)\quad \tilde{h}_{\nu,\sigma}(\xi,k\omega)=\tilde{h}(\xi,k\omega)-\frac{i\pi k^{n-2}}{(2\pi)^n}\int_{k\hat{\eta}\cdot\nu>\sigma}\tilde{h}(\xi,k\hat{\eta})\tilde{h}_{\nu,\sigma}(k\hat{\eta},k\omega)d\hat{\eta},$$

where $|\hat{\eta}|=1$ *and* $d\hat{\eta}$ *is the area element of the sphere* $|\hat{\eta}|=1$.

Proof: It was shown in Example 16.6 that for any $\varphi\in C_0^\infty(\mathbb{R}^n)$,

$$(37.8)\quad \frac{1}{(2\pi)^n}\int_{\mathbb{R}^n}\frac{\varphi(\xi)}{|\xi|^2-k^2+i0(\eta_\nu-\sigma)}d\xi$$
$$=\frac{1}{(2\pi)^n}\int_{\mathbb{R}^n}\frac{\varphi(\xi)}{|\eta|^2-k^2-i0}d\xi-\frac{i\pi k^{n-2}}{(2\pi)^n}\int_{k\hat{\eta}\cdot\nu>\sigma}\varphi(k\hat{\eta})d\hat{\eta}.$$

Therefore, we can rewrite (37.6) in the form

$$(37.9)\quad h_{\nu,\sigma}(x,k\omega)+q(x)\frac{1}{(2\pi)^n}\int_{\mathbb{R}^n}\frac{e^{ix\cdot\eta}\tilde{h}_{\nu,\sigma}(\eta,k\omega)}{|\eta|^2-k^2-i0}d\eta$$
$$=\frac{i\pi k^{n-2}}{(2\pi)^n}\int_{k\hat{\eta}\cdot\nu>\sigma}q(x)e^{ix\cdot k\hat{\eta}}\tilde{h}_{\nu,\sigma}(k\hat{\eta},k\omega)d\hat{\eta}-q(x)e^{ix\cdot k\omega}.$$

Note that (37.9) makes sense for any $\tilde{h}_{\nu,\sigma}(\xi,k\omega)\in H_s(\mathbb{R}^n)$, $s>\frac{1}{2}$. It was proven in §30 that the operator $I+T$, where $T=q(x)E_+h$, has an inverse in $V_{0,s}(\mathbb{R}^n)$. Therefore, we obtain from (37.4) that

$$(37.10)\qquad h(x,k\omega)=-(I+T)^{-1}q(x)e^{ix\cdot k\omega}.$$

Applying $(I+T)^{-1}$ to (37.9) and using (37.10), we get

$$(37.11)\quad h_{\nu,\sigma}(x,k\omega)=h(x,k\omega)-\frac{i\pi k^{n-2}}{(2\pi)^n}\int_{k\hat{\eta}\cdot\nu>\sigma}h(x,k\hat{\eta})\tilde{h}_{\nu,\sigma}(k\hat{\eta},k\omega)d\hat{\eta}.$$

Note that $q(x)e^{ik\hat{\eta}\cdot x}$ continuously depends on $\hat{\eta}$ as an element of $V_{0,s}(\mathbb{R}^n)$. Therefore,

$$
\begin{aligned}
(I+T)^{-1}\int_{k\hat{\eta}\cdot\nu>\sigma} & q(x)e^{ix\cdot k\hat{\eta}}\tilde{h}_{\nu,\sigma}(k\hat{\eta},k\omega)d\hat{\eta} \\
&= \int_{k\hat{\eta}\cdot\nu>\sigma}((I+T)^{-1}q(x)e^{ix\cdot k\hat{\eta}})\tilde{h}_{\nu,\sigma}(k\hat{\eta},k\omega)d\hat{\eta} \\
&= -\int_{k\hat{\eta}\cdot\nu>\sigma} h(x,k\hat{\eta})\tilde{h}_{\nu,\sigma}(k\hat{\eta},k\omega)d\hat{\eta}.
\end{aligned}
$$

Applying the Fourier transform operator to (37.11), we get (37.7). □

Restricting ourselves to $\xi = k\theta$, $|\theta| = 1$, we obtain an integral equation on the sphere S_k of radius k.

Note that (37.7) is a second kind Fredholm equation for $\tilde{h}_{\nu,\sigma}(\cdot,k\omega) \in L_2(S_k)$, since the integral operator in (37.7) is a Hilbert-Schmidt operator in $L_2(S_k)$.

Lemma 37.3. *Equation* (37.7) *is uniquely solvable in* $L_2(S_k)$ *iff equation* (37.6) *is uniquely solvable in* $V_{0,s}(\mathbb{R}^n)$, $s > \frac{1}{2}$.

Proof: Since (37.7) and (37.6) are Fredholm equations, it is enough to prove that (37.7) has a unique solution iff there is no nontrivial solutions of the homogeneous equation corresponding to (37.6):

$$
g(x) + q(x)E_{\nu,\sigma}g = 0. \tag{37.12}
$$

Rewriting (37.12) as in (37.9), applying $(I+T)^{-1}$, and then applying the Fourier transform, we get

$$
\tilde{g}(\xi) = -\frac{i\pi k^{n-2}}{(2\pi)^n}\int_{k\hat{\eta}\cdot\nu>\sigma}\tilde{h}(\xi,k\hat{\eta})\tilde{g}(k\hat{\eta})d\hat{\eta}. \tag{37.13}
$$

Note that if $\tilde{g}(k\hat{\eta}) = 0$ for all $|\hat{\eta}| = 1$, then (37.13) implies that $\tilde{g}(\xi) = 0$ for all $\xi \in \mathbb{R}^n$. Therefore, if $g(x)$ is a nontrivial solution of (37.12), then restricting ourselves to $|\xi| = k$ we obtain a nontrivial solution in $L_2(S_k)$ of the homogeneous equation corresponding to (37.7). Conversely, let $\varphi(k\theta) \in L_2(S_k)$, $|\theta| = 1$, be a nontrivial solution of (37.13). Since $\tilde{h}(\xi,k\hat{\eta})$ is defined for all $\xi \in \mathbb{R}^n$, the right hand side of (37.13) defines $\varphi(\xi)$ for all $\xi \in \mathbb{R}^n$ and we denote $g(x) = F^{-1}\varphi(\xi)$.

Since $q(x)e^{ik\hat{\eta}\cdot x}$ depends continuously on $\hat{\eta}$ as an element of $V_{0,s}(\mathbb{R}^n)$, we conclude that $h(x,k\hat{\eta}) = -(I+T)^{-1}q(x)e^{ix\cdot k\hat{\eta}}$ depends continuously on $\hat{\eta}$ as an element of $V_{0,s}(\mathbb{R}^n)$. Therefore, the right hand side of (37.13) belongs to $H_s(\mathbb{R}^n) = FV_{0,s}(\mathbb{R}^n)$ when $\varphi(k\hat{\eta}) \in L_2(S_k)$, i.e., $\varphi(\xi) \in H_s(\mathbb{R}^n)$ and $g = F^{-1}\varphi(\xi) \in V_{0,s}(\mathbb{R}^n)$. Applying $(I+T)F^{-1}$ to (37.13) and taking into account that $(I+T)h(x,k\hat{\eta}) = -q(x)e^{ix\cdot k\hat{\eta}}$, we see that $g(x)$ is a nontrivial solution of (37.12) in $V_{0,s}(\mathbb{R}^n)$. □

37.2. Analytic continuation of T_r.

Denote by $h(x,\zeta,\tau)$ the solution of the integral equation

$$h_\nu(x,\zeta,\tau) + q(x)E_{\nu,0,\tau}h_\nu = -q(x)e^{ix\cdot\zeta}, \tag{37.14}$$

where $\tau > 0$, $\zeta \in \mathbb{R}^n$, and

$$E_{\nu,0,\tau}h_\nu = \frac{1}{(2\pi)^n}\int_{\mathbb{R}^n}\frac{\tilde{h}_\nu(\eta,\zeta,\tau)e^{ix\cdot\eta}d\eta}{(\eta+i\tau\nu)\cdot(\eta+i\tau\nu)-k^2}, \tag{37.15}$$

$$(\eta+i\tau\nu)\cdot(\eta+i\tau\nu) = |\eta|^2 - \tau^2 + 2i\tau\eta_\nu, \quad \eta_\nu = \eta\cdot\nu. \tag{37.16}$$

Note that $v_\nu = E_{\nu,0,\tau}h_\nu$ is the solution of the differential equation

$$\left[\left(-i\frac{\partial}{\partial x}+i\tau\nu\right)\cdot\left(-i\frac{\partial}{\partial x}+i\tau\nu\right) - k^2 + q(x)\right]v_\nu = -q(x)e^{ix\cdot\zeta}. \tag{37.17}$$

It was proven in §29 (cf. Theorem 29.3) that $E_{\nu,0,\tau}$ is bounded from $V_{0,s}(\mathbb{R}^n)$ to $V_{0,-s}(\mathbb{R}^n)$ and has a small norm when $\tau > 0$ is large (see (29.5)). Therefore, the operator $I + qE_{\nu,0,\tau}$ is invertible in $V_{0,s}(\mathbb{R}^n)$ if $\tau > 0$ is large.

Until now we have not used that $q(x) = O(e^{-\delta|x|})$. Note that Lemmas 37.2 and 37.3 are true for $q(x) = O(\frac{1}{|x|^{\frac{n+1}{2}+\varepsilon}})$, $\varepsilon > 0$. Now we exploit that $q(x) = O(e^{-\delta|x|})$.

Denote by T_τ the following operator:

$$T_\tau h = q(x)E_{\nu,0,\tau}h. \tag{37.18}$$

Denote by $A_{r,\alpha}$, $\alpha \in \mathbb{R}^1$, the Banach space with the norm:

$$\|u\|^2_{A_{r,\alpha}} = \sum_{|p|=0}^{r}\int_{\mathbb{R}^n} e^{2\alpha|x|}\left|\frac{\partial^p u}{\partial x^p}\right|^2 dx. \tag{37.19}$$

For $\alpha > 0$ we denote by $\tilde{A}_{r,\alpha}$ the Fourier transform of $A_{r,\alpha}$. Setting $z = \xi + i\eta$ and differentiating $\int_{\mathbb{R}^n} u(x)e^{-ix\cdot(\xi+i\eta)}dx$ with respect to z for $|\eta| < \alpha$, we get that the functions $\tilde{u}(\xi+i\eta) \in \tilde{A}_{r,\alpha}$ are analytic in the strip $|\eta| < \alpha$.

Parseval's equality gives

$$\frac{1}{(2\pi)^n}\int_{\mathbb{R}^n}|\tilde{u}(\xi+i\eta)|^2 d\xi = \int_{\mathbb{R}^n}|u(x)|^2 e^{2x\cdot\eta}dx \le \|u\|^2_{A_{0,\alpha}}$$

for all $|\eta| < \alpha$. Analogously, for $|\eta| < \alpha$, we have

$$\frac{1}{(2\pi)^n}\int_{\mathbb{R}^n}(1+|\xi|)^{2r}|\tilde{u}(\xi+i\eta)|^2 d\xi \le C\|u\|^2_{A_{r,\alpha}}, \tag{37.20}$$

where C is independent of η.

Let M be a compact set in $\mathbb{C}^n$ such that $|\eta| < \alpha_1 < \alpha$ for $z = \xi + i\eta \in M$. Using the Cauchy integral formula in each z_k, $k = 1, \dots, n$, we get that

$$\max_{\xi + i\eta \in M} |\tilde{u}(\xi + i\eta)| \leq C \|u\|_{A_{0,\alpha}}. \tag{37.21}$$

Let $L(\tau)$ be a family of bounded operators from B_1 to B_2 for any $\tau \in \Omega$. Here B_1 and B_2 are Banach spaces and Ω is an open domain in $\mathbb{C}$. We say that $L(\tau)$ is analytic in Ω if $(L(\tau)h, g)$ is an analytic function in Ω for any $h \in B_1$, $g \in B_2^*$, where B_2^* is dual to B_2, and

$$|(L(\tau)h, g)| \leq C_M \|h\|_{B_1} \|g\|_{B_2^*} \tag{37.22}$$

for any compact set $M \subset \Omega$. Let $\tau_0 \in \Omega$ be such that the disk $|\tau - \tau_0| < \varepsilon$ is contained in Ω. Using the Cauchy integral formula for $(L(\tau)h, g)$ in the disk, we see that the Taylor series $\sum_{m=0}^{\infty} \frac{1}{m!} \frac{d^m}{d\tau^m} L(\tau_0)(\tau - \tau_0)^m$ is convergent in the operator norm for $|\tau - \tau_0| \leq \frac{\varepsilon}{2}$.

Denote by Π_k the half-strip $\{\tau = \mu + i\sigma : \ \mu > 0, \ |\sigma| < k\}$ if $k \leq \delta_0$, and $\Pi_k = \{\tau = \mu + i\sigma : |\sigma| < k \text{ for } 0 < \mu < \mu_0, \ |\sigma| < \delta_0 \text{ for } \mu_0 \leq \mu < +\infty\}$ if $k > \delta_0$. Here $\delta_0 = \delta_1 - \varepsilon$, $\delta_1 < \frac{\delta}{2}$, $\varepsilon > 0$ is small, and δ is the same as in (37.1), $0 < \mu_0 < \frac{\delta_0^2}{\sqrt{k^2 - \delta_0^2}}$.

Lemma 37.4. *Operator T_τ has an analytic extension in τ to the domain Π_k, and T_τ is an analytic operator-valued function in Π_k with values in the space of compact operators in A_{0,δ_1}, $\delta_1 < \frac{\delta}{2}$.*

Proof: It is enough to prove that $\frac{\partial^p}{\partial x^p} E_{\nu,0,\tau}$, $0 \leq |p| \leq 2$, is an analytic operator-valued function in Π_k bounded from A_{0,δ_1} to $A_{0,-\delta_1}$. Then the proof that $q(x)E_{\nu,0,\tau}$ is compact in A_{0,δ_1} is the same as the proof of the compactness of $T = qE_+$ in Theorem 30.2: we represent $T = q(x)E_{\nu,0,\tau}$ in the form $T_\tau = q(x)e^{(\delta_1+\delta_2)|\hat{x}|} T_0 \Lambda^2 e^{-\delta_1|\hat{x}|} E_{\nu,0,\tau}$, where $T_0 = e^{-\delta_2|\hat{x}|}\Lambda^{-2}$, $|\hat{x}| = \sqrt{|x|^2 + 1}$, $0 < \delta_2 < \frac{\delta}{2} - \delta_1$. As in §30, we have that $T_0 = e^{-\delta_2|\hat{x}|}\Lambda^{-2}$ is compact in $L_2(\mathbb{R}^n)$ and, therefore, T_τ is compact in A_{0,δ_1}.

For any $h \in A_{0,\delta_1}$ and $g \in A_{0,\delta_1}$ consider

$$\left(\frac{\partial^p}{\partial x^p} E_{\nu,0,\tau} h, \bar{g}\right) = \frac{1}{(2\pi)^n} \int_{\mathbb{R}^n} \frac{(i\xi)^p \tilde{h}(\xi)\tilde{g}(-\xi) d\xi}{|\xi'|^2 + \xi_\nu^2 - k^2 - \tau^2 + 2i\tau\xi_\nu}. \tag{37.23}$$

We have used that $\overline{F(\bar{g})} = \tilde{g}(-\xi)$. Here $\xi_\nu = \xi \cdot \nu$ and $\xi' = \xi - \xi_\nu \nu$. We introduce polar coordinates $\xi' = r\omega'$, where $r = |\xi'|$, $|\omega'| = 1$. Then

$$\left(\frac{\partial^p}{\partial x^p} E_{\nu,0,\tau} h, \bar{g}\right) = \frac{1}{(2\pi)^n} \left(\int_{\xi_\nu > 0} I_+(\xi_\nu) d\xi_\nu + \int_{\xi_\nu < 0} I_-(\xi_\nu) d\xi_\nu \right), \tag{37.24}$$

where
(37.25)

$$I_{\pm}(\xi_\nu) = \int_0^\infty \int_{|\omega'|=1} \frac{(i\xi_\nu)^{p_\nu}(i\xi')^{p'}\tilde{h}(\xi_\nu, r\omega')\tilde{g}(-\xi_\nu, -r\omega')r^{n-2}drd\omega'}{\xi_\nu^2 + r^2 - k^2 - \tau^2 + 2i\tau\xi_\nu},$$

and $p = (p_\nu, p')$.

For $\xi_\nu > 0, \tau > 0$ we deform the contour of integration in r from the positive real axis to $\Gamma_+ = \{z = r + i\alpha(r)\}$, where $0 \leq r < +\infty$, $\alpha(r) = \frac{\delta_0}{\varepsilon}r$ for $0 \leq r \leq \varepsilon$, and $\alpha(r) = \delta_0$ for $\varepsilon \leq r < \infty$.

Note that $z^2 = (r + i\alpha(r))^2 = r^2 - \alpha^2(r) + 2ir\alpha(r)$ and $r\alpha(r) \geq 0$ for $r \geq 0$. By the Cauchy theorem,
(37.26)

$$I_+(\xi_\nu) = \int_{\Gamma_+} \int_{|\omega'|=1} \frac{(i\xi_\nu)^{p_\nu}(iz\omega')^{p'}\tilde{h}(\xi_\nu, z\omega')\tilde{g}(-\xi_\nu, -z\omega')z^{n-2}dzd\omega'}{\xi_\nu^2 + z^2 - k^2 - \tau^2 + 2i\tau\xi_\nu},$$

since $\frac{1}{\xi_\nu^2+z^2-k^2-\tau^2+2i\tau\xi_\nu}$ is analytic in z between Γ_+ and the positive real axis for $\tau > 0,\ \xi_\nu > 0$.

Analogously,
(37.27)

$$I_-(\xi_\nu) = \int_{\Gamma_-} \int_{|\omega'|=1} \frac{(i\xi_\nu)^{p_\nu}(iz\omega')^{p'}\tilde{h}(\xi_\nu, z\omega')\tilde{g}(-\xi_\nu, -z\omega')z^{n-2}dzd\omega'}{\xi_\nu^2 + z^2 - k^2 - \tau^2 + 2i\tau\xi_\nu},$$

where $\Gamma_- = \{z = r - i\alpha(r),\ 0 \leq r < +\infty\}$ and $\alpha(r)$ is the same as in (37.26).

Until now τ was real and positive. When $\tau = \mu + i\sigma,\ \mu > 0,\ \xi_\nu > 0$, and $z = r + i\alpha(r),\ r \geq 0$, we have the following expression for the denominator in (37.26):

$$(\xi_\nu - \sigma + i\mu)^2 + (r + i\alpha(r))^2 - k^2. \tag{37.28}$$

The curve $\Gamma_0 = \{w(r) = k^2 - (r+i\alpha(r))^2 = k^2 - r^2 + \alpha^2(r) - 2i\alpha(r)r,\ r \geq 0\}$ consists of the segment $k^2 - (1 + i\frac{\delta_0}{\varepsilon})^2 r^2$ for $0 \leq r \leq \varepsilon$ and the parabola $k^2 - (r+i\delta_0)^2$ for $\varepsilon \leq r < +\infty$. Denote by D the range of $w = (\xi_\nu - \sigma + i\mu)^2$, where $\xi_\nu \geq 0,\ \mu + i\sigma \in \Pi_k$. If $k \leq \delta_0$, the region D consists of the upper half-plane $\operatorname{Im} w > 0$, the negative real axis, and the region in the lower half-plane $\operatorname{Im} w < 0$ below the segment $w = t^2,\ 0 \leq t \leq k$ and above the parabola $w = (-k + i\mu)^2,\ 0 \leq \mu < +\infty$. If $k > \delta_0$, the portion of D where $\operatorname{Im} w < 0$ is located above three parabolas: $w = (-k + i\mu)^2,\ 0 \leq \mu \leq \mu_0,\ w = (t + i\mu_0)^2,\ -k \leq t \leq -\delta_0$, and $w = (-\delta_0 + i\mu)^2,\ \mu_0 \leq \mu < +\infty$. In both cases Γ_0 and D do not intersect. The only common point of $\bar{\Gamma}_0$ and $\bar{D}$ is $w = k^2$, i.e., if $|\xi_\nu - \sigma| = k,\ \mu = 0,\ r = 0$. For $\mu > 0$ and $0 < r \leq \varepsilon$ we have:

$$(37.29)\quad |(\xi_\nu + i(\mu + i\sigma))^2 + (r + i\alpha(r))^2 - k^2| \ge C(|(\xi_\nu - \sigma)^2 - k^2| + \mu + r^2) \ge C_1\mu.$$

Therefore, we can differentiate $\int_0^\infty I_+(\xi_\nu)d\xi_\nu$ with respect to $\tau = \mu + i\sigma$, and $\int_0^\infty I_+(\xi_\nu)d\xi_\nu$ is analytic in τ for $\tau \in \Pi_k$. Analogously, $\int_{-\infty}^0 I_-(\xi_\nu)d\xi_\nu$ is analytic in τ for $\tau \in \Pi_k$. Therefore, $(\frac{\partial^p}{\partial x^p}E_{\nu,0,\tau}h, \bar{g})$ is analytic in τ for $\tau \in \Pi_k$. Using (37.21) and (37.20), we obtain that

$$(37.30)\qquad \left|\left(\frac{\partial^p}{\partial x^p}E_{\nu,0,\tau}h, \bar{g}\right)\right| \le C_M\|h\|_{A_{0,\delta_1}}\|\bar{g}\|_{A_{0,\delta_1}}$$

for all τ belonging to some compact set $M \subset \Pi_k$. It follows from (37.30) that $E_{\nu,0,\tau}h$ is a bounded operator from A_{0,δ_1} to $A_{2,-\delta_1}$ for all $\tau \in \Pi_k$. Therefore, $T_\tau = qE_{\nu,0,\tau}$ is a compact operator in A_{0,δ_1} analytically dependent on $\tau \in \Pi_k$.

Remark 37.1. We tried to find the domain of analyticity Π_k as large as possible. However, to prove Theorem 37.1 it is enough to extend T_τ analytically to a smaller domain $\Pi = \{\tau = \mu + i\sigma : \mu > 0,\ |\sigma| < \varepsilon\}$, where $\varepsilon > 0$ is small. This makes the proofs much simpler.

37.3. The limiting values of T_τ and Faddeev's scattering amplitude.

Now we study the limiting values of T_τ, $\tau \in \Pi_k$ in the case where $\mu = \operatorname{Re}\tau \to +0$.

When $\mu = 0$ and σ is fixed, $\xi_\nu \ge 0$, the denominator (37.28) vanishes only if $|\xi_\nu - \sigma| = k,\ r = 0$. Moreover, we have

$$(37.31)\quad |(\xi_\nu - \sigma)^2 + (r + i\alpha(r))^2 - k^2|^{-1} \le C\big|(\xi_\nu - \sigma)^2 - k^2| + r^2\big|^{-1}.$$

Since $\big||(\xi_\nu - \sigma)^2 - k^2| + r^2\big|^{-1}$ is locally integrable in $\mathbb{R}^n$, $n \ge 2$, the estimate (37.30) still holds. Therefore, $T_{0+i\sigma}$ is also a compact operator in A_{0,δ_1}. Moreover, $T_{\tau_2} - T_{\tau_1} \to 0$ in the operator norm as $|\tau_2 - \tau_1| \to 0$, $\tau_j = \mu_j + i\sigma_j$, $\mu_j \ge 0$, $j = 1, 2$, since

$$(37.32)\quad \left|\left(\left(\frac{\partial^p}{\partial x^p}E_{\nu,0,\tau_1} - \frac{\partial^p}{\partial x^p}E_{\nu,0,\tau_2}\right)h, \bar{g}\right)\right| \le C_{\tau_2-\tau_1}\|h\|_{A_{0,\delta_1}}\|\bar{g}\|_{A_{0,\delta_1}},$$

where $C_{\tau_2-\tau_1} \to 0$ as $\tau_2 - \tau_1 \to 0$.

To prove (37.32), we use an elementary estimate

$$\left|\frac{1}{a} - \frac{1}{b}\right| \le \frac{|a-b|}{|a||b|} \le \frac{C(|a|^{1-\varepsilon} + |b|^{1-\varepsilon})|a-b|^\varepsilon}{|a||b|}$$

and estimate (37.29). We have

$$(37.33)\quad \begin{aligned} &|[(\xi_\nu-\sigma_2+i\mu_2)^2+(r+i\alpha(r))^2-k^2]^{-1} \\ &\quad -[(\xi_\nu-\sigma_1+i\mu_1)^2+(r+i\alpha(r))^2-k^2]^{-1}| \\ &\le C_\varepsilon|\tau_2-\tau_1|^\varepsilon\left(\frac{1}{r^{2\varepsilon}(|(\xi_\nu-\sigma_1)^2-k^2|+\mu_1+r^2)}\right. \\ &\qquad\left.+\frac{1}{r^{2\varepsilon}(|(\xi_\nu-\sigma_2)^2-k^2|+\mu_2+r^2)}\right), \end{aligned}$$

where $\varepsilon>0$ is arbitrary.

Since $r^{-2\varepsilon}|(\xi_\nu-\sigma)^2-k^2|+\mu+r^2|^{-1}$ is locally integrable for $\varepsilon<\frac{1}{2}$, the estimate (37.32) holds with $C_{\tau_2-\tau_1}=C|\tau_2-\tau_1|^\varepsilon$.

Therefore, we have obtained that T_τ is a compact operator in A_{0,δ_1} for $\tau\in\overline{\Pi}_k$ and T_τ depends continuously on $\tau\in\overline{\Pi}_k$ in the operator norm. One can simplify the expression for $T_{0+i\sigma}$. We have, for $\tau=i\sigma$, $\xi_\nu>0$,

$$I_+(\xi_\nu)=\int_{\Gamma_+}\int_{|\omega'|=1}\frac{(i\xi_\nu)^{p_\nu}(iz\omega')^{p'}\tilde{h}(\xi_\nu,z\omega')\tilde{g}(-\xi_\nu,-z\omega')z^{n-2}dzd\omega'}{(\xi_\nu-\sigma)^2+r^2+2ir\alpha(r)-\alpha^2(r)-k^2},$$

since $z=r+i\alpha(r)$ on Γ_+. Note that $r\alpha(r)>0$ for $r>0$. Moving the contour Γ_+ back to the positive real axis, we obtain for $\xi_\nu>0$

$$I_+(\xi_\nu)=\int_{\mathbb{R}^{n-1}}\frac{(i\xi)^p\tilde{h}_\nu(\xi)\tilde{g}(-\xi)d\xi'}{(\xi_\nu-\sigma)^2+|\xi'|^2-k^2+i0}.$$

Analogously, for $\xi_\nu<0$ we have

$$I_-(\xi_\nu)=\int_{\mathbb{R}^{n-1}}\frac{(i\xi)^p\tilde{h}_\nu(\xi)\tilde{g}(-\xi)d\xi'}{(\xi_\nu-\sigma)^2+|\xi'|^2-k^2-i0}.$$

Therefore

$$(37.34)\quad \left(\frac{\partial^p}{\partial x^p}E_{\nu,0,i\sigma}h,\overline{g}\right)=\frac{1}{(2\pi)^n}\int_{\mathbb{R}^n}\frac{(i\xi)^p\tilde{h}(\xi)\tilde{g}(-\xi)d\xi}{(\xi_\nu-\sigma)^2+|\xi'|^2-k^2+i0\xi_\nu}.$$

We know that $I+T_\tau$ is invertible in $V_{0,\frac{1}{2}+\varepsilon}$ if $\tau>0$ is large (see Theorem 29.2). Therefore, $I+T_\tau$ is invertible in A_{0,δ_1} for such τ since the null-space of $(I+T_\tau)$ in A_{0,δ_1} is contained in the null-space of $I+T_\tau$ in $V_{0,\frac{1}{2}+\varepsilon}$.

We use the following fact about an analytic family of compact operators (see Rudin [**R**]):

Lemma 37.5. *Assume that T_τ is an analytic family of compact operators in a domain Π_k and assume that $I+T_\tau$ is invertible for some $\tau_0\in\Pi_k$. Then $(I+T_\tau)^{-1}$ exists for all $\tau\in\Pi_k$ with a possible exception of a discrete set Z having no limiting points inside Π_k.*

Moreover, $(I+T_\tau)^{-1}$ depends analytically on τ in $\Pi\setminus Z$. Assuming that T_τ extends continuously to $\overline{\Pi}$, we have that the set of the boundary $\partial\overline{\Pi}_k$ where $I+T_\tau$ has no inverse is closed and nowhere dense in $\partial\overline{\Pi}_k$.

Lemma 37.5 implies that the solution

$$h_\nu(x,\zeta,\tau) = (I+T_\tau)^{-1}(-q(x)e^{ix\cdot\zeta})$$

of the equation (37.14) is analytic in τ when $\tau\in\Pi_k\setminus Z$. Also $h_\nu(x,\zeta,\tau)$ is analytic in ζ when $|\operatorname{Im}\zeta|<\delta_2,\ 0<\delta_2<\delta-\delta_1$ since $q(x)=O(e^{-\delta|x|})$, and $q(x)e^{ix\cdot\zeta}$ is analytic in ζ with values in A_{0,δ_1} when $|\operatorname{Im}\zeta|<\delta_2$. Since $h_\nu(x,\zeta,\tau)\in A_{0,\delta_1}$, we have that $\tilde{h}_\nu(\xi,\zeta,\tau)$ is analytic in (ξ,ζ,τ), where $|\operatorname{Im}\xi|<\delta_1,\ |\operatorname{Im}\zeta|<\delta_2,\ \tau\in\Pi\setminus Z$.

Moreover, there exists an interval $(\sigma_0-\varepsilon,\sigma_0+\varepsilon)$ such that $I+T_{i\sigma}$ is invertible when $\sigma\in(\sigma_0-\varepsilon,\sigma_0+\varepsilon)$. Therefore, there exists a unique solution of the equation

$$h_\nu(x,\zeta,i\sigma)+T_{i\sigma}h_\nu = -q(x)e^{ix\cdot\zeta}. \tag{37.35}$$

It follows from (37.34) that

$$T_{i\sigma}h_\nu = \frac{q(x)}{(2\pi)^n}\int_{\mathbb{R}^n}\frac{e^{ix\cdot\eta}\tilde{h}_\nu(\eta,\zeta,i\sigma)d\eta}{(\eta_\nu-\sigma)^2+|\eta'|^2-k^2+i0\eta_\nu}. \tag{37.36}$$

Changing the variables $\eta_\nu-\sigma$ to η_ν, we get

$$h_\nu(x,\zeta,i\sigma)+q(x)\frac{1}{(2\pi)^n}\int_{\mathbb{R}^n}\frac{e^{ix\cdot\eta+ix_\nu\sigma}\tilde{h}_\nu(\eta_\nu+\sigma,\eta',\zeta,i\sigma)d\eta}{|\eta|^2-k^2+i0(\eta_\nu+\sigma)} = -q(x)e^{ix\cdot\zeta}. \tag{37.37}$$

Multiplying (37.37) by $e^{-ix_\nu\sigma}$ and changing ζ_ν to $\zeta_\nu+\sigma$, we get

$$e^{-ix_\nu\sigma}h_\nu(x,\zeta+\sigma\nu,i\sigma)+q(x)\frac{1}{(2\pi)^n}\int_{\mathbb{R}^n}\frac{e^{ix\cdot\eta}\tilde{h}_\nu(\eta+\sigma\nu,\zeta+\sigma\nu,i\sigma)d\eta}{|\eta|^2-k^2+i0(\eta_\nu+\sigma)} = -q(x)e^{ix\cdot\zeta}. \tag{37.38}$$

Note that $F(e^{-ix_\nu\sigma}h_\nu(x,\zeta+\sigma\nu,i\sigma)) = \tilde{h}_\nu(\xi+\sigma\nu,\zeta+\sigma\nu,i\sigma)$. Comparing (37.38) and (37.6), we get

$$e^{-ix_\nu\sigma}h_\nu(x,\zeta+\sigma\nu,i\sigma) = h_{\nu,-\sigma}(x,\zeta),\quad |\zeta|=k. \tag{37.39}$$

Applying the Fourier transform operator to (37.39), we obtain

$$\tilde{h}_\nu(\xi+\sigma\nu,\zeta+\sigma\nu,i\sigma) = \tilde{h}_{\nu,-\sigma}(\xi,\zeta). \tag{37.40}$$

Note that $C_n\tilde{h}_{\nu,-\sigma}(\xi,\zeta), |\xi|=|\zeta|=k$, is the Faddeev's scattering amplitude.

37.4. Final step: The recovery of $q(x)$.

Now we are ready to prove Theorem 37.1. Consider the integral equation (37.7) for $-\sigma \in (\sigma_0 - \varepsilon, \sigma_0 + \varepsilon)$. The integral equation (37.6) is uniquely solvable when $-\sigma \in (\sigma_0 - \varepsilon, \sigma_0 + \varepsilon)$ since it is equivalent to (37.35) (see (37.36)–(37.39)). Therefore, equation (37.7) is uniquely solvable, i.e., knowing $a(\theta, \omega, k) = C_n \tilde{h}(k\theta, k\omega, k)$, we can recover $\tilde{h}_{\nu,-\sigma}(\xi, \zeta)$ for all $|\xi| = k,\ |\zeta| = k$. It follows from (37.40) that we can recover $\tilde{h}_\nu(\xi, \zeta, i\sigma)$ for $(\xi - \sigma\nu) \cdot (\xi - \sigma\nu) = k^2,\ (\zeta - \sigma\nu) \cdot (\zeta - \sigma\nu) = k^2$.

Since $\tilde{h}_\nu(\xi, \zeta, \tau)$ is analytic in (ξ, ζ, τ) for $|\operatorname{Im} \xi| < \delta_1,\ |\operatorname{Im} \zeta| < \delta_2,\ \tau \in \Pi_k \setminus Z$ and since $\tilde{h}_\nu(\xi, \zeta, i\sigma)$ is the limiting value in τ of $\tilde{h}_\nu(\xi, \zeta, \tau)$, we can recover $\tilde{h}_\nu(\xi, \zeta, \tau)$ by analytic continuation for all (ξ, ζ, τ) that belong to the domain of analyticity of $\tilde{h}_\nu(\xi, \zeta, \tau)$ and are such that $(\xi + i\tau\nu) \cdot (\xi + i\tau\nu) = k^2,\ (\zeta + i\tau\nu) \cdot (\zeta + i\tau\nu) = k^2$.

Let l be an arbitrary vector in $\mathbb{R}^n$. Choose ν and μ such that $|\nu| = 1,\ |\mu| = 1,\ l \cdot \nu = 0,\ \nu \cdot \mu = 0,\ l \cdot \mu = 0$. Take

$$\xi(s) = \frac{1}{2} l + s\mu, \quad \zeta(s) = -\frac{1}{2} l + s\mu, \quad \tau(s) = \sqrt{s^2 + \frac{1}{4} l \cdot l - k^2}, \tag{37.41}$$

where $s \geq s_0 > 0$, and s_0 is large. Note that

$$\begin{aligned} &(\xi(s) + i\tau(s)\nu) \cdot (\xi(s) + i\tau(s)\nu) = k^2, \\ &(\zeta(s) + i\tau(s)\nu) \cdot (\zeta(s) + i\tau(s)\nu) = k^2. \end{aligned} \tag{37.42}$$

Therefore we know $\tilde{h}_\nu(\xi(s), \zeta(s), \tau(s))$. The Fourier transform of (37.14) and the substitute $\xi = \xi(s),\ \zeta = \zeta(s), \tau = \tau(s)$ give (cf. (33.14)):

$$\tilde{h}_\nu(\xi(s), \zeta(s), \tau(s)) + F(qE_{\nu,0,\tau(s)} h_\nu)(\xi(s)) = -\tilde{q}(\xi(s) - \zeta(s)). \tag{37.43}$$

Now take the limit as $s \to +\infty$. We have $\tilde{q}(\xi(s) - \zeta(s)) = \tilde{q}(l)$. We shall show that

$$|F(qE_{\nu,0,\tau(s)} h_\nu)| \leq \frac{C}{\tau(s)}. \tag{37.44}$$

It follows from Theorem 29.3 that

$$\|E_{\nu,0,\tau} h_\nu\|_{0,-\frac{1}{2}-\varepsilon} \leq \frac{C}{\tau} \|h_\nu\|_{0,\frac{1}{2}+\varepsilon}$$

for τ large. Therefore

$$\|T_\tau h_\nu\|_{0,\frac{n+1}{2}} = \|qE_{\nu,0,\tau} h_\nu\|_{0,\frac{n+1}{2}} \leq \frac{C}{\tau} \|h_\nu\|_{0,\frac{1}{2}+\varepsilon}.$$

Since $\|qe^{ix\cdot\zeta}\|_{0,\frac{1}{2}+\varepsilon} \leq C$, where C is independent of ζ, we see that

$$\|h_\nu\|_{0,\frac{1}{2}+\varepsilon} = \|(I + T_\tau)^{-1} qe^{ix\cdot\zeta}\|_{0,\frac{1}{2}+\varepsilon} \leq C$$

for τ large, where C is independent of τ and ζ. Therefore, by the Cauchy-Schwartz inequality,

$$|F(T_\tau h_\nu)| \le \int_{\mathbb{R}^n} |T_\tau h_\nu| dx \le C\|T_\tau h_\nu\|_{0,\frac{n+1}{2}} \le \frac{C_1}{\tau}.$$

Taking the limit as $s \to \infty$, we get

$$\tilde{q}(l) = -\lim_{s\to\infty} h_\nu(\xi(s), \zeta(s), \tau(s)).$$

Since l is arbitrary, we have recovered $q(x)$ from $a(\theta, \omega, k)$ given for fixed k and all $|\theta| = 1,\ |\omega| = 1$.

38. Inverse backscattering

Let $u(x, k\omega)$ be the distorted plane wave solution of the Schrödinger equation (33.1), i.e.,

$$u = e^{ik\omega\cdot x} + v(x, k\omega),$$

where $v = E_+h$, and $h(x, k\omega)$ is the solution of the integral equation (33.5). Performing the Fourier transform in x, we get from (33.5):

$$\tilde{h}(\xi, k\omega) + \frac{1}{(2\pi)^n}\int_{\mathbb{R}^n} \frac{\tilde{q}(\xi-\eta)\tilde{h}(\eta, k\omega)d\eta}{|\eta|^2 - k^2 - i0} = -\tilde{q}(\xi - k\omega). \tag{38.1}$$

Note that the scattering amplitude $a(\theta, \omega, k)$ (see §33) has the following form:

$$a(\theta, \omega, k) = \frac{k^{\frac{n-3}{2}} e^{-i\frac{\pi}{2}(n-3)}}{2^{\frac{n+1}{2}}\pi^{\frac{n-1}{2}}}\tilde{h}(k\theta, k\omega). \tag{38.2}$$

Since the inverse scattering problem of recovering $q(x)$ from $a(\theta, \omega, k)$ is overdetermined when $n \ge 2$, it is reasonable to look for a nonoverdetermined inverse problem. One natural candidate for such an inverse problem is the backscattering inverse problem, i.e., the recovery of $q(x)$ from the backscattering amplitude $b(k\theta) = a(\theta, -\theta, k)$, i.e., when the scattering amplitude is given for all $k > 0$, all $\theta \in S^{n-1}$, and $\omega = -\theta$.

Define a weighted Hölder space $C_{\alpha,N}$ as the closure of $C_0^\infty(\mathbb{R}^n)$ in the norm

(38.3)

$$|f|_{\alpha,N} = \sup_{\xi\in\mathbb{R}^n}(1+|\xi|)^N|f(\xi)| + \sup_{h,\xi\in\mathbb{R}^n}(1+|\xi|)^N\frac{|f(\xi+h) - f(\xi)|}{|h|^\alpha}, 0 < \alpha < 1.$$

Let $C_{\alpha,N}^r$ be the subspace of $C_{\alpha,N}$ consisting of the Fourier transforms of real-valued potentials, i.e., the set of $\tilde{q}(\xi) \in C_{\alpha,N}$ such that $\tilde{q}(\xi) = \overline{\tilde{q}(-\xi)}$.

Theorem 38.1. *Assume that $q(x)$ is such that $\tilde{q}(\xi) \in C_{\alpha,N}$, $0 < \alpha < 1$, $N > n - 2$, $n \ge 3$. There exists a dense open set $O \subset C_{\alpha,N}$ such that the backscattering $b(\xi)$, $\xi = k\omega$, also belongs to $C_{\alpha,N}$. Moreover, the nonlinear* *map* $S(\tilde{q}) = b(\xi)$ *is a continuous Fréchet differentiable map from* $O \subset C_{\alpha,N}$ *to* $C_{\alpha,N}$.

Theorem 38.2. *There exists a dense open set* $O^{(1)} \subset O$ *in* $C_{\alpha,N}$ *such that any* $\tilde{q}_0(\xi) \in O^{(1)}$ *has a neighborhood* $U_{\tilde{q}_0} \subset C_{\alpha,N}$ *for which* $b(\xi) = S(\tilde{q})$ *is a homeomorphism of* $U_{\tilde{q}_0}$ *onto a neighborhood* $U_{b_0} \subset C_{\alpha,N}$ *of* $b_0 = S(q_0)$*; i.e., for any* $b \in U_{b_0}$ *there exists a unique* $\tilde{q}(\xi) = S^{-1}(b) \in U_{\tilde{q}_0}$*, and* $S^{-1}(b)$ *is a continuously Fréchet differentiable map of* U_{b_0} *onto* $U_{\tilde{q}_0}$.

Theorems 38.1 and 38.2 are proven in Eskin-Ralston [**ER2**]. It follows from the proof there that the set $O^{(1)} \cap C^r_{\alpha,N}$ is open and dense in $C^r_{\alpha,N}$ and that the set $O^{(1)}$ contains a neighborhood of $\tilde{q} = 0$. Therefore any $b(\xi)$ with a small $C_{\alpha,N}$ norm is a backscattering for some potential $q(x)$ with $\tilde{q}(\xi) \in C_{\alpha,N}$ and $\tilde{q}(\xi)$ having a small $C_{\alpha,N}$ norm. Moreover, such $\tilde{q}$ is unique in a small ball in $C_{\alpha,N}$ centered at $\tilde{q} = 0$.

38.1. The case of real-valued potentials.

The existence of the continuous inverse of a nonlinear operator S means that the inverse backscattering problem is well posed. Note that in Theorems 38.1 and 38.2 it is not assumed that $q(x)$ is real-valued. In the case of real-valued $q(x)$ we can further reduce the amount of scattering data sufficient for recovering $q(x)$. Let ν be an arbitrary unit vector in $\mathbb{R}^n$. Denote by $\mathbb{R}^+_\nu$ the half-space of all $x \in \mathbb{R}^n$ such that $x \cdot \nu > 0$. We show that the inverse problem of recovering $q(x)$ from the restriction of the backscattering $b(\xi)$ to the half-space $\overline{R^+_\nu}$ is well posed but is still slightly overdetermined.

Let $\chi(t)$ be a $C_0^\infty(\mathbb{R}^1)$ function such that $\chi(t) = \chi(-t)$, $\chi(t) = 0$ for $|t| > 2$, and $\chi(t) = 1$ for $|t| < 1$. For each $\xi \in \mathbb{R}^n$ we have the following decomposition:

$$\xi = (\xi \cdot \nu)\nu + \xi', \tag{38.4}$$

where $\xi' \cdot \nu = 0$. Denote by P_ν the following operation acting on functions defined on $\overline{\mathbb{R}^n_\nu}$:

$$(P_\nu f)(\xi) = f(\xi) + \frac{1}{2}\chi(\xi \cdot \nu)(\overline{f(-\xi')} - f(\xi')). \tag{38.5}$$

Note that
(38.6)
$$(P_\nu f)(\xi') = f(\xi') + \frac{1}{2}\chi(\xi' \cdot \nu)(\overline{f(-\xi')} - f(\xi')) = \frac{1}{2}(f(\xi') + \overline{f(-\xi')}) = \overline{P_\nu f(-\xi')}.$$

Denote by $\hat{P}_\nu$ the extension of $P_\nu f$ to $\mathbb{R}^n$ defined by

$$(\hat{P}_\nu f)(\xi) = \overline{P_\nu f(-\xi)}, \quad \text{where} \quad \xi \cdot \nu < 0. \tag{38.7}$$

Note that for any $f(\xi) \in C_{\alpha,N}$ we have

$$(\hat{P}_\nu)f(\xi) = \overline{\hat{P}_\nu f(-\xi)} \tag{38.8}$$

and

$$(\hat{P}_\nu f)(\xi) = f(\xi) \tag{38.9}$$

if $f(\xi) = \overline{f(-\xi)}$, i.e., if $f(\xi)$ is the Fourier transform of a real-valued function. It follows from (38.7)–(38.9) that $\hat{P}_\nu$ is a bounded linear operator from $C_{\alpha,N}$ onto $C^r_{\alpha,N}$. Let

$$S_r(\tilde{q}) = \hat{P}_\nu S(\hat{q}). \tag{38.10}$$

Let $O \subset C_{\alpha,N}$ be the same as in Theorem 38.1. The following result holds (see Eskin-Ralston [**ER3**]):

Theorem 38.3. *The nonlinear map $S_r(\hat{q})$ is a continuously Fréchet differentiable map of an open dense set $O \cap C^r_{\alpha,N}$ to $C^r_{\alpha,N}$. There exists an open dense set $O^{(2)} \subset (O \cap C^r_{\alpha,N})$ containing $\tilde{q} = 0$ and such that $S_r(\hat{q})$ is a homeomorphism in a neighborhood of any $\hat{q}_0 \in O^{(2)}$.*

Note that the knowledge of $b_r(\xi) = \hat{P}_\nu S(\tilde{q})$ gives less than the knowledge of $b(\xi)$ on $\overline{\mathbb{R}^n_\nu}$ because $b_r(\xi)$ uses only the combination $\frac{b(\xi)+\overline{b(-\xi)}}{2}$ on the plane $\xi \cdot \nu = 0$.

39. Problems

1. Let $u(x)$ be a distorted plane wave (see (33.5)). Prove that $u(x)$ satisfies the following integral equation (the Lippmann-Schwinger equation):

$$u(x) + \int_{\mathbb{R}^n} E_+(x-y)q(y)u(y)dy = e^{ik\omega\cdot x}. \tag{39.1}$$

2. Prove the following formula for the scattering amplitude $a(\theta, \omega, k)$:

$$a(\theta,\omega,k) = -C_n \int_{\mathbb{R}^n} e^{-ik\theta\cdot y} q(y)u(y)dy, \tag{39.2}$$

where C_n is the same as in (33.4), and $u(x)$ and $a(\theta, \omega, k)$ are the same as in (33.5).

3. Assume that $q(x) = 0$ outside of Ω. Using (39.2), prove that

$$a(\theta,\omega,k) = C_n \int_{\partial\Omega} \left(e^{-ik\theta\cdot y}\frac{\partial u(y)}{\partial n} - u(y)\frac{\partial}{\partial n}e^{-ik\theta\cdot y}\right) ds. \tag{39.3}$$

4. The reconstruction procedure (see Nachman [**Na1**]).

a) Using the definition and the selfadjoitness of the Dirichlet-to-Neumann operators, rewrite (39.3) in the form

$$a(\theta,\omega,k) = C_n \int_{\partial\Omega} e^{-ik\theta\cdot y}(\Lambda - \Lambda_0)u_0 ds,$$

where Λ and Λ_0 are the Dirichlet-to-Neumann operators in Ω for $-\Delta + q$ and $-\Delta$, respectively, and u_0 is the restriction of the distorted plane wave to $\partial\Omega$.

b) Applying Green's formula to

$$u(x) + e^{ik\omega\cdot y} = -\int_{\Omega} E_+(x-y)(\Delta_y + k^2)u(y)dy, \ x \notin \Omega,$$

prove that u_0 satisfies the following integral equation on $\partial\Omega$:

$$u_0 + S(\Lambda - \Lambda_0)u_0 = e^{ix\cdot k\omega}|_{\partial\Omega},$$

where $Sf = \int_{\partial\Omega} E_+(x-y')f(y')ds$ is a single layer potential. Note that 4 a), b) give the reconstruction of the scattering amplitude from the Dirichlet-to-Neumann operator.

5. Let $a(\theta,\omega,k)$ be the scattering amplitude. Prove that $a(\theta,\omega,k) = a(-\omega,-\theta,k)$.

6. Prove that if $q(x) = \overline{q(x)}$ and $n = 3$, then

$$a(\theta,\omega,k) - \overline{a(\omega,\theta,k)} = -\frac{ik}{2\pi}\int_{|\omega'|=1} a(\theta,\omega',k)\overline{a(\omega,\omega',k)}d\omega'.$$

In particular,

$$\operatorname{Im} a(\theta,\theta,k) = -\frac{1}{4\pi}\int_{|\omega'|=1} |a(\theta,\omega',k)|^2 d\omega'.$$

7. Consider $\tilde{q}_j(\xi) \in C_{\alpha,N}$, $j = 1,2$, $n = 3$. Let $\tilde{h}_j(\xi, k\omega)$, $j = 1,2$, be the solution of the integral equation (38.1) with $\tilde{q}(\xi)$ replaced by $\tilde{q}_j(\xi)$, $j = 1,2$. Let $b_j(\xi) = \frac{1}{4\pi}\tilde{h}_j(\xi,-\xi)$ be the backscattering amplitudes, $j = 1,2$. The following identity holds (see Eskin-Ralston [**ER2**]):

$$\tilde{h}_1(\xi,-\xi) - \tilde{h}_2(\xi,-\xi) = \tilde{q}_2(2\xi) - \tilde{q}_1(2\xi) + A_3\Big(\frac{\tilde{h}_1 + \tilde{h}_2}{2}\Big)w + A_4(\tilde{h}_1\tilde{h}_2)w, \tag{39.4}$$

where

$$w(\eta) = \tilde{q}_2(2\eta) - \tilde{q}_1(2\eta),$$

$$A_3(\tilde{h})w = \frac{2}{\pi^3}\int_{\mathbb{R}^3} \frac{\tilde{h}(\xi, -\xi + 2t)w(t)dt}{|\xi - 2t|^2 - (|\xi| + i0)^2},$$

$$A_4(\tilde{h}_1\tilde{h}_2)w = \frac{1}{8\pi^6}\int_{\mathbb{R}^3}\int_{\mathbb{R}^3} \frac{\tilde{h}_1(\xi, t)\tilde{h}_2(t - 2\zeta, -\xi)w(t)dtd\zeta}{(|t|^2 - (|\xi| + i0)^2)(|2\zeta - t|^2 - (|\xi| + i0)^2)}.$$

Suppose that the $C_{\alpha,N}$ norms of $\tilde{q}_j$, $j = 1, 2$, are small.

a) Use the identity (39.4) to prove that $b_1(\xi) = b_2(\xi)$ implies $\tilde{q}_1(\xi) = \tilde{q}_2(\xi)$.

b) Use (39.4) to prove that $\hat{P}_\nu b_1 = \hat{P}_\nu b_2$ implies $\tilde{q}_1(\xi) = \tilde{q}_2(\xi)$.

8. Let Ω be a bounded domain with a smooth boundary $\partial\Omega$ and such that $\mathbb{R}^n \setminus \overline{\Omega}$ is connected. Suppose $u = e^{ik\omega\cdot x} + v$ satisfies (36.1), (36.2) in $\mathbb{R}^n \setminus \overline{\Omega}$ and the following boundary condition:

$$\left(\frac{\partial}{\partial n} - a\right) u\Big|_{\partial\Omega} = 0, \tag{39.5}$$

where $\operatorname{Im} a \le 0$. Prove the uniqueness of such a solution.

9. Prove the existence of the solution satisfying (36.1), (36.2), and (39.5) when $\operatorname{Im} a \le 0$.

10. Let B be the space of $u(x)$ locally in L_2 with the norm $\|u\|_B^2 = \lim_{R\to\infty} \sup \frac{1}{R}\int_{|x|\le R} |u(x)|^2 dx$. Prove that if $\varphi(\theta) \in L_2(S^{n-1})$, then

$$\hat{\varphi}(x) = \int_{S^{n-1}} \varphi(\theta) e^{ikx\cdot\theta} d\theta \in B$$

(see Agmon-Hörmander [**AH**]).

11. Let Ω_1 and Ω_2 be two smooth bounded domains, $\overline{\Omega_1} \subset \Omega_2$. Consider two equations in Ω_2:

$$(-\Delta + q_j(x))u_j(x) = 0, \quad j = 1, 2, \quad x \in \Omega_2,$$

where $q_j(x) \in L_\infty(\Omega_2)$. Suppose $q_1 = q_2$ in $\Omega_2 \setminus \Omega_1$. Let $\Lambda_j(\Omega_k)$ be the Dirichlet-to-Neumann operators, $j = 1, 2$, $k = 1, 2$. Prove that $\Lambda_1(\Omega_1) = \Lambda_2(\Omega_1)$ iff $\Lambda_1(\Omega_2) = \Lambda_2(\Omega_2)$.

12. Let $(-\Delta + q(x) - k^2)u = 0$, $x \in \mathbb{R}^n$, $q(x) = O(\frac{1}{(1+|x|)^{\frac{n+1}{2}+\delta}})$.

a) If $\int_{|x|=R} |u(x)|^2 dx = O(1)$, $\int_{|x|=R} \left|\frac{\partial u}{\partial r} - iku\right|^2 dx = o(1)$ as $R \to \infty$, then $u = 0$.

b) If $\frac{1}{R}\int_{|x|\le R} |u(x)|^2 dx = O(1)$ and $\frac{1}{R}\int_{|x|\le R} \left|\frac{\partial u}{\partial r} - iku\right|^2 dx = o(1)$ as $R \to \infty$, then $u = 0$.

13. The case of one direction ω (see Colton-Kress [**CK**]). Consider two scattering obstacles problems of the form (36.1), (36.2), (36.3). Let Ω_j and $a_j(\theta, \omega, k), j = 1, 2$, be the corresponding obstacles and scattering amplitudes. Prove that there exists $\rho > 0$ such that if $\Omega_j \subset B_\rho, j = 1, 2$, and if $a_1(\theta, \omega_0, k) = a_2(\theta, \omega_0, k)$ for fixed $k > 0$, $\omega_0 \in S^{n-1}$ and all $\theta \in S^{n-1}$, then $\Omega_1 = \Omega_2$.

Chapter VI

Pseudodifferential Operators

Introduction to Chapter VI

This chapter starts with the theory of pseudodifferential operators (ψdo's). We choose the simplest class of symbols to make the exposition as accessible as possible.

In §40 we study the boundedness and the composition of ψdo's. In §§41, 42 we deal with the ellipticity and the Fredholm property of ψdo's. In §§43, 45 we consider the adjoints of ψdo's and the change of variables formula.

The rest of Chapter VI is devoted to the applications of ψdo's. In §46 we prove the existence and the uniqueness of the solution of the Cauchy problem for parabolic equations. In §47 we find the asymptotics of the heat kernel as $t \to +0$. The Cauchy problem for the strictly hyperbolic equation of order m is studied in §48. Separately, in §49 we consider the domains of dependence and influence for the strictly hyperbolic equations of any order. In §44 we deal with the pseudolocal property and microlocal regularity of ψdo's.

Finally, in §50 we study propagation of singularities for the equations of real principal type following Hörmander [**H1**], with applications to propagation of singularities of solutions of the Cauchy problem for hyperbolic equations. In the problem section (§51) we indicate an alternative approach to the Cauchy problem for hyperbolic equations following Taylor [**T2**], and we briefly describe a general class of pseudodifferential operators introduced by Beals and Fefferman [**B**], [**BF**].

40. Boundedness and composition of ψdo's

40.1. The boundedness theorem.

Denote by S^α the class of $C^\infty(\mathbb{R}^n \times \mathbb{R}^n)$ functions $A(x,\xi)$ such that

$$\left|\frac{\partial^{k+p}A(x,\xi)}{\partial x^p \partial \xi^k}\right| \leq C_{pk}(1+|\xi|)^{\alpha-|k|}, \ \forall k,p, \tag{40.1}$$

and $A(x,\xi)$ is independent of x for $|x| > R$. Let $A(\infty,\xi) = A(x,\xi)$ for $|x| > R$ and $A'(x,\xi) = A(x,\xi) - A(\infty,\xi)$, i.e., $A'(x,\xi)$ has a compact support in x.

For $A(x,\xi) \in S^\alpha$ we define an operator

$$Au = \frac{1}{(2\pi)^n}\int_{\mathbb{R}^n} A(x,\xi)\tilde{u}(\xi)e^{ix\cdot\xi}d\xi, \ \forall u(x) \in C_0^\infty(\mathbb{R}^n), \tag{40.2}$$

where $\tilde{u}(\xi)$ is the Fourier transform of $u(x)$. If $A(x,\xi) = \sum_{|k|=0}^m a_k(x)\xi^k$ is a polynomial in ξ, then

$$Au = \sum_{|k|=0}^m a_k(x)\left(-i\frac{\partial}{\partial x}\right)^k u \tag{40.3}$$

is a differential operator $A(x,D)u$, $D = -i\frac{\partial}{\partial x}$. We call (40.2) pseudodifferential operator and $A(x,\xi)$ its symbol. We often denote by $A(x,D)$ a pseudodifferential operator with symbol $A(x,\xi)$.

Theorem 40.1. *A pseudodifferential operator $A(x,D)$ with symbol $A(x,\xi)$ $\in S^\alpha$ is bounded from $H_s(\mathbb{R}^n)$ to $H_{s-\alpha}(\mathbb{R}^n)$ for any $s \in \mathbb{R}$:*

$$\|A(x,D)u\|_{s-\alpha} \leq C\|u\|_s, \quad \forall u \in H_s(\mathbb{R}^n). \tag{40.4}$$

Proof: It is clear that

$$\|A(\infty,D)u\|_{s-\alpha}^2 = \int_{\mathbb{R}^n} |A(\infty,\xi)\tilde{u}(\xi)|^2(1+|\xi|)^{2(s-\alpha)}d\xi \leq C_0^2\|u\|_s^2,$$

where $C_0 = \sup_\xi \frac{|A(\infty,\xi)|}{(1+|\xi|)^\alpha}$, $\forall u \in S$.

Let

$$\tilde{A}'(\eta,\xi) = \int_{\mathbb{R}^n} A'(x,\xi)e^{-ix\cdot\eta}dx. \tag{40.5}$$

Taking the Fourier transform of $v(x) = A'(x,D)u$, we get

$$\tilde{v}(\eta) = \frac{1}{(2\pi)^n}\int_{\mathbb{R}^n} \tilde{A}'(\eta-\xi,\xi)\tilde{u}(\xi)d\xi. \tag{40.6}$$

Since $A'(x,\xi) \in C_0^\infty$ in x, we get, using integration by parts in (40.5):

$$|\tilde{A}(\eta,\xi)| \leq C_N(1+|\eta|)^{-N}(1+|\xi|)^\alpha, \ \forall N.$$

Therefore

$$|\tilde{v}(\eta)| \le C_N \int_{\mathbb{R}^n} \frac{(1+|\xi|)^\alpha |\tilde{u}(\xi)|}{(1+|\xi-\eta|)^N} d\xi.$$

It follows from (13.10) that

$$(1+|\xi-\eta|)^{-|t|} \le C_t \frac{(1+|\xi|)^t}{(1+|\eta|)^t}, \quad \forall t \in \mathbb{R}. \tag{40.7}$$

Taking $N \ge n+1+|s-\alpha|$ and using (40.7) with $t = s-\alpha$, we get:

$$(1+|\eta|)^{s-\alpha}|\tilde{v}(\eta)| \le C \int_{\mathbb{R}^n} \frac{(1+|\xi|)^s |\tilde{u}(\xi)|}{(1+|\xi-\eta|)^{n+1}} d\xi.$$

Now as in (13.11), (13.12) we obtain

$$\|v\|_{s-\alpha} \le C\|u\|_s, \ \forall u \in C_0^\infty(\mathbb{R}^n).$$

Since $C_0^\infty(\mathbb{R}^n)$ is dense in $H_s(\mathbb{R}^n)$ (see Theorem 13.2), we can take the closure in $H_s(\mathbb{R}^n)$ and get (40.4).

Definition 40.1. We say that an operator T_α is of order $\le \alpha$ if

$$\|T_\alpha u\|_s \le C_s \|u\|_{s+\alpha} \quad \text{for every} \quad s. \tag{40.8}$$

It follows from Theorem 40.1 that $A(x,D)$ with symbol $A(x,\xi) \in S^\alpha$ has order α.

40.2. Composition of ψdo's.

Theorem 40.2. *Let $A(x,\xi) \in S^\alpha$ and $B(x,\xi) \in S^\beta$. Then for any N the composition $A(x,D)B(x,D)$ has the form*

$$A(x,D)B(x,D) = C_N(x,D) + T_{\alpha+\beta-N-1}, \tag{40.9}$$

where $C_N(x,\xi) \in S^{\alpha+\beta}$, $\operatorname{ord} T_{\alpha+\beta-N-1} \le \alpha+\beta-N-1$, and

$$C_N(x,\xi) = \sum_{|k|=0}^{N} \frac{1}{k!} \frac{\partial^k A(x,\xi)}{\partial \xi^k} D_x^k B(x,\xi), \ D_x = -i\frac{\partial}{\partial x}. \tag{40.10}$$

Proof: Note that $A(x,D)B(\infty,D)$ is a ψdo with symbol $A(x,\xi)B(\infty,\xi)$. Let

$$v(x) = B'(x,D)u = \frac{1}{(2\pi)^n}\int_{\mathbb{R}^n} B'(x,\eta)\tilde{u}(\eta)e^{ix\cdot\eta}d\eta.$$

As in (40.6), we have:

$$\begin{aligned} A(x,D)v &= \frac{1}{(2\pi)^n}\int_{\mathbb{R}^n} A(x,\xi)\tilde{v}(\xi)e^{ix\cdot\xi}d\xi \\ &= \frac{1}{(2\pi)^{2n}}\int_{\mathbb{R}^n}\int_{\mathbb{R}^n} A(x,\xi)\tilde{B}'(\xi-\eta,\eta)e^{ix\cdot\xi}\tilde{u}(\eta)d\xi d\eta, \ \forall u \in C_0^\infty(\mathbb{R}^n). \end{aligned}$$

Make the change of variables $\xi-\eta=\zeta$. Then

$$A(x,D)v=\frac{1}{(2\pi)^{2n}}\int_{\mathbb{R}^n}\int_{\mathbb{R}^n}A(x,\eta+\zeta)\tilde{B}'(\zeta,\eta)e^{ix\cdot(\eta+\zeta)}\tilde{u}(\eta)d\eta d\zeta.$$

Note that $\tilde{B}'(\zeta,\xi)$ is rapidly decreasing in ζ and $\tilde{u}(\eta)$ is rapidly decreasing in η. The Taylor expansion of $A(x,\eta+\zeta)$ in ζ is:

$$A(x,\eta+\zeta)=\sum_{|k|=0}^{N}\frac{1}{k!}\frac{\partial^k A(x,\eta)}{\partial\eta^k}\zeta^k+R_N(x,\eta,\zeta), \tag{40.11}$$

where $R_N(x,\eta,\zeta)$ is the remainder. Note that

$$\frac{1}{(2\pi)^n}\int_{\mathbb{R}^n}\zeta^k\tilde{B}'(\zeta,\eta)e^{ix\cdot\zeta}d\zeta=D_x^kB'(x,\eta).$$

Therefore

$$\begin{aligned}A(x,D)v=&\frac{1}{(2\pi)^n}\int_{\mathbb{R}^n}\sum_{|k|=0}^{N}\frac{1}{k!}\frac{\partial^k A(x,\eta)}{\partial\eta^k}D_{x'}^kB'(x,\eta)\tilde{u}(\eta)e^{ix\cdot\eta}d\eta\\&+T_{\alpha+\beta-N-1}u,\end{aligned}$$

where $T_{\alpha+\beta-N-1}u=\frac{1}{(2\pi)^{2n}}\int_{\mathbb{R}^n}R_N(x,\eta,\zeta)\tilde{B}'(\zeta,\eta)e^{ix\cdot(\eta+\zeta)}\tilde{u}(\eta)d\eta d\zeta$. It remains to prove that $T_{\alpha+\beta-N-1}$ has order $\alpha+\beta-N-1$. We have

$$R_N(x,\eta,\zeta)=R_N(\infty,\eta,\zeta)+R_N'(x,\eta,\zeta), \tag{40.12}$$

where $R_N(\infty,\eta,\zeta)$ is the remainder for $A(\infty,\xi)$ and $R_N'(x,\eta,\zeta)$ is the remainder for $A'(x,\xi)$.

Let $\tilde{R}_N'(\xi,\eta,\zeta)$ be the Fourier transform of $R_N'(x,\eta,\zeta)$ in x. We prove the following estimates for $R_N(\infty,\eta,\zeta)$ and $\tilde{R}_N'(\xi,\eta,\zeta)$:

$$|R_N(\infty,\eta,\zeta)|\le C(1+|\eta|)^{\alpha-N-1}(1+|\zeta|)^{N+|\alpha|+1}, \tag{40.13}$$

$$|\tilde{R}_N'(\xi,\eta,\zeta)|\le C_M(1+|\xi|)^{-M}(1+|\eta|)^{\alpha-N-1}(1+|\zeta|)^{N+|\alpha|+1},\ \forall M. \tag{40.14}$$

To prove (40.13), consider two cases: $|\zeta|\le\frac{1}{2}(1+|\eta|)$ and $|\zeta|>\frac{1}{2}(1+|\eta|)$. For $|\zeta|\le\frac{1}{2}(1+|\eta|)$ we use the Lagrange form of the remainder in the Taylor formula:

$$R_N(\infty,\eta,\zeta)=\sum_{|k|=N+1}\frac{1}{k!}\frac{\partial^k A(\infty,\eta+\theta\zeta)}{\partial\eta^k}\zeta^k,$$

where $|\theta|<1$. Then

$$\left|\frac{\partial^k A(\infty,\eta+\theta\zeta)}{\partial\eta^k}\right|\le C(1+|\eta+\theta\zeta|)^{\alpha-N-1}\le C_1(1+|\eta|)^{\alpha-N-1}, \tag{40.15}$$

since $1+|\eta+\theta\zeta|\ge 1+|\eta|-|\zeta|\ge\frac{1}{2}(1+|\eta|)$ and $(1+|\eta+\theta\zeta|)\le 1+|\eta|+|\zeta|\le\frac{3}{2}(1+|\eta|)$. Therefore (40.13) holds.

If $|\zeta| > \frac{1}{2}(1+|\eta|)$, we get from (40.11):

$$|R_N(\infty,\eta,\zeta)| \le |A(\infty,\eta+\zeta)| + \sum_{|k|=0}^{N} \frac{1}{k!}\left|\frac{\partial^k A(\infty,\eta)}{\partial\eta^k}\right| |\zeta|^{|k|}$$
$$\le C(1+|\eta+\zeta|)^\alpha + C\sum_{|k|=0}^{N}(1+|\eta|)^{\alpha-|k|}|\zeta|^{|k|}.$$

Note that

$$(1+|\eta|)^{\alpha-|k|}|\zeta|^k \le (1+|\eta|)^{\alpha-N-1}(1+|\eta|)^{N-|k|+1}|\zeta|^{|k|}$$
$$\le C(1+|\eta|)^{\alpha-N-1}|\zeta|^{N+1},$$

since $(1+|\eta|)^{N-|k|+1} \le C|\zeta|^{N-|k|+1}$. Using (40.7) with $t=-\alpha,\ \ \xi=\eta+\zeta$ we get

$$(1+|\eta+\zeta|)^\alpha \le C(1+|\eta|)^\alpha(1+|\zeta|)^{|\alpha|} \le C(1+|\eta|)^{\alpha-N-1}(1+|\zeta|)^{|\alpha|+N+1}$$

since $|\zeta| \ge \frac{1}{2}(1+|\eta|)$. Therefore (40.13) holds.

The proof of (40.14) is similar. We replace the estimate $|\frac{\partial^k A(\infty,\eta)}{\partial\eta^k}| \le C(1+|\eta|)^{\alpha-|k|}$ by the estimate $|\frac{\partial^k \tilde{A}'(\xi,\eta)}{\partial\eta^k}| \le C_M(1+|\xi|)^{-M}(1+|\eta|)^{\alpha-|k|}$ for all M. □

Let

$$T_{\alpha+\beta-N-1} = T^{(1)}_{\alpha+\beta-N-1} + T^{(2)}_{\alpha+\beta-N-1},$$

where

(40.16)
$$T^{(1)}_{\alpha+\beta-N-1}u = \frac{1}{(2\pi)^{2n}}\int_{\mathbb{R}^n}\int_{\mathbb{R}^n} R_N(\infty,\eta,\zeta)\tilde{B}'(\zeta,\eta)e^{ix\cdot(\eta+\zeta)}\tilde{u}(\eta)d\eta d\zeta,$$

(40.17)
$$T^{(2)}_{\alpha+\beta-N-1}u = \frac{1}{(2\pi)^{2n}}\int_{\mathbb{R}^n}\int_{\mathbb{R}^n} R'_N(x,\eta,\zeta)\tilde{B}'(\zeta,\eta)e^{ix\cdot(\eta+\zeta)}\tilde{u}(\eta)d\eta d\zeta.$$

Since $(F_xT^{(1)}_{\alpha+\beta-N-1}u)(\xi) = \frac{1}{(2\pi)^n}\int_{\mathbb{R}^n} R_N(\infty,\eta,\xi-\eta)\tilde{B}'(\xi-\eta,\eta)\tilde{u}(\eta)d\eta$ and

$$|\tilde{B}'(\xi-\eta,\eta)| \le C_M(1+|\xi-\eta|)^{-M}(1+|\eta|)^\beta, \tag{40.18}$$

we get, using (40.13):

$$(1+|\xi|)^s\ |(F_xT^{(1)}_{\alpha+\beta-N-1}u)(\xi)| \tag{40.19}$$
$$\le C_M\int_{\mathbb{R}^n}\frac{(1+|\xi|)^s(1+|\xi-\eta|)^{N+|\alpha|+1}(1+|\eta|)^{\alpha+\beta-N-1}|\tilde{u}(\eta)|}{(1+|\xi-\eta|)^M}d\eta.$$

Taking $M > N+1+|\alpha|+|s|+n+1$ and using (40.7) with $t=-s$, we get (40.20)

$$(1+|\xi|)^s \,|(F_x T^{(1)}_{\alpha+\beta-N-1}u)(\xi)| \le C\int_{\mathbb{R}^n} \frac{(1+|\eta|)^{s+\alpha+\beta-N-1}|\tilde{u}(\eta)|}{(1+|\xi-\eta|)^{n+1}}d\eta.$$

The arguments in (13.11), (13.12) applied to (40.20) give

$$\|T^{(1)}_{\alpha+\beta-N-1}u\|_s \le C\|u\|_{s+\alpha+\beta-N-1}.$$

Analogously, we have

$$(F_x T^{(2)}_{\alpha+\beta-N-1}u)(\xi) = \frac{1}{(2\pi)^{2n}}\int_{\mathbb{R}^n} \tilde{R}'_N(\xi-\eta-\zeta,\eta,\zeta)\tilde{B}'(\zeta,\eta)\tilde{u}(\eta)d\eta d\zeta.$$

Using (40.14) with M replaced by M_1, we obtain

$$|(F_x T^{(2)}_{\alpha+\beta-N-1}u)(\xi)| \\ \le C\int_{\mathbb{R}^n}\int_{\mathbb{R}^n} \frac{(1+|\eta|)^{\alpha-N-1}(1+|\zeta|)^{|\alpha|+N+1}(1+|\eta|)^{\beta}|\tilde{u}(\eta)|d\eta d\zeta}{(1+|\xi-\eta-\zeta|)^{M_1}(1+|\zeta|)^M}.$$

For any M_1 we have (cf. (40.7))

$$(1+|\xi-\eta-\zeta|)^{-M_1} \le C_{M_1}(1+|\xi-\eta|)^{-M_1}(1+|\zeta|)^{M_1}. \tag{40.21}$$

Taking $M \ge M_1+N+1+|\alpha|+n+1$ in (40.18) and using (40.21) leads to

$$|(1+|\xi|)^s|(F_x T^{(2)}_{\alpha+\beta-N-1}u)(\xi)| \\ \le C\int_{\mathbb{R}^n} \frac{(1+|\xi|)^s(1+|\eta|)^{\alpha+\beta-N-1}|\tilde{u}(\eta)|}{(1+|\xi-\eta|)^{M_1}}d\eta. \tag{40.22}$$

Treating (40.22) as in (13.11), (13.12), we get

$$\|T^{(2)}_{\alpha+\beta-N-1}u\|_s \le C\|u\|_{\alpha+\beta-N-1+s}.$$

Therefore $T_{\alpha+\beta-N-1}$ is an operator of order $\le \alpha+\beta-N-1$. □

We often call an operator of a negative order a smoothing operator.

Corollary 40.3. *Let $A(x,\xi)\in S^\alpha$ and $B(x,\xi)\in S^\beta$. Then the commutator $[A,B] = A(x,D)B(x,D) - B(x,D)A(x,D)$ has the form*

$$\sum_{k=1}^N C_k(x,D) + T_{\alpha+\beta-N-1},$$

where $\operatorname{ord} T_{\alpha+\beta-N-1} \le \alpha+\beta-N-1$, $C_k(x,\xi)\in S^{\alpha+\beta-k}$, *and, in particular,*

$$C_1(x,\xi) = -i\sum_{j=1}^n\left(\frac{\partial A(x,\xi)}{\partial\xi_j}\frac{\partial B(x,\xi)}{\partial x_k} - \frac{\partial A(x,\xi)}{\partial x_j}\frac{\partial B(x,\xi)}{\partial \xi_j}\right),$$

i.e.,

$$iC_1(x,\xi) = \{A(x,\xi), B(x,\xi)\}. \tag{40.23}$$

Here $\{A(x,\xi),B(x,\xi)\} = \sum_{j=1}^n \left(\frac{\partial A}{\partial \xi_1}\frac{\partial B}{\partial x_j} - \frac{\partial A}{\partial x_j}\frac{\partial B}{\partial \xi_j}\right)$ *is called the Poisson bracket of* $A(x,\xi)$ *and* $B(x,\xi)$.

The proof of the corollary follows immediately from the application of Theorem 40.2 to $A(x,D)B(x,D)$ and $B(x,D)A(x,D)$.

Remark 40.1. Let $|||A(x,D)|||_{(s)}$ be the operator norm of $A(x,D)$ acting from $H_s(\mathbb{R}^n)$ to $H_{s-\alpha}(\mathbb{R}^n)$. It follows from the proof of Theorem 40.1 that

$$|||A(x,D)|||_{(s)} \le \sup_\xi \frac{|A(\infty,\xi)|}{(1+|\xi|)^\alpha} + \sup_{\xi,\eta} \frac{(1+|\eta|)^N|\tilde{A}'(\eta,\xi)|}{(1+|\xi|)^\alpha}, \tag{40.24}$$

where $N \ge n+1+|s-\alpha|$. Note that

$$(1+|\eta|)^N|\tilde{A}'(\eta,\xi)| \le \sum_{|k|=0}^N \int_{\mathbb{R}^n} |D_x^k A'(x,\xi)|dx. \tag{40.25}$$

The estimate (10.25) follows from the equality

$$\eta^k \tilde{A}'(\eta,\xi) = \int_{\mathbb{R}^n} (D_x^k A'(x,\xi))e^{-ix\cdot\eta}d\eta.$$

Remark 40.2. We will later use the following property of the remainder $T_{\alpha+\beta-N-1}$ in (40.9): $(1+|x|^2)^M T_{\alpha+\beta-N-1}$ is an operator of order $\le \alpha+\beta-N-1$ for each M, i.e.,

$$\|(1+|x|^2)^M T_{\alpha+\beta-N-1}u\|_s \le C_M\|u\|_{s+\alpha+\beta-N-1}, \quad \forall s. \tag{40.26}$$

Proof: It follows from (40.17) that $T^{(2)}_{\alpha+\beta-N-1} = 0$ when $|x| > R$ since $R'_N(x,\eta,\zeta) = 0$ when $|x| > R$. Consider now $T^{(1)}_{\alpha+\beta-N-1}$ (see (40.16)). Integrating by parts with respect to ζ in (40.16), we get

$$(1+|x|^2)^M T^{(1)}_{\alpha+\beta-N-1}u = \frac{1}{(2\pi)^{2n}}\int_{\mathbb{R}^n}\int_{\mathbb{R}^n}(1-\Delta_\zeta)^M(R'_N(\infty,\eta,\zeta)\tilde{B}'(\zeta,\eta)) \times e^{ix\cdot(\zeta+\eta)}\tilde{u}(\eta)d\eta d\zeta.$$

We have

$$|(1-\Delta_\zeta)^M(R'_N(\infty,\eta,\zeta)\tilde{B}'(\zeta,\eta))| \le C_{M_1}\frac{(1+|\eta|)^{\alpha+\beta-N-1}}{(1+|\zeta|)^{M_1}}, \quad \forall M_1.$$

Therefore, as in (40.19), (40.20) we get

$$\|(1+|x|^2)^M T^{(1)}_{\alpha+\beta-N-1}u\|_s \le C\|u\|_{\alpha+\beta-N-1+s}.$$

41. Elliptic operators and parametrices

41.1. Parametrix for a strongly elliptic operator.

Let $A(x, D)$ be elliptic differential operator of degree m, i.e., $A(x,\xi) = A_0(x,\xi) + A_1(x,\xi)$, where

$$A_0(x,\xi) = \sum_{|k|=m} a_k(x)\xi^k, \quad A_1(x,\xi) = \sum_{|k|=0}^{m-1} a_k(x)\xi^k,$$

$A_0(x,\xi) \neq 0,\ \forall x \in \mathbb{R}^n,\ \xi \neq 0.$

As in §40, we assume that $A(x,\xi) = A(\infty,\xi)$ when $|x| > R$. An elliptic operator $A(x, D)$ is called strongly elliptic if $\operatorname{Re} A_0(x,\xi) > 0, \forall x \in \mathbb{R}^n,\ \xi \neq 0$. Let $C_\delta = \{\lambda \in \mathbb{C} \setminus \{0\} : -\frac{\pi}{2} - \delta \leq \arg\lambda \leq \frac{\pi}{2} + \delta\},\ \delta > 0$.

Lemma 41.1. *Let $A(x, D)$ be a strongly elliptic operator and let $R_0(x,\xi,\lambda) = (A_0(x,\xi)+\lambda)^{-1}$. Then there exists $\delta > 0$ such that $R_0(x,\xi,\lambda) \in S^{-m}$ and satisfies the estimates*

$$|R_0(x,\xi,\lambda)| \leq C(|\xi|^m + |\lambda|)^{-1} \leq C_1(|\xi| + |\lambda|^{\frac{1}{m}})^{-m},\ \lambda \in C_\delta, \tag{41.1}$$

(41.2)

$$\left|\frac{\partial^{k+p} R_0(x,\xi,\lambda)}{\partial\xi^k \partial x^p}\right| \leq C_{pk}|R_0(x,\xi,\lambda)|\ (|\xi| + |\lambda|^{\frac{1}{m}})^{-|k|},\ \forall p, \forall k,\ \lambda \in C_\delta.$$

Proof: Let $A_{01} = \operatorname{Re} A_0,\ A_{02} = \operatorname{Im} A_0$, and $\lambda = \tau + i\sigma$. Then $|\sigma| = |(\sigma + A_{02}) - A_{02}| \leq |\sigma + A_{02}| + C|\xi|^m$. Therefore, $|\sigma| + |\xi|^m \leq |\sigma + A_{02}| + (C+1)|\xi|^m \leq |\sigma + A_{02}| + C_1 A_{01}$, since $|A_{02}| \leq C|\xi|^m, A_{01} \geq C'|\xi|^m$ for some C, C'. Thus, $|\sigma| + C_1\tau + |\xi|^m \leq |\sigma + A_{02}| + C_1(A_{01} + \tau) \leq |\sigma + A_{02}| + C_1|A_{01} + \tau| \leq C_2|A_0(x,\xi) + \lambda|$. Note that $\tau \geq -|\sigma|\tan\delta$ for $\tau \leq 0,\ \tau + i\sigma \in C_\delta$. Therefore, if $\delta > 0$ is small, then $|\sigma| + C_1\tau \geq C_3(|\sigma| + |\tau|)$ for $\lambda \in C_\delta$. This proves (41.1).

Estimates (41.2) follow from (41.1). □

Suppose that A is a strongly elliptic operator.

Theorem 41.2. *For any $N \geq 0$, there exists a ψdo operator $R^{(N)}(x, D, \lambda)$, $R^{(N)}(x,\xi,\lambda) \in S^{-m}$, such that*

$$(A(x, D) + \lambda I)R^{(N)}(x, D, \lambda) = I + T_{-N-1}, \tag{41.3}$$

where I is the identity operator, $\operatorname{ord} T_{-N-1} \leq -N-1$, *and*

$$|||T_{-N-1}|||_{(s)} \leq C_N|\lambda|^{-\frac{N+1}{m}},\ \lambda \in C_\delta, \tag{41.4}$$

where $|||T_{-N-1}|||_{(s)}$ is the operator norm of T_{-N-1} acting from $H_s(\mathbb{R}^n)$ to $H_s(\mathbb{R}^n)$.

The operator $R^{(N)}(x,D,\lambda)$ is called the parametrix for $A(x,D)+\lambda$.

Proof: In view of Theorem 40.2 we have

$$(A(x,D)+\lambda I)R_0(x,D,\lambda)=I+C_{-1}(x,D,\lambda),$$

where $C_{-1}(x,\xi,\lambda)\in S^{-1}$ and

$$C_{-1}(x,\xi,\lambda)=\sum_{|k|=1}^{m}\frac{1}{k!}\frac{\partial A(x,\xi)}{\partial\xi^k}D_x^kR_0(x,\xi,\lambda)+A_1(x,\xi)R_0(x,\xi,\lambda). \tag{41.5}$$

Note that there is no remainder in (41.5), since $A(x,\xi)$ is a polynomial.

Let

$$R_{-1}(x,\xi,\lambda)=-R_0(x,\xi,\lambda)C_{-1}(x,\xi,\lambda).$$

Then by Theorem 40.2 we have

$$(A(x,D)+\lambda I)(R_0+R_{-1}(x,D,\lambda))=I+C_{-2}(x,D,\lambda), \tag{41.6}$$

where $C_{-2}(x,\xi,\lambda)\in S^{-2}$ and

(41.7)
$$C_{-2}(x,\xi,\lambda)=A_1(x,\xi)R_{-1}(x,\xi,\lambda)+\sum_{|k|=1}^{m}\frac{1}{k!}\frac{\partial^kA(x,\xi)}{\partial\xi^k}D_x^kR_{-1}(x,\xi,\lambda).$$

Analogously defining $R_{-2}(x,\xi,\lambda)=-R_0(x,\xi,\lambda)C_{-2}(x,\xi,\lambda)$, etc., we construct $R_{-p}(x,\xi,\lambda)=-R_0(x,\xi,\lambda)C_{-p}(x,\xi,\lambda)$, where $C_{-p}(x,\xi,\lambda)\in S^{-p}$, $1\le p\le N$, and

(41.8)
$$C_{-p}(x,\xi)=A_1(x,\xi)R_{-p+1}(x,\xi,\lambda)+\sum_{|k|=1}^{m}\frac{1}{k!}\frac{\partial^kA(x,\xi)}{\partial\xi^k}D_x^kR_{-p+1}(x,\xi,\lambda).$$

Therefore, we get:

$$(A+\lambda I)(R_0+R_{-1}+\cdots+R_{-N})=I+T_{-N-1}, \tag{41.9}$$

where $T_{-N-1}(x,\xi,\lambda)\in S^{-N-1}$ and $|D_x^pT_{-N-1}(x,\xi,\lambda)\le\frac{C_{pN}}{|\lambda|^{\frac{N+1}{m}}},\forall|p|\ge 0$. Hence, by Theorem 40.1 and Remark 40.1, the operator T_{-N-1} satisfies (41.4). Note that the sum

$$R^{(N)}(x,\xi,\lambda)=R_0(x,\xi,\lambda)+\cdots+R_{-N}(x,\xi,\lambda) \tag{41.10}$$

can be rewritten in the following form:

$$R^{(N)}(x,\xi,\lambda)=(A_0(x,\xi)+\lambda)^{-1}+\sum_{k=1}^{N}(A_0(x,\xi)+\lambda)^{-(k+1)}p_k(x,\xi), \tag{41.11}$$

where $p_k(x,\xi)$ are polynomials in ξ, $\deg p_k(x,\xi)\le km-\frac{k}{2}$.

41.2. The existence and uniqueness theorem.

Lemma 41.3. *Let $A(x,D)$ be the same as in Theorem 41.2. Let s be arbitrary. Suppose $|\lambda|$ is sufficiently large and $\lambda \in C_\delta$. Then for any $f \in H_{s-m}(\mathbb{R}^n)$ there exists a unique solution $u \in H_s(\mathbb{R}^n)$ of the equation*

$$(A(x,D) + \lambda I)u = f.$$

Proof: Consider (41.3) with $N = 0$. It follows from (41.4) that T_{-1} is an operator with a small norm. Therefore $I+T_{-1}$ is invertible in $H_{s-m}(\mathbb{R}^n)$ and $R = R_0(x,D,\lambda)(I+T_{-1})^{-1}$ is the right inverse to $A+\lambda I$, i.e., $(A+\lambda I)Rf = f$, $\forall f \in H_{s-m}(\mathbb{R}^n)$. Also $u = Rf \in H_s(\mathbb{R}^n)$ since $\operatorname{ord} R = -m$.

Applying Theorem 40.2, we have $R_0(x,D,\lambda)(A(x,D)+\lambda) = I + T_{-1}^{(1)}$, where $T_{-1}^{(1)}$ satisfies (41.4) with $N = 0$. Therefore $(I + T_{-1}^{(1)})^{-1}$ exists when $\lambda \in C_\delta$ is sufficiently large. Thus $R^{(1)} = (I + T_{-1}^{(1)})^{-1} R_0$ is the left inverse of $A(x,D) + \lambda I$. Since $A(x,D) + \lambda I$ has the left and the right inverses, it is invertible as an operator from $H_s(\mathbb{R}^n)$ to $H_{s-m}(\mathbb{R}^n)$, and $R = R^{(1)} = (A(x,D) + \lambda I)^{-1}$. □

41.3. Elliptic regularity.

In the case where $A(x,\xi) \in S^\alpha$ is not a polynomial in ξ we call $A(x,\xi)$ elliptic if $A(x,\xi)$ can be represented in the form

$$A(x,\xi) = A_0(x,\xi)(1 - \chi(\xi)) + A_1(x,\xi),$$

where $A_0(x,\xi)$ is homogeneous in ξ of degree α, $A_0(x,\xi) \neq 0$, $\forall \xi \neq 0$, and $A(x,\xi)$ belongs to C^∞ for $\xi \neq 0$, $\chi(\xi) \in C_0^\infty(\mathbb{R}^n)$, $\chi(\xi) = 1$ for $|\xi| \leq 1$, and $A_1(x,\xi) \in S^{\alpha-1}$.

Lemma 41.4 (Elliptic regularity). *Let $A(x,\xi) \in S^\alpha$ be elliptic and let $u \in H_s(\mathbb{R}^n)$ be a solution of $A(x,D)u = f$ for some $s \in \mathbb{R}$. Suppose $f \in H_{s-\alpha+r}(\mathbb{R}^n)$, $r > 0$. Then $u \in H_{s+r}(\mathbb{R}^n)$. In particular, if $f \in H_N(\mathbb{R}^n)$, $\forall N$, then $u \in H_N(\mathbb{R}^n)$, $\forall N$.*

Proof: Let $R(x,\xi) = A_0^{-1}(x,\xi)(1 - \chi(\xi))$. It follows from Theorem 40.2 that

$$R(x,D)A(x,D)u = u - \chi(D)u + T_{-1}u, \tag{41.12}$$

where $\operatorname{ord} T_{-1} \leq -1$. Therefore

$$u = Rf + \chi(D)u - T_{-1}u.$$

Note that $Rf \in H_{s+r}$ since $\operatorname{ord} R = -\alpha$. Also $T_{-1}u \in H_{s+1}(\mathbb{R}^n)$, $\chi(D)u \in H_N(\mathbb{R}^n), \forall N$. Therefore $u \in H_{s+r_0}$, where $r_0 = \min(1,r)$. If $r > 1$, we can repeat the same arguments with $u \in H_{s+1}(\mathbb{R}^n)$ instead of $u \in H_s(\mathbb{R}^n)$.

After at most $[r] + 1$ steps we prove Lemma 41.4. □

42. Compactness and the Fredholm property

42.1. Compact operators.

Theorem 42.1. *Let* $\operatorname{ord} T' \le -\delta$ *and*

$$\|(1+|x|^2)^M T'u\|_{s+\delta} \le C\|u\|_s, \ \forall s, \tag{42.1}$$

where $M > 0$. *Then* T' *is compact in* $H_s(\mathbb{R}^n)$, $\forall s$.

Proof: We have the identity:

$$T' = (1+|x|^2)^{-M}\Lambda^{-\delta}\Lambda^{\delta}(1+|x|^2)^M T'.$$

Denote $T_1 = \Lambda^\delta(1+|x|^2)^M T'$, where Λ^δ is a ψdo with symbol $\Lambda^\delta(\xi) = (1+|\xi|^2)^{\frac{\delta}{2}}$. It follows from (42.1) that T_1 is bounded in $H_s(\mathbb{R}^n)$, $\forall s$. Therefore it is enough to prove that $T_0 = (1+|x|^2)^{-M}\Lambda^{-\delta}$ is a compact operator in $H_s(\mathbb{R}^n)$ since the product of a compact and of a bounded operator is compact.

The compactness of T_0 in $L_2(\mathbb{R}^n)$ was proven in §30 (see the proof of Theorem 30.3). The proof of the compactness in $H_s(\mathbb{R}^n)$ is similar.

Let $\chi(\xi) \in C_0^\infty(\mathbb{R}^n)$, $|\chi(\xi)| \le 1$, and $\chi(\xi) = 1$ for $|\xi| \le 1$. Denote by T_N the operator $(1+|x|^2)^{-M}\chi(\frac{x}{N})\Lambda^{-\delta}(D)\chi(\frac{D}{N})$. We have

$$\tilde{v}(\xi) = \int_{\mathbb{R}^n} \psi_N(\xi-\eta)\Lambda^{-\delta}(\eta)\chi\left(\frac{\eta}{N}\right)\tilde{u}(\eta)d\eta, \tag{42.2}$$

where

$$v = T_N u, \qquad \psi_N(\xi) = \frac{1}{(2\pi)^n}\int_{\mathbb{R}^n}(1+|x|^2)^{-M}\chi\left(\frac{x}{N}\right)e^{-ix\cdot\xi}d\xi.$$

Therefore

$$\Lambda^s(\xi)\tilde{v}(\xi) = \int_{\mathbb{R}^n} K_N(\xi,\eta)\Lambda^s(\eta)\tilde{u}(\eta)d\eta,$$

where

$$K_N(\xi,\eta) = \Lambda^s(\xi)\psi_N(\xi-\eta)\Lambda^{-s-\delta}(\eta)\chi\left(\frac{\eta}{N}\right). \tag{42.3}$$

Note that

$$\int_{\mathbb{R}^n}\int_{\mathbb{R}^n}|K_N(\xi,\eta)|^2 d\xi d\eta < +\infty,$$

i.e., the integral operator $K_N w = \int_{\mathbb{R}^n} K_N(\xi,\eta)w(\eta)d\eta$ is a Hilbert-Schmidt operator. Since Hilbert-Schmidt operators are compact in $L_2(\mathbb{R}^n)$, we conclude that the operator (42.2) is compact in $\tilde{H}_s(\mathbb{R}^n) = FH_s(\mathbb{R}^n)$, and therefore T_N is compact in $H_s(\mathbb{R}^n)$. We have

$$|||T_0 - T_N|||_{(s)} \to 0 \quad \text{as} \quad N \to \infty, \tag{42.4}$$

where $|||T_0 - T_N|||_{(s)}$ is the operator norm of $T_0 - T_N$ acting in $H_s(\mathbb{R}^n)$. Therefore T_0 is a compact operator in $H_s(\mathbb{R}^n)$ since the limit of compact operators in the operator norm is a compact operator. To prove (42.4) we have used that the operator of multiplication by $(1+|x|^2)^{-M}$ is bounded in $H_s(\mathbb{R}^n)$ (see Problem 15.17) and that the norm of the operator $(1+|x|^2)^{-M}(1-\chi(\frac{x}{N}))$ tends to zero as $N\to\infty$.

Theorem 42.2. *Let Ω be a smooth bounded domain in $\mathbb{R}^n$, and let T be a bounded operator from $H_s(\Omega)$ to $H_{s+\varepsilon}(\Omega)$, $\varepsilon>0$. Then T is a compact operator in $H_s(\Omega)$.*

Proof: Let l be an extension operator from Ω to $\mathbb{R}^n$ such that

$$\|lu\|_{s,\mathbb{R}^n}\le C\|u\|_{s,\Omega}\quad\text{for all}\quad u\in H_s(\Omega)$$

and $lu=0$ for $|x|>R$. We assume that $\overline{\Omega}$ is contained in the ball $\{x : |x|<R\}$.

Denote by $\psi_0(x)$ a $C_0^\infty(\mathbb{R}^n)$ function such that $\psi_0(x)=1$ for $|x|\le R$. Let p be the restriction operator to the domain Ω. We have, for any $u\in H_s(\Omega)$:

$$Tu=pl(Tu)=p\psi_0l(Tu)=p\psi_0\Lambda^{-\varepsilon}\Lambda^{\varepsilon}l(Tu).$$

The operator $\Lambda^{\varepsilon}l(Tu)$ is bounded from $H_s(\Omega)$ to $H_s(\mathbb{R}^n)$ and the operator p is bounded from $H_s(\mathbb{R}^n)$ to $H_s(\Omega)$. It follows from the proof of Theorem 42.1 that $\psi_0(x)\Lambda^{-\varepsilon}$ is compact in $H_s(\mathbb{R}^n)$. Therefore T is compact in $H_s(\Omega)$ as a product of bounded operators and a compact operator. □

Remark 42.1. Analogously to the proof of Theorem 42.2 one can show that the embedding of $H_{s+\varepsilon}(\Omega)$ into $H_s(\Omega)$ is compact: we have, as above, $u=p\psi_0lu=p\psi_0\Lambda^{-\varepsilon}\Lambda^{\varepsilon}lu$ for any $u\in H_{s+\varepsilon}(\Omega)$. The operator $\Lambda^{\varepsilon}lu$ is bounded from $H_{s+\varepsilon}(\Omega)$ to $H_s(\mathbb{R}^n)$. Therefore $Iu=p(\psi_0\Lambda^{-\varepsilon})(\Lambda^{\varepsilon}lu)$ is a compact operator from $H_{s+\varepsilon}(\Omega)$ to $H_s(\Omega)$ as the product of the bounded operators p and $\Lambda^{\varepsilon}lu$ and the compact operator $\psi_0\Lambda^{-\varepsilon}$. □

42.2. Fredholm operators.

Let A be a bounded operator from the Banach space B_1 to the Banach space B_2 and let B_k^* be the dual spaces to B_k, $k=1,2$. Denote by $(u,v^*)_k$ the pairing between B_k and B_k^*, $k=1,2$. For example, when $B=H_s(\mathbb{R}^n)$, $B^*=H_{-s}(\mathbb{R}^n)$ and (u,v) is the extension of the scalar product in $L_2(\mathbb{R}^n)$. The operator A^* is called the adjoint of A if

$$(Au,v^*)_2=(u,A^*v^*)_1\quad\text{for all}\quad u\in B_1,\ v^*\in B_2^*. \tag{42.5}$$

A bounded operator A from B_1 to B_2 is called a Fredholm operator if

a) $\ker A=\{u\in B_1 : Au=0\}$ is finite-dimensional.

b) $\operatorname{Im} A=\{v\in B_2 : v=Au \text{ for some } u\in B_1\}$ is closed in B_2, i.e., the image of A is closed.

c) $\ker A^*=\{v^*\in B_2^* : A^*v^*=0\}$ is finite-dimensional.

It follows from (42.5) that if $v_2^* \in \ker A^*$, i.e., $A^*v_2^* = 0$, then $(Au, v^*)_2 = 0$ for all $u \in B_1$. Vice versa, if $(Au, v^*)_2 = 0$ for all $u \in B_1$, then $(u, A^*v^*) = 0$, $\forall u \in B_1$, and therefore, $A^*v^* = 0$. Thus $\ker A^*$ is "orthogonal" to $\operatorname{Im} A$. We say that $\ker A^*$ is the cokernel of A and that $\operatorname{Im} A$ has a finite codimension if $\dim\ker A^* < +\infty$.

The following proposition is a well-known fact of functional analysis (see Rudin [**R**]).

Proposition 42.3. *Let T be a compact operator in B. Then the operator $I + T$ is Fredholm in B, where I is the identity operator. Moreover,* $\dim\ker(I + T) = \dim\ker(I + T^*)$.

It follows from Proposition 42.3 that the equation $u + Tu = f$ has a solution for any $f \in B$ iff the homogeneous equation $v + Tv = 0$ has only a trivial solution. Indeed, $\dim\ker(I + T) = 0$ implies $\dim\ker(I + T^*) = 0$, and therefore the range of $I + T$ is the whole space B.

Proposition 42.3 is called the Fredholm alternative.

We have already used this proposition in §30.

Proposition 42.4. *A bounded operator A from B_1 to B_2 is Fredholm if and only if there exists a bounded operator R from B_2 to B_1 such that*

$$RA = I + T_1, \quad AR = I + T_2, \tag{42.6}$$

where T_k are compact operators in B_k, $k = 1, 2$.

The operator R is called a regularizer of A.

Proof:

a) Let A be a Fredholm operator. Denote by $B_{02} \subset B_2$ the image of A. Let $f_1^*, \dots, f_{m_-}^*$ be a basis in $\ker A^* \subset B_2^*$. Choose $f_1, \dots, f_{m_-}$ in B_2 such that $(f_j, f_k^*) = \delta_{jk}$, $1 \le j, k \le m_-$. Let $B_{02}^\perp$ be the linear span of $f_1, \dots, f_{m_-}$. Then $B_2 = B_{02} \dot{+} B_{02}^\perp$ is the direct sum of B_{02} and $B_{02}^\perp$, i.e., any $v \in B_2$ has a unique decomposition

$$v = v_1 + v_2, \tag{42.7}$$

where $v_2 = \sum_{k=1}^{m_-}(v, f_k^*)_2 f_k \in B_{02}^\perp$, and $v_1 = (v - v_2) \in B_{02}$. Note that in the case where B_2 is a Hilbert space, (42.7) is the orthogonal decomposition of $\operatorname{Im} A$ and $\ker A^*$. Let

$$P_2 v = \sum_{k=1}^{m_-}(v, f_k^*)_2 f_k. \tag{42.8}$$

Note that $P_2^2 v = P_2 \sum_{k=1}^{m_-}(P_2 v, f_k^*)_2 f_k = P_2 v$, i.e., P_2 is the projector on $B_{02}^\perp$. Therefore, (42.7) has the form:

$$v = (I - P_2)v + P_2 v, \tag{42.9}$$

where P_2 is the projection on $B_{02}^{\perp}$ and $(I - P_2)$ is the projection on B_{02}.

Analogously, let $e_1, \ldots, e_{m_+}$ be a basis in $\ker A \subset B_1$. Let $e_j^* \in B_1^*$ be such that $(e_j, e_k^*)_1 = \delta_{jk}$, $1 \le j, k \le m_+$.

Denote by P_1 the following projection on $\ker A$:

$$P_1 u = \sum_{j=1}^{m_+} (u, e_j^*)_1 e_j, \tag{42.10}$$

and let $B_{01} = \{u \in B_1 : (u, e_j^*)_1 = 0,\ 1 \le j \le m_+\}$. Then we have a direct sum $B_1 = B_{10} \dot{+} \ker A$,

$$u = (I - P_1)u + P_1 u,$$

where $P_1 u \in \ker A$ and $(I - P_1)u \in B_{10}$. The operator A is a bounded one-to-one map of B_{01} onto B_{02}. By the open map theorem (see Rudin [**R**]), there exists a bounded inverse map R_0 of B_{02} onto B_{01}. Denote by R the following bounded operator from B_2 to B_1:

$$Rf = R_0 f \text{ for } f \in B_{02}, \quad Rf = 0 \text{ for } f \in B_{02}^{\perp}. \tag{42.11}$$

Then we have

$$\begin{aligned} ARf &= f - P_2 f, \quad \forall f \in B_2, \\ RAu &= u - P_1 u, \quad \forall u \in B_1. \end{aligned} \tag{42.12}$$

Note that P_1, P_2 are finite rank and hence compact operators. Therefore R is a regularizer of A.

b) Suppose there exists R such that (42.6) holds. Since $\ker A \subset \ker(I+T_1)$ and $\ker(I+T_1)$ is finite-dimensional, we see that $\dim\ker A < +\infty$. Analogously, from $R^*A^* = I + T_2^*$ we conclude that $\dim\ker A^* < +\infty$ since $\ker A^* \subset \ker(I + T_2^*)$.

As in (42.9) we have the following decomposition of B_2:

$$B_2 = \operatorname{Im}(I + T_2) + B_3,$$

where B_3 is finite-dimensional. Since $\operatorname{Im} A \supset \operatorname{Im}(I + T_2)$, we have $\operatorname{Im} A = \operatorname{Im}(I + T_2) + B_{30}$, where B_{30} is a subspace of B_3. Hence $\operatorname{Im} A$ is closed since $\operatorname{Im}(I + T_2)$ is closed and B_{30} is finite-dimensional. Therefore A is Fredholm. □

Remark 42.2. Sometimes it is easier to construct separately the left and the right regularizers, i.e., bounded operators R_1 and R_2 such that $R_1 A = I + T_1, AR_2 = I + T_2$, where T_k are compact operators. Then the proof of Proposition 42.4 implies that A is Fredholm. Moreover, we have $R_1 A R_2 = R_1 + R_1 T_2 = R_2 + T_1 R_2$, and therefore, $R_2 - R_1 = R_1 T_2 - T_1 R_2 = T$, where T is a compact operator from B_2 to B_1. Thus the left and the right regularizers differ by a compact operator.

42.3. Fredholm elliptic operators in $\mathbb{R}^n$.

Not every elliptic operator in $\mathbb{R}^n$ is Fredholm. For example, $-\Delta - k^2$ is not Fredholm from $H_s(\mathbb{R}^n)$ to $H_{s-2}(\mathbb{R}^n)$ for any s.

We consider a class of symbols $A(x,\xi) \in S^\alpha$ such that

$$(42.13) \qquad |A(x,\xi)| \geq C(1+|\xi|)^\alpha \quad \text{if} \quad |x|^2+|\xi|^2 \geq R^2 \quad \text{for some} \quad R>0.$$

Theorem 42.5. *Let $A(x,\xi) \in S^\alpha$ and let (42.13) holds. Then $A(x,D)$ is a Fredholm operator from $H_s(\mathbb{R}^n)$ to $H_{s-\alpha}(\mathbb{R}^n)$, $\forall s$.*

Proof: Suppose that $\chi(\xi)$ is as in Theorem 42.1. Define $R_0(x,\xi) = A^{-1}(x,\xi)(1-\chi(\frac{x}{R})\chi(\frac{\xi}{R}))$. Note that $R_0(x,\xi) \in S^{-\alpha}$. Applying Theorem 40.2 we have:

$$(42.14) \qquad A(x,D)R_0(x,D) = \Big(I - \chi\Big(\frac{x}{R}\Big)\chi\Big(\frac{D}{R}\Big)\Big) + T_{-1}^{(1)},$$

where $\operatorname{ord} T_{-1}^{(1)} \leq -1$. Moreover, by Remark 40.2, $\operatorname{ord}\,(1+|x|^2)^M T_{-1}^{(1)} \leq -1$. By Theorem 42.1, the operator $T_1^{(1)} - \chi(\frac{x}{R})\chi(\frac{D}{R})$ is compact in $H_{s-\alpha}(\mathbb{R}^n)$. Therefore $R_0(x,D)$ is the right regularizer of $A(x,D)$. Analogously it follows from Theorem 42.1 that $R_0(x,D)$ is a left regularizer of $A(x,D)$ since $R_0(x,D)A(x,D) = I - \chi(\frac{x}{R})\chi(\frac{D}{R}) + T_{-1}$ and $-\chi(\frac{x}{R})\chi(\frac{D}{R}) + T_{-1}$ is a compact operator in $H_s(\mathbb{R}^n)$. Therefore $A(x,D)$ is Fredholm.

Remark 42.3. If $A(x,\xi) \in S^\alpha$ and (42.13) holds, then, starting from $R_0(x,\xi) = A^{-1}(x,\xi)(1-\chi(\frac{x}{R})\chi(\frac{\xi}{R}))$ instead of $R_0(x,\xi,\lambda) = (A_0(x,\xi)+\lambda)^{-1}$, we can construct, similarly to the proof of Theorem 41.2, an operator $R^{(N)}(x,D)$ such that

$$A(x,D)R^{(N)}(x,D) = I + T_{-N-1}, \quad \forall N,$$

where $R^{(N)}(x,\xi) \in S^{-\alpha}$ and $\operatorname{ord} T_{-N-1} \leq -N-1$. Analogously, we can construct $R_1^{(N)}(x,D)$ such that

$$R_1^{(N)}(x,D)A(x,D) = I + T_{-N-1}^{(1)},$$

where $R_1^{(N)}(x,\xi) \in S^{-\alpha}$ and $\operatorname{ord} T_{-N-1}^{(1)} \leq -N-1$. As in Remark 42.2, we have $R^{(N)}(x,D) - R_1^{(N)}(x,D) = T_{-\alpha-N-1}^{(2)}$, where $\operatorname{ord} T_{-\alpha-N-1}^{(2)} \leq -\alpha-N-1$.

43. The adjoint of a pseudodifferential operator

43.1. A general form of ψdo's.

We rewrite expression (40.2) for a ψdo as follows:

$$(43.1) \qquad Au = \frac{1}{(2\pi)^n}\int_{\mathbb{R}^n}\int_{\mathbb{R}^n} A(x,\xi)e^{i(x-y)\cdot\xi}u(y)dyd\xi, \quad \forall u \in C_0^\infty(\mathbb{R}^n),$$

where the integral is understood as a repeated integral: first integration is performed with respect to y and then, with respect to ξ. It is convenient to consider, instead of (43.1), a more general form of a pseudodifferential operator:

$$Au = \frac{1}{(2\pi)^n} \int_{\mathbb{R}^n} \int_{\mathbb{R}^n} a(x,y,\xi) e^{i(x-y)\cdot\xi} u(y) dy d\xi, \tag{43.2}$$

where the integral is understood as a repeated integral and $a(x,y,\xi) \in C^\infty(\mathbb{R}^n \times \mathbb{R}^n \times \mathbb{R}^n)$ with the following properties:

$$\left| \frac{\partial^{p+k+r} a(x,y,\xi)}{\partial x^p \partial y^k \partial \xi^r} \right| \le C_{pkr}(1+|\xi|)^{\alpha-|r|}, \ \forall p,k,r, \tag{43.3}$$

$$\begin{aligned} a(x,y,\xi) &= a(\infty,y,\xi) \quad \text{for} \quad |x| \ge R, \\ a(x,y,\xi) &= a(x,\infty,\xi) \quad \text{for} \quad |y| \ge R. \end{aligned} \tag{43.4}$$

We represent $a(x,y,\xi)$ in the following form:

$$a(x,y,\xi) = a(x,\infty,\xi) + a'(x,y,\xi), \tag{43.5}$$

where $a'(x,y,\xi) = 0$ for $|y| \ge R$.

The following theorem shows that an operator of the form (43.2) can be represented in the form (40.2) up to an operator of an arbitrary low order.

Theorem 43.1. *Let A be an operator of the form (43.2). Then for any N,*

$$Au = \sum_{|k|=0}^{N} A_k(x,D)u + T_{\alpha-N-1}u, \tag{43.6}$$

where the $A_k(x,D)$ are operators of the form (40.2),

$$A_0(x,\xi) = a(x,\infty,\xi) + a'(x,x,\xi) = a(x,x,\xi), \tag{43.7}$$

(43.8)

$$A_k(x,\xi) = \frac{1}{k!} D_y^k \frac{\partial^k}{\partial \xi^k} a'(x,y,\xi)|_{y=x}, \ |k| \ge 1, \quad \operatorname{ord} T_{\alpha-N-1} \le \alpha - N - 1.$$

Proof: We have, for any $u \in C_0^\infty(\mathbb{R}^n)$:

$$Au = a(x,\infty,D)u + A'u, \tag{43.9}$$

where

$$A'u = \frac{1}{(2\pi)^n} \int_{\mathbb{R}^n} \int_{\mathbb{R}^n} a'(x,y,\xi) u(y) e^{i(x-y)\cdot\xi} dy d\xi. \tag{43.10}$$

Compute the integral in y using the convolution formula:

$$\int_{\mathbb{R}^n} a'(x,y,\xi) u(y) e^{-iy\cdot\xi} dy = \frac{1}{(2\pi)^n} \int_{\mathbb{R}^n} \tilde{a}'(x,\xi-\eta,\xi) \tilde{u}(\eta) d\eta, \tag{43.11}$$

where $\tilde{a}'(x,\eta,\xi) = F_y a'(x,y,\xi)$. We get

$$(43.12)\qquad A'u = \frac{1}{(2\pi)^{2n}} \int_{\mathbb{R}^n}\int_{\mathbb{R}^n} \tilde{a}'(x,\xi-\eta,\xi)\tilde{u}(\eta)e^{ix\cdot\xi}d\eta d\xi.$$

Making the change of variables $\xi - \eta = \zeta$, we obtain

$$(43.13)\qquad A'u = \frac{1}{(2\pi)^{2n}} \int_{\mathbb{R}^n}\int_{\mathbb{R}^n} \tilde{a}'(x,\zeta,\eta+\zeta)\tilde{u}(\eta)e^{ix\cdot(\eta+\zeta)}d\eta d\zeta.$$

The Taylor expansion of $\tilde{a}'(x,\zeta,\eta+\zeta)$ gives

$$(43.14)\qquad \tilde{a}'(x,\zeta,\eta+\zeta) = \sum_{|k|=0}^{N} \frac{1}{k!}\frac{\partial^k \tilde{a}'(x,\zeta,\eta)}{\partial\eta^k}\zeta^k + R_N(x,\zeta,\eta).$$

Taking into account that

$$(43.15)\qquad \frac{1}{(2\pi)^n}\int_{\mathbb{R}^n} \frac{\partial^k \tilde{a}'(x,\zeta,\eta)}{\partial\eta^k}\zeta^k e^{ix\cdot\zeta}d\zeta = \frac{\partial^k}{\partial\eta^k}D_y^k a'(x,y,\eta)|_{y=x},$$

we get (43.6) with $T_{\alpha-N-1}u$ of the form

$$(43.16)\qquad T_{\alpha-N-1}u = \frac{1}{(2\pi)^{2n}} \int_{\mathbb{R}^n}\int_{\mathbb{R}^n} R_N(x,\zeta,\eta)e^{ix\cdot(\eta+\zeta)}\tilde{u}(\eta)d\eta d\zeta.$$

It remains to show that $\operatorname{ord} T_{\alpha-N-1} \le \alpha - N - 1$. Note that

$$(43.17)\qquad \tilde{a}'(x,\zeta,\eta) = \tilde{a}'(\infty,\zeta,\eta) + \tilde{a}''(x,\zeta,\eta),$$

where $\tilde{a}''(x,\zeta,\eta) = 0$ for $|x| \ge R$. Then

$$(43.18)\qquad R_N(x,\zeta,\eta) = R_{N1}(\infty,\zeta,\eta) + R_{N2}(x,\zeta,\eta),$$

where R_{N1} corresponds to $\tilde{a}'(\infty,\zeta,\eta)$ and R_{N2} corresponds to $\tilde{a}''(x,\zeta,\eta)$. The Fourier transform in x of $T_{\alpha-N-1}u$ is

$$(43.19)\qquad (F_x(T_{\alpha-N-1}u))(\xi) = \frac{1}{(2\pi)^n}\int_{\mathbb{R}^n} R_{N1}(\infty,\xi-\eta,\eta)\tilde{u}(\eta)d\eta + \frac{1}{(2\pi)^{2n}} \int_{\mathbb{R}^n}\int_{\mathbb{R}^n} \tilde{R}_{N2}(\xi-\eta-\zeta,\zeta,\eta)\tilde{u}(\eta)d\eta d\zeta.$$

It follows from (40.13), (40.14), and (43.14) that

$$(43.20)\qquad |R_{N1}(\infty,\xi-\eta,\eta)| \le C_M(1+|\xi-\eta|)^{-M}(1+|\eta|)^{\alpha-N-1}(1+|\xi-\eta|)^{N+1+|\alpha|},\ \forall M,$$

$$(43.21)\qquad |\tilde{R}_{N2}(\xi-\eta-\zeta,\zeta,\eta)| \le C_{M_1,M}(1+|\xi-\eta-\zeta|)^{-M_1}(1+|\zeta|)^{-M} \times (1+|\eta|)^{\alpha-N-1}(1+|\zeta|)^{N+1+|\alpha|},\quad \forall M_1,\ \forall M.$$

Estimating the term in (43.19) containing R_{N1} as in (40.19), (40.20) and the term in (43.19) containing R_{N2} as in (40.21), (40.22), we get

$$\|T_{\alpha-N-1}u\|_s \leq C_s\|u\|_{s+\alpha-N-1}, \ \forall s. \tag{43.22}$$

□

43.2. The adjoint operator.

Let A^* be the formally adjoint operator to the ψdo $A(x,D)$, i.e.,

$$(Au,v) = (u, A^*v), \ \forall u \in C_0^\infty(\mathbb{R}^n), \ v \in C_0^\infty(\mathbb{R}^n),$$

where (u,v) is the L_2 scalar product.

Writing $A(x,D)u$ in the form (43.1), we get

$$(Au,v) = \frac{1}{(2\pi)^n}\int_{\mathbb{R}^n}\left(\int_{\mathbb{R}^n}\int_{\mathbb{R}^n} A(x,\xi)u(y)e^{i(x-y)\cdot\xi}dyd\xi\right)\overline{v(x)}dx. \tag{43.23}$$

We can rewrite (43.23) in the form

$$(Au,v) = \frac{1}{(2\pi)^n}\int_{\mathbb{R}^n} u(y)\left(\int_{\mathbb{R}^n}\int_{\mathbb{R}^n} A(x,\xi)e^{i(x-y)\cdot\xi}\overline{v(x)}dxd\xi\right)dy. \tag{43.24}$$

Therefore

$$(A^*v)(y) = \frac{1}{(2\pi)^n}\int_{\mathbb{R}^n}\int_{\mathbb{R}^n}\overline{A(x,\xi)}v(x)e^{i(y-x)\cdot\xi}dxd\xi. \tag{43.25}$$

Changing the notation from y to x and vice versa, we get

$$(A^*v)(x) = \frac{1}{(2\pi)^n}\int_{\mathbb{R}^n}\int_{\mathbb{R}^n}\overline{A(y,\xi)}v(y)e^{i(x-y)\cdot\xi}dyd\xi, \tag{43.26}$$

i.e., A^*v has the form (43.2). Theorem 43.1 applied to (43.26) gives

Theorem 43.2. *Let $A(x,D)$ be a ψdo with symbol $A(x,\xi) \in S^\alpha$. Then the adjoint operator A^* is also a ψdo operator and*

$$(A^*v)(x) = \sum_{|k|=0}^{N} A_k(x,D)v + T_{\alpha-N-1}v, \tag{43.27}$$

where

$$A_0(x,\xi) = \overline{A(x,\xi)}, \tag{43.28}$$

$$A_k(x,\xi) = \frac{1}{k!}D_x^k\frac{\partial^k}{\partial\xi^k}\overline{A(x,\xi)}, \quad 1 \leq |k| \leq N, \tag{43.29}$$

$$\operatorname{ord} T_{\alpha-N-1} \leq \alpha - N - 1.$$

□

43.3. Weyl's ψdo's.

Let $A(x,\xi) \in S^\alpha$. Denote by $A_w(x,D)$ an operator of the form (43.2):

$$A_w(x,D)u = \frac{1}{(2\pi)^n}\int_{\mathbb{R}^n} A\left(\frac{x+y}{2},\xi\right) e^{i(x-y)\cdot\xi}u(y)dyd\xi. \tag{43.30}$$

The operators of the form (43.30) are called Weyl's pseudodifferential operators.

Applying Theorem 43.1 to (43.30), we obtain

$$A_w(x,D) = A(x,D) + A_1(x,D) + T_{\alpha-N-1}, \tag{43.31}$$

where $\operatorname{ord} T_{\alpha-N-1} \le \alpha - N - 1$ and

$$A_1(x,\xi) = \sum_{|k|=1}^{N} \frac{1}{2^k k!} D_x^k \frac{\partial^k}{\partial \xi^k} A(x,\xi).$$

Therefore $A_w(x,D)$ and $A(x,D)$ differ by the terms of order $\alpha - 1$.

Consider the case where $A(x,\xi)$ is real-valued. Then the ψdo $A(x,D)$ is not selfadjoint if the terms (43.29) are not zero. However, $A_w(x,D)$ is selfadjoint if $A(x,\xi)$ is real since

$$\begin{aligned}(A_w(x,D)u,v) &= \frac{1}{(2\pi)^n}\int_{\mathbb{R}^n}\int_{\mathbb{R}^n}\int_{\mathbb{R}^n} A\left(\frac{x+y}{2},\xi\right) u(y)e^{i(x-y)\cdot\xi}dyd\xi\overline{v(x)}dx \\ &= \frac{1}{(2\pi)^n}\int_{\mathbb{R}^n}\int_{\mathbb{R}^n}\int_{\mathbb{R}^n} A\left(\frac{x+y}{2},\xi\right) u(y)\overline{e^{i(y-x)\cdot\xi}v(x)}dydxd\xi \\ &= (u, A_w(x,D)v).\end{aligned}$$

44. Pseudolocal property and microlocal regularity

44.1. The Schwartz kernel.

Let A be a linear continuous operator from $\mathcal{D}(\mathbb{R}^n)$ to $\mathcal{D}'(\mathbb{R}^n)$. The Schwartz kernel of A is the distribution $E \in \mathcal{D}'(\mathbb{R}^n \times \mathbb{R}^n)$ such that $(Au,v) = E(u(y)\overline{v(x)})$ for all $u \in C_0^\infty(\mathbb{R}^n)$, $v \in C_0^\infty(\mathbb{R}^n)$. Let Au be a ψdo of the form (43.2). Consider

$$E_\varepsilon(x,y) = \frac{1}{(2\pi)^n}\int_{\mathbb{R}^n} a(x,y,\xi)e^{i(x-y)\cdot\xi}\chi(\varepsilon\xi)d\xi, \tag{44.1}$$

where $\chi(\xi) \in C_0^\infty(\mathbb{R}^n)$, $\chi(\xi) = 1$ when $|\xi| < 1$, $\chi(\xi) = 0$ when $|\xi| > 2$. Note that $E_\varepsilon(x,y) \in C^\infty(\mathbb{R}^n \times \mathbb{R}^n)$. Let $w(x,y) \in C_0^\infty(\mathbb{R}^n \times \mathbb{R}^n)$. Then

$$E_\varepsilon(w(x,y)) = \int_{\mathbb{R}^n}\int_{\mathbb{R}^n} E_\varepsilon(x,y)w(x,y)dxdy$$

converges in $\mathcal{D}'(\mathbb{R}^n \times \mathbb{R}^n)$ to $E \in \mathcal{D}'(\mathbb{R}^n \times \mathbb{R}^n)$, where

$$E(w(x,y)) = \frac{1}{(2\pi)^n} \int_{\mathbb{R}^n} \int_{\mathbb{R}^n} \int_{\mathbb{R}^n} a(x,y,\xi) e^{i(x-y)\cdot\xi} w(x,y)\,dx\,dy\,d\xi. \tag{44.2}$$

The integral in (44.2) is understood as a repeated integral: first integration is with respect to x and y and then with respect to ξ. Since $(Au, v) = E(u(y)\overline{v(x)})$, the distribution E is the Schwartz kernel of A. When $\alpha < -n$, the integral

$$E(x,y) = \frac{1}{(2\pi)^n} \int_{\mathbb{R}^n} a(x,y,\xi) e^{i(x-y)\cdot\xi} d\xi \tag{44.3}$$

converges absolutely and defines the Schwartz kernel of A. When $\alpha \geq -n$, we understand (44.3) as a limit in $\mathcal{D}'(\mathbb{R}^n \times \mathbb{R}^n)$ of (44.1) as $\varepsilon \to 0$.

Theorem 44.1. *For an arbitrary $\delta > 0$, the distribution $(1 - \chi(\frac{x-y}{\delta}))E$ belongs to $C^\infty(\mathbb{R}^n \times \mathbb{R}^n)$.*

Proof: Since $\frac{\partial}{\partial \xi_k} e^{i(x-y)\cdot\xi} = i(x_k - y_k) e^{i(x-y)\cdot\xi}$, we have the following identity:

$$e^{i(x-y)\cdot\xi} = (-\Delta_\xi)^N \frac{e^{i(x-y)\cdot\xi}}{|x-y|^{2N}}, \tag{44.4}$$

where $\Delta_\xi = \sum_{j=1}^n \frac{\partial^2}{\partial \xi_j^2}$. Substituting (44.4) into (44.1) and integrating by parts with respect to ξ, we get

$$\begin{aligned}(44.5)\quad &\Big(1 - \frac{\chi(x-y)}{\delta}\Big) E_\varepsilon(x,y) \\ &= \frac{1}{(2\pi)^n} \int_{\mathbb{R}^n} (-\Delta_\xi)^N (a(x,y,\xi)\chi(\varepsilon\xi)) \frac{(1 - \chi(\frac{x-y}{\delta}))}{|x-y|^{2N}} e^{i(x-y)\cdot\xi} d\xi.\end{aligned}$$

Let

$$E^{(\delta)}(x,y) = \frac{1}{(2\pi)^n} \int_{\mathbb{R}^n} ((-\Delta_\xi)^N a(x,y,\xi)) \frac{(1 - \chi(\frac{x-y}{\delta}))}{|x-y|^{2N}} e^{i(x-y)\cdot\xi} d\xi. \tag{44.6}$$

We will study the limit of (44.5) as $\varepsilon \to 0$. Fix $M > 0$ and assume that $2N > 2M + n + 1 + \alpha$. Note that $|\frac{\partial^k}{\partial \xi^k} \chi(\varepsilon\xi)| \leq C\varepsilon^{|k|}$ and $\frac{\partial^k}{\partial \xi^k} \chi(\varepsilon\xi)$ differ from zero only when $\frac{1}{\varepsilon} < |\xi| < \frac{2}{\varepsilon}$ for $|k| \geq 1$. Also $1 - \chi(\varepsilon\xi) = 0$ as $|\xi| < \frac{1}{\varepsilon}$. Therefore

$$\begin{aligned}&\left|\Big(1 - \frac{\chi(x-y)}{\delta}\Big) E_\varepsilon(x,y) - E^{(\delta)}(x,y)\right| \\ &\qquad\leq C \sum_{|k|=0}^{2N} \int_{\frac{1}{\varepsilon} < |\xi|} \varepsilon^{|k|} (1 + |\xi|)^{\alpha - 2N + |k|} d\xi \to 0\end{aligned}$$

as $\varepsilon \to 0$, because $2N > n+1+\alpha$. Since $(1 - \frac{\chi(x-y)}{\delta})E_\varepsilon(x,y)$ also converges to $(1 - \frac{\chi(x-y)}{\delta})E$ in the distribution sense, we see that

$$\left(1 - \frac{\chi(x-y)}{\delta}\right)E = E^{(\delta)}, \tag{44.7}$$

where $E^{(\delta)}$ is the same as in (44.6). Note that $\frac{(1-\chi(\frac{x-y}{\delta}))}{|x-y|^{2N}} \in C^\infty(\mathbb{R}^n \times \mathbb{R}^n)$ since $\frac{(1-\chi(\frac{x-y}{\delta}))}{|x-y|^{2N}} = 0$ for $|x-y| < \delta$. We show that $E^{(\delta)} \in C^\infty(\mathbb{R}^n \times \mathbb{R}^n)$. Note that $|(-\Delta_\xi)^N a(x,y,\xi)| \le C(1+|\xi|)^{\alpha-2N} \le C(1+|\xi|)^{-(n+1)-2M}$. Therefore, differentiating (44.6) with respect to x and y, we obtain that $E^{(\delta)}$ has M continuous derivatives in x and y. Since M is arbitrary we get that $E^{(\delta)} = (1 - \frac{\chi(x-y)}{\delta})E \in C^\infty(\mathbb{R}^n \times \mathbb{R}^n)$.

44.2. Pseudolocal property of ψdo's.

Corollary 44.2 (Pseudolocal property of ψdo). *Assume that $\varphi \in C_0^\infty(\mathbb{R}^n)$, $\psi(x) \in C_0^\infty(\mathbb{R}^n)$, and $\operatorname{supp}\varphi \cap \operatorname{supp}\psi = \emptyset$. Let $A(x,D)$ be a ψdo of the form (43.2). Then $\psi(x)A\varphi u$ is an integral operator with $C^\infty(\mathbb{R}^n \times \mathbb{R}^n)$ kernel.*

Proof: Since $\operatorname{supp}\varphi \cap \operatorname{supp}\psi = \emptyset$, there exists $\delta > 0$ such that $\psi(x)E(x,y)\varphi(y) = 0$ when $|x-y| < \delta$. Therefore, $\psi(x)A\varphi$ is an integral operator with $C^\infty(\mathbb{R}^n \times \mathbb{R}^n)$ kernel, i.e., $\operatorname{ord}\psi A\varphi = -\infty$. Note that $\psi A \varphi = 0$ if A is a differential operator.

Lemma 44.3. *Let $A(x,\xi) \in S^\alpha$ and let $u \in H_s(\mathbb{R}^n)$ for some $s \in \mathbb{R}$. We have*

$$\operatorname{sing\,supp} Au \subset \operatorname{sing\,supp} u.$$

Proof: Recall that $\operatorname{sing\,supp} u$ is the complement of the largest open set where $u \in C^\infty$ (cf. §14). Let $x_0 \notin \operatorname{sing\,supp} u$. Then there exists $\varphi(x) \in C_0^\infty(\mathbb{R}^n)$ such that $\varphi(x) = 1$ near x_0 and $\operatorname{supp}\varphi \cap \operatorname{sing\,supp} u = \emptyset$. Let $\psi(x) \in C_0^\infty(\mathbb{R}^n)$, $\psi(x_0) \neq 0$, and $\operatorname{supp}\psi$ is contained in the interior of the set where $\varphi = 1$. Then $\operatorname{supp}\psi \cap \operatorname{supp}(1-\varphi) = \emptyset$. We have:

$$\psi A u = \psi A \varphi u + \psi A (1-\varphi) u. \tag{44.8}$$

Note that $\varphi u \in C_0^\infty(\mathbb{R}^n)$, since $\operatorname{supp}\varphi \cap \operatorname{sing\,supp}\ u = \emptyset$. Then, by Theorem 40.1, $\psi A \varphi u \in C^\infty$. Also $\psi A(1-\varphi)u \in C^\infty$ by Corollary 44.2. Therefore, $\psi A u \in C^\infty$ and $x_0 \notin \operatorname{sing\,supp} Au$ since $\psi(x_0) \neq 0$. □

The following theorem gives a refinement of Lemma 44.3.

Theorem 44.4. *Let $A(x,\xi) \in S^\alpha$ and let $u \in H_s(\mathbb{R}^n)$ for some $s \in \mathbb{R}$. We have*

$$WF(Au) \subset WF(u),$$

where $WF(v)$ is the wave front set of v (see §14 for the definition of the wave front set).

Proof: Suppose $(x_0, \xi_0) \notin WF(u)$. There exists $\varphi(x) \in C_0^\infty(\mathbb{R}^n)$ with $\varphi(x) = 1$ in a small neighborhood of x_0, and there exists $\alpha(\xi) \in C^\infty(\mathbb{R}^n \setminus \{0\})$ such that $\alpha(\xi)$ is homogeneous in ξ of degree zero, $\alpha(\xi) = 1$ in a small conic neighborhood of ξ_0, and $\alpha(D)\varphi(x)u \in C^\infty(\mathbb{R}^n)$ (cf. Definition 14.2).

Suppose $\psi(x) \in C_0^\infty(\mathbb{R}^n)$, $\psi(x_0) \neq 0$, and $\operatorname{supp}\psi$ is contained in the interior of $\{x : \varphi(x) = 1\}$. Let $\beta(\xi) \in C^\infty(\mathbb{R}^n \setminus \{0\})$, and let $\beta(\xi)$ be homogeneous of degree 0, where $\beta(\xi_0) \neq 0$ and $\operatorname{supp}\beta \cap \{|\xi| = 1\}$ is contained in the interior of $\{\xi : \alpha(\xi) = 1\}$.

To prove Theorem 44.4 we need to show that $\beta(D)\psi(x)Au \in C^\infty(\mathbb{R}^n)$. Since $\beta(\xi)$ is not smooth at $\xi = 0$, we replace $\beta(D)$ by $(1 - \chi(D))\beta(D)$, where $\chi(\xi)$ is the cutoff function as above. Note that $\chi(D)w \in C^\infty(\mathbb{R}^n)$ if $w \in H_s, \forall s$. Since $\operatorname{supp}\psi \cap \operatorname{supp}(1 - \varphi(x)) = \emptyset$, we see that $\psi A(1 - \varphi)u \in C^\infty$ by the pseudolocality of $A(x, D)$. Consider now $(1 - \chi(D))\beta(D)\psi A\varphi u$. Applying Theorem 40.2 to $(1 - \chi(D))\beta(D)$ and ψA, we get

$$(1 - \chi(D))\beta(D)\psi A\varphi u = \sum_{|k|=0}^{N} C_k(x, D)\varphi u + T_{\alpha-N-1}\varphi u, \tag{44.9}$$

where $\operatorname{ord} T_{\alpha-N-1} \leq \alpha - N - 1$, and the symbols $C_k(x, \xi)$ contain either $\beta(\xi)$ or the derivatives $\frac{\partial^k \beta(\xi)}{\partial \xi^k}$. Since $\alpha(\xi) = 1$ on $\operatorname{supp}\beta(\xi)$, we have $C(x, \xi) = \sum_{|k|=0}^{N} C_k(x, \xi) = C(x, \xi)\alpha(\xi)$. Therefore

$$(1 - \chi(D))\beta(D)\psi A\varphi u = C\alpha(D)\varphi u + T_{\alpha-N-1}\varphi u.$$

Since $\alpha(D)\varphi u \in C^\infty$, we have $C\alpha(D)\varphi u \in C^\infty$ (cf. Theorem 40.1). Therefore $(1 - \chi(D))\beta(D)\psi A\varphi u \in H_{s-\alpha+N+1}(\mathbb{R}^n)$. Since N is arbitrary, we conclude that $\beta(D)\psi A\varphi u \in C^\infty$. □

44.3. Microlocal regularity.

Theorem 44.5 (Microlocal regularity). *Suppose $A(x, \xi) \in S^m$ and there exists $A_0(x, \xi)$ that is homogeneous of degree m in ξ and such that $A(x, \xi) - A_0(x, \xi)\chi(\xi) \in S^{m-1}$. Suppose $A(x, \xi)$ is microelliptic at (x_0, ξ_0), i.e., $A_0(x_0, \xi_0) \neq 0$. Let $u \in H_s(\mathbb{R}^n)$ be the solution of $Au = f$ and let f be C^∞ at (x_0, ξ_0), i.e., $(x_0, \xi_0) \notin WF(f)$. Then $(x_0, \xi_0) \notin WF(u)$.*

Proof: Let $\varphi(x), \alpha(\xi), \psi(x), \beta(\xi)$ be the same as in the proof of Theorem 44.4. We assume that $A_0(x, \xi) \neq 0$ when $x \in \operatorname{supp}\varphi(x)$, $|\xi| = 1$ and $\xi \in \operatorname{supp}\alpha(\xi)$, and that $\alpha(D)\varphi(x)f \in C^\infty(\mathbb{R}^n)$.

Let $B(x, \xi)$ be the extension of $A(x, \xi)$ to $\mathbb{R}^n \times \mathbb{R}^n$ from a neighborhood of $\operatorname{supp}\varphi(x)\alpha(\xi)(1 - \chi(\xi))$, and let $B(x, \xi) \in S^m$ and $|B(x, \xi)| \geq C(1 + |\xi|)^m$ for $|x|^2 + |\xi|^2 \geq R^2$. As in Remark 42.3, one can construct a ψdo $R^{(N)}(x, D)$, $R^{(N)}(x, \xi) \in S^{-m}$, such that $R^{(N)}(x, D)B(x, D) = I + T_{-N-1}$, where ord $T_{-N-1} \leq -N-1$. Since $Au = f$, we have

$$Bu = f + (B - A)u. \tag{44.10}$$

Applying $R^{(N)}$ to (44.10), we get

$$u = R^{(N)}f + R^{(N)}(B - A))u - T_{-N-1}u. \tag{44.11}$$

Now we apply $(1 - \chi(D))\beta(D)\psi(x)$ to (44.11). Since $R^{(N)}(x, \xi) \in S^{-m}$ and $\alpha(D)\varphi(x)f \in C^\infty$, we conclude from the proof of Theorem 44.4 that $(1 - \chi(D))\beta(D)\psi R^{(N)}f \in C^\infty$. We now apply Theorem 40.2 to the composition C of the pseudodifferential operators $(1 - \chi(D)\beta(D)\psi(x)$, $R^{(N)}(x, D)$, and $B(x, D) - A(x, D)$. We obtain:

$$Cu = \sum_{|k|=0}^{N} C_k(x, D)u + T^{(1)}_{-N-1}u,$$

where ord $T^{(1)}_{-N-1} \leq -N-1$ and $C_k(x, \xi) \in S^{-|k|}$. Note that all $C_k(x, \xi)$ contain either $\psi(x)\beta(\xi)$ or its derivatives with respect to ξ and x. On the other hand, $B(x, \xi) - A(x, \xi) = 0$ on supp $\varphi(x)\alpha(\xi)(1 - \chi(\xi))$ for $|x|^2 + |\xi|^2 \geq R^2$, i.e., $B(x, \xi) - A(x, \xi) = 0$ on a neighborhood of supp $\psi(x)\beta(\xi)(1 - \chi(\xi))$. Therefore $C_k(x, xi) = 0$ for $|x|^2 + |\xi|^2 \geq R^2$ for all k, $0 \leq |k| \leq N$. This implies that $(1 - \chi(D))\psi(x)u \in H_{s+N+1}(\mathbb{R}^n)$ for an arbitrary N. Thus $(1 - \chi(D))\beta(D)\psi(x)u \in C^\infty$. □

Remark 44.1. It follows from (44.3) that a ψdo with symbol S^α, $\alpha < -n$, has a continuous kernel. We shall prove the same result for any operator of order $\alpha < -n$.

Lemma 44.6. *Let T_α be an operator of order $\alpha < -n$. Then the kernel $T_\alpha(x, y)$ of T_α is continuous in (x, y).*

Proof: Take $s = -\frac{n}{2} - \varepsilon$, $\varepsilon > 0$. Since ord $T_\alpha = \alpha$, we have

$$\|v\|_{-\frac{n}{2}-\varepsilon-\alpha} \leq |||T_\alpha||| \, \|u\|_{-\frac{n}{2}-\varepsilon}, \tag{44.12}$$

where $v = T_\alpha u$, $|||T_\alpha|||$ is the operator norm of T_α from $H_{-\frac{n}{2}-\varepsilon}(\mathbb{R}^n)$ to $H_{-\frac{n}{2}-\varepsilon-\alpha}(\mathbb{R}^n)$. Note that the space $H_{\frac{n}{2}+\varepsilon}$ is dual to $H_{-\frac{n}{2}-\varepsilon}$ with respect to the extension of the L_2-scalar product (cf. §13) and that $H_{\frac{n}{2}+\varepsilon} \supset H_{-\frac{n}{2}-\varepsilon-\alpha}$, since $-\frac{n}{2} - \varepsilon - \alpha > \frac{n}{2} + \varepsilon$ when ε is small. Therefore T_α is a bounded operator from $H_{-\frac{n}{2}-\varepsilon}$ to $H_{\frac{n}{2}+\varepsilon}$.

Let δ_x be the delta-function at x, i.e., $\delta_x(\varphi) = \varphi(x)$. Analogously, $\delta_y(\varphi) = \varphi(y)$. Note that δ_x and δ_y belong to $H_{-\frac{n}{2}-\varepsilon}(\mathbb{R}^n)$ and depend continuously on x and y, respectively. Let

$$T_\alpha(x, y) = (T_\alpha\delta(y), \delta(x)), \tag{44.13}$$

i.e., $T_\alpha(x,y)$ is the kernel of T_α. Note that $(T_\alpha u_1, u_2)$ is a continuous bilinear form in $H_{-\frac{n}{2}-\varepsilon}(\mathbb{R}^n)\times H_{-\frac{n}{2}-\varepsilon}(\mathbb{R}^n)$. Therefore, $T_\alpha(x,y)$ is continuous in (x,y). Moreover,

$$|T_\alpha(x,y)| \le C\,|||T_\alpha|||\,. \tag{44.14}$$

Example 44.1. Let $E(x,y)$ be the Schwartz kernel of the operator (43.2). We find the wave front set $WF(E(x,y)) \subset \mathbb{R}^{2n}\times\mathbb{R}^{2n}\setminus\{0\}$. It follows from Theorem 44.1 that $\operatorname{sing\,supp} E \subset \{(x,y) : x = y\}$. Fix any point (x_0,x_0) and let $\psi(x,y)\in C_0^\infty(\mathbb{R}^{2n})$, $\psi(x_0,x_0)\neq 0$. Take the Fourier transform of $\psi(x,y)E(x,y)$ in x and in y. We get:

(44.15)
$$\begin{aligned}F(\psi E) &= \frac{1}{(2\pi)^n}\int_{\mathbb{R}^n}\int_{\mathbb{R}^n}\int_{\mathbb{R}^n}\psi(x,y)a(x,y,\theta)e^{-ix\cdot\xi-iy\cdot\eta}e^{i(x-y)\cdot\theta}dxdyd\theta\\ &= \frac{1}{(2\pi)^n}\int_{\mathbb{R}^n}\tilde b(\xi-\theta,\eta+\theta,\theta)d\theta,\end{aligned}$$

where $b(x,y,\theta)=\psi(x,y)a(x,y,\theta)$, and $\tilde b(\xi,\eta,\theta)$ is the Fourier transform of $b(x,y,\theta)$ in x and y. To justify (44.15) we introduce $\chi(\varepsilon\theta)$ and take the limit in the distribution sense as $\varepsilon\to 0$ (cf. the proof of Theorem 44.1).

Since $b(x,y,\theta)$ is C_0^∞ in x and in y, we have

$$|\tilde b(\xi-\theta,\eta+\theta,\theta)| \le C_N(1+|\xi-\theta|)^{-N}(1+|\eta+\theta|)^{-N}(1+|\theta|)^m, \forall N. \tag{44.16}$$

Make a change of variables $\theta+\eta=\theta'$. Then

$$F(\psi(x,y)E(x,y)) = \frac{1}{(2\pi)^n}\int_{\mathbb{R}^n}\tilde b(\xi+\eta-\theta',\theta',\theta'-\eta)d\theta'.$$

Note that
(44.17)
$$|\tilde b(\xi+\eta-\theta',\theta',\theta'-\eta)| \le C_N(1+|\xi+\eta-\theta'|)^{-N}(1+|\theta'|)^{-N}(1+|\theta'|+|\eta|)^m.$$

We show that

$$|F(\psi E)(\xi,\eta)) \le C_N(1+|\xi+\eta|)^{-N}(1+|\eta|)^m, \quad \forall N. \tag{44.18}$$

We have

$$F(\psi E) = I_1 + I_2, \tag{44.19}$$

where

$$I_1 = \int_{|\theta'|<\frac{1}{2}|\xi+\eta|}\tilde b(\xi+\eta-\theta',\theta',\theta'-\eta)d\theta' \quad \text{and} \quad I_2 = F(\psi E) - I_1.$$

If $|\theta'|<\frac12|\xi+\eta|$, we get $1+|\xi+\eta-\theta'|\ge 1+|\xi+\eta|-|\theta'|\ge\frac12(1+|\xi+\eta|)$. Therefore

$$|I_1(\xi,\eta)| \le C_N(1+|\xi+\eta|)^{-N}(1+|\eta|)^m, \ \forall N.$$

If $|\theta'| > \frac{1}{2}|\xi+\eta|$, we get

$$|I_2(\xi,\eta)| \le C_N(1+|\xi+\eta|)^{-\frac{N}{2}}(1+|\eta|)^m, \ \forall N.$$

Therefore (44.18) holds. Let $\omega_1 = \frac{\xi}{\sqrt{|\xi|^2+|\eta|^2}}$, $\omega_2 = \frac{\eta}{\sqrt{|\xi|^2+|\eta|^2}}$, provided $(\xi,\eta) \ne (0,0)$. Then

$$|F(\psi E)| \le \frac{C_N(1+|\eta|)^m}{(1+|\omega_1+\omega_2|\sqrt{|\xi|^2+|\eta|^2})^N}, \ \forall N.$$

Therefore, for any small conic neighborhood of (ω_1,ω_2) such that $\omega_1+\omega_2 \ne 0$, we have

$$|F(\psi E)| \le C_N(1+\sqrt{|\xi|^2+|\eta|^2})^{-N}, \ \forall N.$$

Thus $WF(E) \subset \{(x,y,\xi,\eta) \in \mathbb{R}^{2n} \setminus (\mathbb{R}^{2n} \setminus \{0\}) : x = y, \xi = -\eta\}$.

45. Change-of-variables formula for ψdo's

Let

$$x = s(\hat{x}) \tag{45.1}$$

be a one-to-one diffeomorphism of $\mathbb{R}^n$ onto $\mathbb{R}^n$, $s(\hat{x}) = \hat{x}$ for $|\hat{x}| > R$, and $J(\hat{x}) = \det\frac{\partial s(\hat{x})}{\partial \hat{x}} \ne 0$, $\forall \hat{x} \in \mathbb{R}^n$, where $\frac{\partial s(\hat{x})}{\partial \hat{x}}$ is the Jacobian matrix of $x = s(\hat{x})$. Let $A(x,D)$ be a ψdo operator with symbol $A(x,\xi) \in S^\alpha$. We have

$$A(x,D)u = Bu + T_{-\infty}u, \tag{45.2}$$

where

$$Bu = \frac{1}{(2\pi)^n}\int_{\mathbb{R}^n} A(x,\xi)\chi\Big(\frac{x-y}{\delta}\Big)e^{i(x-y)\cdot\xi}u(y)dyd\xi, \tag{45.3}$$

$$T_{-\infty}u = \frac{1}{(2\pi)^n}\int_{\mathbb{R}^n} A(x,\xi)\Big(1-\chi\Big(\frac{x-y}{\delta}\Big)\Big)e^{i(x-y)\cdot\xi}u(y)dyd\xi. \tag{45.4}$$

It follows from Theorem 44.1 that $T_{-\infty}$ is an integral operator with $C^\infty(\mathbb{R}^n \times \mathbb{R}^n)$ kernel.

Consider $v(x) = A(x,D)u$, where $u(x) \in C_0^\infty(\mathbb{R}^n)$. Make the change of variables (45.1). Let $\hat{u}(\hat{x}) = u(s(\hat{x}))$ and $\hat{v}(\hat{x}) = v(s(\hat{x}))$ be $u(x)$ and $v(x)$ in the new coordinates and let $\hat{v}(\hat{x}) = \hat{A}\hat{u}$ be the image of $A(x,D)$ in the new coordinates. Then

$$\hat{A}\hat{u} = \hat{B}\hat{u} + \hat{T}_{-\infty}\hat{u},$$

where $\hat{B}$ is the image of B and $\hat{T}_{-\infty}$ is the image of $T_{-\infty}$ in new coordinates. Since $T_{-\infty}$ is an operator with $C^\infty(\mathbb{R}^n \times \mathbb{R}^n)$ kernel, the operator $\hat{T}_{-\infty}$ is also an integral operator with C^∞ kernel.

We show that $\hat{B}$ is also a ψdo modulo $T_{\alpha-N-1}$, where ord $T_{\alpha-N-1} \leq \alpha - N - 1$, N is arbitrary, and we compute the symbol of $\hat{B}$. Substituting $x = s(\hat{x}), y = s(\hat{y})$ in (45.3), we get

(45.5)
$$Bu = \frac{1}{2\pi^n}\int_{\mathbb{R}^n} A(s(\hat{x}),\xi)\chi\Big(\frac{s(\hat{x})-s(\hat{y})}{\delta}\Big)e^{i(s(\hat{x})-s(\hat{y}))\cdot\xi}u(s(\hat{y}))|J(\hat{y})|d\hat{y}d\xi.$$

Note that

$$s(\hat{x})-s(\hat{y}) = \int_0^1 \frac{d}{dt}s(\hat{y}+t(\hat{x}-\hat{y}))dt = H(\hat{x},\hat{y})(\hat{x}-\hat{y}), \tag{45.6}$$

where $H(\hat{x},\hat{y})$ is a C^∞ matrix and $H(\hat{x},\hat{x}) = \frac{\partial s(\hat{x})}{\partial \hat{x}}$. Since det $\frac{\partial s(\hat{x})}{\partial \hat{x}} \neq 0, \forall \hat{x}$, we see that $\det H(\hat{x},\hat{y}) \neq 0$ if $|s(\hat{x})-s(\hat{y})| \leq \delta$, where δ is small. We have

$$(s(\hat{x})-s(\hat{y}))\cdot\xi = H(\hat{x},\hat{y})(\hat{x}-\hat{y})\cdot\xi = (x-y)\cdot H^T(\hat{x},\hat{y})\xi, \tag{45.7}$$

where $H^T(\hat{x},\hat{y})$ is a matrix transpose to $H(\hat{x},\hat{y})$. We substitute (45.7) into (45.5) and make the change of variables

$$\eta = H^T(\hat{x},\hat{y})\xi. \tag{45.8}$$

In order to justify the change of variables (45.8), one must introduce the cutoff factor $\chi(\varepsilon\xi)$ into (45.5), make the change of variables, and then take the limit as $\varepsilon \to 0$. We get

(45.9)
$$\hat{B}\hat{u} = \frac{1}{(2\pi)^n}\int_{\mathbb{R}^n} A(s(\hat{x}),(H^T(\hat{x},\hat{y}))^{-1}\eta)\chi\Big(\frac{s(\hat{x})-s(\hat{y})}{\delta}\Big)e^{i(\hat{x}-\hat{y})\cdot\eta}\hat{u}(\hat{y}) \times |J(\hat{y})|\,|\det H(\hat{x},\hat{y})|^{-1}d\hat{y}d\eta.$$

Therefore $\hat{B}$ is a ψdo of the form (43.2) with symbol

(45.10)
$$\hat{b}(\hat{x},\hat{y},\eta) = A(s(\hat{x}),(H^T(\hat{x},\hat{y}))^{-1}\eta)\chi(\frac{s(\hat{x})-s(\hat{y})}{\delta})|J(\hat{y})|\,|\det H(\hat{x},\hat{y})|^{-1}.$$

Using Theorem 43.1, we can represent (45.9) in the form

$$\hat{B}\hat{u} = \sum_{|k|=0}^{N}\hat{B}_k(\hat{x},\hat{D})\hat{u} + \hat{T}_{\alpha-N-1}\hat{u},$$

where

$$\hat{B}_0(\hat{x},\eta) = \hat{b}(\hat{x},\hat{x},\eta) = A\Big(s(\hat{x}),\Big(\Big(\frac{\partial s(\hat{x})}{\partial \hat{x}}\Big)^T\Big)^{-1}\eta\Big),$$

$$B_k(\hat{x},\eta) = \frac{1}{k!}D_{\hat{y}}^k\frac{\partial^k}{\partial\eta^k}b(\hat{x},\hat{y},\eta)|_{\hat{y}=\hat{x}},$$

ord $\hat{T}_{\alpha-N-1} \leq \alpha - N - 1$, and N is arbitrary. □

46. The Cauchy problem for parabolic equations

46.1. Parabolic ψdo's.

Consider a differential equation in $\mathbb{R}^{n+1}_+ = \{t > 0, x \in \mathbb{R}^n\}$ of the form:

$$\frac{\partial u(x,t)}{\partial t} + A(x,t,D)u(x,t) = f, \ t > 0, \ x \in \mathbb{R}^n, \tag{46.1}$$

where $A(x,t,\xi) = A_0(x,t,\xi) + A_1(x,t,\xi)$, $A_0(x,t,\xi) = \sum_{|k|=m} a_k(x,t)\xi^k$, $A_1(x,t,\xi) = \sum_{|k|\le m-1} a_k(x,t)\xi^k$, $D = -i\frac{\partial}{\partial x}$.

We assume that $a_k(x,t) \in C^\infty(\mathbb{R}^{n+1})$, $a_k(x,t) = a_k^\infty$ for $|x| > R$, and $a_k(x,t) = a_k(x,\infty)$ when $|t| > R$. Equation (46.1) is called parabolic if $\operatorname{Re} A_0(x,t,\xi) \ge C|\xi|^m$, $\forall(x,t)$. We study the Cauchy problem, i.e., $u(x,t)$ satisfies (46.1) and the initial condition

$$u(x,0) = u_0(x), \quad x \in \mathbb{R}^n. \tag{46.2}$$

Let $v(x,t) = e^{-t\tau}u(x,t)$, $g = e^{-t\tau}f$, $\tau > 0$. Then $v(x,t)$ satisfies the equation:

$$\frac{\partial v(x,t)}{\partial t} + \tau v + A(x,t,D)v(x,t) = g, \quad t > 0, \quad x \in \mathbb{R}^n, \tag{46.3}$$

and

$$v(x,0) = u_0(x). \tag{46.4}$$

We prove the existence and the uniqueness of the solution of the Cauchy problem (46.3), (46.4) provided τ is large.

We introduce the Sobolev spaces that are adapted to the form of the parabolic equation (46.1). Let $\Pi_{\frac{s}{m},s}(\mathbb{R}^{n+1})$ be the space of distributions with finite norm

$$\|u\|^2_{\frac{s}{m},s} = \int_{\mathbb{R}^{n+1}} |\tilde{u}(\xi,\sigma)|^2 |i\sigma + (|\xi|^2+1)^{\frac{m}{2}}|^{\frac{2s}{m}} d\xi d\sigma, \tag{46.5}$$

where

$$\tilde{u}(\xi,\sigma) = \int_{\mathbb{R}^{n+1}} u(x,t)e^{-ix\cdot\xi - i\sigma t} dxdt.$$

As in $ 22, we denote by $\overset{\circ}{\Pi}_{\frac{s}{m},s}(\mathbb{R}^{n+1})$ the subspace of $\Pi_{\frac{s}{m},s}(\mathbb{R}^{n+1})$ consisting of distributions with supports in $\mathbb{R}^{n+1}_+ = \{t > 0, x \in \mathbb{R}^n\}$ and denote by $\Pi_{\frac{s}{m},s,\tau}(\mathbb{R}^{n+1}_+)$ the space of restrictions of $u \in \Pi_{\frac{s}{m},s}(\mathbb{R}^{n+1})$ to the half-space $t > 0$ with the norm

$$\|u\|^+_{\frac{s}{m},s} = \inf_l \|lu\|_{\frac{s}{m},s}, \tag{46.6}$$

where lu is an arbitrary extension of u to $\mathbb{R}^{n+1}$ (cf. §22).

Denote by $P_{\alpha,m}$ the class of symbols $A(x,t,\xi,\sigma)$ such that A is C^∞ in all variables, $A(x,t,\xi,\sigma) = A(\infty,\xi,\sigma)$ when $|x| \geq R$, $A(x,t,\xi,\sigma) = A(x,\infty,\xi,\sigma)$ when $|t| > R$, and

$$\left|\frac{\partial^{k_1+k_2+k_3+k_4} A(x,t,\xi,\sigma)}{\partial x^{k_1}\partial t^{k_2}\partial\xi^{k_3}\partial\sigma^{k_4}}\right| \leq C_k|\Lambda_m^+|^{\frac{\alpha}{m}-\frac{k_3}{m}-k_4}, \quad \forall k, \tag{46.7}$$

where

$$\Lambda_m^+ = i\sigma + (|\xi|^2+1)^{\frac{m}{2}}. \tag{46.8}$$

Note that if $m=1$, then $P_{\alpha,m}$ coincides with S^α (cf. §40). Note also that the norm (46.5) has the form $\|(\Lambda_m^+)^{\frac{s}{m}}u\|_0$, where $\|\ \|_0$ is the norm in $L_2(\mathbb{R}^{n+1})$, and Λ_m^+ is the ψdo with symbol $\Lambda_m^+(\xi,\sigma)$.

Analogously to the proof of Theorems 40.1 and 40.2, we have

Lemma 46.1. *A pseudodifferential operator A with symbol $A(x,t,\xi,\sigma) \in P_{\alpha,m}$ is bounded from $\Pi_{\frac{s}{m},s}(\mathbb{R}^{n+1})$ to $\Pi_{\frac{s-\alpha}{m},s-\alpha}(\mathbb{R}^{n+1})$. If $A(x,t,\xi,\sigma) \in P_{\alpha,m}$, $B(x,t,\xi,\sigma) \in P_{\beta,m}$, then*

$$AB = C + T_{\alpha+\beta-1}, \tag{46.9}$$

where $C(x,t,\xi,\sigma) = A(x,t,\xi,\sigma)B(x,t,\xi,\sigma) \in P_{\alpha+\beta,m}$, and

$$\|T_{\alpha+\beta-1}u\|_{\frac{s}{m},s} \leq C\|u\|_{\frac{\alpha+\beta-1+s}{m},\alpha+\beta-1+s} \quad \forall u \in C_0^\infty(\mathbb{R}^{n+1}),\ \forall s. \tag{46.10}$$

The following lemma is crucial for this section:

Lemma 46.2. *Suppose $A(x,t,\xi,\sigma-i\tau) \in P_{\alpha,m}$ is analytic in $z=\sigma-i\tau$ for $\tau>0$ and continuous for $\tau \geq 0$. Suppose also that $A(x,t,\xi,\sigma-i\tau)$ satisfies estimates (46.7) with σ replaced by $z=\sigma-i\tau, \tau\geq 0$. Then the ψdo $A(x,t,D_x,D_t)$ maps $\overset{\circ}{\Pi}_{\frac{s}{m},s}(\mathbb{R}^{n+1})$ into $\overset{\circ}{\Pi}_{\frac{s-\alpha}{m},s-\alpha}(\mathbb{R}^{n+1})$ for each $s \in \mathbb{R}$.*

Proof: For any $u(x,t) \in C_0^\infty(\mathbb{R}_+^{n+1})$ we have:

$$Au = \frac{1}{(2\pi)^{n+1}}\int_{\mathbb{R}^{n+1}} A(x,t,\xi,\sigma)\tilde{u}(\xi,\sigma)e^{ix\cdot\xi+it\sigma}dxdt. \tag{46.11}$$

Since A and $\tilde{u}(\xi,z)$ are analytic in $z=\sigma-i\tau$ for $\tau>0$ and since $\tilde{u}(\xi,z)$ decays rapidly as $|\operatorname{Re} z| \to \infty$, we can move the line of integration in z using the Cauchy theorem

$$Au = \frac{1}{(2\pi)^n}\int_{\mathbb{R}^n}\int_{-\infty}^{\infty} A(x,t,\xi,\sigma-i\tau)\tilde{u}(\xi,\sigma-i\tau)e^{ix\cdot\xi+it(\sigma-i\tau)}d\sigma d\xi, \tag{46.12}$$

where $\tau>0$ is arbitrary. Therefore

$$|(Au)(x,t)| \leq Ce^{t\tau}, \quad \forall\tau>0, \tag{46.13}$$

since $|\tilde{u}(\xi,\sigma-i\tau)| \leq C_N(1+|\xi|+|\sigma-i\tau|)^{-N}$, $\forall N$, for $\xi\in\mathbb{R}^n$, $\sigma\in\mathbb{R}^1$, $\tau\geq 0$.

Fix any $t_0 < 0$. Taking the limit in (46.13) as $\tau \to +\infty$, we get $(Au)(x,t_0) = 0$. Therefore, $\operatorname{supp} Au \subset \overline{\mathbb{R}^{n+1}_+}$.

It follows from Lemma 46.1 that $Au \in \Pi_{\frac{s-\alpha}{m},s-\alpha}(\mathbb{R}^{n+1})$. Therefore $Au \in \overset{\circ}{\Pi}_{\frac{s-\alpha}{m},s-\alpha}$. Since $C_0^\infty(\mathbb{R}^{n+1}_+)$ is dense in $\overset{\circ}{\Pi}_{\frac{s}{m},s}(\mathbb{R}^{n+1})$, we have that $Au \in \overset{\circ}{\Pi}_{\frac{s-\alpha}{m},s-\alpha}$ for any $u \in \overset{\circ}{\Pi}_{\frac{s}{m},s}$. □

46.2. The Cauchy problem with zero initial conditions.

We shall denote by $P^+_{\alpha,m}$ the class of symbols in $P_{\alpha,m}$ that are analytic in $z = \sigma - i\tau$ for $\tau > 0$ and satisfy estimates (46.7) with σ replaced by $\sigma - i\tau, \tau \geq 0$.

Theorem 46.3. *Consider the equation (46.3) in $\mathbb{R}^{n+1}$. Let s be arbitrary. If τ is sufficiently large, then for any $g \in \overset{\circ}{\Pi}_{\frac{s}{m},s}(\mathbb{R}^{n+1})$ there exists a unique $v \in \overset{\circ}{\Pi}_{\frac{s+m}{m},s+m}(\mathbb{R}^{n+1})$ that solves (46.3) in $\mathbb{R}^{n+1}$.*

Proof: Let $R_0(x,t,\xi,\sigma - i\tau) = (i\sigma + \tau + A_0(x,t,\xi))^{-1}$, $\tau > 0$. We have $|R_0(x,t,\xi,\sigma - i\tau)| \leq C(|\sigma| + \tau + |\xi|^m)^{-1}$. Moreover, $R_0 \in P^+_{-m,m}$ (cf. (41.1), (41.2)).

It follows from Lemmas 46.1 and 46.2 that the ψdo R_0 maps $\overset{\circ}{\Pi}_{\frac{s}{m},s}(\mathbb{R}^{n+1})$ to $\overset{\circ}{\Pi}_{\frac{s+m}{m},s+m}(\mathbb{R}^{n+1})$ for any s. Note that

$$(i\sigma + \tau + A(x,t,\xi))R_0(x,t,\xi,\sigma - i\tau) = 1 + A_1(x,\xi)(i\sigma + \tau + A_0)^{-1}.$$

We have

$$|A_1(x,\xi)|\,|R_0| \leq \frac{C(1+|\xi|)^{m-1}}{|\sigma| + \tau + |\xi|^m} \leq C_1(|\sigma| + \tau + |\xi|^m)^{-\frac{1}{m}}. \tag{46.14}$$

Therefore, $A_1(x,\xi)R_0 \in P^+_{-1,m}$. Applying (46.9) to the composition of $\frac{\partial}{\partial t} + \tau + A$ and R_0, we get

$$\left(\frac{\partial}{\partial t} + \tau + A\right) R_0 = I + T^{(1)}_{-1}, \tag{46.15}$$

where $T^{(1)}_{-1} = A_1(x,D)R_0 + T_{-1}$, $\operatorname{ord} T^{(1)}_{-1} \leq -1$. Moreover, the presence of the parameter τ implies that (cf. (46.14) and (41.4))

$$\|T^{(1)}_{-1}u\|_{\frac{s}{m},s} \leq \frac{C}{\tau^{\frac{1}{m}}}\|u\|_{\frac{s}{m},s}. \tag{46.16}$$

Note that $T^{(1)}_1$ acts in $\overset{\circ}{\Pi}_{\frac{1}{m},s}(\mathbb{R}^{n+1})$ since $((\frac{\partial}{\partial t} + \tau + A)R_0 - I)f = 0$ when $t < 0$ if $f = 0$ when $t < 0$. Therefore $T^{(1)}_1$ has a small norm in $\overset{\circ}{\Pi}_{\frac{s}{m},s}(\mathbb{R}^{n+1})$ when $\tau \geq \tau_0$, τ_0 is large and $R = R_0(I + T^{(1)}_{-1})^{-1}$ is the right inverse of

$\frac{\partial}{\partial t}+\tau+A$. Analogously (cf. $ 41) one can prove that $\frac{\partial}{\partial t}+\tau+A$ has the left inverse when $\tau \geq \tau_0$. □

Theorem 46.4. *Let $\tau \geq \tau_0$, where τ_0 is large. Then for any $f(x,t)$ in $\mathbb{R}^{n+1}$ such that $e^{-t\tau}f \in \overset{\circ}{\Pi}_{\frac{s}{m},s}(\mathbb{R}^{n+1})$, there exists a unique solution of (46.1) in $\mathbb{R}^{n+1}$ such that $e^{-t\tau}u \in \overset{\circ}{\Pi}_{\frac{s+m}{m},s+m}(\mathbb{R}^{n+1})$.*

Proof: Let $v(x,t) \in \overset{\circ}{\Pi}_{\frac{s}{m},s}(\mathbb{R}^{n+1})$ be the solution of (46.3) in $\mathbb{R}^{n+1}$, where $\tau \geq \tau_0$, $g = e^{-t\tau}f$. We will write $v(x,t,\tau)$ instead of $v(x,t)$ to emphasize the dependence of v on τ. For $\tau = \tau_0$ we have $\frac{\partial v(x,t,\tau_0)}{\partial t} + \tau_0 v(x,t,\tau_0) + Av(x,t,\tau_0) = e^{-t\tau_0}f(x,t)$ in $\mathbb{R}^{n+1}$. Then $e^{-t(\tau-\tau_0)}v(x,t,\tau_0)$ satisfies (46.3) with $g = e^{-t\tau}f$. The uniqueness of the solution of (46.3) in $\overset{\circ}{\Pi}_{\frac{s}{m},s}(\mathbb{R}^{n+1})$ implies that $e^{-t(\tau-\tau_0)}v(x,t,\tau_0) = v(x,t,\tau)$. Set $u(x,t) = e^{t\tau_0}v(x,t,\tau_0)$. Then $v(x,t,\tau) = e^{-t\tau}u(x,t)$, and $u(x,t)$ solves (46.1) in $\mathbb{R}^{n+1}$. □

46.3. The Cauchy problem with nonzero initial conditions.

Now we study the Cauchy problem (46.3) with nonzero initial data (46.4). Analogously to the proof of Theorem 13.6 we have:

Lemma 46.5. *Suppose $s > \frac{m}{2}$. Then $\Pi_{\frac{s}{m},s}(\mathbb{R}^{n+1})$ is embedded into the space $C(\mathbb{R}, H_{s-\frac{m}{2}}(\mathbb{R}^n))$ and*

$$\sup_t \|u(\cdot,t)\|_{s-\frac{m}{2}} \leq C\|u\|_{\frac{s}{m},s}. \tag{46.17}$$

Theorem 46.6. *Suppose $\frac{1}{2} < \frac{s}{m} < \frac{3}{2}$ and τ is large. Then for any $u_0(x) \in H_{s-\frac{m}{2}}(\mathbb{R}^n)$ and for any $g \in \Pi_{\frac{s-m}{m},s-m}(\mathbb{R}^{n+1}_+)$ there exists a unique solution $v \in \Pi_{\frac{s}{m},s}(\mathbb{R}^{n+1}_+)$ of (46.3) such that (46.4) holds.*

Proof: Analogously to the proof of Example 13.3, for any $u_0(x) \in H_{s-\frac{m}{2}}(\mathbb{R}^n)$ there exists $v_0(x,t) \in \Pi_{\frac{s}{m},s}(\mathbb{R}^{n+1})$ such that $v_0(x,0) = u_0(x)$ and

$$\|v_0\|_{\frac{s}{m},s} \leq C\|u_0\|_{s-\frac{m}{2}}. \tag{46.18}$$

If $w(x,t) = v(x,t) - v_0(x,t)$ for $t \geq 0$, then

$$w(x,0) = 0 \tag{46.19}$$

and $w(x,t)$ satisfies

$$\left(\frac{\partial}{\partial t}+\tau+A\right)w = g_1 \quad \text{for} \quad t>0, \tag{46.20}$$

where $g_1 = g - p(\frac{\partial}{\partial t}+\tau+A)v_0 \in \Pi_{\frac{s-m}{m},s-m}(\mathbb{R}^{n+1}_+)$. Here p is the restriction operator to the half-space $\mathbb{R}^{n+1}_+$.

Let $g_+ = g_1$ for $t > 0$, $g_+ = 0$ for $t < 0$. Since $-\frac{1}{2} < \frac{s-m}{m} < \frac{1}{2}$, one can prove (see, for example, Eskin [**E1**], §5) that $g_+ \in \overset{\circ}{\Pi}_{\frac{s-m}{m},s-m}(\mathbb{R}^{n+1})$. Applying Theorem 46.3, we conclude that there exists $w_+ \in \overset{\circ}{\Pi}_{\frac{s}{m},s}(\mathbb{R}^{n+1})$ that satisfies $(\frac{\partial}{\partial t} + \tau + A)w_+ = g_+$ and

$$\|w_+\|_{\frac{s}{m},s} \leq C\|g_+\|_{\frac{s-m}{m},s-m}. \tag{46.21}$$

Note that $w_+(x,0) = 0$. Choosing $v(x,t) = w_+(x,t) + v_0(x,t)$ for $t > 0$, we prove Theorem 46.6. □

It is convenient to consider the spaces $\Pi_{\frac{s}{m},s,s'}(\mathbb{R}^{n+1})$ with the norm

$$\|u\|_{\frac{s}{m},s,s'} = \|\Lambda^{s'}(D)u\|_{\frac{s}{m},s}, \tag{46.22}$$

where $\Lambda^{s'}(D)$ is a ψdo with symbol $(1+|\xi|^2)^{\frac{s'}{2}}$, $s' \in \mathbb{R}$. Repeating the proof of Theorem 46.6, we get:

Theorem 46.7. *Let $\tau \geq \tau_0$, τ_0 be large. For any $g \in \Pi_{\frac{s-m}{m},s-m,s'}(\mathbb{R}^{n+1}_+)$ and any $u_0 \in H_{s-\frac{m}{2}+s'}(\mathbb{R}^n)$, $\frac{m}{2} < s < \frac{3m}{2}$, there exists a unique solution $v \in \Pi_{\frac{s}{m},s,s'}(\mathbb{R}^{n+1}_+)$ of the Cauchy problem* (46.3), (46.4) *and*

$$\|v\|^+_{\frac{s}{m},s,s'} \leq C\|g\|^+_{\frac{s-m}{m},s-m,s'} + C\|u_0\|_{s-\frac{m}{2}+s'}. \tag{46.23}$$

It follows from (46.23) and (46.17) that

$$\sup_t \|v(\cdot,t)\|_{s'+s-\frac{m}{2}} \leq C\|g\|^+_{\frac{s-m}{m},s-m,s'} + C\|u_0\|_{s-\frac{m}{2}+s'}. \tag{46.24}$$

Remark 46.1. As in Theorem 46.4, we see that for τ large there exists a unique solution $u(x,t)$ of the Cauchy problem (46.1), (46.2) such that $e^{-\tau t}u \in \Pi_{\frac{s}{m},s}(\mathbb{R}^{n+1}_+)$ provided $\frac{m}{2} < s < \frac{3m}{2}$, $u_0(x) \in H_{s-\frac{m}{2}}(\mathbb{R}^n)$, and $e^{-\tau t}f \in \Pi_{\frac{s-m}{m},s-m}(\mathbb{R}^{n+1}_+)$.

The following estimate holds:

$$\|e^{-\tau t}u\|^+_{\frac{s}{m},s} \leq C\|e^{-\tau t}f\|^+_{\frac{s}{m}-1,s-m} + C\|u_0\|_{s-\frac{m}{2}}, \tag{46.25}$$

where $\|\ \|^+_{\frac{s}{m},s}$ is the norm in $\Pi_{\frac{s}{m},s}(\mathbb{R}^{n+1}_+)$.

Using the norms (46.22), we also have (cf. (46.23))

$$\|e^{-\tau t}u\|^+_{\frac{s}{m},s,s'} \leq C\|e^{-\tau t}f\|^+_{\frac{s-m}{m},s-m,s'} + C\|u_0\|_{s-\frac{m}{2}+s'}, \tag{46.26}$$

where $\frac{m}{2} < s < m$, and s' is arbitrary.

Remark 46.2. Consider the Cauchy problem (46.1), (46.2) on a finite time interval $R_T = \mathbb{R}^n \times (0,T)$. Suppose $f \in \Pi_{\frac{s}{m}-1,s-m}(R_T)$. Let f_1 be an extension of $f(x,t)$ such that $e^{-\tau t}f_1 \in \Pi_{\frac{s-m}{m},s-m}(\mathbb{R}^{n+1}_+)$ and $\|e^{-\tau t}f_1\|_{\frac{s}{m}-1,s-m} \leq C\|f\|_{\frac{s}{m}-1,s-m,R_T}$.

By Remark 46.1, there exists a unique solution u_1 of the Cauchy problem (46.1), (46.2) in $\mathbb{R}^{n+1}_+$ such that $e^{-\tau t}u_1 \in \Pi_{\frac{s}{m},s}(\mathbb{R}^{n+1}_+)$ and the estimate (46.25) holds.

Let f_2 be another extension of $f(x,t)$ from $R_T = \mathbb{R}^n \times (0,T)$ to $\mathbb{R}^{n+1}_+$ such that $\|e^{-\tau t}f_2\|_{\frac{s}{m}-1,s-m} \le C\|f\|_{\frac{s}{m}-1,s-m,R_T}$ and u_2 is the corresponding unique solution of the Cauchy problem (46.1), (46.2). Denote $u_+ = u_1 - u_2$ for $t > 0$, $u_+ = 0$ for $t < 0$, and analogously $f_+ = f_1 - f_2$ for $t > 0$ and $f_+ = 0$ for $t < 0$. Then

$$\left(\frac{\partial}{\partial t} + A\right) u_+ = f_+ \quad \text{in} \quad \mathbb{R}^{n+1}. \tag{46.27}$$

Note that $f_+ = 0$ for $t < T$. Then analogously to the proof of Lemma 46.2 and Theorem 46.3 one can show that $u_+ = 0$ for $t < T$, i.e., the restriction of $u_1(x,t)$ to R_T does not depend on the choice of the extension f_1.

Note that $e^{-\tau t}$ and $e^{\tau t}$ are C^∞ bounded functions on $[0,T]$. Thus $u_1(x,t)$ solves the Cauchy problem (46.1), (46.2) on $\mathbb{R}^n \times (0,T)$ and

$$\begin{aligned}\|u_1\|_{\frac{s}{m},s,R_T} \le C\|e^{-\tau t}u_1\|^+_{\frac{s}{m},s} &\le C\|u_0\|_{s-\frac{m}{2}} + C\|e^{-\tau t}f_1\|^+_{\frac{s}{m}-1,s-m} \\ &\le C\|u_0\|_{s-\frac{m}{2}} + C_1\|f\|_{\frac{s}{m}-1,s-m,R_T}.\end{aligned}$$

47. The heat kernel

47.1. Solving the Cauchy problem by Fourier-Laplace transform.

Consider the Cauchy problem (46.1), (46.2) provided the coefficients of A are independent of t and $\operatorname{Re} A_0(x,\xi) \ge C|\xi|^m$. Let $f = 0$ and let $u(x,t)$ be the solution of

$$\frac{\partial u}{\partial t} + A(x,D)u = 0, \quad t > 0,\ x \in \mathbb{R}^n, \tag{47.1}$$

$$u(x,0) = u_0(x), \quad x \in \mathbb{R}^n. \tag{47.2}$$

Let s_0 be arbitrary. For any $\frac{m}{2} < s < \frac{3m}{2}$, $s' = s_0 - s + \frac{m}{2}$, and for any $u_0(x) \in H_{s_0}(\mathbb{R}^n)$, there exists a unique solution $u(x,t)$ of (47.1), (47.2) such that $e^{-\tau t}u(x,t) \in H_{\frac{s}{m},s,s_0-s+\frac{m}{2}}(\mathbb{R}^{n+1}_+)$ and (cf. (46.23), (46.24))

$$\sup_{t\ge 0} \|e^{-\tau t}u(\cdot,t)\|_{s_0} \le C\|e^{-\tau t}u\|^+_{\frac{s}{m},s,s_0-s+\frac{m}{2}} \le C\|u_0\|_{s_0}. \tag{47.3}$$

Here $\tau \ge \tau_0 > 0$ is large.

Let $u_+ = u(x,t)$ for $t > 0$, $u_+ = 0$ for $t < 0$. Since $e^{-\tau t}u(x,t) \in \Pi_{\frac{s}{m},s,s_0-s+\frac{m}{2}}(\mathbb{R}^{n+1}_+)$, we have that $e^{-\tau t}u(x,t) \in L_2(\mathbb{R}^1_+, H_{\frac{m}{2}+s_0}(\mathbb{R}^n))$, i.e., $e^{-\tau t}u(x,t)$ is an L_2-function of t with values in $H_{\frac{m}{2}+s_0}(\mathbb{R}^n)$. Suppose

$\tilde{u}_+(x, \sigma - i\tau)$, $\tau \geq \tau_0$, is the Fourier-Laplace transform of $u_+(x,t)$ in t only. The Parseval equality gives

$$\int_0^\infty e^{-2t\tau}\|u(\cdot,t)\|^2_{\frac{m}{2}+s_0}dt = \frac{1}{2\pi}\int_{-\infty}^\infty \|\tilde{u}_+(\cdot,\sigma - i\tau)\|^2_{\frac{m}{2}+s_0}d\sigma \tag{47.4}$$

for each $\tau \geq \tau_0$. We have

$$\frac{\partial u_+}{\partial t} + A(x,D)u_+ = \delta(t)u_0(x). \tag{47.5}$$

Performing the Fourier transform in (47.5), we get

$$i(\sigma - i\tau)\hat{u}_+(x,\sigma - i\tau) + A(x,D)\hat{u}_+(x,\sigma - i\tau) = u_0(x).$$

It follows from Lemma 41.3 that for $\tau \geq \tau_0$,

$$\hat{u}(x,\sigma - i\tau) = (A + i(\sigma - i\tau)I)^{-1}u_0(x).$$

Therefore

$$u(x,t) = \frac{1}{2\pi}\int_{-\infty}^\infty (A + i(\sigma - i\tau)I)^{-1}u_0(x)e^{it(\sigma - i\tau)}d\sigma, \quad \tau \geq \tau_0. \tag{47.6}$$

Note that the integral (47.6) does not depend on τ by the Cauchy integral theorem, and for $\tau = \tau_0$ it is understood as the Fourier transform of a function in $L_2(\mathbb{R}^1, H_{\frac{m}{2}+s_0}(\mathbb{R}^n))$ (cf. Parseval's formula (47.4)).

Applying $(A + i(\sigma - i\tau)I)^{-1}$ to (41.9) from the left, we get

$$(A + \lambda I)^{-1} = R^{(N)} + T^{(1)}_{-m-N-1}, \tag{47.7}$$

where $\lambda = i(\sigma - i\tau)$, $R^{(N)} = \sum_{k=0}^N R_{-k}$, $T^{(1)}_{-(m+N+1)} = -(A+\lambda I)^{-1}T_{-N-1}$. Note that $\operatorname{ord} T^{(1)}_{-(m+N+1)} \leq -(m+N+1)$ since $\operatorname{ord}(A+\lambda I)^{-1} = -m$. Substituting (47.7) into (47.6), we get

$$u(x,t) = u_1(x,t) + u_2(x,t),$$

where

$$u_1(x,t) = \frac{1}{2\pi}\int_{-\infty}^\infty R^{(N)}(x,D,i(\sigma - i\tau))u_0(x)e^{it(\sigma - i\tau)}d\sigma, \tag{47.8}$$

$$u_2(x,t) = \frac{1}{2\pi}\int_{-\infty}^\infty (T^{(1)}_{-(m+N+1)}u_0)(x)e^{it(\sigma - i\tau)}d\sigma, \tag{47.9}$$

and $R^{(N)}(x,\xi,\lambda)$ has the form (41.11).

The integral with respect to σ in (47.8) can be computed easily by residue theory using the Jordan lemma (cf. Proposition 16.1):

$$\frac{1}{2\pi}\int_{-\infty}^\infty \frac{e^{it(\sigma - i\tau)}d\sigma}{(A_0(x,\xi) + i(\sigma - i\tau))^{k+1}} = \frac{t^k}{k!}e^{-tA_0(x,\xi)}. \tag{47.10}$$

Therefore

$$u_1(x,t) = \frac{1}{(2\pi)^n}\int_{\mathbb{R}^n}\sum_{k=0}^{N} p_k(x,\xi)\frac{t^k}{k!}e^{-tA_0(x,\xi)}\tilde{u}_0(\xi)e^{ix\cdot\xi}d\xi, \tag{47.11}$$

where $p_0 = 1$, $\deg p_k \le mk - \frac{k}{2}$, $k \ge 1$ (cf. (41.11)).

It follows from Lemma 44.6 that $T^{(1)}_{-(m+N+1)}$ is an integral operator with a continuous kernel, $T^{(1)}_{-(m+N+1)}(x,y,\sigma - i\tau)$, when $-m-N-1 < -n$.

47.2. Asymptotics of the heat kernel as $t \to +0$.

Denote by e^{-tA} the solution operator of the Cauchy problem (47.1), (47.2), i.e.,

$$u(x,t) = e^{-tA}u_0(x). \tag{47.12}$$

Let $G(x,y,t)$ be the kernel of e^{-tA}. Note that $G(x,y,t)$ is the solution of the Cauchy problem

$$\begin{aligned} &\left(\frac{\partial}{\partial t} + A\right)G(x,y,t) = 0, \quad t>0,\ x\in\mathbb{R}^n,\ y\in\mathbb{R}^n,\\ &G(x,y,0) = \delta(x-y), \end{aligned} \tag{47.13}$$

i.e., $u_0(x) = \delta(x-y) \in H_{-\frac{n}{2}-\varepsilon}(\mathbb{R}^n)$.

$G(x,y,t)$ is called the heat kernel.

We have from (47.9)–(47.11) that

$$G(x,y,t) = \sum_{k=0}^{N}\frac{1}{(2\pi)^n}\int_{\mathbb{R}^n} p_k(x,\xi)\frac{t^k}{k!}e^{-tA_0(x,\xi)+i(x-y)\cdot\xi}d\xi + G_N(x,y,t), \tag{47.14}$$

where

$$G_N(x,y,t) = \frac{1}{2\pi}\int_{-\infty}^{\infty} T^{(1)}_{-m-N-1}(x,y,\sigma-i\tau)e^{it(\sigma-i\tau)}d\sigma. \tag{47.15}$$

Note that $T^{(1)}(x,y,\sigma-i\tau)$ is analytic in $\sigma - i\tau$ for $\tau > \tau_0$ since all the other terms in (47.7) are analytic in $z = \sigma - i\tau$. It follows from estimates of the form (44.14), (46.16) that

$$|T^{(1)}_{-m-N-1}(x,y,\sigma)| \le C|\sigma - i\tau|^{-N_1-2}, \tag{47.16}$$

where $-m-N-1 < -n-N_1-2$. Therefore, (47.16) implies that $G_N(x,y,t)$ has N_1 continuous derivatives with respect to t. Also $G_N(x,y,t) = 0$ for $t<0$ because of the analyticity of $T^{(1)}_{-m-N-1}$. Therefore

$$|G_N(x,y,t)| \le C_N t^{N_1}, \quad t>0. \tag{47.17}$$

Take $x = y$ in (47.14) and make a change of variables

$$\eta = \frac{\xi}{t^{\frac{1}{m}}}. \tag{47.18}$$

We get

$$\frac{1}{(2\pi)^n} \int_{\mathbb{R}^n} e^{-tA_0(x,\xi)} d\xi = \frac{c_0(x)}{t^{\frac{n}{m}}}, \tag{47.19}$$

where $c_0(x) = \frac{1}{(2\pi)^n} \int_{\mathbb{R}^n} e^{-A_0(x,\eta)} d\eta$. Making the change of variables (47.18) for $1 \le k \le N$ in (47.14), taking $x = y$, and collecting the terms having the same power of t, we get

$$G(x,x,t) = \frac{1}{t^{\frac{n}{m}}} \left(c_0(x) + \sum_{k=1}^{N_2-1} c_k(x) t^{\frac{k}{m}} \right) + O(t^{\frac{N_2-n}{m}}), \tag{47.20}$$

where the coefficients $c_k(x)$ have an explicit form. Note that N_2 in (47.20) is arbitrary since N in (47.14), (47.15) is arbitrary and we take $N_1 > \frac{N_2-n}{m}$.

If $A_0(x,\eta)$ is even in η, then the integrals of the form

$$\int_{\mathbb{R}^n} \eta_1^{r_1} \eta_2^{r_2} \cdots \eta_n^{r_n} e^{-A_0(x,\eta)} d\eta$$

are equal to zero if the $r_1 + r_2 + \cdots + r_n$ are odd. In this case (see (47.20)) $c_k(x) = 0$ if the k are odd.

Further simplifications occur when $A_0(x,D)$ is a second order elliptic operator, $A_0(x,\xi) = \sum_{j,k=1}^n g^{jk}(x)\xi_j\xi_k \ge C|\xi|^2$, i.e., $\operatorname{Im} A_0 = 0$. We have

$$c_0(x) = \frac{1}{(2\pi)^n} \int_{\mathbb{R}^n} e^{-\sum_{j,k=1}^n g^{jk}(x)\eta_j\eta_k} d\eta = \frac{1}{(2\sqrt{\pi})^n} \sqrt{g(x)}, \tag{47.21}$$

where $g(x) = (\det[g^{jk}(x)]_{j,k=1}^n)^{-1}$.

To compute (47.21), one can make an orthogonal transformation $\eta = O\zeta$ such that $\sum_{j,k=1}^n g^{jk}(x)\eta_j\eta_k = \sum_{j=1}^n \lambda_j \zeta_j$ and then make the changes of coordinates $\zeta_j = \frac{1}{\sqrt{\lambda_j}}\zeta_j', 1 \le j \le n$.

48. The Cauchy problem for strictly hyperbolic equations

Let $H(x,t,\xi,\sigma)$ be a polynomial of degree m,

$$H(x,t,\xi,\sigma) = H_0(x,t,\xi,\sigma) + H_1(x,t,\xi,\sigma),$$

where

$$H_0(x,t,\xi,\sigma) = \sum_{|k|+j=m} a_{jk}(x,t)\xi^k\sigma^j, \quad a_{m0} \ne 0,$$

$$H_1(x,t,\xi,\sigma) = \sum_{|k|+j\le m-1} a_{jk}(x,t)\xi^k\sigma^j.$$

We assume that $a_{jk}(x,t) \in C^\infty(\mathbb{R}^{n+1})$ and $a_{jk}(x,t) = a_{jk}^\infty$ for $|x| > R$, and $a_{jk}(x,t) = a_{jk}(x,\infty)$ for $|t| > R$. The polynomial $H(x,t,\xi,\sigma)$ is called strictly hyperbolic if $H_0(x,t,\xi,\sigma) = 0$ has m real roots $\sigma_k(x,t,\xi)$ such that $\sigma_j(x,t,\xi) \neq \sigma_k(x,t,\xi)$ when $j \neq k$ for all $(x,t) \in \mathbb{R}^{n+1}$ and $\xi \neq 0$. Note that $\sigma_k(x,t,\xi)$ is a homogeneous function of ξ of degree 1 and $\sigma_k(x,t,\xi) \in C^\infty(\mathbb{R}^{n+1} \times (\mathbb{R}^n \setminus \{0\}))$.

Consider the Cauchy problem

$$(48.1) \quad H(x,t,D_x,D_t)u(x,t) = f(x,t), \quad (x,t) \in \mathbb{R}_+^{n+1} = \{t > 0, x \in \mathbb{R}^n\},$$

where $D_x = -i\frac{\partial}{\partial x}$, $D_t = -i\frac{\partial}{\partial t}$, with the initial conditions

$$(48.2) \qquad \left.\frac{\partial^k u(x,t)}{\partial t^k}\right|_{t=0} = g_k(x), \quad 0 \le k \le m-1.$$

Suppose first that $g_k = 0$, $0 \le k \le m-1$. As in the parabolic case, let $v(x,t) = e^{-\tau t}u(x,t)$, $g(x,t) = e^{-\tau t}f(x,t)$. Then $v(x,t)$ satisfies the equation

$$(48.3) \qquad H(x,t,D_x,D_t - i\tau)v(x,t) = g(x,t), \quad (x,t) \in \mathbb{R}_+^{n+1},$$

and

$$(48.4) \qquad \left.\frac{\partial^k v(x,t)}{\partial t^k}\right|_{t=0} = 0, \quad 0 \le k \le m-1.$$

Denote by $H_{p,s}(\mathbb{R}^{n+1})$ the Sobolev space with the norm

$$(48.5) \qquad \|v\|_{p,s}^2 = \int_{\mathbb{R}^{n+1}} (1 + |\xi|^2 + \sigma^2)^p (1 + |\xi|^2)^s |\tilde{u}(\xi,\sigma)|^2 d\xi d\sigma,$$

where $p \ge 0$ is an integer, $s \in \mathbb{R}$. The spaces $\overset{\circ}{H}_{p,s}(\mathbb{R}^{n+1})$ and $H_{p,s}(\mathbb{R}_+^{n+1})$ are defined as in the parabolic case.

Let $f_+ = f$ for $t > 0$, and $f_+ = 0$ for $t < 0$. Analogously, let $u_+ = u$, $v_+ = v(x,t)$, $g_+ = g(x,t)$ for $t > 0$, and $u_+ = v_+ = g_+ = 0$ for $t < 0$.

Theorem 48.1. *Let $\tau \ge \tau_0$, where τ_0 is large, and let s be arbitrary. Then for any f_+ such that $e^{-\tau t}f_+ \in \overset{\circ}{H}_{0,s}(\mathbb{R}^{n+1})$ there exists a unique solution u_+ of $H(x,t,D_x,D_t)u_+ = f_+$ in $\mathbb{R}^{n+1}$ such that $e^{-\tau t}u_+ \in \overset{\circ}{H}_{m-1,s}(\mathbb{R}^{n+1})$.*

Proof: Assume, for simplicity, that $a_{m0} = 1$. Otherwise, divide (48.1) by a_{m0}. Then

$$H_0(x,t,\xi,\sigma - i\tau) = \prod_{j=1}^m (\sigma - i\tau - \sigma_j(x,t,\xi)) \neq 0$$

if $\tau \neq 0$. We have

$$\frac{\partial H_0(x,t,\xi,\sigma - i\tau)}{\partial \sigma} = \sum_{k=1}^m q_k(x,t,\xi,\sigma - i\tau),$$

where

$$q_k(x,t,\xi,\sigma-i\tau)=\prod_{j\neq k}(\sigma-i\tau-\sigma_j(x,t,\xi)),\quad 1\le k\le m. \tag{48.6}$$

Therefore

$$\begin{aligned} H_0\overline{\frac{\partial H_0}{\partial\sigma}} &= \prod_{j=1}^m(\sigma-i\tau-\sigma_j)\sum_{k=1}^m\prod_{j\neq k}(\sigma+i\tau-\sigma_j) \\ &= \sum_{k=1}^m(\sigma-i\tau-\sigma_k)\prod_{j\neq k}((\sigma-\sigma_j)^2+\tau^2) = -i\tau Q_0+Q_1, \end{aligned} \tag{48.7}$$

where

$$Q_0=\sum_{k=1}^m q_k\overline{q_k}=\sum_{k=1}^m\prod_{j\neq k}((\sigma-\sigma_j)^2+\tau^2),\qquad Q_1=\sum_{k=1}^m(\sigma-\sigma_k)q_k\overline{q_k}. \tag{48.8}$$

Since $Q_0(x,t,\xi,\sigma,\tau)>0$ for $(\xi,\sigma,\tau)\neq(0,0,0)$, we get by the homogeneity in (ξ,σ,τ),

$$Q_0(x,t,\xi,\sigma,\tau)\ge C_0(|\xi|^2+|\sigma|^2+\tau^2)^{m-1}. \tag{48.9}$$

Note that Q_0 and Q_1 are symmetric polynomials in the roots $\sigma_1,\dots,\sigma_m$. It is a well-known theorem in algebra that any symmetric polynomial in $\sigma_1,\dots,\sigma_n$ can be represented as a polynomial in the coefficients of H_0. Therefore Q_0 and Q_1 are polynomials in (ξ,σ,τ).

48.1. The main estimate.

Lemma 48.2. *For any $v_+\in\overset{\circ}{H}_{m,s}(\mathbb{R}^{n+1})$ the following estimate holds:*

$$\|\Gamma^{m-1}(D_x,D_t,\tau)v_+\|_{0,s}\le\frac{C}{\tau}\|H(x,t,D_x,D_t-i\tau)v_+\|_{0,s}, \tag{48.10}$$

where $\Gamma^{m-1}(D_x,D_t,\tau)$ is a pseudodifferential operator with symbol

$$\Gamma^{m-1}(\xi,\sigma,\tau)=(|\xi|^2+\sigma^2+\tau^2)^{\frac{m-1}{2}}$$

and $\tau\ge\tau_0$, τ_0 is large.

Proof: Let $\Lambda^s(\xi)=(1+|\xi|^2)^{\frac{s}{2}}$. Since $a_{m0}=1$, we have, using Theorem 40.2:

(48.11)

$$\Lambda^s(D)H(x,t,D_x,D_t-i\tau)=H(x,t,D_x,D_t-i\tau)\Lambda^s(D)+Q_2\Lambda^s(D),$$

where $\|Q_2\Lambda^s v_+\|_0\le C\|\Gamma^{m-1}(D_x,D_t,\tau)v_+\|_{0,s}$.

Consider the L_2 scalar product in $\mathbb{R}^{n+1}$:

$$(H\Lambda^s v_+,H_{0\sigma}\Lambda^s v_+)=(H_{0\sigma}^*H\Lambda^s v_+,\Lambda^s v_+),$$

where $H_{0\sigma}^*$ is formally adjoint to $H_{0\sigma}$. Using the theorems analogous to Theorems 40.2 and 43.2 for the case of differential operators, we get:

$$(48.12)\quad (H\Lambda^s v_+, H_{0\sigma}\Lambda^s v_+) = -i\tau(Q_0\Lambda^s v_+, \Lambda^s v_+) + (Q_1\Lambda^s v_+, \Lambda^s v_+) + (Q_3\Lambda^s v_+, \Lambda^s v_+),$$

where Q_0, Q_1 are differential operators with symbols (48.8) and

$$(48.13)\quad |(Q_3\Lambda^s v_+, \Lambda^s v_+)| \le C\|\Gamma^{m-1}\Lambda^s v_+\|_0^2.$$

Note that the symbol $Q_1(x,t,\xi,\tau)$ is real-valued. Therefore, $\operatorname{ord}(Q_1 - Q_1^*) \le 2m-2$, and we see that $\operatorname{Im}(Q_1\Lambda^s v_+, \Lambda^s v_+)$ satisfies the estimate of the form (48.13). Thus, taking the imaginary part of (48.12), we get

$$(48.14)\quad -\operatorname{Im}(H\Lambda^s v_+, H_{0\sigma}\Lambda^s v_+) \ge \tau\sum_{k=1}^{m}(Q_0\Lambda^s v_+, \Lambda^s v_+) - C\|\Gamma^{m-1}\Lambda^s v_+\|_0^2.$$

We will show that

$$(48.15)\quad (Q_0\Lambda^s v_+, \Lambda^s v_+) \ge C\|\Gamma^{m-1}\Lambda^s v_+\|_0^2 - C_1\|\Gamma^{m-\frac{3}{2}}\Lambda^s v_+\|_0^2.$$

The inequality (48.15) is called the Gårding inequality. We have

$$(48.16)\quad Q_0(x,t,\xi,\sigma,\tau) = Q_0 - \frac{C_0}{2}(|\xi|^2 + |\sigma|^2 + \tau^2)^{m-1} + \frac{C_0}{2}(|\xi|^2 + |\sigma|^2 + \tau^2)^{m-1},$$

where Q_0 and C_0 are the same as in (48.9).

Let

$$(48.17)\quad B(x,t,\xi,\sigma,\tau) = \sqrt{Q_0 - \frac{C_0}{2}\Gamma^{2(m-1)}(\xi,\sigma,\tau)}.$$

Note that $B > 0$ and $B(x,t,\xi,\sigma,\tau) \in S^{m-1}$. Denote by B^* the adjoint operator to B. By Theorems 40.2 and 43.2 we have

$$(48.18)\quad Q_0 - \frac{C_0}{2}\Gamma^{2(m-1)} = B^*B + Q_4,$$

where $\operatorname{ord} Q_4 \le 2m-3$, and

$$(48.19)\quad |(Q_4\Lambda^s v_+, \Lambda^s v_+)| \le C_1\|\Gamma^{m-\frac{3}{2}}\Lambda^s v_+\|_0^2.$$

Note that $(B^*B\Lambda^s v_+, \Lambda^s v_+) = \|B\Lambda^s v_+\|^2 \ge 0$. Therefore, (48.18) and (48.19) imply that

$$(48.20)\quad \tau(Q_0\Lambda^s v_+, \Lambda^s v_+) \ge \frac{C_0}{2}\tau\|\Gamma^{m-1}\Lambda^s v_+\|_0^2 - C_1\tau\|\Gamma^{m-\frac{3}{2}}\Lambda^s v_+\|_0^2$$
$$\ge \left(\frac{C_0}{2}\tau - C_1\right)\|\Gamma^{m-1}\Lambda^s v_+\|_0^2,$$

since $\tau\|\Gamma^{m-\frac{3}{2}}\Lambda^s v_+\|_0^2 \le \|\Gamma^{m-1}\Lambda^s v_+\|_0^2$. Combining (48.14) and (48.20), we obtain

$$\operatorname{Im}(H\Lambda^s v_+, H_{0\sigma}\Lambda^s v_+) \ge C\tau\|\Gamma^{m-1}\Lambda^s v_+\|_0^2.$$

It follows from (48.11) that

$$(\Lambda^s H v_+, H_{0\sigma}\Lambda^s v_+) = (H\Lambda^s v_+, H_{0\sigma}\Lambda^s v_+) + (Q_2\Lambda^s v_+, H_{0\sigma}\Lambda^s v_+).$$

Therefore

$$\begin{aligned} C\tau\|\Gamma^{m-1}\Lambda^s v_+\|_0^2 &\le |(\Lambda^s H v_+, H_{0\sigma}\Lambda^s v_+)| + |(Q_2\Lambda^s v_+, H_{0\sigma}\Lambda^s v_+)| \\ &\le C_1\|\Lambda^s H v_+\|_0\ \|\Gamma^{m-1}\Lambda^s v_+\|_0 + C_2\|\Gamma^{m-1}\Lambda^s v_+\|_0^2. \end{aligned}$$

Hence (48.10) holds. □

48.2. Uniqueness and parabolic regularization.

Now we prove the uniqueness result.

Lemma 48.3. *Suppose $Hu_+ = 0$ in $\mathbb{R}^{n+1}$, where*

$$e^{-\tau t}u_+ = v_+ \in \overset{\circ}{H}_{m-1,s}(\mathbb{R}^{n+1}).$$

Then $u_+ = 0$.

Proof: Since $v_+ \in \overset{\circ}{H}_{m-1,s}(\mathbb{R}^{n+1})$ and

$$Hv_+ = D_t^m v_+ + \sum_{k=0}^{m-1} b_k(x,t,D_x,\tau)D_t^k v_+ = 0,$$

where $\operatorname{ord} b_k \le m-k$, we see that

$$D_t^m v_+ = -\sum_{k=0}^{m-1} b_k(x,t,D_x,\tau)D_t^k v_+ \in \overset{\circ}{H}_{0,s-1}(\mathbb{R}^{n+1}). \tag{48.21}$$

Therefore $v_+ \in \overset{\circ}{H}_{m,s-1}(\mathbb{R}^{n+1})$. Applying the inequality (48.10) with $Hv_+ = 0$ and s replaced by $s-1$, we obtain that $v_+ = 0$. □

The next step is the proof of the existence result.

Lemma 48.4. *For any $g_+ \in \overset{\circ}{H}_{0,s}(\mathbb{R}^{n+1})$ there exists a unique solution of (48.3) in $\mathbb{R}^{n+1}$, where $v_+ \in \overset{\circ}{H}_{m-1,s}(\mathbb{R}^{n+1})$.*

Proof: We use the method of "parabolic" regularization.

Let $H_\varepsilon(x,t,\xi,\sigma-i\tau) = H(x,t,\xi,\sigma-i\tau-i\varepsilon\Lambda(\xi))$, $\varepsilon > 0$. The operator $H(x,t,D_x,D_t-i\tau-\varepsilon\Lambda(D_x))$ is a "parabolic" ψdo in the sense that $H_\varepsilon = H(x,t,\xi,z-i\varepsilon\Lambda(\xi))$ and the H_ε^{-1} are analytic when $\operatorname{Im} z < 0$ and

$$C_{1\varepsilon}(|\xi|+|z|)^m \ge |H(x,t,\xi,z-i\varepsilon\Lambda(\xi))| \ge C_\varepsilon(|\xi|+|z|)^m \tag{48.22}$$

for all $\operatorname{Im} z > 0$ large (cf. the class $P_{\alpha,m}^+$ in §46).

Repeating the proof of Theorem 46.3, we conclude that for any $g_+ \in \overset{\circ}{H}_{0,s}(\mathbb{R}^{n+1})$, there exists $v_\varepsilon \in \overset{\circ}{H}_{m,s}(\mathbb{R}^{n+1})$ such that

$$H(x,t,D_x,D_t - i\tau - i\varepsilon\Lambda(D_x))v_\varepsilon = g_+, \quad (x,t) \in \mathbb{R}^{n+1}. \tag{48.23}$$

Note that the proof of the estimate (48.10) can be repeated for the equation (48.23), and we get

$$\|\Gamma^{m-1}\Lambda^s v_\varepsilon\|_0 \le C\|\Lambda^s g_+\|_0, \tag{48.24}$$

where the constant C is independent of ε. Therefore, $\{v_\varepsilon\}$, $\varepsilon > 0$, is bounded in $\overset{\circ}{H}_{m-1,s}(\mathbb{R}^{n+1})$. It follows from the weak compactness of a bounded set in $\overset{\circ}{H}_{m-1,s}(\mathbb{R}^{n+1})$ that there exists a sequence v_{ε_k} that converges weakly to some $v_+ \in \overset{\circ}{H}_{m-1,s}(\mathbb{R}^{n+1})$ (see Rudin [**R**]) . For any $\varphi \in C_0^\infty(\mathbb{R}^{n+1})$, we have

$$(g_+,\varphi) = (H_{\varepsilon_k}v_{\varepsilon_k},\varphi) = (v_{\varepsilon_k}, H^*_{\varepsilon_k}\varphi). \tag{48.25}$$

Passing to the limit as $\varepsilon_k \to 0$, we get $(g_+,\varphi) = (v_+, H^*\varphi)$, i.e., $Hv_+ = g_+$ in $\mathbb{R}^{n+1}$. □

Lemmas 48.3 and 48.4 imply that for $\tau \ge \tau_0$, where $\tau_0 > 0$ is large, and for any f such that $e^{-t\tau}f \in \overset{\circ}{H}_{0,s}(\mathbb{R}^{n+1})$, there exists a unique solution $v(x,t,\tau) \in \overset{\circ}{H}_{m-1,s}(\mathbb{R}^{n+1})$ of (48.3) with $g_+ = e^{-t\tau}f_+$.

Analogously to the proof of Theorem 46.4 one can show that there exists $u_+(x,t)$ solving (48.1) and such that

$$v_+(x,t,\tau) = e^{-t\tau}u_+(x,t) \in \overset{\circ}{H}_{m-1,s}(\mathbb{R}^{n+1}).$$

Remark 48.1. It follows from Theorem 48.1 that for any $g_+ \in \overset{\circ}{H}_{0,s}(\mathbb{R}^{n+1})$ there exists a unique $v_+ \in \overset{\circ}{H}_{m-1,s}(\mathbb{R}^{n+1})$. Then it follows from (48.21) that $D_t^m v_+ \in \overset{\circ}{H}_{0,s-1}(\mathbb{R}^{n+1})$. Therefore, $v_+ \in \overset{\circ}{H}_{m,s-1}(\mathbb{R}^{n+1})$ and consequently the $\frac{\partial^k v_+(x,t)}{\partial t^k}$ are continuous functions of t with values in $H_{m+s-k-\frac{3}{2}}(\mathbb{R}^n)$, $0 \le k \le m-1$ (cf. Theorem 13.6). Since $v_+ = 0$ for $t < 0$, we see that $\frac{\partial^k v_+(x,0)}{\partial t^k} = 0$, $0 \le k \le m-1$, i.e., $v_+(x,t)$ satisfies the zero initial conditions (48.4). □

One may consider also the Cauchy problem (48.1), (48.2) with nonzero initial conditions. If $g_k(x) \in H_{m-k+s-\frac{1}{2}}(\mathbb{R}^n)$, $0 \le k \le m-1$, there exists $v_0 \in H_{m,s}(\mathbb{R}^{n+1}_+)$ such that $\frac{\partial^k v_0(x,+0)}{\partial t^k} = g_k(x)$, $0 \le k \le m-1$, and $v_0 = 0$ for large t (cf. Example 13.3). We look for the solution of (48.1) in the form $u = v_0 + w$, where $w(x,t)$ satisfies

$$H(x,t,D_x,D_t)w(x,t) = f_0, \ t > 0, \quad f_0 = f(x,t) - Hv_0. \tag{48.26}$$

Let f_+ be the extension of f_0 by zero for $t<0$. Note that $e^{-t\tau}f_+ \in \overset{\circ}{H}_{0,s}(\mathbb{R}^{n+1})$. By Theorem 48.1, there exists w_+ such that $Hw_+ = f_+$ in $\mathbb{R}^{n+1}$ and $e^{-t\tau}w_+ \in \overset{\circ}{H}_{m-1,s}(\mathbb{R}^{n+1})$. Let $w = w_+$ for $t>0$. It was shown in Remark 48.1 that $e^{-t\tau}w_+ \in H_{m,s-1}(\mathbb{R}^{n+1}_+)$ and $\frac{\partial^k w(x,0)}{\partial t^k} = 0$, $0\le k\le m-1$. Therefore, $u = v_0 + w$ satisfies (48.1) for $t>0$, $e^{-t\tau}u \in H_{m,s-1}(\mathbb{R}^{n+1}_+)$, and $\frac{\partial^k u(x,0)}{\partial t^k} = g_k(x)$, $0\le k\le m-1$.

We have proven the following result:

Lemma 48.5. *For any $e^{-\tau t}f \in H_{0,s}(\mathbb{R}^{n+1}_+)$ and any $g_k(x)\in H_{m+s-k-\frac{1}{2}}(\mathbb{R}^n)$, $0\le k\le m-1$, there exists a unique $e^{-\tau t}u(x,t)\in H_{m,s-1}(\mathbb{R}^{n+1}_+)$ such that (48.1), (48.2) hold.* □

48.3. The Cauchy problem on a finite time interval.

A modification of the proof of Theorem 48.1 gives an existence and uniqueness theorem in $R_T = \mathbb{R}^n \times (0,T)$ with less restrictive requirements on the smoothness of the initial data than those in Lemma 48.5.

Denote by $[u,v]$ the L_2 scalar product in $\mathbb{R}^n$ for t fixed, and let $[u]_s = [\Lambda^s u]_0$ be the norm in $H_s(\mathbb{R}^n)$. Denote by $C_{m-1,s}$ the space of functions such that $\frac{\partial^k u(x,t)}{\partial t^k}$, $0\le k\le m-1$, are continuous in $t\in[0,T]$ with values in $H_{m-k-1+s}(\mathbb{R}^n)$. The norm in $C_{m-1,s}$ is

$$\max_{0\le t\le T}\sum_{k=0}^{m-1}[D_t^k u]_{m-1-k+s}.$$

Also denote by $L_1[(0,T),H_s(\mathbb{R}^n)]$ the space with the norm $\int_0^T [f(x,t)]_s dt$.

Theorem 48.6. *Let s be arbitrary. For any $g_k(x)\in H_{s+m-1-k}(\mathbb{R}^n)$, $0\le k\le m-1$, and any $f(x,t)\in L^1[(0,T),H_s(\mathbb{R}^n)]$, there exists a unique solution $u(x,t)\in C_{m-1,s}$ of the Cauchy problem (48.1), (48.2) in R_T such that*

$$\max_{0\le t\le T}\sum_{k=0}^{m-1}[D_t^k u(x,t)]_{m-k-1+s} \le C\sum_{k=0}^{m-1}[g_k]_{m-1-k+s} + C\int_0^T [f]_s dt. \tag{48.27}$$

Proof: Take any $u(x,t)\in H_{m,s}(R_T)$. Note that $H_{m,s}(R_T)\subset C_{m-1,s}$. Let $w = e^{-t\tau}u(x,t)$, where τ is large, and let $F(x,t) = H(x,t,D_x,D_t - i\tau)w$, $\varphi_k(x) = \frac{\partial^k w(x,0)}{\partial t^k}$, $0\le k\le m-1$. Note that $\sigma_k(x,t,\xi)$ have bounded first derivatives in ξ. Therefore, as in (48.12), we have, using Theorem 40.2 with $N=0$:

$$\begin{aligned}(48.28)\quad &[\Lambda^s H(x,t,D_x,D_t-i\tau)w, H_{0\sigma}\Lambda^s w]\\ &\qquad=\sum_{k=1}^{m}[(D_t-i\tau-\sigma_k(x,t,D_x))q_k\Lambda^s w, q_k\Lambda^s w]+[Q_2\Lambda^s w,\Lambda^s w],\end{aligned}$$

where

$$(48.29)\qquad |[Q_2\Lambda^s w,\Lambda^s w]|\le C\sum_{k=0}^{m-1}[\Lambda_\tau^{m-1-k}D_t^k w]_s^2\overset{\text{def}}{=}C|[w]|_{m-1,s}^2,$$

and $\Lambda_\tau=(|\xi|^2+\tau^2+1)^{\frac{1}{2}}$. Integrating (48.28) from 0 to t, multiplying by -1 and taking the imaginary part, we get as in (48.14):

$$\begin{aligned}(48.30)\quad &\sum_{k=1}^{m}[q_k(x,t,D_x,D_t-i\tau)\Lambda^s w]_0^2\\ &\qquad-\sum_{k=1}^{m}[q_k(x,0,D_x,D_t-i\tau)\Lambda^s w]_0^2+\tau\int_0^t\sum_{k=1}^{m}[q_k\Lambda^s w]_0^2dt'\\ &\qquad\qquad\le C\int_0^t|[w]|_{m-1,s}^2dt'+C\int_0^t[F]_s|[w]|_{m-1,s}dt'.\end{aligned}$$

We need a variant of the Gårding inequality (cf. (48.15)):

$$(48.31)\qquad \sum_{k=1}^{m}[q_k(x,t,D_x,D_t-i\tau)\Lambda^s w]_0^2\ge C|[w]|_{m-1,s}^2-\frac{C_1}{\tau}|[w]|_{m-1,s}^2.$$

Let

$$(48.32)\qquad w_k(x,t)=\Lambda_\tau^{m-1-k}\Lambda^s D_t^k w(x,t),\quad 0\le k\le m-1.$$

We have

$$q_k(x,t,D_x,D_t-i\tau)\Lambda^s w=\sum_{j=0}^{m-1}a_{kj}(x,t,D_x,\tau)w_j(x,t),$$

where the $a_{kj}(x,t,D_x,\tau)$ are ψdo's of order zero. Note that

$$(48.33)\qquad \sum_{k=1}^{m}[q_k\Lambda^s w]_0^2=\Big[\sum_{j,p=0}^{m-1}a_{jp}^{(1)}w_p,w_j\Big],$$

where $a_{jp}^{(1)}=\sum_{k=1}^m a_{kj}^*a_{kp}$, and the a_{kj}^* are the adjoint operators to a_{kj}.

Denote by $A(x,t,\xi,\tau)$ the matrix with elements

$$\sum_{k=1}^{m}\overline{a_{kj}(x,t,\xi,\tau)}a_{kp}(x,t,\xi,\tau).$$

Fix any (x_0, t_0). Then we have

$$(A(x_0, t_0, D_x, D_t)\vec{w}(x,t), \vec{w}(x,t)) = (Q_0(x_0, t_0, D_x, D_t, \tau)w(x,t), w(x,t)),$$

where $\vec{w} = (w_0, \dots, w_{m-1})$. Therefore (48.9) implies that $A(x,t,\xi,\sigma,\tau)$ is positive definite for any (x,t) and

$$A(x,t,\xi,\tau) \geq C_0 I.$$

Let $B(x,t,\xi,\tau) = (A(x,t,\xi,\tau) - \frac{C_0}{2}I)^{\frac{1}{2}}$ and let $B(x,t,D_x,\tau)$ be the matrix ψdo with symbol $B(x,t,\xi,\tau)$. Using Theorems 40.2, 43.2 with $N=0$, we get (cf. (48.18))

$$\sum_{j,p=0}^{m-1} [a_{jp}^{(1)} w_p, w_j] = \frac{C_0}{2} \sum_{k=0}^{m-1} [w_k]_0^2 + [B^* B\vec{w}, \vec{w}] + [C_2 \vec{w}, \vec{w}], \tag{48.34}$$

where $\operatorname{ord} C_2 \leq -1$, i.e., $|[C_2\vec{w}, \vec{w}]| \leq C \sum_{k=0}^{m-1} [\Lambda_\tau^{-1} w_k]_0^2$. Therefore

$$|[C_2 \vec{w}, \vec{w}]| \leq \frac{C}{\tau} \sum_{k=0}^{m-1} [w_k]_0^2. \tag{48.35}$$

Since $[B^*B\vec{w}, \vec{w}] = [B\vec{w}, B\vec{w}] \geq 0$ and since $\sum_{k-0}^{m-1} [w_k]_0^2 = |[w]|_{m-1,s}^2$ we get (48.31). □

Taking τ large and combining (48.30) and (48.31), we get

$$\begin{aligned} |[w(x,t)]|_{m-1,s}^2 \leq C \sum_{k=1}^{m} [q_k(x,0,D_x,D_t - i\tau)\Lambda^s w]_0^2 \\ + C \int_0^T [F]_s dt \max_{0 \leq t \leq T} |[w]|_{m-1,s}. \end{aligned} \tag{48.36}$$

Take a t_0 that satisfies

$$|[w(x,t_0)]|_{m-1,s}^2 = \max_{0 \leq t \leq T} |[w(x,t)]|_{m-1,s}.$$

Choosing such t_0 in (48.36) and using the inequality

$$\begin{aligned} \sum_{k=1}^{m} [q_k(x,0,D_x,D_t - i\tau)\Lambda^s w]_s^2 \\ \leq C \sum_{k=0}^{m-1} [\Lambda_\tau^{m-k-1} \varphi_k]_s^2 \leq C_1 \max_{0 \leq t \leq T} |[w(x,t)]|_{m-1,s}^2, \end{aligned} \tag{48.37}$$

we get, dividing by $\max_{0 \leq t \leq T} |[w(x,t)]|_{m-1,s}$:

$$\max_{0 \leq t \leq T} |[w(x,t)]|_{m-1,s} \leq C \sum_{k=0}^{m-1} [\Lambda_\tau^{m-1-k} \varphi_k]_s + C \int_0^T [F]_s dt. \tag{48.38}$$

We have $w = e^{-t\tau}u(x,t), u(x,t) \in H_{m,s}(R_T)$ is arbitrary. Note that $e^{-t\tau}$ and $e^{t\tau}$ are bounded on $[0,T]$. Estimating $u = e^{t\tau}w$ using (48.38) we get

$$(48.39)\qquad \max_{0\le t\le T}\sum_{k=0}^{m-1}\left[\frac{\partial^k u(x,t)}{\partial t^k}\right]_{m-1-k+s} \le C\sum_{k=0}^{m-1}\left[\frac{\partial^k u(x,0)}{\partial t^k}\right]_{m-1-k+s} + C\int_0^T [H(x,t,D_x,D_t)u]_0 dt. \qquad \square$$

The proof of the uniqueness for the solution of the Cauchy problem in R_T is the same as in Lemma 48.3: Let $u \in C_{m-1,s}$ be a solution of $H(x,t,D_x,D_t)u = 0$ in R_T, with zero initial conditions. Note that $u \in H_{m-1,s}(R_T)$ since $C_{m-1,s} \subset H_{m-1,s}$. It follows from $Hu = 0$ that $\frac{\partial^m u}{\partial t^m} \in H_{0,s-1}(R_T)$ (cf. (48.21). Therefore $w \in H_{m,s-1}(R_T)$. Applying the estimate (48.39) with s replaced by $s-1$, we get $u = 0$.

To prove the existence, choose any $F \in H_{0,s+1}(R_T)$. Note that $F \in L^1[(0,T), H_s(\mathbb{R}^n)]$ since $(\int_0^T [F]_s dt)^2 \le T\|F\|_{0,s}^2$. Choose also any $\varphi_k \in H_{m-k+\frac{1}{2}+s}(\mathbb{R}^n)$, $0 \le k \le m-1$. It follows from Lemma 48.5 that there exists $w \in H_{m,s}(R_T)$ such that $Hw = F$, $\frac{\partial^k w(x,0)}{\partial t^k} = \varphi_k(x)$, $0 \le k \le m-1$. Note that the estimate (48.39) holds for $w(x,t)$ since $w \in H_{m,s}(R_T)$. Take a sequence $F_n \in H_{0,s+1}(R_T)$ that converges to $f(x,t)$ in the norm of $L^1[(0,T), H_s(\mathbb{R}^n)]$. Also assume that $\varphi_k^{(n)}(x) \in H_{m-k+\frac{1}{2}+s}(\mathbb{R}^n)$ is convergent to $g_k \in H_{m-1-k+s}(\mathbb{R}^n)$ in $H_{m-k-1+s}(\mathbb{R}^n)$, $0 \le k \le m-1$. Let $w_n(x) \in H_{m,s}(R_T)$ be the solution of the Cauchy problem $Hw_n = F_n$ in R_T, $\frac{\partial^k w_n(x,0)}{\partial t^k} = \varphi_k^{(n)}(x)$, $0 \le k \le m-1$. It follows from (48.39) that w_n converges in $C_{m-1,s}$ to the solution $u(x,t) \in C_{m-1,s}$ of the Cauchy problem (48.1), (48.2) in R_T.

48.4. Strictly hyperbolic equations of second order.

Let

$$(48.40)\quad H(x,t,\xi,\sigma) = g^{00}(x,t)\sigma^2 + 2\sum_{j=1}^n g^{j0}(x,t)\xi_j\sigma + \sum_{j,k=1}^n g^{jk}(x,t)\xi_j\xi_k + b_0(x,t)\sigma + \sum_{j=1}^n b_j(x,t)\xi_j + c(x,t)$$

be a strictly hyperbolic polynomial of second order. Assume that $g^{00}(x,t) = 1$. Let

$$\sigma_\pm(x,t,\xi) = -a(x,t,\xi) \pm \sqrt{b(x,t,\xi)}$$

be the roots of $H_0(x,t,\xi,\sigma)=0$, where

$$a=\sum_{j=1}^{n} g^{jo}(x,t)\xi_j, \quad b=a^2-\sum_{j,k=1}^{n} g^{jk}(x,t)\xi_j\xi_k.$$

The strict hyperbolicity means that $\sigma_+-\sigma_-=2\sqrt{b(x,t\xi)}>0$ for all x,t, and $\xi\neq 0$. In the notation of (48.7), (48.8) we have

$$\begin{aligned}Q_0(x,t,\xi,\sigma,\tau)&=(\sigma-\sigma_+)^2+\tau^2+(\sigma-\sigma_-)^2+\tau^2\\&=(\sigma+a-\sqrt{b})^2+\tau^2+(\sigma+a+\sqrt{b})^2+\tau^2=2[(\sigma+a)^2+\tau^2+b]\\&=2\Bigg[\Big(\sigma+\sum_{j=1}^{n} g^{j0}\xi_j\Big)^2+\Big(\sum_{j=1}^{n} g^{j0}\xi_j\Big)^2-\sum_{j,k=1}^{n} g^{jk}\xi_j\xi_+\tau^2\Bigg].\end{aligned}$$

The proof of Theorem 48.6 works, obviously, in the case $m=2$. In this subsection we present another proof of the estimate (48.27).

Consider the Cauchy problem:

$$H\left(x,t,\frac{\partial}{\partial x},\frac{\partial}{\partial t}\right)u(x,t)=f(x,t), \quad x\in\mathbb{R}^n,\ 0<t<T, \tag{48.41}$$

$$u(x,0)=g_0(x), \quad \frac{\partial u(x,0)}{\partial t}=g_1(x),\ x\in\mathbb{R}^n. \tag{48.42}$$

Assume, for simplicity, that all coefficients in (48.40) are real-valued. We will look for a real-valued solutions $u(x,t)$ of the Cauchy problem (48.41), (48.42).

Let $w(x,t)\in H_2(R_T)$ be arbitrary. Consider the identity

(48.43)

$$\left[H\left(x,t,\frac{\partial}{\partial x},\frac{\partial}{\partial t}\right)w(x,t),\frac{\partial w}{\partial t}+\sum_{j=0}^{n} g^{j0}\frac{\partial w}{\partial x_j}\right]=\left[F(x,t),\frac{\partial w}{\partial t}+\sum_{j=1}^{n} g^{j0}\frac{\partial w}{\partial x_j}\right],$$

where $F(x,t)=H(x,t,\frac{\partial}{\partial x},\frac{\partial}{\partial t})w$ and $[u,v]$ is the the scalar product in $L_2(\mathbb{R}^n)$. We have

$$\begin{aligned}&\left[\frac{\partial^2 w}{\partial t^2}+2\sum_{j=1}^{n} g^{j0}(x,t)\frac{\partial^2 w}{\partial x_j\partial t},\frac{\partial w}{\partial t}+\sum_{j=1}^{n} g^{j0}\frac{\partial w}{\partial x_j}\right]\\&=\frac{1}{2}\frac{\partial}{\partial t}\left[\frac{\partial w}{\partial t}+\sum_{j=1}^{n} g^{j0}\frac{\partial w}{\partial x_j}\right]_0^2+\frac{1}{2}\frac{\partial}{\partial t}\left[\sum_{j=1}^{n} g^{j0}\frac{\partial w}{\partial x_j}\right]_0^2\\&\qquad+\frac{1}{2}\int_{\mathbb{R}^n}\sum_{j=1}^{n}\frac{\partial}{\partial x_j}\left(g^{j0}\left(\frac{\partial w}{\partial t}\right)^2\right)dx+R_1,\end{aligned} \tag{48.44}$$

where

$$|R_1| \le C\left([w]_1^2 + \left[\frac{\partial w}{\partial t}\right]_0^2\right), \tag{48.45}$$

and $[w]_s$ is the norm in $H_s(\mathbb{R}^n)$. Note that the third term in the right hand side of (48.44) is equal to zero. Analogously, after the integration by parts with respect to x_j, $1 \le j \le n$, we get

$$\begin{aligned}\left[\sum_{j,k=1}^n g^{jk}\frac{\partial^2 w}{\partial x_j \partial x_k}, \frac{\partial w}{\partial t}\right] &= -\left[\sum_{j,k=1}^n g^{jk}\frac{\partial w}{\partial x_k}, \frac{\partial^2 w}{\partial x_j \partial t}\right] + R_2 \\ &= -\frac{1}{2}\frac{\partial}{\partial t}\left[\sum_{j,k=1}^n g^{jk}\frac{\partial w}{\partial x_k}, \frac{\partial w}{\partial x_j}\right] + R_3,\end{aligned} \tag{48.46}$$

where R_2, R_3 satisfy estimates of the form (48.45).

Finally,

$$\begin{aligned}&\left[\sum_{j,k=1}^n g^{jk}\frac{\partial^2 w}{\partial x_j \partial x_k}, \sum_{p=1}^n g^{p0}\frac{\partial w}{\partial x_p}\right] \\ &\quad = -\int_{\mathbb{R}^n}\sum_{p=1}^n \frac{\partial}{\partial x_p}\left(g^{p0}\sum_{j,k=1}^n g^{jk}\frac{\partial w}{\partial x_j}\frac{\partial w}{\partial x_k}\right)dx + R_4 = R_4,\end{aligned} \tag{48.47}$$

because the integral in the right hand side of (48.47) is equal to zero. Note that R_4 has an estimate of the form (48.45).

Substituting (48.44), (48.46), (48.47) in (48.43), we get

$$\frac{d}{dt}E_1(t) \le C|[w]|_1^2 + C[F]_0|[w]|_1, \tag{48.48}$$

where

$$\begin{aligned}E_1(t) = \frac{1}{2}\int_{\mathbb{R}^n}\Bigg\{&\left(\frac{\partial w}{\partial t} + \sum_{j=1}^n g^{j0}\frac{\partial w}{\partial x_j}\right)^2 + \left(\sum_{j=1}^n g^{j0}\frac{\partial w}{\partial x_j}\right)^2 \\ &- \sum_{j,k=1}^n g^{jk}\frac{\partial w}{\partial x_j}\frac{\partial w}{\partial x_k}\Bigg\}dx,\end{aligned} \tag{48.49}$$

$$|[w]|_1^2 = [w]_1^2 + \left[\frac{\partial w}{\partial t}\right]_0^2.$$

Note that

$$C_2|[w]|_1 \le E_1(t) \le C_1|[w]|_1, \quad \forall t. \tag{48.50}$$

Using (48.50) and the Gronwall's inequality, we get

$$|[w(\cdot,t)]|_1^2 \le C\int_0^t [F(\cdot,t')]_0|[w(\cdot,t')]|_1 dt' + C|[w(\cdot,0)]|_1^2. \tag{48.51}$$

Now we can continue the proof of the estimate (48.17), using (48.51) instead of (48.36), and we get (cf. (48.38))

$$\max_{0\le t\le T} |[w(\cdot,t)]|_1 \le C[g_0]_1 + C[g_1]_0 + C\int_0^T [F(\cdot,t)]_0 dt. \tag{48.52}$$ □

Having estimate (48.52) one can complete the proof of the uniqueness and existence theorem for the Cauchy problem (48.41)), (48.42) as in Subsection 48.3.

49. Domain of dependence

We shall study now the domain of dependence and the domain of influence of the solutions to hyperbolic equations. We give here only an estimate of the domains of influence and dependence.

Denote by $K_{a,-t_0}$ the half-cone

$$K_{a,-t_0} = \{|x|^2 < a^2(t+t_0)^2,\ t > -t_0\}. \tag{49.1}$$

Theorem 49.1. *Let $f_+(x,t)$ and $u_+(x,t)$ be the same as in Theorem* 48.1. *Suppose* $\operatorname{supp} f_+(x,t) \subset \overline{K}_{a_0,0}$, $a_0 > 0$. *There exists $a > a_0$ such that* $\operatorname{supp} u_+(x,t) \subset K_{a,-t_0}$, *where $t_0 > 0$.*

Proof: We have

$$\operatorname{supp} f_+ \subset K_{a_0,0} \subset \overline{K}_{a,-t_0},$$

for any $a > a_0, t_0 > 0$. We shall define a one-to-one map

$$(y, y_0) = s(x,t) \tag{49.2}$$

of $K_{a,-t_0}$ onto the half-space $y_0 > -t_0$, $y \in \mathbb{R}^n$ such that $s(x,t) \in C^\infty$ when $(x,t) \ne (0,-t_0)$ and the Jacobian matrix $\frac{\partial(y,y_0)}{\partial(x,t)}$ is close to I.

To describe explicitly the map (49.2), we choose spherical coordinates (r,ψ,ω) in $K_{a,-t_0}$, where

$$\omega = \frac{x}{|x|} \in S^{n-1}, \quad r = \sqrt{|x|^2 + (t+t_0)^2},$$

$$\cos\psi = \frac{t+t_0}{\sqrt{|x|^2+(t+t_0)^2}}, \quad 0 \le \psi \le \frac{\pi}{2} - \alpha, \quad a = \cot\alpha.$$

Analogously, we introduce spherical coordinates in the half-space $y_0 > -t_0$, $y \in \mathbb{R}^n$:

$$\omega' = \frac{y}{|y|}, \quad r' = \sqrt{|y|^2 + (y_0 + t_0)^2},$$

$$\cos\psi' = \frac{y_0 + t_0}{\sqrt{|y|^2 + (y_0 + t_0)^2}}, \qquad 0 \le \psi \le \frac{\pi}{2}.$$

The change of coordinates (49.2) has the following form in spherical coordinates:

$$\omega' = \omega, \quad r' = r, \quad \psi' = \frac{1}{1 - \frac{2\alpha}{\pi}}\psi.$$

In the Cartesian coordinates we have

$$y = x, \ y_0 + t_0 = \sqrt{|x|^2 + (t + t_0)^2} \cos \frac{\psi}{1 - \frac{2\alpha}{\pi}}, \tag{49.3}$$

where $\psi = \cos^{-1} \frac{t+t_0}{\sqrt{|x|^2 + (t+t_0)^2}}$.

Let $\chi(x_1) \in C_0^\infty(\mathbb{R}^1)$, $\chi(x_1) = 1$ for $|x_1| < 1$, $\chi(x_1) = 0$ for $|x_1| > 2$, and $|\chi(x_1)| \le 1$. Instead of (49.3) we consider the following change of variables $(y, y_0) = s_N(x, t)$:

$$y = x, \quad y_0 + t_0 = \sqrt{|x|^2 + (t + t_0)^2} \cos \frac{\psi}{1 - \frac{2\alpha}{\pi}\chi(\frac{|x|}{N})}. \tag{49.4}$$

Then (49.4) coincides with (49.3) if $|x| < N$, and (49.4) is the identity map if $|x| > 2N$. Note that $\frac{\partial s_N(x,t)}{\partial(x,t)}$ is close to the identity when α is small.

Make the change of variables (49.4). Then $H(x, t, D_x, D_t)$ has the form $H_N(y, y_0, D_y, D_{y_0}) = H(s_N^{-1}(y, y_0), ((\frac{\partial s_N^{-1}(y,y_0)}{\partial(y,y_0)})^T)^{-1}(D_y, D_{y_0}))$ (cf. §45). We assume that $H(x, t, D_x, D_t)$ is strictly hyperbolic with respect to D_t, i.e., $H_0(x, t, \xi, \sigma) = 0$ has m distinct real roots $\sigma_k(x, t, \xi)$, $1 \le k \le m$, for all $x, t, \xi \ne 0$. By the implicit function theorem, $H_N(y, y_0, D_y, D_{y_0})$ will also be strictly hyperbolic with respect to D_{y_0}, since the Jacobian matrix $\frac{\partial s_N(x,t)}{\partial(x,t)}$ is close to the identity.

Consider the equation

$$H_N(y, y_0, D_y, D_{y_0})\hat{w}_N(y, y_0) = \hat{f}_+(y, y_0) \tag{49.5}$$

in the strip $\{-t_0 + \varepsilon < y_0 < T, \ y \in \mathbb{R}^n\}$, where $\hat{f}_+(y, y_0)$ is the image of $f_+(x, t) \in \overset{\circ}{H}_{0,s}(\mathbb{R}^{n+1})$ under the map (49.4). We look for

$$\hat{w}_N(y, y_0) \in H_{m,s-1}(\mathbb{R}^n \times (-t_0 + \varepsilon, T))$$

with zero initial conditions when $t = -t_0 + \varepsilon$. It follows from Theorem 48.6 that such $\hat{w}_N(y, y_0)$ exists and is unique.

Denote by $\Sigma^{(N)}_{-t_0+\varepsilon}$ and $\Sigma^{(N)}_T$ the images of the planes $y_0 = -t_0 + \varepsilon$ and $y_0 = T$ under the map $s_N^{-1}(y, y_0)$. Let $\Omega^{(N)}_{-t_0+\varepsilon,T}$ be the domain between $\Sigma^{(N)}_{-t_0+\varepsilon}$ and $\Sigma^{(N)}_T$. Denote by $w_N(x,t)$ the image of $\hat{w}_N(y, y_0)$ restricted to the strip $-t_0 + \varepsilon < y_0 < T$, $y \in \mathbb{R}^n$. Then $w_N(x,t)$ satisfies the equation $Hw_N = f_+(x,t)$ in $\Omega^{(N)}_{-t_0+\varepsilon,T}$ and has zero Cauchy data on $\Sigma^{(N)}_{-t_0+\varepsilon}$. Let $w_N^+(x,t)$ be the extension of $w_N(x,t)$ by zero below $\Sigma^{(N)}_{-t_0+\varepsilon}$. Then

$$H(x,t,D_x,D_t)w_N^+(x,t) = f_+(x,t) \quad \text{for} \quad (x,t) \in \Omega^{(N)}_{-\infty,T}, \tag{49.6}$$

where $(x,t) \in \Omega^{(N)}_{-\infty,T}$ if (x,t) is below $\Sigma^{(N)}_T$.

Compare $w_N^+(x,t)$ with $u_+(x,t)$ that solves $H(x,t,D_x,D_t)u_+ = f_+(x,t)$ in $\mathbb{R}^{n+1}$. Let T_1 be such that the plane $t = T_1$ is below $\Sigma^{(N)}_T$, i.e., $t = T_1$ is contained in $\Omega^{(N)}_{-\infty,T}$. Then by the uniqueness part of Theorem 48.1 we have that $u_+ = w_N^+(x,t)$ for $t < T_1$. Therefore $u_+(x,t) = 0$ if $t < T_1$ and $(x,t) \in \Omega^{(N)}_{-\infty,-t_0+\varepsilon}$. Taking the limit as $N \to \infty$, we see that $u_+(x,t) = 0$ for $|x|^2 > a^2(t+t_0-\varepsilon)^2$, $t < T_1$, since $\Sigma^{(N)}_{-t_0+\varepsilon}$ coincides with $\{(x,t) : |x|^2 = a^2(t+t_0-\varepsilon)^2\}$ for $|x| < N$. Here $a = \cot\alpha$. Since we can choose T (and consequently T_1) arbitrarily large, and since $\varepsilon > 0$ is arbitrary, we prove that $u_+ = 0$ for $a^2|x|^2 > |t+t_0|^2$, $\forall t > -t_0$. □

Corollary 49.2 (Domain of influence). *Let $G(x,y,t,t')$ be the forward Green function, i.e.,*

$$\begin{aligned} &G(x,y,t,t') = 0 \quad \textit{for} \quad t < t',\ x \in \mathbb{R}^n, \\ &H(x,t,D_x,D_t)G(x,y,t,t') = 0 \quad \textit{for} \quad t > t',\ x \in \mathbb{R}^n, \\ &\frac{\partial^k G(x,y,t,t')}{\partial t^k}\Big|_{t=t'+0} = 0, \quad 0 \le k \le m-2, \\ &\frac{\partial^{m-1} G(x,y,t,t')}{\partial t^{m-1}}\Big|_{t=t'+0} = \delta(x-y), \end{aligned} \tag{49.7}$$

where $t'+0$ means the limit as $t \to t'$, $t > t'$.

We use Theorem 49.1 to show that $\operatorname{supp} G(x,y,t,t') \subset \overline{K}_{a,y,t'}$, where $K_{a,y,t'}$ is the half-cone $\{(x,t) : |x-y|^2 < a^2(t-t')^2, t \ge t'\}$, and $a > 0$ is large.

Let $h_+(x,t) = \frac{(t-t')^{m-1}}{(m-1)!}\delta(x-y)$ for $t \ge t'$, $h_+(x,t) = 0$ for $t < t'$, and let

$$G^{(1)}(x,t) = G(x,y,t,t') - h_+(x,t) \quad \text{for} \quad t \ge t'. \tag{49.8}$$

We have

$$HG^{(1)} = f_+, \quad t > t',$$

where $f_+ = -Hh_+$. Note that $G^{(1)}$ has zero initial data if $t = t'$.

Let $G_+^{(1)} = G^{(1)}$ for $t \geq t'$, $G_+^{(1)} = 0$ for $t < t'$. Note that $\operatorname{supp} f_+ = \{(x,t) : x = y,\ t \geq t'\}$ and $e^{-t\tau} f_+ \in \overset{\circ}{H}_{0,s}(\mathbb{R}^{n+1})$, where $s < -\frac{n}{2} - m$, $\tau > 0$. By Theorem 48.1, there exists $G_+^{(1)}$, $e^{-t\tau} G_+^{(1)} \in \overset{\circ}{H}_{m,s-1}(\mathbb{R}^{n+1})$, that satisfies $HG_+^{(1)} = f_+$ in $\mathbb{R}^{n+1}$. Therefore, we have proven that $G(x,y,t,t') = G_+^{(1)} + h_+(x,t)$ is the forward Green function. To prove that $\operatorname{supp} G(x,y,t,t') \subset K_{a,y,t'}$, it is enough to prove that $\operatorname{supp} G_+^{(1)} \subset \overline{K}_{a,y,t'}$. Consider the cone $K_{a,y,t'-t_0}$, where $t_0 > 0$ is arbitrary. Then $K_{a,y,t'-t_0}$ contains the support of f_+ for $a > 0$ and each $t_0 > 0$. By Theorem 49.1, $\overline{K}_{a,y,t'-t_0} \supset \operatorname{supp} G_+^{(1)}(x,t)$ when $a > 0$ is large, and since t_0 is arbitrary, we see that $\overline{K}_{a,y,t'} \supset \operatorname{supp} G_+^{(1)}(x,t)$. Therefore the domain of influence of the point (y,t') is contained in $\overline{K}_{a,y,t'}$. □

Consider the Cauchy problem $Hu_+ = f_+$ with zero initial conditions when $t = 0$, where $\operatorname{supp} f_+ \subset D$, $\overline{D} \subset \mathbb{R}_+^{n+1}$ is a bounded domain, and f_+ is smooth. We have $u_+ = \int_D G(x,y,t,t') f_+(y,t') dy dt'$.

This shows that that the domain of influence of $\overline{D}$ is contained in the closure of $\Delta = \bigcup_{(y,t') \in \overline{D}} \overline{K}_{a,y,t'}$. If D is a bounded domain then the intersection $\overline{\Delta} \cap \{t = t_0\}$ is bounded for each t_0. This means that the speed of propagation of disturbances originated in D is finite.

Corollary 49.3 (Domain of dependence). *Denote by $G^*(x,y,t,t')$ the backward Green function, i.e.,*

$$
\begin{aligned}
&G^*(x,y,t,t') = 0 \quad \text{for} \quad t > t',\ x \in \mathbb{R}^n, \\
&H^*(x,t,D_x,D_t) G^*(x,y,t,t') = 0 \quad \text{for} \quad t < t',\ x \in \mathbb{R}^n, \\
&\left.\frac{\partial^k G^*(x,y,t,t')}{\partial t^k}\right|_{t=t'-0} = 0,\ 0 \leq k \leq m-2, \\
&\left.\frac{\partial^{m-1} G^*(x,y,t,t')}{\partial t^{m-1}}\right|_{t=t'-0} = \delta(x-y),
\end{aligned}
\tag{49.9}
$$

where $t' - 0$ means the limit as $t \to t'$, $t < t'$.

Here H^* is the operator formally adjoint to H.

The proof of the existence of the backward Green function is the same as for the forward Green function. Let $K_{a,y,t'}^- = \{(x,t) : |x-y|^2 < a^2(t-t')^2,\ t < t'\}$ is the backward half-cone. Analogously to the proof of Theorem 49.1 and Corollary 49.2, we show that $\operatorname{supp} G^*(x,y,t,t')$ is contained in $K_{a,y,t'}^-$. Therefore the domain of dependence of the point (y,t') is contained in $\overline{K}_{a,y,t'}^-$. Let $Hu_+ = f_+$, where $f_+ = u_+ = 0$ for $t < 0$, f_+ is smooth, and

$\operatorname{supp} f_+ \subset D \subset \mathbb{R}^{n+1}_+$. Since $H^*G^* = \delta(x-y)\delta(t-t')$, we get $(f_+, G^*) = (Hu_+, G^*) = (u_+, H^*G^*) = u_+(y,t')$, where (u,v) is the extension of L_2 scalar product in $\mathbb{R}^{n+1}$. Therefore, $u_+(y,t') = \int_D G^*(x,y,t,t') f_+(x,t)dxdt$, i.e., $u_+(y,t')$ depends only on the values of $f_+(x,t)$ in $\bar{D} \cap \bar{K}^-_{a,y,t'}$.

50. Propagation of singularities

50.1. The null-bicharacteristics.

Let $y = (x, x_0)$, $\eta = (\xi, \xi_0)$, where $(x,\xi) \in \mathbb{R}^n \times \mathbb{R}^n$, $x_0 \in \mathbb{R}$, $\xi_0 \in \mathbb{R}$. Let $P_0(y,\eta)$ be a symbol homogeneous in η, $\deg_\eta P_0(y,\eta) = m$. We assume that $P_0(y,\eta) \in C^\infty$ when $\eta \neq 0$, $P_0(y,\eta)$ is real-valued, and $\frac{\partial P_0(y,\eta)}{\partial \eta} \neq 0$ when $P_0(y,\eta) = 0$, $\eta \neq 0$.

Such symbols P_0 will be called the symbols of real principal type.

The bicharacteristics of $P_0(y,\eta)$ are the solutions of the system

$$\begin{aligned} \frac{d\hat{y}}{ds} &= \frac{\partial P_0(\hat{y}(s), \hat{\eta}(s))}{\partial \eta}, \quad \hat{y}(0) = y, \\ \frac{d\hat{\eta}}{ds} &= -\frac{\partial P_0(\hat{y}(s), \hat{\eta}(s))}{\partial y}, \quad \hat{\eta}(0) = \eta, \ s \in \mathbb{R}. \end{aligned} \tag{50.1}$$

Note that

$$\frac{d}{ds} P_0(\hat{y}(s), \hat{\eta}(s)) = \frac{\partial P_0(\hat{y}(s), \hat{\eta}(s))}{\partial y} \frac{d\hat{y}}{ds} + \frac{\partial P_0(\hat{y}(s), \hat{\eta}(s))}{\partial \eta} \frac{d\hat{\eta}}{ds} = 0, \tag{50.2}$$

where $\hat{y}(s), \hat{\eta}(s)$ is the solution of (50.1), i.e., $P_0(\hat{y}(s), \hat{\eta}(s)) = \text{const}$ along a bicharacteristic. If the initial value (y,η) of (50.1) is such that $P_0(y,\eta) = 0$, then $P_0(\hat{y}(s), \hat{\eta}(s)) = 0$ for all s.

In this case the bicharacteristic is called the null-bicharacteristic.

We write the solution of (50.1) in the form:

$$\begin{aligned} \hat{y} &= \hat{y}(s, y, \eta), \\ \hat{\eta} &= \hat{\eta}(s, y, \eta). \end{aligned} \tag{50.3}$$

50.2. Operators of real principal type.

Let

$$P(y, D_y) = \sum_{k=0}^{N} P_k(y, D_y)(1 - \chi(D_y)) + T_{m-N-1}, \tag{50.4}$$

where $P_k(y,\eta)$ are homogeneous and smooth when $\eta \neq 0$, $\deg_\eta P_k(y,\eta) = m-k$, $P_0(y,\eta)$ is a symbol of real principal type, $\operatorname{ord} T_{m-N-1} \leq m-N-1$, N is arbitrary, $\chi(\eta) \in C_0^\infty(\mathbb{R}^{n+1})$, $\chi(\eta) = 1$ for $|\eta| < 1$, and $P(y,\eta) = P(\infty, \eta)$ for large $|x|$.

We call $P(y, D_y)$ of the form (50.4) a ψdo of real principal type.

The following theorem holds.

Theorem 50.1. *Let $u(y) \in H_s(\mathbb{R}^{n+1})$ be the solution of $P(y, D_y)u = f$ in $\mathbb{R}^{n+1}$ for some $s \in \mathbb{R}$. Assume that $P(y, D_y)$ is a ψdo of real principal type. If $P(y^{(0)}, \eta^{(0)}) = 0$, $(y^{(0)}, \eta^{(0)}) \notin WF(u)$ and if the null-bicharacteristic $\hat{y} = \hat{y}(s, y^{(0)}, \eta^{(0)})$, $\hat{\eta} = \hat{\eta}(s, y^{(0)}, \eta^{(0)})$, $\hat{y}(0, y^{(0)}, \eta^{(0)}) = y^{(0)}$, $\hat{\eta}(0, y^{(0)}, \eta^{(0)}) = \eta^{(0)}$ does not belong to $WF(f)$ for any $s \in [0, L]$, then this null-bicharacteristic does not belong to $WF(u)$ for $s \in [0, L]$.*

We denote by $\gamma(y, \eta)$ the null-bicharacteristic given by (50.1).

Proof: Let Λ^s be a ψdo with the symbol $(1 + |\xi|^2 + |\xi_0|^2)^{\frac{s}{2}}$. If $s < 0$, we apply Λ^s to $Pu = f$. Using Theorem 40.2, we get $\hat{P}\hat{u} = \hat{f}$, where $\hat{f} = \Lambda^s f$, $\hat{u} = \Lambda^s u$, and $\hat{P}$ is again a ψdo of the form (50.4) with the same principal part. Note that $\hat{u} \in H_0(\mathbb{R}^n)$ and $\gamma(y^{(0)}, \eta^{(0)})$ does not belong to $WF(\hat{f})$ since Λ^s is an elliptic operator (see §44). Therefore, without loss of generality we may assume that $u \in H_s(\mathbb{R}^{n+1})$, where $s \geq 0$.

Let $P_0(y^{(0)}, \eta^{(0)}) = 0$, $\eta^{(0)} \neq 0$. Since $P_0(y, \eta)$ is of real principal type, there exists ξ_j such that $\frac{\partial P_0(y^{(0)}, \eta^{(0)})}{\partial \xi_j} \neq 0$, $0 \leq j \leq n$. Assume, for the definiteness, that $\frac{\partial P_0(y^{(0)}, \eta^{(0)})}{\partial \xi_0} > 0$, where $y^{(0)} = (x^{(0)}, x_0^{(0)})$, $\eta^{(0)} = (\xi^{(0)}, \xi_0^{(0)})$. Note that $\xi^{(0)} \neq 0$. Otherwise we have that $P_0(y^{(0)}, 0, \xi_0) = 0$ for any $\xi_0 = \tau \xi_0^{(0)}, \tau > 0$, and hence $\frac{\partial P_0(y^{(0)}, 0, \xi_0^{(0)})}{\partial \xi_0} = 0$.

By the implicit function theorem, there exists a conic neighborhood U_0 of $(y^{(0)}, \eta^{(0)})$ such that

$$P_0(y, \eta) = q_0(y, \eta)(\xi_0 - \lambda_0(y, \xi)), \tag{50.5}$$

where $q_0(y, \eta) \neq 0$ in U_0, $\lambda_0(y, \xi)$ is real-valued, $\xi \neq 0$ in U_0 since $\xi^{(0)} \neq 0$, $\deg_\eta q_0(y, \eta) = m - 1$, $\deg_\xi \lambda_0(y, \xi) = 1$, and q_0 and λ_0 are C^∞ in U_0.

For simplicity of notation assume that $x_0^{(0)} = 0$. Denote by U_1 a neighborhood of $(x^{(0)}, 0, \eta^{(0)})$ in $U_0 \cap \{x_0 = 0\}$.

We often write $\hat{y}(s) = (\hat{x}(s), \hat{x}_0(s))$, $\hat{\eta}(s) = (\hat{\xi}(s), \hat{\xi}_0(s))$ instead of $\hat{y}(s, y, \eta)$, $\hat{\eta}(s, y, \eta)$. Since $\hat{y}(s), \hat{\eta}(s)$ is a null-bicharacteristic and $q_0 \neq 0$ in U_0, we have

$$\hat{\xi}_0(s) - \lambda_0(\hat{x}(s), \hat{x}_0(s), \hat{\xi}(s)) = 0. \tag{50.6}$$

Consider the null-bicharacteristics in U_0 with initial conditions on U_1, i.e., $\hat{x}(0) = x$, $\hat{x}_0(0) = 0$, $\hat{\eta}(0) = \eta$, $(x, 0, \eta) \in U_1$.

Since $\frac{\partial P_0}{\partial \xi_0} > 0$ in U_0, we can take

$$\hat{x}_0 = \int_0^s \frac{\partial P_0(\hat{y}(s'), \hat{\eta}(s'))}{\partial \xi_0} ds' \tag{50.7}$$

as a parameter on $\gamma(x,0,\eta)$. Note that

$$(50.8)\quad \frac{\partial P_0(\hat{y}(s),\hat{\eta}(s))}{\partial\xi_0} = \frac{\partial q_0(\hat{y}(s),\hat{\eta}(s))}{\partial\xi_0} \times \left(\hat{\xi}_0(s) - \lambda_0(\hat{x}(s),\hat{x}_0(s),\hat{\xi}(s))\right) + q_0(\hat{y}(s),\hat{\eta}(s)) = q_0(\hat{y}(s),\hat{\eta}(s)).$$

Analogously,

$$(50.9)\quad \frac{\partial P_0(\hat{y}(s),\hat{\eta}(s))}{\partial\xi} = q_0(\hat{y}(s),\hat{\eta}(s))\left(-\frac{\partial\lambda_0(\hat{y}(s),\hat{\xi}(s))}{\partial\xi}\right),$$

$$(50.10)\quad \frac{\partial P_0(\hat{y}(s),\hat{\eta}(s))}{\partial y} = q_0(\hat{y}(s),\hat{\eta}(s))\left(-\frac{\partial\lambda_0(\hat{y}(s),\hat{\xi}(s))}{\partial y}\right).$$

For simplicity of notation we write x_0 instead of $\hat{x}_0$. Therefore, the system (50.1) takes the following equivalent form when x_0 is a parameter:

$$(50.11)\quad \begin{aligned} \frac{d\hat{x}(x_0)}{dx_0} &= -\frac{\partial\lambda_0(\hat{x}(x_0),x_0,\hat{\xi}(x_0))}{\partial\xi}, \quad \hat{x}(0)=x,\\ \frac{d\hat{\xi}(x_0)}{dx_0} &= \frac{\partial\lambda_0(\hat{x}(x_0),x_0,\hat{\xi}(x_0))}{\partial x}, \quad \hat{\xi}(0))=\xi, \end{aligned}$$

and

$$(50.12)\quad \frac{d\hat{\xi}_0(x_0)}{dx_0} = \frac{\partial\lambda_0(\hat{x}(x_0),x_0,\hat{\xi}(x_0))}{\partial x_0}, \quad \hat{\xi}_0(0)=\xi_0.$$

Taking into account (50.11) that the solution of (50.12) is

$$\hat{\xi}_0(x_0) = \lambda_0(\hat{x}(x_0),x_0,\hat{\xi}(x_0))$$

and the system (50.11) supplemented with the equations $x_0 = x_0$ and $\hat{\xi}(x_0) - \lambda_0(\hat{x}(x_0),x_0,\hat{\xi}(x_0)) = 0$ is the equation for the null-bicharacteristics of $P_0(y,D_y)$ and $D_{x_0} - \lambda_0(y,D_x)$.

50.3. Propagation of singularities for operators of real principal type.

Let $V(-3\varepsilon, L+3\varepsilon)$ be the union of all null-bicharacteristics starting at $(x,0,\eta)\in U_1$ and having $-3\varepsilon < x_0 < L+3\varepsilon$. We assume that U_1 and L are such that $V(-3\varepsilon,L+3\varepsilon)\subset U_0$ and $WF(f)\cap\overline{V(-3\varepsilon,L+3\varepsilon)}=\emptyset$.

Let $\chi_{2\varepsilon}(x_0)\in C_0^\infty(-3\varepsilon,L+3\varepsilon)$ and $\chi_{2\varepsilon}=1$ for $-2\varepsilon<x_0<L+2\varepsilon$ and let $\chi_\varepsilon(x_0)\in C_0^\infty(-2\varepsilon,L+2\varepsilon)$, $\chi_\varepsilon(x_0)=1$ for $-\varepsilon<x_0<L+\varepsilon$.

Lemma 50.2. *There exists $\varphi(y,\eta)\in S^0$ such that $\varphi(y^{(0)},\eta^{(0)})\neq 0$,*

$$\operatorname{supp}\varphi(y,\eta)\subset V(-3\varepsilon,L+3\varepsilon),$$

and

$$\varphi(y, D_y)P(y, D_y)u = P(y, D_y)\varphi(y, D_y)u + T_{m-N-1}u + T_\varepsilon u, \tag{50.13}$$

where $T_\varepsilon u = 0$ *for* $-\varepsilon < x_0 < L + \varepsilon$, $\operatorname{ord} T_{m-N-1} \le m - N - 1$, *and* N *is arbitrary.*

Note that T_ε will play no role, since later we will consider the restriction of (50.13) to the interval $(-\varepsilon, L+\varepsilon)$. Note also that $\varphi(y,\eta)$ depends on N, but we did not indicate this for simplicity of notation.

Proof of Lemma 50.2. We are looking for $\varphi(y, \eta)$ in the form:

$$\varphi(y,\eta) = \sum_{k=0}^{N} \varphi_{-k}(y,\eta)\left(1 - \chi\left(\frac{\eta}{R}\right)\right)\chi_\varepsilon(x_0), \tag{50.14}$$

where the φ_{-k} are homogeneous in η of degree $-k$, $\operatorname{supp}\varphi_{-k}\chi_\varepsilon(x_0) \subset V(-2\varepsilon, L+2\varepsilon)$, $\chi(\eta) \in C_0^\infty(\mathbb{R}^n)$, $\chi(\eta) = 1$ for $|\eta| < 1$, and R is large. Compute the commutator $P\varphi - \varphi P$. Using Theorem 40.2 and equating to zero symbols of the same degree of homogeneity in $\eta = (\xi, \xi_0)$, we get:

$$\varphi_0(y,\eta)P_0(y,\eta) = P_0(y,\eta)\varphi_0(y,\eta),$$

$$\sum_{j=0}^{n} \frac{\partial\varphi_0(y,\eta)}{\partial\xi_j}\frac{\partial P_0(y,\eta)}{\partial x_j} - \sum_{j=0}^{n} \frac{\partial\varphi_0(y,\eta)}{\partial x_j}\frac{\partial P_0(y,\eta)}{\partial\xi_j} = 0, \tag{50.15}$$

(50.16)
$$\sum_{j=0}^{n}\left(\frac{\partial\varphi_{-k}(y,\eta)}{\partial\xi_j}\frac{\partial P_0(y,\eta)}{\partial x_j} - \frac{\partial\varphi_{-k}(y,\eta)}{\partial x_j}\frac{\partial P_0(y,\eta)}{\partial\xi_j}\right) = b_k(y,\eta), 1 \le k \le N,$$

where $\deg_\eta b_k(y,\eta) = m-k-1$, and $b_k(y,\eta)$ depends on $\varphi_0, \varphi_{-1}, \dots, \varphi_{-k+1}$.

The left hand side of (50.15) is called the Poisson bracket of φ_0 and P_0 (cf. Corollary 40.3). Substituting (50.1) or (50.11), (50.12) into (50.15), we get

$$\frac{d}{dx_0}\varphi_0(\hat{x}(x_0,x,\eta), x_0, \hat{\eta}(x_0,x,\eta)) = 0. \tag{50.17}$$

Let $\psi(x,\eta) \in C_0^\infty(U_1)$, $\psi(x^{(0)},\eta^{(0)}) \ne 0$, $\deg_\eta\psi(x,\eta) = 0$. We assume that the $\varphi_{-k}(x, x_0, \eta)$, $0 \le k \le n$, satisfy the following initial conditions for $x_0 = 0$:

$$\begin{aligned} &\varphi_0(x,0,\eta) = \psi(x,\eta), \\ &\varphi_{-k}(x,0,\eta) = 0, \quad 1 \le k \le N. \end{aligned} \tag{50.18}$$

Then

$$\varphi_0(\hat{x}(x_0,x,\eta), x_0, \hat{\eta}(x_0,x,\eta)) = \psi(x,\eta)$$

for $-2\varepsilon \le x_0 \le L + 2\varepsilon$, where $\hat{x}, \hat{\eta}$ are the solutions of (50.11), (50.12), $\hat{x}(0) = x$, $\hat{\eta}(0) = \eta$. Let

(50.19) $$x = x(x_0, \hat{x}, \hat{\eta}), \qquad \eta = \eta(x_0, \hat{x}, \hat{\eta})$$

be the inverse to (50.11), (50.12). Then

(50.20)
$$\varphi_0(\hat{x}, x_0, \hat{\eta}) = \psi(x(x_0, \hat{x}, \hat{\eta}), \eta(x_0, \hat{x}, \hat{\eta})), \quad (\hat{x}, x_0, \hat{\eta}) \in V(-2\varepsilon, L + 2\varepsilon).$$

Using x_0 as a parameter, we get from (50.16)

$$\frac{d}{dx_0}\varphi_{-k}(\hat{x}(x_0), x_0, \hat{\eta}(x_0)) = b_k(\hat{x}(x_0), x_0, \hat{\eta}(x_0)) \left(\frac{\partial P_0(\hat{x}(x_0), x_0, \hat{\eta}(x_0))}{\partial \xi_0} \right)^{-1}.$$

Integrating with respect to x_0 from 0 to x_0 and using the initial conditions (50.18) and the inverse map (50.19), we obtain, analogously to (50.20), that

$$\operatorname{supp} \varphi_{-k}\chi_\varepsilon(x_0) \subset V(-2\varepsilon, L + 2\varepsilon).$$

Note that $\varphi(x, x_0, \eta) \neq 0$ on $\gamma(x^{(0)}, 0, \eta^{(0)})$ since $\varphi_0 \neq 0$, $\deg_\eta \varphi_{-k} \le -k$, $k \ge 1$, $\chi(\frac{\eta}{R}) = 1$ for $|\eta| < R$, where R can be taken large enough.

Applying $\varphi(y, D_y)$ to $P(y, D_y)u = f$ and using (50.15), (50.16), we get

(50.21) $$P(y, D_y)\varphi(y, D_y)u = \varphi(y, D_y)f - T_{m-N-1}u - T_\varepsilon u.$$

The contribution to $T_\varepsilon u$ comes from the terms containing derivatives of $\chi_\varepsilon(x_0)$ when we compute the commutator $P\varphi - \varphi P$. Note that $\varphi(y, D_y)f \in C^\infty$ since $\operatorname{supp} \varphi \subset V(-2\varepsilon, L + 2\varepsilon)$ and therefore $\operatorname{supp} \varphi \cap WF(f) = \emptyset$. □

Lemma 50.3. *There exists $Q_N(y, \eta) \in S^{m-1}$ such that after applying $Q_N(y, D_y)$ to (50.21), we get*

(50.22) $$(D_{x_0} - \lambda_0(y, D_x) + a_N(y, D_x))\varphi u = -T_\varepsilon^{(N)}u - T_{-N-1}^{(N)}u + Q_N\varphi f,$$

where $a_N(y, \xi) \in S^0$, $T_\varepsilon^{(N)}u = 0$ for $-\varepsilon < x_0 < L + \varepsilon$, $\operatorname{ord} T_{-N-1}^{(N)} \le -N - 1$, and N is arbitrary.

Proof: Let $\varphi^{(1)}(y, \eta) \in C^\infty(\mathbb{R}^{n+1} \times (\mathbb{R}^{n+1} \setminus \{0\}))$, $\deg_\eta \varphi^{(1)} = 0$ and let $\operatorname{supp} \varphi^{(1)}$ be contained in U_0. We assume that $\varphi^{(1)} = 1$ in a conic neighborhood of $\operatorname{supp} \varphi_{-k}(y, \eta)$, $0 \le k \le N$. Therefore, $\operatorname{supp}(1 - \varphi^{(1)}) \cap \operatorname{supp} \varphi_{-k} = \emptyset$, $0 \le k \le N$.

Let Q_1 be a ψdo with symbol $\varphi^{(1)}(y, \eta)q_0^{-1}(y, \eta)(1 - \chi(\eta))\chi_{2\varepsilon}(x_0)$. Applying $Q_1(y, D_y)$ to (50.21) and using Theorem 40.2, we get

(50.23) $$(D_{x_0} - \lambda_0(y, D_x))\varphi(y, D_y)u = C_0(y, D_y)\varphi u - T_{-N-1}^{(1)}u - T_\varepsilon^{(1)}u + Q_1\varphi f,$$

where $\operatorname{ord} T_{-N-1}^{(1)} \le -N - 1$, $C_0(y, \xi) \in S^0$, and $T_\varepsilon^{(1)} = 0$ for $-\varepsilon \le x_0 \le L + \varepsilon$. Since $\varphi f \in C^\infty$, we conclude that $g = Q_1\varphi f \in C^\infty(\mathbb{R}^{n+1})$. In

(50.23), we have used that the composition of a ψdo with symbol $(\xi_0 - \lambda_0)(1 - \varphi^{(1)}(y,\eta)\chi_{2\varepsilon}(x_0))$ and of $\varphi(y, D_y)$ is an operator of order $-\infty$, and that $Q_1 T_\varepsilon = T_\varepsilon^{(1)} + T'_{-N-1}$, where $\operatorname{ord} T'_{-N-1} \le -N-1$.

By the Taylor theorem,

$$C_0(y,\xi,\xi_0) = C_0(y,\xi,\lambda_0(y,\xi)) + C_{-1}^{(1)}(y,\xi,\xi_0)(\xi_0 - \lambda_0(y,\xi)). \tag{50.24}$$

Denote $\lambda_{-1}(y,\xi) = C_0(y,\xi,\lambda_0(y,\xi))$. We can rewrite (50.24) in the form

$$\begin{aligned} C_0(y,\xi,\xi_0) = \lambda_{-1}(y,\xi) + C_{-1}^{(1)}(y,\xi,\xi_0)(\xi_0 - \lambda_0(y,\xi) - \lambda_{-1}(y,\xi)) \\ + C_{-1}^{(1)}(y,\xi,\xi_0)\lambda_{-1}(y,\xi). \end{aligned} \tag{50.25}$$

Since $\xi \neq 0$ on $\gamma(x^{(0)}, 0, \xi^{(0)}, \xi_0^{(0)})$ and $\operatorname{supp}\varphi^{(1)}$ is contained in a small conic neighborhood of γ, we have that $|\xi|^{-1} \le C(|\xi| + |\xi_0|)^{-1}$ on $\operatorname{supp}\varphi^{(1)}$. Therefore, since $C_0(y,\xi,\xi_0)$ contains $\varphi^{(1)}$ or derivatives of $\varphi^{(1)}$, we have that $C_0(y,\xi,\xi_0) = 0$ when $\frac{|\xi|}{|\xi|+|\xi_0|}$ is small. Hence $C_{-1}^{(1)}(y,\xi,\xi_0) \in S^{-1}$ and $\lambda_{-1}(y,\xi) \in S^0(\mathbb{R}^{n+1} \times \mathbb{R}^n)$.

Applying Theorem 40.2 to (50.25), we get

$$\begin{aligned} C_0(y, D_y)\varphi u = \lambda_{-1}(y, D_x)\varphi u \\ + C_{-1}^{(1)}(y, D_y)(D_{x_0} - \lambda_0(y, D_x) - \lambda_{-1}(y, D_x))\varphi u \\ + C_{-1}^{(2)}(y, D_y)\varphi u + T_{-N-1}^{(2)} u + T_\varepsilon^{(2)}\varphi u, \end{aligned} \tag{50.26}$$

where $\operatorname{ord} T_{-N-1}^{(2)} \le -N-1$, $C_{-1}^{(2)}(y,\eta) \in S^{-1}$. Substituting (50.26) into (50.23), we obtain:

$$\begin{aligned} (D_{x_0} - (\lambda_0 + \lambda_{-1}))\varphi u - C_{-1}^{(1)}(y, D_y)(D_{x_0} - (\lambda_0 + \lambda_{-1}))\varphi u \\ - C_{-1}^{(2)}(y, D_y)\varphi u = -T_\varepsilon^{(3)} u - T_{-N-1}^{(3)} u + Q_1\varphi f, \end{aligned} \tag{50.27}$$

Since $C_{-1}^{(1)} \in S^{-1}$, there exists $Q_2 = I + Q_2'$, $Q_2'(y,\eta) \in S^{-1}$ such that

$$Q_2(I - C_{-1}^{(1)}) = I + T_{-N-1}^{(3)} \tag{50.28}$$

(cf. §42). Multiplying (50.27) by Q_2, we get

$$\begin{aligned} (D_{x_0} - (\lambda_0(y, D_x) + \lambda_{-1}(y, D_x)))\varphi u \\ = C_{-1}^{(3)}(y, D_y) - T_\varepsilon^{(4)} - T_{-N-1}^{(4)}\varphi u + Q_2 Q_1 \varphi f, \end{aligned} \tag{50.29}$$

where $C_{-1}^{(3)}(y,\eta) \in S^{-1}$.

Now again by the Taylor formula (cf. (50.24)) we have:

$$
\begin{aligned}
(50.30)\quad C_{-1}^{(3)}(y,\xi,\xi_0) &= C_{-1}^{(3)}(y,\xi,\lambda_0+\lambda_{-1}) + C_{-2}^{(3)}(y,\xi,\xi_0)(\xi_0-\lambda_0-\lambda_1) \\
&= \lambda_{-2}(y,\xi) + C_{-2}^{(3)}(y,\xi,\xi_0)(\xi_0-\lambda_0-\lambda_{-1}-\lambda_{-2}) + C_{-2}^{(3)}(y,\xi)\lambda_{-2}(y,\xi),
\end{aligned}
$$

where $C_{-2}^{(3)}(y,\eta) \in S^{-2}$, $\lambda_{-2}(y,\xi) = C_{-1}^{(3)}(y,\xi,\lambda_0+\lambda_1)$. Repeating (50.27) with $C_0, C_{-1}^{(1)}$ replaced by $C_{-1}^{(3)}, C_{-2}^{(3)}$, we get, as in (50.27), (50.29), that

$$
\begin{aligned}
(50.31)\quad &(D_{x_0} - (\lambda_0(y,D_x) + \lambda_{-1}(y,D_x) + \lambda_{-2}(y,D_x)))\varphi u \\
&- C_{-2}^{(3)}(y,D_y)(D_{x_0} - (\lambda_0(y,D_x) + \lambda_{-1}(y,D_x) + \lambda_{-2}(y,D_x)))\varphi u \\
&\qquad = C_{-2}^{(4)}(y,D_y)\varphi u - T_\varepsilon^{(5)}\varphi u - T_{-N-1}^{(5)}\varphi u + Q_2Q_1\varphi f,
\end{aligned}
$$

where $C_{-2}^{(4)} \in S^{-2}$.

Let $Q_3(y,D_y) = I + Q_3'(y,D_y)$ be such that $Q_3(I - C_{-2}^{(3)}) = I + T_{-N-1}$. Multiplying (50.31) by $Q_3(y,D_y)$, we get, as in (50.29):

$$
(D_{x_0} - (\lambda_0+\lambda_{-1}+\lambda_{-2}))\varphi u = C_{-2}^{(5)}\varphi u - T_\varepsilon^{(6)}u - T_{-N-1}^{(6)}u + Q_3Q_2Q_1\varphi f,
$$

where $C_{-2}^{(5)}(y,\eta) \in S^{-2}$.

After N such steps we get (50.22) with $a_N = -\sum_{k=1}^{N}\lambda_{-k}(y,D_x)$. □

Now we can conclude the proof of Theorem 50.1.

Take the restriction of (50.22) to $R_L = \mathbb{R}^n \times (0,L)$. Then $T_\varepsilon^{(N)}u = 0$. We assumed that $(x^{(0)},0,\eta^{(0)}) \notin WF(u)$. Therefore

$$
(50.32)\qquad \varphi u|_{x_0=0} = h \in C^\infty.
$$

Also the right hand side of (50.22) belongs to $H_{s+N+1}(R_L)$. By Theorem 48.6, there exists a unique solution of the Cauchy problem (50.22), (50.32) belonging to $H_{0,s+N+1}(R_L)$. Here $H_{0,s}(R_L)$ is the same as in §48. Since, by assumption, $s \geq 0$, we have $\varphi u \in H_s(R_L) \subset H_{0,s}(R_L)$. Therefore, by the uniqueness part of Theorem 48.6, $\varphi u \in H_{0,s+N+1}(R_L)$.

In view of (50.22), $D_{x_0}\varphi u \in H_{0,s+N}(R_L)$. Differentiating (50.22) with respect to x_0, we obtain $D_{x_0}^2\varphi u \in H_{0,s+N-1}(R_L)$. Since N is arbitrary, we conclude, repeatedly differentiating (50.22) with respect to x_0, that $\varphi u \in H_N(R_L)$ and therefore $\varphi u \in C^\infty$. □

Remark 50.1. We show that the solution of (50.11) exists for all $-\infty < x_0 < +\infty$, provided that $\lambda_0(x,x_0,\xi) = \lambda(\infty,x_0,\xi)$ for large $|x|$ and $\lambda(x,x_0,\xi) = \lambda(x,\infty,\xi)$ for large $|x_0|$. Since $\deg_\xi \lambda_0(x,x_0,\xi) = 1$, we have that $|\frac{\partial\lambda_0(x,x_0,\xi)}{\partial\xi}| \leq C$ for all $(x,x_0) \in \mathbb{R}^{n+1}, \xi \neq 0$. Therefore

$$
(50.33)\qquad |\hat{x}(x_0) - x| \leq C|x_0 - x_0^{(0)}|.
$$

Considering the dot product $\hat{\xi}(x_0) \cdot \frac{d\hat{\xi}(x_0)}{dx_0}$, we get

$$\left|\hat{\xi}(x_0) \cdot \frac{d\hat{\xi}_0}{dx_0}\right| = \left|\hat{\xi}(x_0) \cdot \frac{\partial \lambda_0}{\partial x}\right| \leq C|\hat{\xi}(x_0)|^2.$$

Therefore $\frac{1}{2}\frac{d}{dx_0}|\hat{\xi}(x_0)|^2 \leq C|\hat{\xi}(x_0)|^2$, and we obtain

$$|\hat{\xi}(x_0)|^2 \leq |\xi^{(0)}|^2 \exp(2C|x_0 - x_0^{(0)}|). \tag{50.34}$$

Estimates (50.33), (50.34) imply that the solution of (50.11), (50.12) exists on $(-\infty, +\infty)$. □

We prove that the solution of (50.1) also exists for all $\infty < s < +\infty$. Note that $|\frac{\partial P_0(y,\eta)}{\partial y}| \neq 0$ for all (y, η), $\eta \neq 0$. For $P_0(y, \eta) = 0$ this follows from the definition of the symbol of real principal type. For $P_0(y, \eta) \neq 0$ this follows from Euler's formula $mP_0(y, \eta) = \eta \cdot \frac{\partial P_0(y,\eta)}{\partial \eta}$.

Introduce a new parameter in (50.1):

$$\tau = \int_0^s \left|\frac{\partial P_0(\hat{y}(s), \hat{\eta}(s))}{\partial \eta}\right| ds.$$

When τ is a parameter, the system (50.1) takes the form:

$$\frac{d\hat{y}}{d\tau} = \frac{\frac{\partial P_0(\hat{y},\hat{\eta})}{\partial \eta}}{|\frac{\partial P_0(\hat{y},\hat{\eta})}{\partial \eta}|}, \qquad \frac{d\hat{\eta}}{d\tau} = \frac{\frac{\partial P_0(\hat{y},\hat{\eta})}{\partial y}}{|\frac{\partial P_0(\hat{y},\hat{\eta})}{\partial \eta}|}. \tag{50.35}$$

The right hand sides in (50.35) have the same degree of homogeneity in $\hat{\eta}$ as the right hand sides of (50.11). Therefore we can repeat without changes the proof of the existence of a global solution for the system (50.11). □

Using Remark 50.1 we get the global version of Theorem 50.1:

Theorem 50.4. *Let $P(x, D)u = f$, where $P(x, D)$ is a ψdo of real principal type. Let $\gamma(s) = \{\hat{y} = y(s),\ \hat{\eta} = \eta(s),\ -\infty < s < +\infty\}$ be the null-bicharacteristic* (50.1). *Suppose $WF(f) \cap \gamma(s) = \emptyset$ for all $s \in \mathbb{R}$. Then either $WF(u) \cap \gamma(s) = \emptyset$ for all $s \in \mathbb{R}$, or $\gamma(s) \in WF(u)$ for all $s \in \mathbb{R}$.*

Proof: Suppose $\gamma(0) \notin WF(u)$. Then, by Theorem 50.1, there exists a conic neighborhood $U_0 \subset \mathbb{R}^n \times \mathbb{R}^n \setminus \{0\}$ of $(\hat{y}(0), \hat{\eta}(0))$ such that $(U_0 \cap \gamma) \cap WF(u) = \emptyset$. Since $P_0(y, \eta)$ is homogeneous in η and $P_0(y, \eta) = P_0(\infty, \eta)$ for $|y|$ large, there exists a finite cover of $\mathbb{R}^n \times \mathbb{R}^n \setminus \{0\}$ by conic neighborhoods $U_j, 1 \leq j \leq N$, such that (50.5) holds. Using Theorem 50.1 and the connectedness of $\mathbb{R}^n \times (\mathbb{R}^n \setminus \{0\})$, we get that $(\gamma \cap U_j) \cap WF(u) = \emptyset$ for all j, $1 \leq j \leq N$. Therefore $\gamma \cap WF(u) = \emptyset$ for all $s \in \mathbb{R}$.

Suppose now that $\gamma(s_1) = (\hat{y}(s_1), \hat{\eta}(s_1)) \in WF(u)$. Then all $\gamma(s) = (y(s), \eta(s)) \in WF(u)$ because if some point $\gamma(s_2) = (\hat{y}(s_2), \hat{\eta}(s_2)) \notin WF(u)$, then, as proven above, $\gamma(s) \notin W(u)$ for all $s \in \mathbb{R}$. Therefore, either

$\gamma(s)$, $-\infty < s < +\infty$, is contained in $WF(u)$, or the intersection of $\gamma(s)$ with $WF(u)$ is empty. □

50.4. Propagation of singularities in the case of a hyperbolic Cauchy problem.

Let $H(x,t,D_x,D_t)$ be a strictly hyperbolic operator. We describe the wave front set $WF(u)$ of the Cauchy problem (48.1), (48.2) in terms of $WF(f)$ and the wave front sets of initial conditions $g_k(x)$, $0 \le k \le m-1$. Since $H_0(x,t,\xi,\sigma)$ has m distinct real roots $\sigma = \lambda_k(x,x_0,\xi)$, $1 \le k \le m$, there are m null-bicharacteristics passing through an arbitrary point $(x^{(0)},t_0,\xi^{(0)}) \in \mathbb{R}^{n+1} \times (\mathbb{R}^n \setminus \{0\})$ with $\xi_0^{(k)} = \lambda_k(x^{(0)},t_0,\xi^{(0)})$, $1 \le k \le m$. Consider first the Cauchy problem (48.1) with zero initial conditions, i.e., $Hu_+ = f_+$ in $\mathbb{R}^{n+1}$, $e^{-\tau t}u_+ \in \overset{\circ}{H}_{m,s}(\mathbb{R}^{n+1})$ for some $s \in \mathbb{R}$ and $\tau > 0$, and $u_+ = 0$ for $t < 0$.

If $(x^{(0)},t_0,\xi^{(0)},\xi_0^{(0)}) \in WF(f_+)$, $t_0 \ge 0$, then, obviously,

$$(x^{(0)},t_0,\xi^{(0)},\xi_0^{(0)}) \in WF(u_+),$$

since $WF(f_+) \subset WF(u_+)$.

If $\xi_0^{(0)} \neq \lambda_k(x^{(0)},t_0,\xi^{(0)})$, $1 \le k \le m$, i.e., H_0 is microelliptic at $(x^{(0)},t_0,\xi^{(0)},\xi_0^{(0)})$, then there is no propagation of singularities (Theorem 44.5). If $\xi_0^{(0)} = \lambda_{k_0}(x^{(0)},t_0,\xi^{(0)})$ for some $1 \le k_0 \le m$, then the singularity may propagate or not (cf. Example 50.1 below).

Now consider the case of the Cauchy problem (48.1), (48.2). Assume that $u \in H_{m,s}(\mathbb{R}^{n+1}_+)$ for some $s \in \mathbb{R}$ and let $u_+ = u$ for $t \ge 0$, $u_+ = 0$ for $t < 0$. Then $u_+(x,t)$ satisfies the following equation in $\mathbb{R}^{n+1}$:

$$H(x,t,D_x,D_t)u_+ = f_+ + \sum_{k=0}^{m-1} b_k(x,0,D)g_k(x)\delta^{(m-1-k)}(t), \tag{50.36}$$

where $\operatorname{ord} b_k \le k$, $f_+ = f$ for $t > 0$, and $f_+ = 0$ for $t < 0$.

Suppose $(x^{(0)},\xi^{(0)}) \in \mathbb{R}^n \times (\mathbb{R}^n \setminus \{0\})$ belongs to $WF(g_{k_0}(x))$ for some k_0, $0 \le k_0 \le m-1$. Then $(x^{(0)},0,\xi^{(0)},\sigma) \in WF(g_{k_0}(x)\delta^{(m-1-k)}(t))$, where σ is arbitrary. In particular, $(x^{(0)},0,\xi^{(0)},\lambda_j(x^{(0)},0,\xi^{(0)}))$ belongs to $WF(g_{k_0}(x)\delta^{(m-1-k)}(t))$ for each $1 \le j \le m$.

There are m null-bicharacteristics $\gamma_{jk_0}(x^{(0)},0,\xi^{(0)},\lambda_j(x^{(0)},0,\xi^{(0)}))$ that start at $(x^{(0)},0,\xi^{(0)},\lambda_j(x^{(0)},0,\xi^{(0)}))$, $1 \le j \le m$.

The following theorem holds:

Theorem 50.5. *Let $u(x,t)$ be the solution of the Cauchy problem* (48.1), (48.2). *Assume, for simplicity, that $f(x,t)$ is smooth for $0 \le t \le \delta$. Let* *the wave front sets* $WF(f_+) \subset \mathbb{R}^{n+1}_+ \times (\mathbb{R}^{n+1} \setminus \{0\})$ *and* $WF(g_k(x)) \subset \mathbb{R}^n \times (\mathbb{R}^n \setminus \{0\})$, $0 \le k \le m-1$, *be given. Denote by* $\Sigma(f, g_0, \dots, g_{m-1}) \subset \mathbb{R}^{n+1}_+ \times (\mathbb{R}^{n+1} \setminus \{0\})$ *the set of points satisfying the following conditions:*

a) $\Sigma \supset WF(f_+) \cap \{t > 0\}$.

b) *If* $(x^{(0)}, t_0, \xi^{(0)}, \sigma_0) \in WF(f_+)$, $t_0 \ge 0$, $\xi^{(0)} \ne 0$, *and* $\sigma_0 = \lambda_{k_0}(x^{(0)}, t_0, \xi^{(0)})$ *for some* k_0, $1 \le k_0 \le m$, *then the forward null-bicharacteristic* $\gamma^+_{k_0}(x^{(0)}, t_0, \xi^{(0)}, \sigma_0)$, $t \ge t_0$, *belongs to* Σ.

c) *If* $(x^{(0)}, \xi^{(0)}) \in WF(g_k)$, $0 \le k \le m-1$, *then the union of* m *forward null-bicharacteristics* $\gamma^+_{jk}(x^{(0)}, 0, \xi^{(0)}, \lambda_j(x^{(0)}, 0, \xi^{(0)}))$, $t \ge 0$, $1 \le j \le m$, *belongs to* Σ.

Then $WF(u) \cap \{t > 0\}$ *is contained in* Σ.

Proof: Let $\mathcal{P}_0 = (x^{(0)}, t_0, \xi^{(0)}, \sigma_0) \notin \Sigma$, where $t_0 > 0$. If $\xi^{(0)} = 0$ or $\sigma_0 \ne \lambda_k(x^{(0)}, t_0, \xi^{(0)})$, $1 \le k \le m$, then H_0 is microelliptic at $\mathcal{P}_0$. Since $\mathcal{P}_0 \notin \Sigma$, we have that $\mathcal{P} \notin WF(f_+)$ and therefore $\mathcal{P} \notin WF(u)$. Suppose $\xi^{(0)} \ne 0$ and $\sigma_0 = \lambda_{j_0}(x^{(0)}, t_0, \xi^{(0)})$ for some $1 \le j_0 \le m$. Denote by γ_- the backward null-bicharacteristic starting at $(x^{(0)}, t_0, \xi^{(0)}, \sigma_0)$ provided that $\sigma_0 = \lambda_{j_0}$. Let $(x^{(1)}, 0, \xi^{(1)}, \lambda_{j_0}(x^{(1)}, 0, \xi^{(1)}))$ be the point on γ_- where $t = 0$. Since $\mathcal{P} \notin \Sigma$, we have that $(x^{(1)}, \xi^{(1)}) \notin WF(g_k(x))$ for $0 \le k \le m-1$.

We show that $(x^{(1)}, 0, \xi^{(1)}, \lambda_{j_0}(x^{(1)}, 0, \xi^{(1)})) \notin WF(b_k g_k(x) \delta^{(m-k-1)}(t))$ for all $0 \le k \le m-1$.

Let $\beta(x,t) \in C_0^\infty(\mathbb{R}^{n+1})$, $\beta(x^{(1)}, 0) \ne 0$, and $\beta(x,t)$ is zero outside of a small neighborhood of $(x^{(1)}, 0)$. Then $F(\beta b_k g_k \delta^{(m-k-1)}(t)) = \sum_{p=0}^{m-k-1} \sigma^p \beta_p(\xi)$, where $|\beta_p(\xi)| \le C_N (1 + |\xi|)^{-N}$, $\forall N$, in a conic neighborhood of $\xi^{(1)}$, since $(x^{(1)}, \xi^{(1)}) \notin WF(g_k), k = 1, \dots, m$. Therefore $|\sum_{p=0}^{m-k-1} \sigma^p \beta_p(\xi)| \le C(1 + |\xi| + |\sigma|)^{-N}$, $\forall N$, in a conic neighborhood of $(\xi^{(1)}, \lambda_{j_0}(x^{(1)}, 0, \xi^{(1)}))$ since $|\xi| \ge C(|\xi| + |\sigma|)$ in this neighborhood. This proves that

$$(x^{(1)}, 0, \xi^{(1)}, \lambda_{j_0}(x^{(1)}, 0, \xi^{(1)})) \notin WF(b_k g_k(x) \delta^{(m-k-1)}(t))$$

for all $0 \le k \le m-1$. Therefore, the wave front set of the right hand side of (50.36) does not intersect γ_- for $-\infty < t \le t_0$. Since $u_+ = 0$ for $t < 0$, we get from Theorem 50.1 that $WF(u_+) \cap \gamma_- = \emptyset$ for $t \le t_0$, i.e., $\mathcal{P}_0 \notin WF(u)$. □

Remark 50.2. Consider the Cauchy problem (48.1), (48.2) assuming that $f(x,t) \in C^\infty(\overline{\mathbb{R}^{n+1}_+})$. Then, by the partial hypoellipticity (cf. Proposition 7.6), the solution $u(x,t)$ is a smooth function of $t \ge 0$ with the distribution values in $\mathcal{D}'(\mathbb{R}^n)$. We describe the wave front set of $u(x,t) \subset \mathbb{R}^n \times (\mathbb{R}^n \setminus \{0\})$ for a fixed $t > 0$. It follows from Theorems 50.1 and 50.5 that if

$(x^{(0)}, t_0, \xi^{(0)}, \sigma^{(0)}) \in WF(u)$, $t_0 > 0$, then $\xi^{(0)} \neq 0$ and $\sigma_0 = \lambda_{j0}(x^{(0)}, t_0, \xi^{(0)})$ for some $1 \leq j_0 \leq m$.

Denote by $W(t_0) \subset \mathbb{R}^n \times (\mathbb{R}^n \setminus \{0\})$ the set consisting of all $(x^{(0)}, \xi^{(0)})$ such that $(x^{(0)}, t_0, \xi^{(0)}, \sigma_0) \in WF(u)$ for some $\sigma_0 \in \mathbb{R}^1$.

Lemma 50.6. *The wave front set $WF(u(\cdot, t_0))$ is contained in $W(t_0)$, $t_0 > 0$.*

Proof. Suppose $(x^{(0)}, \xi^{(0)}) \notin W(t_0)$, i.e., $(x^{(0)}, t_0, \xi^{(0)}, \sigma) \notin WF(u)$ for all $\sigma \in \mathbb{R}$. Note that $(x^{(0)}, t_0, 0, \pm 1) \notin WF(u)$ since $H_0(x^{(0)}, t_0, 0, \pm 1) \neq 0$ and $f \in C^\infty$ for $t > 0$. Then there exists $\varphi(x) \in C_0^\infty(\mathbb{R}^n)$, $\varphi(x_0) \neq 0$, $\alpha(\xi) \in C^\infty(\mathbb{R}^n \setminus \{0\})$, $\deg_\xi \alpha(\xi) = 0$, $\alpha(\xi^{(0)}) \neq 0$, and a small $\delta > 0$ such that $\alpha(D_x)\varphi(x)\chi(\frac{t-t_0}{\delta})u \in C^\infty(\mathbb{R}^{n+1})$. In particular $\alpha(D_x)\varphi(x)u(x, t_0) \in C^\infty(\mathbb{R}^n)$, i.e., $(x^{(0)}, \xi^{(0)}) \notin WF(u(\cdot, t_0))$. Therefore $WF(u(\cdot, t_0))$ is contained in $W(t_0)$. $\square$

When $m \geq 2$, it may happen that $WF(u(., t_0)) \neq W(t_0)$. For example, let $\frac{\partial^2 u(x,t)}{\partial t^2} - \Delta u(x, t) = 0$ in $\mathbb{R}^{n+1}$, $u(x, t_0) = 0$, $\frac{\partial u(x,t_0)}{\partial t} = \delta(x)$. Then $WF(u(\cdot, t_0)) = \emptyset$ and $(0, t_0, \xi^{(0)}, \pm|\xi^{(0)}|) \in WF(u)$ for any $\xi^{(0)} \neq 0$, i.e., $(0, \xi^{(0)}) \in W(t_0)$. However, in the case $m = 1$, i.e., when $H(x, t, D_x, D_t) = D_t - \lambda(x, t, D_x)$, we have $WF(u(\cdot, t_0)) = W(t_0)$. To prove this take any $(x^{(0)}, \xi^{(0)}) \notin WF(u(\cdot, t_0))$. Then there exist $\psi(x) \in C_0^\infty(\mathbb{R}^n)$ and $\alpha(\xi) \in C^\infty(\mathbb{R}^n \setminus \{0\})$ such that $\psi(x^{(0)}) \neq 0$, $\alpha(\xi^{(0)}) \neq 0$, $\alpha(\xi)$ is homogeneous of degree zero, and

$$\psi(x)\alpha(D_x)(1 - \chi(D_x))u(x, t_0) \in C^\infty(\mathbb{R}^n).$$

When $m = 1$, one can take the solution $\varphi(x, t, \xi)$ of (50.15) independent of σ. Therefore, taking $\varphi|_{t=t_0} = \psi(x)\alpha(\xi)(1 - \chi(\xi))$, we get by Theorem 50.1 that $\chi(\frac{t-t_0}{\delta})\varphi(x, t, D_x)u \in C^\infty(\mathbb{R}^{n+1})$. Thus, $(x^{(0)}, t_0, \xi^{(0)}, \sigma) \notin WF(u)$ for any $\sigma \in \mathbb{R}$, i.e., $(x^{(0)}, \xi^{(0)}) \notin W(t_0)$. Hence, $W(t_0) \subset WF(u(\cdot, t_0))$. Combining this with Lemma 50.6, we get $WF(u(\cdot, t_0)) = W(t_0)$ for $m = 1$.

Example 50.1. Consider the Cauchy problem

$$\frac{\partial u(x,t)}{\partial t} = f(x,t), \quad t > 0,\ x \in \mathbb{R}^n, \qquad u(x, 0) = 0,$$

where $f(x, t) = h(t)\delta(x)$, $h(t) \in C_0^\infty(\mathbb{R}_+^1)$, $\operatorname{supp} h(t) = [1, 2] \cup [3, 4]$, and $\int_1^2 h(t)dt \neq 0$, $\int_1^2 h(t)dt + \int_3^4 h(t)dt = 0$. Then $u(x, t) = \delta(x)\int_{-\infty}^t h(\tau)d\tau$ is the solution of the Cauchy problem. We have $WF(f) = \{0, t, \xi, 0\}$, $\forall \xi \neq 0$ and $\forall t \in [1, 2] \cup [3, 4]$, and $WF(u) = \{(0, t, \xi, 0)\}$, $\forall \xi \neq 0$ and $\forall t \in [1, 4]$. Note that here the set Σ defined in Theorem 50.5 is equal to $\{0, t, \xi, 0\}$, $\forall \xi \neq 0$ and $\forall t \in [1, +\infty)$, i.e., $WF(u) \neq \Sigma$.

51. Problems

1. Consider a boundary value problem in the half-space $\mathbb{R}^n_+$:

(51.1)
$$\sum_{k=1}^{n} \frac{\partial^2 u(x', x_n)}{\partial x_k^2} - u(x', x_n) = 0, \quad x_n > 0,$$

(51.2)
$$-\frac{\partial u(x, x_n)}{\partial x_n} + \sum_{k=1}^{n-1} b_k(x') \frac{\partial u(x', x_n)}{\partial x_k} + \lambda u \Big|_{x_n=0} = g(x'),$$

where $x' = (x_1, \ldots, x_{n-1})$ and $b_k(x') \in C_0^\infty(\mathbb{R}^{n-1})$ are real-valued functions, $1 \le k \le n-1$, $\lambda > 0$.

a) Reduce the boundary value problem (51.1), (51.2) to a pseudodifferential equation $A(x', D')v(x') = g(x')$ in $\mathbb{R}^{n-1}$, where $v(x') = u(x', 0)$.

b) Find the adjoint operator $A^*(x', D')$.

c) Show that $A(x', D')v = g$ has a unique solution in $H_s(\mathbb{R}^{n-1})$, $s \in \mathbb{R}^1$, for any $g \in H_{s-1}(\mathbb{R}^{n-1})$, assuming that $\lambda > 0$ is sufficiently large.

d) Find the first two terms of the parametrix to $A(x', D')$.

2. Denote by L^m the class of symbols $A(x, \xi) \in C^\infty(\mathbb{R}^n \times \mathbb{R}^n)$ such that

(51.3)
$$\left| \frac{\partial^{p+k} A(x, \xi)}{\partial x^p \partial \xi^k} \right| \le C_{pk}(1 + |\xi| + |x|)^{m-|p|-|k|}, \quad \forall p, k.$$

An example of a symbol of class L^m is the symbol of the harmonic oscillator $-\frac{d^2}{dx_1^2} + x_1^2$.

a) Let $E(x, y)$ be the Schwartz kernel of $A(x, D)$ (cf. §44), where $A(x, \xi) \in L^{-N}$ and $N > 0$ is large. Prove that

(51.4)
$$\left| \frac{\partial^{p_1+p_2} E(x, y)}{\partial x^{p_1} \partial y^{p_2}} \right| \le C_N(1 + |x|)^{-N_1}(1 + |y|)^{-N_1},$$

$0 \le |p_1| \le N_1, 0 \le |p_2| \le N_1$, where N_1 is large if N is large.

b) Prove that operator $A(x, D)$ is compact in $L_2(\mathbb{R}^2)$ if $A(x, \xi) \in L^{-\delta}, \delta > 0$.

c) We say that $A(x, \xi) \in L^m$ is elliptic if $|A(x, \xi)| \ge C(|x| + |\xi|)^m$ for $|x|^2 + |\xi|^2 \ge R^2$, $R > 0$. Prove that an elliptic operator with symbol $A(x, \xi) \in L^0$ is Fredholm in $L_2(\mathbb{R}^n)$.

d) Let $A(x, \xi) \in L^m$ be elliptic. Construct a parametrix to $A(x, D)$, i.e., a ψdo $R_N(x, D), R_N(x, \xi) \in L^{-m}$ such that $A(x, D)R_N(x, D) = I + T_{-N}$, where the kernel of T_{-N} satisfies estimates (51.4).

3. Suppose $A(x,\xi) = \sum_{k=0}^{2m} A_k(x,\xi)$ is a polynomial in x and in ξ. Let $\zeta = (x,\xi)$ and let $A_k(x,\xi) = \sum_{|p|=k} a_{pk}\zeta^p$, $0 \le k \le 2m$. Suppose $A_{2m}(x,\xi) = \sum_{|p|=2m} a_{p0}\zeta^p > 0$ when $\zeta = (x,\xi) \ne (0,0)$. Let $G(x,y,t)$ be the heat kernel for $A(x,D)$, i.e.,

$$\frac{\partial G}{\partial t} + A(x,D)G = 0 \quad \text{for} \quad t > 0, \tag{51.5}$$

$$G(x,y,0) = \delta(x-y). \tag{51.6}$$

Find the principal term of the asymptotics of the heat trace $\int_{\mathbb{R}^n} G(x,x,t)dx$ as $t \to 0$. Consider the particular case where $m = 1$, i.e., $A(x,\xi) = Q(\zeta,\zeta)$, where $Q(\zeta,\zeta)$ is a positive definite quadratic form in ζ.

4. Parabolic equations of higher order in D_t. Let $P_0(x,t,\xi,\sigma)$ be a quasi-homogeneous polynomial in (ξ,σ):

$$P_0(x,t,\alpha\xi,\alpha^b\sigma) = \alpha^m P_0(x,t,\xi,\sigma), \quad \forall\alpha > 0,$$

where $b > 1$, $r = \frac{m}{b}$ is an integer and r is the degree of P_0 in σ.

Such a P_0 is called a parabolic symbol if

$$|P_0(x,t,\xi,\sigma - i\tau)| \ge C(|\xi|^m + |\sigma|^{\frac{m}{b}} + \tau^{\frac{m}{b}}) \tag{51.7}$$

for all $\tau \ge 0$.

An equivalent definition of a parabolic symbol is that all the roots $\sigma = \mu_j(x,t,\xi)$, $1 \le j \le \frac{m}{b}$ of $P_0(x,t,\xi,\sigma) = 0$ have an estimate

$$\operatorname{Im}\mu_j(x,t,\xi) \ge C|\xi|^b, \ 1 \le j \le r, \quad r = \frac{m}{b}.$$

An example of parabolic symbol with $r > 1$ is $(i\sigma + |\xi|^2)^2$.

Consider the Cauchy problem

$$P(x,t,D_x,D_t)u(x,t) = f(x,t), \ \ t > 0, \ x \in \mathbb{R}^n, \tag{51.8}$$

$$\frac{\partial^k u(x,0)}{\partial t^k} = g_k(x), \ x \in \mathbb{R}^n, \ 0 \le k \le r-1, \tag{51.9}$$

where $P(x,t,\xi,\sigma) = P_0(x,t,\xi,\sigma) + P_1(x,t,\xi,\sigma)$, P_0 is parabolic, $r = \frac{m}{b}$, and $|P_1(x,t,\xi,\sigma - i\tau)| \le C(1 + |\xi| + |\sigma - i\tau|^{\frac{1}{b}})^{m-1}$.

Extend the results of §46 to this class of parabolic equations of order $r \ge 2$.

5. Elliptic pseudodifferential systems. Let $A(x,D)u(x) = f(x)$ be an $m \times m$ system of pseudodifferential equations where

$$u(x) = (u_1(x),\dots,u_m(x)), \ f = (f_1,\dots,f_m(x)), \quad x \in \mathbb{R}^n,$$

$$A(x,D) = [a_{jk}(x,D)]_{j,k=1}^m, \ a_{jk}(x,\xi) \in S^r(\mathbb{R}^n \times \mathbb{R}^n).$$

Such an $A(x, D)$ is called elliptic in $\mathbb{R}^n$ if

$$|\det A(x,\xi)| \geq C(1+|\xi|)^{mr} \quad \text{when } |x|^2+|\xi|^2 \geq R^2.$$

Prove that for each $s \in \mathbb{R}$, $A(x, D)$ is a Fredholm operator from $H_s(\mathbb{R}^n)$ to $H_{s-r}(\mathbb{R}^n)$.

6. **Symmetric first order hyperbolic system.** Consider the Cauchy problem for the first order symmetric hyperbolic system

$$D_t u(x,t) + A(x,t,D_x)u(x,t) = f(x,t), \quad x \in \mathbb{R}^n, \ t \in [0,T], \tag{51.10}$$

$$u(x,0) = g(x), \quad x \in \mathbb{R}^n, \tag{51.11}$$

where $A(x,t,\xi) = [a_{ij}(x,t,\xi)]_{i,j=1}^m$, $a_{ij} \in S^1(\mathbb{R}^{n+1} \times \mathbb{R}^n)$ and

$$A^*(x,t,\xi) - A(x,t,\xi) = A_1 \in S^0. \tag{51.12}$$

a) Assuming that $u(x,t) \in H_{1,s}(\mathbb{R}^n \times (0,T))$, $g \in H_s(\mathbb{R}^n)$ and $f \in L^1((0,T), H_s(\mathbb{R}^n)$, prove the estimate

$$\max_{0\leq t\leq T}[u(\cdot,t)]_s \leq C[g]_s + C\int_0^T [f(\cdot,t')]_s dt' \tag{51.13}$$

b) Prove the uniqueness for the Cauchy problem (51.10), (51.11) in $H_{0,s}(\mathbb{R}^n)$, $\forall s \in \mathbb{R}$.

c) Prove the existence of the solution of (51.10), (51.11) in $C((0,T), H_s(\mathbb{R}^n))$ using the Friedrichs regularization (see Taylor [**T2**]):
Let $A_\varepsilon(x,t,D_x) = A(x,t,D_x)\chi(\varepsilon D_x)$, $\chi(\xi) \in C_0^\infty(\mathbb{R}^n)$, and $\chi(\xi) = 1$ for $|\xi| \leq 1$. Prove the existence of u_ε and the estimate (51.13) with A replaced by A_ε. Then use the weak compactness to pass to the limit as $\varepsilon \to 0$.

7. **Strictly hyperbolic system of the first order.** Consider the Cauchy problem

$$D_t u(x,t) + L(x,t,D_x)u = f(x,t), \quad x \in \mathbb{R}^n, \ t \in [0,T], \tag{51.14}$$

$$u(x,0) = g(x), \quad x \in \mathbb{R}^n. \tag{51.15}$$

The system (51.14) is called strictly hyperbolic if $L(x,t,\xi) \in S^1$, $L(x,t,\xi) = L_0(x,t,\xi) + L_1(x,t,\xi)$ for $|\xi| \geq 1$, where $L_0(x,t,\xi)$ is homogeneous in ξ of order one, $L_1(x,t,\xi) \in S^0$, and $L_0(x,t,\xi)$ has m distinct real eigenvalues $\lambda_j(x,t,\xi)$, $1 \leq j \leq m$, for all $(x,t) \in \mathbb{R}^{n+1}$, $\xi \neq 0$.

a) Show that there exists $B_0(x,t,\xi)$, homogeneous of degree zero, $\det B_0(x,\xi) \neq 0, \forall (x,t)$, $\xi \neq 0$, such that $L_0(x,\xi) = B_0^{-1}D_0B_0$, where $D_0 = [\lambda_j(x,t,\xi)\delta_{jk}]_{j,k=1}^m$ is a diagonal matrix. Prove

that $Q_0(x,t,\xi) = B_0^*(x,t,\xi)B_0(x,t,\xi)$ has the following property:
$Q_0(x,t,\xi)L_0(x,t,\xi) = L_0^*(x,t,\xi)Q_0(x,t,\xi)$, i.e., Q_0L_0 is a symmetric matrix.

b) Using the Gårding inequality (cf. §48), prove that

$$Q_0(x,t,D_x)(1-\chi(D_x)) + C\Lambda^{-1} \tag{51.16}$$

is a positive definite operator when $C > 0$ is large, $\Lambda^{-1}(\xi) = (1+|\xi|^2)^{-\frac{1}{2}}$.

c) Use (51.16) to reduce the study of the Cauchy problem (51.14), (51.15) for the strictly hyperbolic system to the case of symmetric hyperbolic system (51.10), (51.11).

d) Consider the strictly hyperbolic equation of order m (cf. §48):

$$H(x,t,D_x,D_t)u = f. \tag{51.17}$$

Setting $v_j = D_t^j\Lambda^{m-j}u, 1 \le j \le m$, reduce (51.17) to the strictly hyperbolic system of the first order.

8. **Beals-Fefferman classes of ψdo's** (see Beals [**B**] and Beals-Fefferman [**BF**]). Let $\Phi(x,\xi), \varphi(x,\xi)$ be C^∞ functions with the following properties:

a) $c \le \Phi(x,\xi) \le C(1+|\xi|),\ C \ge |\varphi(x,\xi)| \ge C(1+|\xi|)^{\varepsilon-1}, \varepsilon > 0.$

b) $\Phi\varphi \ge C.$

c) $\frac{\Phi(x,\xi)}{\varphi(x,\xi)} \sim \frac{\Phi(y,\eta)}{\varphi(y,\eta)}$ if $|\xi| \sim |\eta|$, where $A \sim B$ means $c_1 \le \frac{A}{B} \le c_2$.

d) $\left|\frac{\partial^{p+q}\Phi(x,\xi)}{\partial x^p\partial\xi^q}\right| \le C_{pq}\Phi^{1-|q|}\varphi^{-|p|}, \left|\frac{\partial^{p+q}\varphi(x,\xi)}{\partial x^p\partial\xi^q}\right| \le C_{pq}\Phi^{-|q|}\varphi^{1-|p|},$ $\forall p,q.$

We say that the symbol $A(x,\xi)$ belongs to $S_{\Phi,\varphi}^{M,m}$ if $A(x,\xi) \in C^\infty(\mathbb{R}^n\times\mathbb{R}^n)$ and $\left|\frac{\partial^{p+q}A(x,\xi)}{\partial x^p\partial\xi^q}\right| \le C_{pq}\Phi^{M-|q|}\varphi^{m-|p|}$, $\forall p,q$. A particular case of the class $S_{\Phi,\varphi}^{M,m}$ is Hörmander's class $S_{\rho,\delta}^m$ of symbols $A(x,\xi) \in C^\infty(\mathbb{R}^n\times\mathbb{R}^n)$ satisfying

$$\left|\frac{\partial^{p+q}A(x,\xi)}{\partial x^p\partial\xi^q}\right| \le C_{pq}(1+|\xi|)^{m-\rho|q|+\delta|p|},$$

where $0 \le \delta < \rho \le 1$. Here $\Phi = (1+|\xi|^2)^{\frac{\rho}{2}}$, $\varphi = (1+|\xi|^2)^{-\frac{\delta}{2}}$. The following theorem is proven in Beals-Fefferman [**BF**]:

Theorem 51.1. *If $A(x,\xi) \in S_{\Phi,\varphi}^{0,0}$, then*

$$\|A(x,D)u\|_0 \le C\|u\|_0, \quad \forall u \in C_0^\infty(\mathbb{R}^n). \tag{51.18}$$

Prove that Theorems 40.2 and 43.2 hold for $A(x,\xi) \in S_{\Phi,\varphi}^{M_1,m_1}$, $B(x,\xi) \in S_{\Phi,\varphi}^{M_2,m_2}$

9. The sharp Gårding inequality (see Beals [**B**] and Beals-Fefferman [**BF**]). Let $A(x,\xi) \in S^m$ and $A(x,\xi) \geq C > 0,\ m > 0$. Then

$$\operatorname{Re}(A(x,D)u,u) \geq -C\|u\|^2_{\frac{m-1}{2}}, \forall u \in C_0^\infty(\mathbb{R}^n). \tag{51.19}$$

One way to prove the sharp Gårding inequality is to use the Beals-Fefferman calculus.

Suppose $m = 1$. Define $\Phi(x,\xi) = (1+|\xi|^2)^{\frac{1}{4}}\sqrt{A(x,\xi)}$, $\varphi(x,\xi) = (1+|\xi|^2)^{-\frac{1}{4}}\sqrt{A(x,\xi)}$.

a) Prove that $\Phi(x,\xi)$ and $\varphi(x,\xi)$ satisfy a)–d) and that $A(x,\xi) \in S^{1,1}_{\Phi,\varphi}$ and $\sqrt{A(x,\xi)} \in S^{\frac{1}{2},\frac{1}{2}}_{\Phi,\varphi}$.

b) Complete the proof of (51.19).

Chapter VII

Elliptic Boundary Value Problems and Parametrices

Introduction to Chapter VII

We start with the Fredholm theory of ψdo's on a manifold without boundary (§52). In §§53 and 54 we study the elliptic boundary value problem for the elliptic equations of order m following the author's book (see Eskin [**E1**]). In §55 we construct the parametrix for such boundary value problems using "plus" and "minus" ψdo's as in [**E1**] and [**E2**]. The approach of §55 is used in §56 for the derivation of the heat trace asymptotics as $t \to 0$ in the case of elliptic boundary value problems. For the Laplace-Beltrami equations with either Dirichlet or Neumann boundary condition we find explicitly the two leading terms of the heat trace asymptotics. In §57 another application of the parametrix construction of §55 is the parametrix for the Dirichlet-to-Neumann operator in the case of the Laplace-Beltrami operator (see Lee-Uhlmann [**LU**]). In §58 we study the heat trace expansions for the strongly elliptic operators in the case of boundary value problems and in the case of a manifold without boundary. In the selfadjoint case we give conditions for the uniform convergence of the eigenfunction expansions. Finally, §59 is devoted to the Atiyah-Singer index theorem in $\mathbb{R}^n$. We follow the works of Atiyah-Singer [**AtS1**], [**AtS2**], and Seeley [**Se3**], who proved this result independently. At the end of §59 we present, without proof, Fedosov's explicit formula for the index (see Fedosov [**F1**]).

52. Pseudodifferential operators on a manifold

52.1. Manifolds and vector bundles.

Let M be an n-dimensional C^∞ manifold, i.e., M is a topological space such that any point $x_0 \in M$ has a neighborhood U_0 that is homeomorphic to a neighborhood $B_0 \subset \mathbb{R}^n$. Let $y = s_0(x)$ be the map of U_0 onto B_0. We call $y = (y_1, \ldots, y_n)$ local coordinates in U_0. If U_j and U_k are two neighborhoods in M such that $U_j \cap U_k \neq \emptyset$, and if $y^{(j)} = s_j(x)$, $y^{(k)} = s_k(x)$ are local coordinates in U_j and U_k respectively, then $y^{(j)} = s_{jk}(y^{(k)}) = s_j(s_k^{-1}(y^{(k)}))$ is the map of $s_k(U_j \cap U_k) \subset \mathbb{R}^n$ onto $s_j(U_j \cap U_k) \in \mathbb{R}^n$.

The manifold M is called a C^∞ manifold if $s_{jk}(y^{(k)}) \in C^\infty$ and the Jacobian $\det \frac{Ds_{jk}(y^{(k)})}{Dy^{(k)}} \neq 0$. In this section we assume that M is compact. A function $\varphi(x)$ on M is called a C^∞ function if $\varphi(s_j^{-1}(y^{(j)})) \in C^\infty$ for any neighborhood U_j.

By definition, the vector bundle V on M is a topological space with the following properties. There exists a continuous projection p of V onto M, and for each $x_0 \in M$ there is a neighborhood U_0 such that the preimage $p^{-1}(U_0)$ is homeomorphic to $U_0 \times \mathbb{R}^m$ for some $m \geq 1$: $\Phi_0(p^{-1}(U_0)) = U_0 \times \mathbb{R}^m$. Moreover, if U_j and U_k are two such neighborhoods, $U_j \cap U_k \neq \emptyset$, and Φ_j and Φ_k are the corresponding maps on $p^{-1}(U_j)$ and $p^{-1}(U_k)$, respectively, then for any $(x, \xi) \in (U_j \cap U_k) \times \mathbb{R}^m$ we have

$$(52.1) \qquad (y, \eta) = \Phi_j \Phi_k^{-1}(x, \xi),$$

where $y = x$, $\eta = c_{jk}(x)\xi$, and $c_{jk}(x)$ is an invertible linear map in $\mathbb{R}^m$. The matrices $c_{jk}(x)$ are called transition matrices of the bundle V. Note that the $c_{jk}(x)$, $x \in U_j \cap U_k$, have the following properties: $c_{kj}(x) = c_{jk}^{-1}(x)$, $c_{jk}(x)c_{kp}(x) = c_{jp}(x)$ when $x \in U_j \cap U_k \cap U_p$. In the local coordinates $y^{(j)} = s_j(x)$, $x \in U_j$, and $y^{(k)} = s_k(x)$, $x \in U_k$, we get a map of $s_k(U_j \cap U_k) \times \mathbb{R}^m$ onto $s_j(U_j \cap U_k) \times \mathbb{R}^m$ of the form

$$(52.2) \qquad y^{(j)} = s_{jk}(y^{(k)}), \quad \eta^{(j)} = \hat{c}_{jk}(y^{(k)})\eta^{(k)},$$

where $s_{jk}(y^{(k)}) = s_j(s_k^{-1}(y^{(k)}))$, $\hat{c}_{jk}(y^{(k)}) = c_{jk}(s_k^{-1}(y^{(k)})) \in C^\infty$. We call $\xi_x \in p^{-1}(x)$ a fiber at the point $x \in M$ and we write the elements of V in the form (x, ξ_x).

An example of a vector bundle is the cotangent bundle $T^*(M)$, where $m = n$ and the transition matrices $\hat{c}_{jk}(y^{(k)})$ (cf. §45) have the form

$$(52.3) \qquad \hat{c}_{jk}(y^{(k)}) = \left(\left(\frac{\partial s_{jk}(y_k)}{\partial y^{(k)}} \right)^T \right)^{-1}.$$

Since M is assumed to be compact, there exists a finite cover of M by neighborhoods $U_j, 1 \le j \le N$.

Let $\{\varphi_j(x), U_j\}_{j=1}^N$ be a partition of unity, that is, $\operatorname{supp}\varphi_j \subset U_j$, $\sum_{j=1}^N \varphi_j = 1$ on M and $\varphi_j \in C^\infty(M)$. For $u(x) \in C^\infty(M)$ we introduce a Sobolev norm

$$\|u\|_s^2 = \sum_{j=1}^N \|\varphi_j(s_j^{-1}(y^{(j)}))u(s_j^{-1}(y^{(j)}))\|_s^2, \tag{52.4}$$

where $\|\varphi_j(s_j^{-1}(y^{(j)}))u(s_j^{-1}(y^{(j)}))\|_s$ is the norm in $H_s(\mathbb{R}^n)$.

We denote by $H_s(M)$ the closure of $C^\infty(M)$ in the norm (52.4). It is easy to prove (cf. §13) that if we take another partition of unity and another set of local coordinates, we get a norm that is equivalent to (52.4).

52.2. Definition of a pseudodifferential operator on a manifold. We say that a linear operator T is of order $\le \alpha$ if

$$\|Tu\|_s \le C\|u\|_{s+\alpha}, \quad u \in C^\infty(M), \tag{52.5}$$

for any $s \in \mathbb{R}$.

We call A a pseudodifferential operator (ψdo) on M if the following two conditions hold:

a) If $\psi(x) \in C^\infty(M)$, $\varphi(x) \in C^\infty(M)$ and $\operatorname{supp}\psi \cap \operatorname{supp}\varphi = \emptyset$, then $\psi A\varphi$ is an operator of order $-\infty$.

b) If $\operatorname{supp}\varphi(x)$ and $\operatorname{supp}\psi(x)$ belong to the same neighborhood U_0, then, for any N,

$$\psi A\varphi u = \sum_{j=0}^N \psi A^{(j)}\varphi u + \psi T_{\alpha-N-1}\varphi u,$$

where $\operatorname{ord} T_{\alpha-N-1} \le \alpha - N - 1$ and $A^{(j)}$ are ψdo's in local coordinates $y^{(0)} = s_0(x)$, $x \in U_0$. We assume that the symbol of $A^{(j)}$ is equal to $A^{(j)}(y^{(0)}, \eta^{(0)})(1 - \chi(\eta^{(0)}))$ in local coordinates, $A^{(j)}(y^{(0)}, \eta^{(0)})$ is homogeneous in $\eta^{(0)}$ of degree $\alpha - j$ and is C^∞ when $\eta^{(0)} \ne 0$. We call $A^{(0)}(y^{(0)}, \eta^{(0)})$ the principal symbol of A in local coordinates in U_0.

Now we define the principal symbol of a ψdo on M. If $U_1 \cap U_2 \ne \emptyset$ and if $A_p^{(0)}$, $p = 1, 2$, are the principal symbols of A in U_p, $p = 1, 2$, then we have (cf. §45):

$$\psi(s_1^{-1}(y^{(1)}))A_1^{(0)}(y^{(1)}, \xi^{(1)})\varphi(s_1^{-1}(y^{(1)})) \\ = \psi(s_2^{-1}(y^{(2)}))A_2^{(0)}(y^{(2)}, \xi^{(2)})\varphi(s_2^{-1}(y^{(2)}))$$

in $U_1 \cap U_2$, where

$$y^{(2)} = s_{12}(y^{(1)}),\ \xi^{(2)} = \Big(\Big(\frac{\partial s_{12}(y^{(1)})}{\partial y^{(1)}}\Big)^T\Big)^{-1}\xi^{(1)},\ s_{12}(y^{(1)}) = s_2(s_1^{-1}(y^{(1)})).$$

Therefore, the operator $A^{(0)}$ determines a function $A_0(x, \xi_x)$ on $T_0^*(M) = \{(x, \xi_x) \in T^*(M), \xi_x \neq 0\}$. The symbol $A_j^{(0)}(y^{(j)}, \xi^{(j)})$ is the realization of this function in local coordinates in U_j.

We call $A^{(0)}(x, \xi_x)$ the principal symbol of the ψdo on M. Note that $A^{(0)}(x, \xi_x)$ is homogeneous in ξ_x of degree α. Vice versa, given such $A^{(0)}(x, \xi_x)$, we can construct a ψdo on M whose principal symbol is $A^{(0)}(x, \xi_x)$. The operator A has the form $Au = \sum_{j=1}^N \psi_j A_j^{(0)} \varphi_j u$, where $A_j^{(0)}$ is a ψdo in $\mathbb{R}^n$ with symbol $A_j^{(0)}(y^{(j)}, \xi^{(j)})(1 - \chi(\xi^{(j)}))$, $A_j^{(0)}(y^{(j)}, \xi^{(j)})$ is the realization of $A^{(0)}(x, \xi_x)$ in the local system of coordinates $y^{(j)}$ in U_j, $\psi_j(x) \in C_0^\infty(U_j)$, $\operatorname{supp}(1 - \psi_j) \cap \operatorname{supp} \varphi_j = \emptyset$, and $\sum_{j=1}^N \varphi_j = 1$.

52.3. Elliptic ψdo's on a manifold.

It follows from Theorems 40.1, 40.2 and §45 that A is an operator of order α. Also if A and B are two ψdo's with principal symbols $A^{(0)}(x, \xi_x)$, $B^{(0)}(x, \xi_x)$, $\deg_{\xi_x} A^{(0)} = \alpha$, $\deg_{\xi_x} B^{(0)} = \beta$, then

$$AB = C + T,$$

where $\operatorname{ord} T \leq \alpha + \beta - 1$ and C is a ψdo with principal symbol $C^{(0)}(x, \xi_x) = A^{(0)}(x, \xi_x) B^{(0)}(x, \xi_x)$.

Theorem 52.1. *Let A be an elliptic ψdo on M, i.e., $A^{(0)}(x, \xi_x) \neq 0$, $\forall (x, \xi_x) \in T_0^*(M)$. Then A is a Fredholm operator from $H_s(M)$ to $H_{s-\alpha}(M)$ for all s.*

Proof: Denote by R the ψdo with the principal symbol $R^{(0)}(x, \xi_x) = (A^{(0)}(x, \xi_x))^{-1}$. Then $AR = I + T^{(1)}$, $RA = I + T^{(2)}$, where $\operatorname{ord} T^{(p)} \leq -1$, $p = 1, 2$.

It follows from §42 that any operator of negative order is compact in $H_s(M)$ for all s. Therefore, R is the left and the right regularizer of A and hence A is Fredholm.

53. Boundary value problems in the half-space

53.1. Factorization of an elliptic symbol.

Let $\mathbb{R}_+^n = \{(x', x_n) : x_n > 0, x' \in \mathbb{R}^{n-1}\}$. Let $A_0(\xi', \xi_n)$ be a homogeneous polynomial of degree m. The polynomial A_0 is elliptic if $A_0(\xi', \xi_n) \neq 0$ for all $(\xi', \xi_n) \neq (0, 0)$. Assuming that $A_0(0, +1) = 1$, we obtain that

$A_0(\xi',\xi_n) = \prod_{j=1}^m(\xi_n - \lambda_j(\xi'))$, where $\deg_{\xi'}\lambda_j(\xi') = 1$ and $\operatorname{Im}\lambda_j(\xi') \neq 0$ when $\xi' \neq 0$. Denote

$$
\begin{aligned}
A_+(\xi',\xi_n) &= \prod_{j=1}^{m_+}(\xi_n - \lambda_j^+(\xi')),\\
A_-(\xi',\xi_n) &= \prod_{j=1}^{m_-}(\xi_n - \lambda_j^-(\xi')),
\end{aligned}
\tag{53.1}
$$

where $\operatorname{Im}\lambda_j^+(\xi')) > 0$, $1 \le j \le m_+$, $\operatorname{Im}\lambda_j^-(\xi')) < 0$, $1 \le j \le m_-$, $m_+ + m_- = m$, $\xi' \neq 0$. Note that $\lambda_j^\pm$ are continuous in ξ', $\xi' \neq 0$. If $n \ge 3$, then $\operatorname{Im}\lambda_j^+(\xi') > 0$ for all $\xi' \in \mathbb{R}^{n-1}\setminus\{0\}$ since $\mathbb{R}^{n-1}\setminus\{0\}$ is connected. Therefore m_+ (and analogously m_-) does not depend on ξ'. Since $A_0(-\xi',-\xi_n) = (-1)^m A_0(\xi',\xi_n)$, we have that $-\lambda_j^+(-\xi') = \lambda_{k(j)}^-(\xi')$ for some $k(j)$. Therefore $m_+ = m_-$ and $m = m_+ + m_-$ is even. If $n = 2$, this is not always true. For example, $A_0(\xi_1,\xi_2) = \xi_1 + i\xi_2$ is an elliptic polynomial. In this and the following sections we assume that $\deg A_0(\xi)$ is even also when $n = 2$, and that $m_+ = m_- = \frac{m}{2}$.

Since $A_0(\xi',\xi_n)$ vanishes when $(\xi',\xi_n) = (0,0)$, we replace it by $\hat{A}_0(\xi',\xi_n) = A_0((1+|\xi'|)\frac{\xi'}{|\xi'|},\xi_n)$.

Note that $\hat{A}_0(\xi',\xi_n) - A_0(\xi',\xi_n) = A_1(\xi',\xi_n)$, where $\operatorname{ord} A_1(\xi) \le m-1$ and $C_1(1+|\xi|)^m \le |\hat{A}_0(\xi)| \le C_2(1+|\xi|)^m$. Also for any homogeneous function $C(\xi',\xi_n)$ we let $\hat{C}(\xi',\xi_n) = C((1+|\xi'|)\frac{\xi'}{|\xi'|},\xi_n)$.

Let $B_{j0}(\xi',\xi_n)$ be homogeneous polynomials of degree m_j and let

$$\hat{B}_{j0}(\xi',\xi_n) = B_{j0}\Big((1+|\xi'|)\frac{\xi'}{|\xi'|},\xi_n\Big).$$

We consider the following boundary value problem in $\mathbb{R}^n_+$:

$$\hat{A}_0(D',D_n)u(x',x_n) = f(x',x_n), \quad x_n > 0, \tag{53.2}$$

$$\hat{B}_{j0}(D',D_n)\,u(x',x_n)\big|_{x_n=0} = g_j(x'), \quad 1 \le j \le m_+ = \frac{m}{2}, \tag{53.3}$$

where $f \in H_{s-m}(\mathbb{R}^n_+)$, $g_j(x') \in H_{s-m_j-\frac12}(\mathbb{R}^{n-1})$, $s > \max_{1\le j\le m_+}(m_j + \frac12)$. We require that $s > \max_{1\le j\le m_+}(m_j+\frac12)$ to ensure the existence of the restrictions (traces) $B_{j0}u|_{x_n=0}$ (cf. §22). We look for the solution $u \in H_s(\mathbb{R}^n_+)$. Let

$$\mathcal{H}_{(s)} = H_{s-m}(\mathbb{R}^n_+) \times \prod_{j=1}^{m_+} H_{s-m_j-\frac12}(\mathbb{R}^{n-1}).$$

The operator $\mathcal{A}_0 u = (\hat{A}_0 u|_{x_n>0}, \hat{B}_{j0}\,u|_{x_n=0}\,, 1 \le j \le m_+)$ is bounded from $H_s(\mathbb{R}^n_+)$ to $\mathcal{H}_{(s)}$.

53.2. Explicit solution of the boundary value problem. We find conditions for $\mathcal{A}_0$ to be invertible.

Let $lf \in H_{s-m}(\mathbb{R}^n)$ be an extension of $f(x', x_n)$ such that

$$\|lf\|_{s-m} \le C\|f\|_{s-m}^+,$$

where $\|f\|_{s-m}^+$ is the norm in $H_{s-m}(\mathbb{R}_+^n)$ (cf. §22). Then

$$u_0(x) = F^{-1}\frac{\widetilde{lf}(\xi)}{\hat{A}_0(\xi', \xi_n)} \tag{53.4}$$

is a solution of $\hat{A}_0(D', D_n)u_0 = f$ in $\mathbb{R}_+^n$. Let

$$v(x) = u(x) - u_0(x), \quad x_n > 0.$$

Then $v(x)$ satisfies

$$\hat{A}_0(D', D_n)v(x', x_n) = 0, \quad x_n > 0, \tag{53.5}$$

$$\hat{B}_{j0}(D', D_n)\, v(x', x_n)\big|_{x_n=0} = h_j(x'), \quad 1 \le j \le m_+, \tag{53.6}$$

where $h_j(x') = g_j(x') - \hat{B}_{j0}\, u_0|_{x_n=0} \in H_{s-m_j-\frac{1}{2}}(\mathbb{R}^{n-1})$, since $u_0 \in H_s(\mathbb{R}^n)$. Performing the Fourier transform in x', we get

$$\hat{A}_0(\xi', D_n)\tilde{v}(\xi', x_n) = 0, \tag{53.7}$$

$$\hat{B}_{j0}(\xi', D_n)\, \tilde{v}(\xi', x_n)\big|_{x_n=0} = \tilde{h}_j(\xi'). \tag{53.8}$$

Note that (53.7), (53.8) is the boundary value problem on $\mathbb{R}_+^1$ for an ordinary differential equation. The general solution of (53.7) belonging to $H_s(\mathbb{R}_+)$ has the form

$$\begin{aligned}\tilde{v}(\xi', x_n) &= \frac{1}{2\pi}\int_{-\infty}^{\infty} \frac{\sum_{k=1}^{m_+} c_k(\xi')\xi_n^{k-1}e^{ix_n\xi_n}}{\hat{A}_+(\xi', \xi_n)}d\xi_n \\ &= \frac{1}{2\pi}\int_{\gamma_+}\sum_{k=1}^{m_+}\frac{z^{k-1}c_k(\xi')e^{ix_nz}}{\hat{A}_+(\xi', z)}dz,\end{aligned} \tag{53.9}$$

where γ_+ is a simple closed contour in $\operatorname{Im} z > 0$ enclosing all zeros of $\hat{A}_+(\xi', z)$ (cf. Proposition 16.1 (the Jordan lemma)). If ξ' is such that all $\lambda_j^+(\xi')$ are simple roots, $1 \le j \le m_+ = \frac{m}{2}$, then computing (53.9) with the help of residues, we get

$$\tilde{v}(\xi', x_n) = \sum_{j=1}^{m_+} c_{j1}(\xi')e^{ix_n\hat{\lambda}_j^+(\xi')},$$

where $\hat{\lambda}_j^+(\xi') = \lambda_j^+((1+|\xi'|)\frac{\xi'}{|\xi'|})$. The advantage of representation (53.9) is that it does not require any assumptions on the multiplicity of roots. Substitution of (53.9) into (53.8) gives the following linear system for $c_k(\xi')$:

$$\sum_{k=1}^{m_+} \hat{b}_{jk}(\xi')c_k(\xi') = \tilde{h}_j(\xi'), \quad 1 \le j \le m_+, \tag{53.10}$$

where

$$b_{jk}(\xi') = \frac{1}{2\pi}\int_{\gamma_+} \frac{B_{j0}(\xi', z)z^{k-1}}{A_+(\xi', z)}dz, \tag{53.11}$$

$\hat{b}_{jk}(\xi') = b_{jk}((1+|\xi'|)\frac{\xi'}{|\xi'|})$. Note that $b_{jk}(\xi')$ are homogeneous functions of degree $m_j + k - m_+$.

We assume that

$$\det[b_{jk}(\xi')]_{j,k=1}^{m_+} \neq 0, \quad \forall \xi' \neq 0. \tag{53.12}$$

Let $[d_{kj}(\xi')]_{k,j=1}^{m_+}$ be the inverse matrix to $[b_{jk}(\xi')]_{j,k=1}^{m_+}$. Then

$$c_k(\xi') = \sum_{j=1}^{m_+} \hat{d}_{kj}(\xi')\tilde{h}_j(\xi'), \quad 1 \le k \le m_+, \tag{53.13}$$

and (53.9), (53.13) give the unique solution of the boundary value problem (53.7), (53.8). Note that $\deg_{\xi'} d_{kj}(\xi) = m_+ - k - m_j$. We show that

$$\|v\|_s^+ \le C\sum_{j=1}^{m_+}[h_j]_{s-m_j-\frac{1}{2}}, \tag{53.14}$$

where $[h_j]_t$ is the norm in $H_t(\mathbb{R}^{n-1})$. To prove (53.14), choose an integer $N \ge s$. Note that $\Lambda_-^N(\xi', z) = (z + i|\xi'|)^N$ has no zeros inside the contour γ_+.

We have

$$\tilde{v}(\xi', x_n) = \frac{1}{2\pi}\int_{\gamma_+}\sum_{k=1}^{m_+}\frac{\hat{\Lambda}_-^N(\xi', z)z^{k-1}c_k(\xi')e^{ix_nz}}{\hat{\Lambda}_-^N(\xi', z)\hat{A}_+(\xi', z)}dz,$$

where $\hat{\Lambda}_-^N(\xi', z) = (z + i(|\xi'|+1))^N$.

Denote by $p_k(\xi', z)$ the remainder of the division of $\Lambda_-^N(\xi', z)z^{k-1}$ by $A_+(\xi, z)$. Note that $p_k(\xi', z)$ is homogeneous in (ξ', z) of degree $N + k - 1$ and $p_k(\xi, z)$ is a polynomial in z of degree $\le m_+ - 1$:

$$\Lambda_-^N(\xi', z)z^{k-1} = q_k(\xi', z)A_+(\xi', z) + p_k(\xi', z), \tag{53.15}$$

where q_k is a polynomial in z. Note that

$$\int_{\gamma_+}\frac{\hat{q}_k(\xi', z)c_k(\xi')e^{ix_nz}}{\hat{\Lambda}_-^N(\xi', z)}dz = 0$$

since $\hat{\Lambda}_-^N(\xi', z)$ has no zeros inside γ_+. Using the Jordan lemma, we get for $x_n > 0$,

$$\tilde{v}(\xi', x_n) = \frac{1}{2\pi}\sum_{k=1}^{m_+}\int_{-\infty}^{\infty}\frac{\hat{p}_k(\xi', \xi_n)c_k(\xi')e^{ix_n\xi_n}}{\hat{\Lambda}_-^N(\xi', \xi_n)\hat{A}_+(\xi', \xi_n)}d\xi_n. \tag{53.16}$$

Extend $\tilde{v}(\xi', x_n)$ for $x_n < 0$ by the right hand side of (53.16) and perform the Fourier transform in x_n. We get

$$(\|v\|_s^+)^2 \leq C\sum_{k=1}^{m_+}\int_{\mathbb{R}^n}\frac{(1+|\xi'|+|\xi_n|)^{2s}|\hat{p}_k(\xi', \xi_n)|^2|c_k(\xi')|^2}{|\hat{\Lambda}_-^N(\xi', \xi_n)|^2|\hat{A}_+(\xi', \xi_n)|^2}d\xi' d\xi_n.$$

Making the change of variables $\xi_n = (1+|\xi'|)\eta_n$ and taking into account that $N \geq s$, we obtain

$$(\|v\|_s^+)^2 \leq C\sum_{k=1}^{m_+}\int_{\mathbb{R}^{n-1}}(1+|\xi'|)^{2s-2m_++2k-1}|c_k(\xi')|^2 d\xi'. \tag{53.17}$$

Since $\deg_{\xi'} d_{kj}(\xi') = m_+ - k - m_j$, we get (53.14) from (53.17) and (53.13). Therefore $u = u_0(x', x_n) + v(x', x_n)$, where u_0 is given by (53.4), is the unique solution in $H_s(\mathbb{R}_+^n)$ of the boundary value problem (53.2), (53.3), and the following estimate holds:

$$\|u\|_s^+ \leq C\|f\|_{s-m}^+ + C\sum_{j=1}^{m_+}[g_j(x')]_{s-m_j-\frac{1}{2}},$$

where $s > \max_{1\leq j\leq m_+}(m_j + \frac{1}{2})$. Denote by R_0 the bounded operator from $\mathcal{H}_s$ to $H_s(\mathbb{R}_+^n)$ given by $u = u_0(x', x_0) + v(x', x_n)$, $x_n > 0$. We have proven the following theorem:

Theorem 53.1. *Let $s > \max_{1\leq j\leq m}(m_j + \frac{1}{2})$. Suppose the condition* (53.12) *is satisfied. Then R_0 is the inverse to the operator $\mathcal{A}_0$:*

$$\mathcal{A}_0 R_0 = I_1, \quad R_0\mathcal{A}_0 = I_2, \tag{53.18}$$

where I_1 is the identity operator in $\mathcal{H}_{(s)}$ and I_2 is the identity operator in $H_s(\mathbb{R}_+^n)$.

54. Elliptic boundary value problems in a bounded domain

54.1. The method of "freezing" coefficients.

Let Ω be a bounded domain with a smooth boundary $\partial\Omega$. Consider the boundary value problem

$$A(x, D)u(x) = f(x), \qquad x \in \Omega, \tag{54.1}$$

$$B_j(x, D)u(x)|_{\partial\Omega} = g_j(x'), \qquad x' \in \partial\Omega, \quad 1 \leq j \leq m_+, \tag{54.2}$$

where $A(x, D)$ is an elliptic operator of order $m = 2m_+$, and the $B_j(x, D)$ are differential operators of order m_j, $1 \leq j \leq m_+$.

Denote by $A_0(x, D), B_{j0}(x, D)$ the principal terms of $A(x, D), B_j(x, D)$, $1 \leq j \leq m_+$. We have $A_0(x, \xi) \neq 0$, $\forall x \in \overline{\Omega}$, $\xi \neq 0$.

Denote by $\mathcal{H}_{(s)}(\Omega)$ the space of the right hand sides of (54.1), (54.2):

(54.3)
$$\mathcal{H}_{(s)}(\Omega) = H_{s-m}(\Omega) \times \prod_{j=1}^{m_+} H_{s-m_j-\frac{1}{2}}(\partial\Omega), \quad \text{where} \quad s > \max_{1\leq j\leq m_+}\left(m_j + \frac{1}{2}\right).$$

The operator $\mathcal{A}u = (A(x, D)u|_\Omega, B_j u|_{\partial\Omega}, 1 \leq j \leq m_+)$ is bounded from $H_s(\Omega)$ to $\mathcal{H}_{(s)}(\Omega)$. We study the condition for the operator $\mathcal{A}$ to be Fredholm.

Let $x_0 \in \Omega$ and let U_0 be a neighborhood of x_0, such that $\overline{U}_0 \cap \partial\Omega = \emptyset$. Let $\psi(x) \in C_0^\infty(U_0)$, $|\psi(x)| \leq 1$.

Lemma 54.1. *For any $\varepsilon > 0$, there exists U_0 such that*

$$\psi(x)A(x, D)u = \psi(x)(\hat{A}_0(x_0, D) + K + T_0)u, \tag{54.4}$$

where $A_0(x_0, D) = \sum_{|k|=m} a_k(x_0)D^k$, $\hat{A}_0(x_0, \xi) = A_0(x_0, (1 + |\xi|)\frac{\xi}{|\xi|})$, and

$$\|Ku\|_{s-m} \leq C\varepsilon\|u\|_s, \tag{54.5}$$

$\operatorname{ord} T_0 \leq m - 1$.

Proof: Let $\psi_1(x) \in C_0^\infty(U_0)$, $\operatorname{supp}(1 - \psi_1(x)) \cap \operatorname{supp}\psi(x) = \emptyset$. Let

$$Ku = \sum_{|k|=m} \Lambda^{-(s-m)}\psi_1(x)(a_k(x) - a_k(x_0))\Lambda^{(s-m)}D^k u, \tag{54.6}$$

where $\Lambda(D)$ is a ψdo with symbol $\Lambda(\xi) = (1 + |\xi|^2)^{\frac{1}{2}}$.

Let $\varepsilon = \max_{x,k} |\psi_1(x)|\, |a_k(x) - a_k(x_0)|$.

We have

$$\begin{aligned} \|Ku\|_{s-m} &\leq \sum_{|k|=m} C\|\psi_1(x)(a_k(x) - a_k(x_0))\Lambda^{s-m}D^k u\|_0 \\ &\leq C\varepsilon \sum_{|k|=m} \|\Lambda^{s-m}D^k u\|_0 \leq C_1\varepsilon\|u\|_s. \end{aligned} \tag{54.7}$$

We have used in (54.7) that $\|\psi_1(x)(a_k(x) - a_k(x_0))v\|_0 \leq \varepsilon\|v\|_0$.

Commuting Λ^{s-m} and $\psi_1(x)(a_k(x) - a_k(x_0))$, we conclude that $T_0 = \psi_2(x)(A(x, D) - \hat{A}_0(x_0, D) - K)$ has order $\leq m - 1$, where $\psi_2(x) \in C_0^\infty(U_0)$, $\operatorname{supp}(1 - \psi_2) \cap \operatorname{supp}\psi_1 = \emptyset$. □

Take any $x_0 \in \partial\Omega$ and let $U^{(1)}$ be a neighborhood of x_0. Introduce in $U^{(1)}$ a local system of coordinates $y = s_1(x)$ such that $y_n = 0$ is the equation of $\partial\Omega \cap U^{(1)}$, $y_n \geq 0$ when $x \in U^{(1)} \cap \Omega$ and $(y_0', 0) = s_1(x_0)$. Let $A^{(1)}(y,\eta), B_j^{(1)}(y,\eta)$ be the symbols of $A(x,D), B_j(x,D)$ in y-coordinates. Denote by $\mathcal{A}^{(1)}$ the operator

$$\mathcal{A}^{(1)}u = (A^{(1)}(y,D_y)u(y)\big|_{U_+^{(1)}},\ B_j^{(1)}(y,D_y)u\big|_{U^{(1)}\cap\{y_n=0\}},\ 1 \leq j \leq m_+),$$

where $U_+^{(1)} = U^{(1)} \cap \{y_n > 0\}$ and $A_0^{(1)}, B_{j0}^{(1)}$ are the principal parts of $A^{(1)}, B_j^{(1)}$. Let $\hat{\mathcal{A}}_0^{(1)}$ be the operator $\mathcal{A}_0^{(1)}$ with the coefficients "frozen" at the point $(y_0', 0)$ and ξ' replaced by $(1+|\xi'|)\frac{\xi'}{|\xi'|}$. Let

$$\varepsilon = \max\{|\psi_1(y)|\,|a_k^{(1)}(y',y_n) - a_k^{(1)}(y_0',0)|, |\psi_1(y',0)|\,|b_{jk}^{(1)}(y',0) - b_{jk}^{(1)}(y_0',0)|\},$$

where $\psi_1(y) \in C_0^\infty(U^{(1)})$, $\psi_1(y) = 1$ near $(y_0',0)$, and $a_k^{(1)}, b_{jk}^{(1)}$ are the coefficients of $A_0^{(1)}, B_{j0}^{(1)}$. The maximum is taken over all $y \in U^{(1)}$, k and j.

Denote by $\psi(y)$ a $C_0^\infty(U^{(1)})$ function with $\operatorname{supp}\psi \cap \operatorname{supp}(1-\psi_1) = \emptyset$.

Lemma 54.2. *For any $\varepsilon > 0$, one can choose $U^{(1)}$ and $\psi_1(y) \in C_0^\infty(U^{(1)})$ such that*

$$\psi(y)\mathcal{A}^{(1)}u = \psi(y)(\hat{\mathcal{A}}_0^{(1)} + K_1 + T^{(0)}), \tag{54.8}$$

where $|||K_1 u|||_{(s)} \leq C\varepsilon\|u\|_s$, $\operatorname{ord} T^{(0)} \leq \operatorname{ord}\hat{\mathcal{A}}_0^{(1)} - 1$.

Here $|||K_1 u|||_{(s)}$ is the norm in $\mathcal{H}_{(s)}$, and $\psi(y)(f(y), g_1(y'), \dots, g_{m_+}(y')) = (\psi(y)f(y), \psi(y',0)g_1(y'), \dots, \psi(y',0)g_{m_+}(y'))$.

The proof of Lemma 54.2 is the same as the proof of Lemma 54.1 with two adjustments:

1) We use $\Lambda_-^{s-m}(\xi) = (\xi_n + i(1+|\xi'|^2)^{\frac{1}{2}})^{s-m}$ instead of $\Lambda^{s-m}(\xi) = ((1+|\xi|^2)^{\frac{1}{2}})^{\frac{s-m}{2}}$ in the definition of K_1 (cf. (54.6)). The operator $\Lambda_-^{s-m}(D', D_n)$ has the property that $p_+\Lambda_-^{s-m}l_1u = p_+\Lambda_-^{s-m}l_2u$, where p_+ is the restriction to the half-space $y_n > 0$ and l_1u, l_2u are two arbitrary extensions of u from $\mathbb{R}_+^n$ to $\mathbb{R}^n$.

The proof of this property is similar to the proof of Lemma 46.2.

Since $l_1u - l_2u = 0$ for $y_n < 0$ and since $\Lambda_-^{s-m}(\xi', z)$ is analytic in z when $\operatorname{Im} z > 0$, we see that $\Lambda_-^{s-m}(l_1u - l_2u) = 0$ for $y_n < 0$. Therefore, the norm in $H_{s-m}(\mathbb{R}_+^n)$ is equivalent to $\|p_+\Lambda_-^{s-m}lu\|_0$ (cf. §55).

2) The second adjustment is the use of $(\Lambda')^{s-m_j} = (1+|\xi'|^2)^{\frac{s-m_j}{2}}$ instead of $\Lambda^{s-m}(\xi) = (1+|\xi|^2)^{\frac{s-m}{2}}$ when dealing with $\psi(y',0)B_j^{(1)}(y',0,D)u|_{y_n=0}$.

54.2. The Fredholm property.

Theorem 54.3. *Suppose $A(x,D)$ is elliptic in $\overline{\Omega}$. Assume that for any $x_0 \in \partial\Omega$ the condition (53.12) is satisfied for $\hat{\mathcal{A}}_0^{(1)}$, and also that $s > \max_{1\le j\le m_+}(m_j + \frac{1}{2})$. Then the operator $\mathcal{A}$ is a Fredholm operator from $H_s(\Omega)$ to $\mathcal{H}_{(s)}(\Omega)$. The following estimate holds:*

$$\|u\|_s^+ \le C\Big(\|f\|_{s-m}^+ + \sum_{j=1}^{m_+}[g_j]_{s-m_j-\frac{1}{2}} + \|u\|_{s-1}^+\Big), \quad \forall s > \max_j\Big(m_j + \frac{1}{2}\Big).$$

Proof: Let x_0 be any interior point of Ω. Denote by R_0 the pseudo-differential operator with symbol $(\hat{A}_0(x_0,\xi))^{-1}$. It follows from Lemma 54.1 that $\psi(x)A(x,D)u = \psi(\hat{A}_0(x_0,D) + K + T)$. Therefore $\psi(x)A(x,D)R_0 = \psi(I + KR_0 + TR_0)$. We choose $\varepsilon > 0$ in Lemma 54.1 such that $|||KR_0||| < 1$, where $|||KR_0|||$ is the norm of the operator KR_0 in $H_{s-m}(\mathbb{R}^n)$. Then we have

$$\psi(x)A(x,D)R_0(I+KR_0)^{-1} = \psi(I+T_1), \tag{54.9}$$

where $T_1 = TR_0(I+KR_0)^{-1}$, $\operatorname{ord} T_1 \le -1$.

Consider now an arbitrary point $x_0 \in \partial\Omega$. It follows from Lemma 54.2 that the representation (54.8) holds in some local system of coordinates. By Theorem 53.1, there exists $R^{(0)}$ such that $\hat{\mathcal{A}}_0^{(0)}R^{(0)} = I$. Choose $U^{(1)}$ in Lemma 54.2 such that $\big|\big|\big|K_1R^{(0)}\big|\big|\big| < 1$. Then

$$\psi\mathcal{A}R^{(0)}(I+K_1R^{(0)})^{-1} = \psi(I+T_1^{(0)}), \tag{54.10}$$

where $T_1^{(0)} = T^{(0)}R^{(0)}(I+K_1R^{(0)})^{-1}$, $\operatorname{ord} T_1^{(0)} \le -1$. Therefore we have an open cover of $\overline{\Omega}$ by neighborhoods of the form U_0 when $x_0 \notin \partial\Omega$ and by neighborhoods of the form $U^{(1)}$ when $x_0 \in \partial\Omega$. Choose a finite subcover $\{U_j\}_{j=1}^N$ and let $\varphi_j(x) \in C_0^\infty(U_j)$ be the corresponding partition of unity. Denote by R the following operator bounded from $\mathcal{H}_{(s)}$ to $H_s(\Omega)$:

$$RF = \sum_{j=1}^N \psi_j R_j \varphi_j F, \tag{54.11}$$

where $F = (f, g_1, \dots, g_{m_+}) \in \mathcal{H}_{(s)}$, $\varphi_j F = (\varphi_j f, \varphi_j g_1\big|_{\partial\Omega}, \dots, \varphi_j g_{m_+}\big|_{\partial\Omega})$, $\psi_j(x) \in C_0^\infty(U_j)$, $\operatorname{supp}\varphi_j \cap \operatorname{supp}(1-\psi_j) = \emptyset$, and R_j has the form $R_0(I+KR_0)^{-1}$ (cf. (54.9)) in the case where $U_j \cap \partial\Omega = \emptyset$ and the form $R^{(0)}(I+K_1R^{(0)})^{-1}$ (cf. (54.10)) in the local coordinates in U_j when $U_j \cap \partial\Omega \ne \emptyset$. We have

$$\mathcal{A}RF = \mathcal{A}\sum_{j=1}^N \psi_j R_j \varphi_j F = \sum_{j=1}^N (\psi_j\mathcal{A} + \mathcal{A}_j)R_j\varphi_j F,$$

where $\operatorname{ord} \mathcal{A}_j \le \operatorname{ord} \mathcal{A} - 1$. Since $\psi_j \mathcal{A} R_j = \psi_j(I + T_j)$, $\operatorname{ord} T_j \le -1$, and since $\psi_j \varphi_j = \varphi_j$, we get

$$\mathcal{A}RF = F + TF,$$

where $T = \sum_j (\psi_j T_j \varphi_j + \mathcal{A}_j R_j \varphi_j)$, $\operatorname{ord} T \le -1$. Since $\operatorname{ord} T \le -1$, the operator T is compact (cf. §42). Therefore R is the right regularizer of $\mathcal{A}$. Analogously, we can construct the left regularizer in the form:

$$R^{(1)}F = \sum_{j=1}^{N} \varphi_j R_j^{(1)} \psi_j F, \tag{54.12}$$

where $\varphi_j(x)$ and $\psi_j(x)$ are the same as in (54.11), $R_j^{(1)}$ has the form $(I + R_0 K)^{-1} R_0$ in the case $U_j \cap \partial\Omega = \emptyset$ (cf. (54.9)), and $R_j^{(1)}$ has the form $(I + R^{(0)} K_1)^{-1} R^{(0)}$ in the case of $U_j \cap \partial\Omega \neq \emptyset$. We get

$$R^{(1)} \mathcal{A} u = \sum_{j=1}^{N} \varphi_j R_j^{(1)} \psi_j \mathcal{A} u = \sum_{j=1}^{N} \varphi_j R_j^{(1)} \psi_j (\mathcal{A}_j^{(0)} + K_j + T_j).$$

Since $\psi_j(\mathcal{A}_j^{(0)} + K_j) = (\mathcal{A}_j^{(0)} + K_j)\psi_j + T_{j1}$, where $\operatorname{ord} T_{j1} \le \operatorname{ord} \mathcal{A} - 1$, and since $R_j^{(1)}(\mathcal{A}_j^{(0)} + K_j) = I$, we get

$$R^{(1)} \mathcal{A} u = u + T^{(1)} u, \tag{54.13}$$

where $\operatorname{ord} T^{(1)} \le -1$, i.e., $T^{(1)}$ is compact.

Since $\mathcal{A}$ has the right and the left regularizers, it is Fredholm.

Remark 54.1. The proof of the Fredholm property given in this section is simpler than the proofs based on the construction of parametrices for the elliptic boundary value problem (cf. §55). It can also be extended to more general cases such as boundary value problems in domains with corners and wedges, where the construction of parametrices is much more complicated.

Remark 54.2. Theorem 54.3 was proven under the assumption that the condition (53.12) is satisfied for each $x_0 \in \partial\Omega$ in some local system of coordinates near x_0. We call the boundary conditions (54.2) that satisfy (53.12) for all $x_0 \in \partial\Omega$ the elliptic boundary conditions. The condition (53.12) is also called the Shapiro-Lopatinskii condition, or the coercivity condition. We now show that if (53.12) is satisfied at $x_0 \in \partial\Omega$ in some local system of coordinates $y = s(x)$, then it is satisfied in any other local system of coordinates $y^{(1)} = s^{(1)}(x)$.

Denote by $y^{(1)} = t(y)$ the transformation $t(y) = s^{(1)}(s^{-1}(y))$. Let $\frac{\partial t(y)}{\partial y}$ be the Jacobian matrix. Since $y_n^{(1)} = t_n(y', 0) = 0$, we have $\frac{\partial t_n(y',0)}{\partial y_k} = 0$, $1 \le k \le n - 1$.

Let $y^{(0)} = s(x_0)$, $y_0^{(1)} = s^{(1)}(x_0)$. Denote by $c(y) = [c_{jk}(y)]_{j,k=1}^n$ the matrix $[(\frac{\partial t(y)}{\partial y})^T]^{-1}$. If $A_0(y,\xi), B_{j0}(y,\xi)$ are the principal symbols of $A(x,D)$, $B_j(x,D)$ in y-coordinates and $A_0^{(1)}(y^{(1)},\eta), B_{j0}^{(1)}(y^{(1)},\eta)$ are the principal symbols in $y^{(1)}$-coordinates, then (cf. §45)

$$A_0^{(1)}(t(y),c(y)\xi) = A_0(y,\xi), \quad B_{j0}^{(1)}(t(y),c(y)\xi) = B_{j0}(y,\xi), \tag{54.14}$$

where $1 \le j \le m_+$.

We have at $y = y^{(0)}$:

$$\eta_j = \sum_{k=1}^{n-1} c_{jk}(y^{(0)})\xi_k, \quad 1 \le j \le n-1, \tag{54.15}$$

$$\eta_n = \sum_{k=1}^{n} c_{nk}(y^{(0)})\xi_k, \quad \text{where} \quad c_{nn}(y^{(0)}) > 0. \tag{54.16}$$

Analogously to (53.1) we have

$$A_+(y^{(0)},\xi',\xi_n) = \prod_{j=1}^{m_+}(\xi_n - \lambda_j^+(y^{(0)},\xi')),$$

$$A_+^{(1)}(y_0^{(1)},\eta',\eta_n) = \prod_{j=1}^{m_+}(\eta_n - \lambda_{j1}^+(y_0^{(1)},\eta')),$$

where $\operatorname{Im}\lambda_j > 0$, $\operatorname{Im}\lambda_{j1} > 0$. Therefore

$$A_+(y^{(0)},\xi) = \prod_{j=1}^{m_+}\left(c_{nn}(y^{(0)})\xi_n + \sum_{k=1}^{n-1} c_{nk}(y^{(0)})\xi_k - \lambda_{j1}^+(y_0^{(1)},c'(y^{(0)})\xi')\right),$$

where c' is the $(n-1)\times(n-1)$ matrix in (54.15). Analogously to (53.11) we have

$$b_{jk}^{(1)}(y_0^{(1)},\eta') = \frac{1}{2\pi}\int_{\gamma_+} \frac{B_{j0}^{(1)}(y_0^{(1)},\eta',z^{(1)})(z^{(1)})^{k-1}}{A_+^{(1)}(y_0^{(1)},\eta',z^{(1)})}dz^{(1)}. \tag{54.17}$$

Making the change of variables $z^{(1)} = c_{nn}z + \sum_{k=1}^{n-1} c_{nk}\xi_k$, $\eta' = c'\xi'$, we get that

$$b_{jk}^{(1)}(y_0^{(1)},\eta') = \frac{c_{nn}(y^{(0)})}{2\pi}\int_{\gamma_+} \frac{B_{j0}(y^{(0)},\xi',z)(c_{nn}z + \sum_{k=1}^{n-1} c_{nk}\xi_k)^{k-1}dz}{A_+(y^{(0)},\xi',z)}.$$

Therefore

$$\det[b_{jk}^{(1)}(y_0^{(1)},\eta')]_{j,k=1}^{m_+} = c_{nn}^{\frac{m_+(m_++1)}{2}} \det[b_{jk}(y^{(0)},\xi')]_{j,k=1}^{m_+},$$

and hence the condition (53.12) is satisfied in $y^{(1)}(= s^{(1)}(x))$-coordinates if it is satisfied in $y(= s(x))$-coordinates, since $c_{nn}(y^{(0)}) > 0$ and $\det c'(y^{(0)}) \ne 0$.

54.3. Invariant form of the ellipticity of boundary conditions.

We now give a more invariant description of the condition (53.12).

For any $x' \in \partial\Omega$ denote by $\nu(x')$ the inward unit normal to $\partial\Omega$ at x'. Let Ω_ε be the subdomain of Ω consisting of points $x' + x_n\nu(x')$ where $0 \leq x_n < \varepsilon$, i.e., Ω_ε is diffeomorphic to $\partial\Omega \times [0, \varepsilon)$. In a neighborhood U_0 of $x' \in \partial\Omega$ we introduce local coordinates (y', y_n), where $y_n = x_n$ and the y' are local coordinates in $U_0 \cap \partial\Omega$. The operators $A(x, D), B_j(x, D)$ have the principal terms $A_0(y', y_n, D_{y'}, D_{y_n}), B_{j0}(y', y_n, D_{y'}, D_{y_n})$ in these local coordinates. Let $(y^{(1)}, y_n')$ be another local system of coordinates in U_1, $x' \in U_1$, where $y_n' = x_n$ and $y^{(1)}$ are local coordinates in $U_1 \cap \partial\Omega$. We have in $U_0 \cap U_1$:

$$y^{(1)} = t'(y') \quad \text{and} \quad y_n' = y_n.$$

The principal symbols of $A(x, D), B_j(x, D)$ in $(y^{(1)}, y_n)$- and (y', y_n)-coordinates are related by formula (54.15), where $c'(y') = \left(\left(\frac{\partial t'(y')}{\partial y'}\right)^T\right)^{-1}$, and (54.16) reduces to $\eta_n = \xi_n$. Therefore, the operators $A(x, D), B_j(x, D)$ determine a family of operators $A_0(x', \xi'_{x'}, 0, D_n)$, $B_{j0}(x', \xi_{x'}, 0, D_n)$, $1 \leq j \leq m_+$ on $\mathbb{R}^1_+$, depending smoothly on $(x', \xi'_{x'}) \in T_0^*(\partial\Omega)$.

Consider the family of boundary value problems on the half-line $x_n > 0$ depending smoothly on $(x', \xi'_x) \in T_0^*(\partial\Omega)$,

$$A_0(x', \xi'_{x'}, 0, D_n)u(x_n) = 0, \quad x_n > 0, \tag{54.18}$$

$$B_{j0}(x', \xi'_{x'}, 0, D_n)u(x_n)\,|_{x_n=0} = h_j, \quad h_j \in \mathbb{C}, \ 1 \leq j \leq m_+. \tag{54.19}$$

The condition (53.12), $\forall x_0 \in \partial\Omega$, i.e., the condition that the boundary value problem (54.1), (54.2) is elliptic, is equivalent to the condition that the boundary value problem (54.18), (54.19) has a unique solution in $L_2(\mathbb{R}_+)$. Note that the kernel of $A_0(x', \xi'_{x'}, 0, D_n)$ in $L_2(\mathbb{R}^1)$ is m_+-dimensional for any $(x', \xi'_{x'}) \in T_0^*(\partial\Omega)$, and the number of boundary conditions (54.19) is also m_+. Therefore the solvability of (54.18), (54.19) is equivalent to the uniqueness in $L_2(\mathbb{R}_+)$ of the solution of (54.18), (54.19), with $h_j = 0$, $1 \leq j \leq m_+$.

54.4. Boundary value problems for elliptic systems of differential equations.

Let $A(x, D)$ be an $r \times r$ system of differential operators. An operator $A(x, D)$ is called elliptic if $A(x, \xi) = A_0(x, \xi) + A_1(x, \xi)$, where $A_0(x, \xi)$ is homogeneous of degree p, $\deg A_1 \leq p - 1$, and $\det A_0(x, \xi)$ is an elliptic operator of degree $m = pr$, i.e., $\det A_0(x, \xi) \neq 0$, $\forall \xi \neq 0$, $\forall x \in \overline{\Omega}$. Consider a boundary value problem

$$A(x, D)u = f(x) \quad \text{in} \quad \Omega, \tag{54.20}$$

$$B_j(x, D)u\,|_{\partial\Omega} = g_j(x'), \quad 1 \le j \le m_+ = \frac{m}{2}. \tag{54.21}$$

Here $u(x) = (u_1(x), \dots, u_r(x))$, $f(x) = (f_1(x), \dots, f_r(x))$, the $g_j(x)$ are scalar functions and the $B_j(x, D)$ are $1 \times r$ matrices.

Let $B_j(x, \xi) = B_{j0}(x, \xi) + B_{j1}(x, \xi)$ where $B_{j0}(x, \xi)$ is a homogeneous polynomial of degree m_j, $1 \le j \le m_+$, $\deg B_{j1} \le m_j - 1$. Denote by $B(x, D)$ the $m_+ \times r$ matrix whose rows are B_j. Then the boundary conditions (54.21) can be written in the form

$$B(x, D)u\,|_{\partial\Omega} = g(x'), \tag{54.22}$$

where $g = (g_1, \dots, g_{m_+})$. Let $B_0(x, D)$ be the principal part of $B(x, D)$. Analogously to (54.18), (54.19) we can associate with $A(x, D), B(x, D)$ a family $A_0(x', \xi'_{x'}, 0, D_n)$, $B_0(x', \xi'_{x'}, 0, D_n)$ of differential operators on $\mathbb{R}^1_+$ depending smoothly on $(x', \xi'_{x'}) \in T^*_0(\partial\Omega)$. Consider the boundary value problem in $\mathbb{R}^1_+$:

$$A(x', \xi'_{x'}, 0, D_n)u(x_n) = 0, \quad x_n > 0, \tag{54.23}$$

$$B(x', \xi'_{x'}, 0, D_n)u\,|_{x_n=0} = h, \quad \forall h \in \mathbb{C}^{m_+}. \tag{54.24}$$

The ellipticity of the boundary conditions (54.22) means that for any $(x', \xi'_{x'}) \in T^*_0(\partial\Omega)$, the boundary value problem (54.23), (54.24) has a unique solution in $L_2(\mathbb{R}'_+)$.

Similarly to Theorem 54.3 we have the following theorem:

Theorem 54.4. *Let $A(x, D)$ be elliptic in $\overline{\Omega}$ and let the family* (54.23), (54.24) *is uniquely solvable in $L_2(\mathbb{R}^1_+)$ for any $(x', \xi'_{x'}) \in T^*_0(\partial\Omega)$. Let $s > \max_{1 \le j \le m_+}(m_j + \frac{1}{2})$, where $m_j = \deg B_j$. Then the boundary value problem* (54.20), (54.21) *defines a Fredholm operator from $H_s(\Omega)$ to $H_{s-p}(\Omega) \times \prod_{j=1}^{m_+} H_{s-m_j-\frac{1}{2}}(\partial\Omega)$.*

Proof: Let x_0 be any point inside Ω. Then, as in proof of Theorem 54.3, we take $R_0 = \hat{A}_0^{-1}(x_0, D)$, where $\hat{A}_0^{-1}$ is the ψdo with symbol $A_0^{-1}(x_0, (|\xi| + 1)\frac{\xi}{|\xi|})$. If $x'_0 \in \partial\Omega$, we introduce local coordinates in a neighborhood of x_0, write the symbols $A_0(x'_0, \xi', 0, \xi_n), B_0(x'_0, \xi', \xi_n)$ in these coordinates, and consider the boundary value problem of the form (53.5), (53.6) in the half-space $x_n > 0$:

$$\hat{A}_0(x'_0, D', D_n)u(x', x_n) = 0, \tag{54.25}$$

$$\hat{B}_0(x'_0, D', D_n)u(x', x_n)\,|_{x_n=0} = h(x'). \tag{54.26}$$

Performing the Fourier transform in x', we obtain (cf. (53.7), (53.8)) a family of boundary value problems on $\mathbb{R}_+^1$ depending on a parameter ξ':

$$A_0\left(x_0', (1+|\xi'|)\frac{\xi'}{|\xi'|}, D_n\right)\tilde{u}(\xi', x_n) = 0, \tag{54.27}$$

$$B_0\left(x_0', (1+|\xi'|)\frac{\xi'}{|\xi'|}, D_n\right)\tilde{u}(\xi', x_n)\Big|_{x_n=0} = \tilde{h}(\xi'). \tag{54.28}$$

The system (54.27), (54.28) coincides, for ξ' fixed, with the system (54.23), (54.24). By assumption the system (54.23), (54.24) is uniquely solvable for any fixed $(x', \xi_{x'},)$. Therefore, there exists an operator $\tilde{u}(\xi', x_n) = \tilde{R}(x_0', \xi')\tilde{h}(\xi')$ that is the inverse to (54.27), (54.28). Performing the inverse Fourier transform in ξ', we get an operator $R_{x_0'}^{(0)}$ similar to the operator R_0 in (53.18). The continuation of the proof of Theorem 54.4 is similar to the proof of Theorem 54.3. Note that we gave an explicit construction of the operator R_0 in §53, which will be used in §55. However, to prove Theorem 54.4 (and Theorem 54.3 too), one needs only the existence of the operator $R_{x_0'}^{(0)}$.

Remark 54.3. Let $\ker A_0(x', \xi_{x'}')$ be the kernel of $A_0(x', \xi_{x'}', 0, D_n)$ in $L_2(\mathbb{R}_+^1)$. It has the dimension m_+ for all $(x', \xi_{x'}') \in T_0^*(\Omega)$. Locally we can find a basis of $\ker A_0(x', \xi_{x'}')$, smooth in $(x', \xi_{x'}')$. Therefore, $\ker A_0(x', \xi_{x'}')$ is a vector bundle on $T_0^*(\partial\Omega)$. Let e_j, $1 \le j \le m_+$, be a basis in $\mathbb{C}^{m_+}$. Solving (54.23), (54.24) with h replaced by e_j, $1 \le j \le m_+$, we get a basis in $\ker A_0(x', \xi_{x'}')$, $\forall (x', \xi_{x'}') \in T_0^*(\partial\Omega)$, smoothly dependent on $(x', \xi_{x'}')$. Therefore $\ker A_0(x', \xi_{x'}')$ is isomorphic to the direct product $T_0^*(\partial\Omega) \times \mathbb{C}^{m_+}$, i.e., $\ker A_0(x', \xi_{x'}')$ is a trivial bundle on $T_0^*(\partial\Omega)$.

This fact shows that the triviality of the vector bundle $\ker A_0(x', \xi_{x'}')$ is a necessary condition for the existence of an elliptic boundary value problem (see Atiyah-Bott [**AtB**]). For the necessary and sufficient condition for the existence of the pseudodifferential boundary value problems for $A(x, D)$ see also Eskin [**E1**], §16.

55. Parametrices for elliptic boundary value problems

55.1. Plus-operators and minus-operators.

We construct the parametrix for the boundary value problem (54.1), (54.2) using the partition of unity $\{\varphi_j, U_j\}$ in $\overline{\Omega}$. When $U_j \cap \partial\Omega = \emptyset$, we construct the parametrix as in Theorem 41.2 with minor changes.

Since we have no parameter λ in (41.3), we define $\hat{R}_0(x, \xi) = A_0^{-1}(x, \xi)(1 - \chi(\xi))$, where $\chi(\xi) \in C_0^\infty(\mathbb{R}^n)$, $\chi(\xi) = 1$ for $|\xi| < 1$. Then

$\hat{R}_0(x,\xi) \in S^{-m}$. Defining $C_{-p}(x,\xi), R_{-p}(x,\xi)$ as in (41.5)–(41.8) with $R_0(x,\xi,\lambda) = (A_0(x,\xi)+\lambda)^{-1}$ replaced by $\hat{R}_0(x,\xi)$, we get

$$A(x,D)R(x,D) = I + T_{-N-1}, \tag{55.1}$$

where $\operatorname{ord} T_{-N-1} \le -N-1$, and $R(x,D)$ has the form (41.10).

If $U_j \cap \partial\Omega \ne \emptyset$, we introduce a local system of coordinates (y', y_n) in U_j (cf. §54). Let $A_j(y', y_n, D', D_n)$, $B_{jk}(y', y_n, D', D_n)$ be the operators $A(x,D), B_k(x,D)$ in y-coordinates in U_j. We extend $A_j(y', y_n, D', D_n)$, $B_k^{(j)}(y', y_n, D', D_n)$ to $\mathbb{R}^n$ and consider the boundary value problem in $\mathbb{R}^n_+$:

$$A_j(y', y_n, D', D_n)u(y', y_n) = f(y', y_n), \quad y_n > 0, \tag{55.2}$$

$$B_k^{(j)}(y', y_n, D', D_n)u(y', y_n)|_{y_n=0} = g_k(y'), \quad 1 \le k \le m_+. \tag{55.3}$$

Let lf be an arbitrary extension of f to $\mathbb{R}^n$. Denote by R_{j0} the parametrix for $A_j u = lf$ in $\mathbb{R}^n$ constructed in (55.1). We look for the parametrix of (55.2), (55.3) in the form

$$u = R_{j0} lf + R_{j1}(h_1, \dots, h_{m_+}), \tag{55.4}$$

where

$$h_k(y') = g_k - B_k^{(j)} R_{j0} lf|_{y_n=0}, \quad 1 \le k \le m_+, \tag{55.5}$$

and R_{j1} is the parametrix of the boundary value problem

$$A_j(y', y_n, D', D_n)v(x', x_n) = 0, \quad (y', y_n) \in \mathbb{R}^n_+, \tag{55.6}$$

$$B_k^{(j)}(y', y_n, D', D_n)v(y', y_n)|_{y_n=0} = h_k(y'), \quad 1 \le k \le m_+. \tag{55.7}$$

Now we shall construct R_{j1}. For simplicity of notation, we drop the subscript j in (55.6), (55.7) and replace (y', y_n) by (x', x_n).

Let A_0, B_{k0} be the principal parts of $A(x', x_n, \xi', \xi_n), B_k(x', x_n, \xi', \xi_n)$. Let $A_0(x, 0, +1) = 1$ and let

$$A_0(x, \xi', \xi_n) = A_+(x, \xi', \xi_n)A_-(x, \xi', \xi_n) \tag{55.8}$$

be the factorization of A_0, i.e.,

$$A_+(x, \xi', \xi_n) = \prod_{j=1}^{m_+}(\xi_n - \sigma_j^+(x, \xi')),$$

$$A_-(x, \xi', \xi_n) = \prod_{j=1}^{m_+}(\xi_n - \sigma_j^-(x, \xi')),$$

where $\operatorname{Im}\sigma_j^+ > 0$, $\operatorname{Im}\sigma_j^- < 0$, $\sigma_j^\pm(x,\xi')$ are homogeneous functions of ξ' of degree 1, $\xi' \ne 0$.

Let $\chi_1(\xi') \in C_0^\infty(\mathbb{R}^{n-1})$ and $\chi_1 = 1$ when $|\xi'| < \frac{1}{2}$, $\chi_1 = 0$ when $|\xi'| > 1$. Denote

$$R_0^\pm(x,\xi) = (1-\chi_1(\xi'))A_\pm^{-1}(x,\xi). \tag{55.9}$$

We call the symbol $C^+(x,\xi',\xi_n)$ a plus-symbol if $C^+(x,\xi',\xi_n)$ is analytic in ξ_n for $\operatorname{Im}\xi_n < 0$, is C^∞ in (x,ξ',ξ_n), and is homogeneous in (ξ',ξ_n) for $|\xi'| > 1$. Analogously, $C^-(x,\xi',\xi_n)$ is a minus-symbol if it is analytic in ξ_n for $\operatorname{Im}\xi_n > 0$ and smooth and homogeneous for $|\xi'| > 1$. For example, R_0^+ is a plus-symbol and R_0^- is a minus-symbol. We consider in this section only symbols that are rational functions of ξ_n with no poles when ξ_n is real.

Any such symbol $C(x,\xi',\xi_n)$ that decays as $\xi_n \to \infty$ can be uniquely represented in the form

$$C(x,\xi',\xi_n) = C^+(x,\xi',\xi_n) + C^-(x,\xi',\xi_n), \tag{55.10}$$

where C^+ is a plus-symbol and C_- is a minus-symbol.

The uniqueness follows from Liouville's theorem.

The existence follows from the explicit formulas

$$C^+ = \Pi^+ C, \quad C^- = \Pi^- C, \tag{55.11}$$

where

$$\Pi^\pm C = \mp\frac{i}{2\pi}\int_{-\infty}^{\infty} \frac{C(x,\xi',\eta_n)}{\xi_n - \eta_n \mp i0}d\eta_n$$

(cf. Example 5.2).

We can compute the integral Π^+C using the residues of $C(x,\xi',\eta_n)$ in the half-plane $\operatorname{Im}\eta_n > 0$. Analogously Π^-C can be computed by using residues of $C(x,\xi',\eta_n)$ in the half-plane $\operatorname{Im}\eta_n < 0$.

We call ψdo's plus-operators or minus-operators if their symbols are plus-symbols or minus-symbols, respectively.

Remark 55.1. The main property of plus- and minus-operators is the following: If C^+ is a plus-operator and $\operatorname{supp} u_+ \subset \overline{\mathbb{R}^n_+}$, then $\operatorname{supp} C^+u_+ \subset \overline{\mathbb{R}^n_+}$. Analogously if C^- is a minus-operator and $\operatorname{supp} u_- \subset \overline{\mathbb{R}^n_-}$, then $\operatorname{supp} C^-u_- \subset \overline{\mathbb{R}^n_-}$.

The proof of these properties is the same as the proof of Lemma 46.2.

Therefore, if $C^+ \in S^\alpha$, then C^+ is a bounded operator from $\overset{\circ}{H}_s(\mathbb{R}^n)$ to $\overset{\circ}{H}_{s-\alpha}(\mathbb{R}^n)$. Note that if $C^- \in S^\alpha$, then p_+C^-lu is independent of the extension lu, where p_+ is the restriction to $\mathbb{R}^n_+$. If l_1u and l_2u are two extensions, then $l_1u - l_2u = u_-$ has support in $\overline{\mathbb{R}^n_-}$ and so $p_+C^-u_- = 0$. Hence pC^-lu defines a bounded operator from $H_s(\mathbb{R}^n_+)$ to $H_{s-\alpha}(\mathbb{R}^n_+)$.

Let Λ_-^s be the minus-operator with the symbol

$$\Lambda_-(\xi', \xi_n) = \left(\xi_n + i\sqrt{|\xi'|^2 + 1}\right)^s.$$

Since $p_+\Lambda_-^s lu$ does not depend of the extension, we have that the norm $\|u\|_s^+ = \inf_l \|lu\|_s$, where the infimum is taken over all extensions lu, is equivalent to the norm $\|p_+\Lambda_-^s l_0 u\|_0$, where $l_0 u$ is a fixed extension. □

55.2. Construction of the parametrix in the half-space.

For any N we construct plus-operators with symbols $R_p^+(x, \xi', \xi_n)$, $\deg R_p^+ = -m_+ - p$ for $|\xi'| > 1$, and minus-operators with symbols $A_p^-(x, \xi', \xi_n)$, $\deg A_p^- = m_+ - p$ for $|\xi'| > 1$, $0 \le p \le N$, such that

$$A(x, D', D_n) \sum_{p=0}^{N} R_p^+(x, D', D_n) = \sum_{p=0}^{N} A_p^-(x, D', D_n) + T_{m_+ - N - 1}, \tag{55.12}$$

where

$$\|T_{m_+ - N - 1} u\|_{0,s} \le C\|u\|_{m_+ - 1, s - N}, \quad \forall s. \tag{55.13}$$

Here the $\|u\|_{p,s}$ are the same norms as in §48. Since A is a differential operator, the product $A(x, D)R_p^+(x, D', D_n)$ has the form

$$A(x, D)R_p^+ = \sum_{j=0}^{m} B_{jp}(x, D), \tag{55.14}$$

where

$$\begin{aligned} B_{0p}(x, \xi) &= A_0(x, \xi)R_p^+(x, \xi', \xi_n), \\ B_{jp}(x, \xi) &= \sum_{r=0}^{j-1} \sum_{|k|=j-r} \frac{1}{k!} \frac{\partial^k A_r(x, \xi)}{\partial \xi^k} D_x^k R_p^+ + A_j(x, \xi)R_p^+, \quad j \ge 1, \\ A(x, \xi) &= A_0(x, \xi) + \sum_{j=1}^{m} A_j(x, \xi), \ \operatorname{ord} A_j = m - j. \end{aligned} \tag{55.15}$$

Taking the product of A and $\sum_{p=0}^N R_p^+$ and collecting symbols of the same degree of homogeneity in (ξ', ξ_n) for $|\xi'| > 1$, we get

$$A_0(x, \xi)R_0^+(x, \xi', \xi_n) = A_0^-(x, \xi', \xi_n), \tag{55.16}$$

$$A_0(x, \xi)R_p^+(x, \xi', \xi_n) + T_p(x, \xi', \xi_n) = A_p^-(x, \xi', \xi_n), \quad p \ge 1, \tag{55.17}$$

where T_p depends on R_j^+, $0 \le j \le p - 1$:

$$T_p = \sum_{j+r=p} B_{jr}. \tag{55.18}$$

Note that $\deg T_p = m_+ - p$ for $|\xi'| > 1$.

It follows from (55.16) that

$$A_0^- = A_-(x,\xi',\xi_n)(1-\chi_1(\xi')). \tag{55.19}$$

To solve (55.17), multiply it by A_-^{-1} and use the decomposition (55.10),

$$A_-^{-1}T_p = \Pi^+ A_-^{-1}T_p + \Pi^- A_-^{-1}T_p.$$

As a result we get

$$A_+R_p^+ + \Pi^+ A_-^{-1}T_p = A_-^{-1}A_p^- - \Pi^- A_-^{-1}T_p. \tag{55.20}$$

Since the decomposition (55.10) is unique, we obtain

$$R_p^+ = -A_+^{-1}\Pi^+ A_-^{-1}T_p, \quad A_p^- = A^-\Pi^- A_-^{-1}T_p, \quad \forall p \geq 1. \tag{55.21}$$

Note that the degree of homogeneity in (ξ',ξ_n) for $|\xi'|>1$ decreases as p increases, but the decay in ξ_n of R_p^+ is $O(\frac{1}{\xi_n^{m_++1}})$ for all p. Also T_{m_+-N-1} is a ψdo with symbol (55.18) for $p=N+1$. Therefore, the estimate (55.13) holds, since $T_p(x,\xi',\xi_n) = O(\xi_n^{m_+-1})$.

We look for the parametrix of (55.6) in the form

$$u = \left(\sum_{p=0}^{N} R_p^+(x,D',D_n)\right)\left(\sum_{k=1}^{m_+} c_k(x')\delta^{(k-1)}(x_n)\right), \quad x_n>0. \tag{55.22}$$

Since

$$p_+\left(\sum_{p=0}^{N} A_p^-\right)\left(\sum_{k=1}^{m_+} c_k(x')\delta^{(k-1)}\right) = 0,$$

we get

$$A(x,D)u = T_{m_+-N-1}\left(\sum_{k=1}^{m_+} c_k(x')\delta^{(k-1)}(x_n)\right) \quad \text{for} \quad x_n>0. \tag{55.23}$$

Note that

$$\left\|T_{m_+-N-1}\left(\sum_{k=1}^{m_+} c_k\delta^{(k-1)}\right)\right\|_s^+ \leq C\sum_{k=1}^{m_+}[(\Lambda')^{s+m_++k-N-\frac{3}{2}}c_k]_0, \tag{55.24}$$

where $\Lambda' = (|\xi'|^2+1)^{\frac{1}{2}}$.

To prove (55.24), note that the symbol $T_{m_+-N-1}(x,\xi',\xi_n)\xi_n^{k-1}$ is a rational function of ξ_n. Take $M>s$, where M is an integer. The composition of ψdo's $\Lambda_-^M(D',D_n)$ and $T_{m_+-N-1}D_n^{k-1}$ has the form:

$$\Lambda_-^M T_{m_+-N-1}D_n^{k-1} = P_k(x',D',D_n) + r_k(x,D',D_n) + T'_{-N-1},$$

where $P_k(x,\xi',\xi_n)$ is a polynomial in ξ_n, $r_k(x,\xi',\xi_n) = O(\frac{1}{\xi_n})$, $\operatorname{ord} T'_{-N-1} \leq -N-1$. Note that $p_+\Lambda_-^{s-M}P_k(c_k\delta) = 0$ since $\Lambda_-^{s-M}P_k$ is a minus-symbol.

Letting $v_k = T_{m_+-N-1}(c_k\delta^{(k-1)}(x_n))$, we get

$$(55.25)\quad \begin{aligned} \|v_k\|_s^+ &\le C\|\Lambda_-^s v_k\|_0 = C\|\Lambda_-^{s-M}\Lambda_-^M v_k\|_0 \\ &\le C\|\Lambda_-^{s-M} r_k(x,D',D_n)(c_k\delta)\|_0 + C\|\Lambda_-^{s-M}T'_{-N-1}(c_k\delta)\|_0. \end{aligned}$$

Now repeating the proof of Theorem 40.1 we get (55.24). □

Substitute (55.22) into the boundary conditions (55.7).

Let p' be the restriction operator to $x_n = 0$. Let $D(x,\xi',\xi_n)$ be a symbol rational in ξ_n, $D = O(\frac{1}{\xi_n})$. Then

$$p'D(c(x')\delta(x_n)) = \lim_{x_n\to+0}\frac{1}{(2\pi)^n}\int_{\mathbb{R}^n} D(x',x_n,\xi',\xi_n)\tilde{c}(\xi')e^{ix_n\xi_n+ix'\cdot\xi'}d\xi_n d\xi'.$$

Applying the Jordan lemma we get

$$(55.26)\qquad p'D(c(x')\delta(x_n)) = d(x',D')c(x'),$$

where

$$(55.27)\qquad d(x',\xi') = \frac{1}{2\pi}\int_{\gamma_+} D(x',0,\xi',\xi_n)d\xi_n$$

and γ_+ is a closed contour in $\operatorname{Im}\xi_n > 0$ encircling all the poles of $D(x,\xi',\xi_n)$ in $\operatorname{Im}\xi_n > 0$.

Let $\Pi' D(x',0,\xi',\xi_n) = \frac{1}{2\pi}\int_{\gamma_+} Dd\xi_n$. Note that $\Pi' D_+ = \frac{1}{2\pi}\int_{\gamma_+} Dd\xi_n = 0$ if $D_+ = O(\frac{1}{\xi_n^2})$ and D_+ is a plus-symbol. We have, as in (55.26), (55.27),

$$(55.28)\quad p'B_j(x_1',0,D',D_n)u_k(x) = b_{jk}(x',D')c_k(x') + T_{m_j+k-1-m_+-N-1}\,c_k,$$

where $u_k = R^+\frac{\partial^{k-1}}{\partial x_n^{k-1}}(c_k\delta)$, $R^+ = \sum_{p=0}^N R_p^+$,

$$(55.29)\qquad b_{jk}(x',\xi') = \Pi' B_{jk}(x',0,\xi',\xi_n),\quad 1\le j,k\le m_+,$$

and B_{jk} is the symbol of the composition of B_j and $R^+\frac{\partial^{k-1}}{\partial x_n^{k-1}}$.

Let $b_{jk0}(x',\xi')$ be the principal part of b_{jk}. Since the boundary value problem (55.2), (55.3) is elliptic, we have that $\det[b_{jk0}(x',\xi')]_{j,k=1}^{m_+} \ne 0$, $\forall(x',\xi')$, $\xi'\ne 0$ (cf. (53.12)). Therefore, as in (55.1), we can construct a matrix pseudodifferential operator $[d_{kj}(x',D')]$ such that

$$(55.30)\qquad [b_{jk}][d_{kp}] = I + [t_{jp}],\quad \operatorname{ord} t_{jk} < -N,\quad 1\le j,k\le m_+.$$

Let

$$(55.31)\qquad c_k(x') = \sum_{p=1}^{m_+} d_{kp}(x',D')h_p,\quad 1\le k\le m_+.$$

Substituting (55.31) into (55.22) and (55.28) we get:

$$p_+A(x', x_n, D', D_n)u = T_{-N-1}(h_1, \dots, h_{m_+}), \tag{55.32}$$

$$p'B_j(x', 0, D', D_n)u = h_j + T^{(j)}_{-N-1}(h_1, \dots, h_{m_+}), \tag{55.33}$$

where u is defined by (55.22) with $c_k(x')$, $1 \le k \le m_-$, given by (55.31), i.e., (55.22), (55.31) is the parametrix of the boundary value problem (55.6), (55.7). □

55.3. Parametrix in a bounded domain.

Restoring the subscript j in (55.6), (55.7), we denote this parametrix by R_{j1}. Then (55.4), (55.5) is the parametrix of the boundary value problem (55.2), (55.3), in the local coordinates in U_j, when $U_j \cap \partial\Omega \neq \emptyset$. We have (cf. (55.22), (55.31))

$$p_+A_jR_{j1}(h_1, \dots, h_{m_+}) = T^{(j)}_{-N-1}(h_1, \dots, h_{m_+}), \tag{55.34}$$

$$p'B^{(j)}_k R_{j1}(h_1, \dots, h_{m_+}) = h_k + T^{(jk)}_{-N-1}(h_1, \dots, h_{m_+}),\ 1 \le k \le m_+, \tag{55.35}$$

where $\operatorname{ord} T^{(j)}_{-N} \le -N$, $\operatorname{ord} T^{(jk)}_{-N} \le -N$.

Now construct the parametrix of the boundary value problem (54.1), (54.2). Let $\{\varphi_j, U_j\}$ be a partition of unity, $\psi_j(x) \in C_0^\infty(U_j)$, and $\operatorname{supp}(1 - \psi_j) \cap \operatorname{supp}\varphi_j = \emptyset$.

Denote $F = (f(x), g_1(x'), \dots, g_{m_+}(x'))$. We look for the parametrix in the form

$$\mathcal{R}F = \sum_j \psi_j R_j \varphi_j F. \tag{55.36}$$

When $\operatorname{supp} U_j \cap \partial\Omega = \emptyset$, we assume that R_j is the parametrix for $Au = \varphi_j f$ in $\mathbb{R}^n$ (cf. (55.1)). When $U_j \cap \partial\Omega \neq \emptyset$, we use a local system of coordinates in U_j, and R_j is the parametrix for (55.2), (55.3) in these local coordinates, i.e., R_j has the form (55.4), (55.5). If $U_j \cap \partial\Omega = \emptyset$, then $A(x, D)R_j = I + T_{-N}$. Therefore

$$A\psi_jR_j\varphi_jf = \psi_jAR_j\varphi_jf + [A, \psi_j]R_j\varphi_jf = \varphi_jf + \psi_jT_{-N}\varphi_jf + [A, \psi_j]R_j\varphi_jf,$$

where $[A, \psi_j]$ is the commutator of A and ψ_j. Since $\psi_j = 1$ in a neighborhood of $\operatorname{supp}\varphi_j$ we have that $\operatorname{ord}[A, \psi_j]R_j\varphi_j = -\infty$ by the pseudolocal property (cf. Corollary 44.2). Therefore $A\psi_jR_j\varphi_jf = \varphi_jf + T^{(1)}_{-N}f$, $\operatorname{ord} T^{(1)}_{-N} \le -N$. If U_j is such that $U_j \cap \partial\Omega \neq \emptyset$, then $R_j = R_{j0} + R_{j1}$. For R_{j0} we have, as above, $A_j\psi_jR_{j0}\varphi_jF = \varphi_jf + T^{(1)}_{-N}f$ in the local system of coordinates. Note that R_{j1} also has the pseudolocal property since it holds for the operators (55.22), (55.31). Therefore $\operatorname{ord}[A_j, \psi_j]R_{j1}\varphi_jF = T_{-\infty}F$. Using (55.34) and (55.35) we get

(55.37) $$p_\Omega A\mathcal{R}F = f + T_{-N}F,$$

(55.38) $$p_{\partial\Omega}B_k\mathcal{R}F = g_k + T_{-N}^{(k)}F,$$

where $\operatorname{ord} T_{-N} \le -N$, $\operatorname{ord} T_{-N}^{(k)} \le -N$.

56. The heat trace asymptotics

56.1. The existence and the estimates of the resolvent.

Let

(56.1) $$(A(x,D)+\lambda)u(x) = f(x), \quad x\in\Omega,$$

(56.2) $$B_j(x,D)u|_{\partial\Omega} = g_j(x'), \quad 1\le j\le m_+,$$

be an elliptic boundary value problem. We assume that $A(x,D)$ is a strongly elliptic operator of order $m = 2m_+$, i.e., $\operatorname{Re} A_0(x,\xi) > 0$ for $x\in\overline{\Omega}$, $\xi\ne 0$, $\operatorname{ord} B_j = m_j$, $1\le j\le m_+$. It follows from Theorem 54.3 that the left hand sides of (56.1), (56.2) define a Fredholm operator $\mathcal{A}(\lambda)$ acting from $H_s(\Omega)$ to $\mathcal{H}_{(s)} = H_s(\Omega)\times\prod_{j=1}^{m_+} H_{s-m_j-\frac12}(\partial\Omega)$ for any $s > m_0 = \max_{1\le j\le m_+}(m_j+\frac12)$.

It follows from (41.1) that there exists $\delta > 0$ such that $(A_0(x,\xi)+\lambda)^{-1}$ is analytic when $\lambda\in C_\delta = \{\lambda\in\mathbb{C}\setminus\{0\}, -\frac{\pi}{2}-\delta\le\arg\lambda\le\frac{\pi}{2}+\delta\}$, and $|A_0(x,\xi)+\lambda|^{-1}\le C(|\xi|+|\lambda|^{\frac1m})^{-m}$ when $\lambda\in C_\delta$, $\forall x\in\overline{\Omega}$, $\forall\xi$. Instead of condition (53.12) we impose a stronger condition.

Let $\tilde{A}_0(\tilde{x},D,D_n), \tilde{B}_{j0}(\tilde{x},D',D_n)$ be the principal parts of $A(\tilde{x},D)$, $B_j(\tilde{x},D)$ in a local system of coordinates in a neighborhood of an arbitrary $\tilde{x}\in\partial\Omega$. Let $\tilde{A}_0(\tilde{x},\xi',\xi_n)+\lambda = \tilde{A}_-(\tilde{x},\xi',\lambda^{\frac1m},\xi_n)\tilde{A}_+(\tilde{x},\xi',\lambda^{\frac1m},\xi_n)$ be the factorization of $\tilde{A}_0+\lambda$ (cf. (53.1)). Here $\lambda^{\frac1m}$ is the branch of the mth root that is positive when $\lambda>0$. We assume that

(56.3)
$$\det[b_{jk}(\tilde{x},\xi',\lambda^{\frac1m})]_{j,k=1}^{m_+}\ne 0 \text{ for all } \xi'\in\mathbb{R}^n,\ \lambda\in C_\delta,\ |\xi'|+|\lambda^{\frac1m}|>0,\ \tilde{x}\in\partial\Omega,$$

where the $b_{jk}(\tilde{x},\xi',\lambda^{\frac1m})]$ are the same as in (53.11) with $A_+(\xi',z)$ replaced by $\tilde{A}_+(\tilde{x},\xi',\lambda^{\frac1m},z)$.

The presence of parameter λ allows us to obtain a stronger result (cf. §41).

Theorem 56.1. *The operator $\mathcal{A}(\lambda)$ is an invertible operator from $H_s(\Omega)$ to $\mathcal{H}_{(s)}$, $s>m_0$, when $\lambda\in C_\delta = \{\lambda\in\mathbb{C}\setminus\{0\}, |\lambda|>\lambda_0\}$, λ_0 is large, assuming that the condition* (56.3) *is satisfied.*

Proof: Consider the boundary value problem (53.2), (53.3) in $\mathbb{R}^n_+$, where $\hat{A}_0$ is replaced by $A_0(x,D)+\lambda$ and $\hat{B}_{j0}$ by $B_{j0}(D',D_n)$. Let $\mathcal{A}_0(\lambda)$ be the operator defined by this boundary value problem. Let $A_0(\xi', \xi_n) + \lambda = A_+(\xi', \lambda^{\frac{1}{m}}, \xi_n) A_-(\xi', \lambda^{\frac{1}{m}}, \xi_n)$ be the factorization of $A_0(\xi', \xi_n) + \lambda$. Note that $A_\pm(\xi', \lambda^{\frac{1}{m}}, \xi_n)$ are homogeneous in $(\xi', \lambda^{\frac{1}{m}}, \xi_n)$ and analytic in λ when $\lambda \in C_\delta$.

Let $R_0(\lambda)$ be the inverse to $\mathcal{A}_0(\lambda)$ constructed in §53, where we change $\hat{A}_0$ to $A_0(\xi) + \lambda$, $\hat{B}_{j0}$ to $B_{j0}(\xi', \xi_n)$, and $\Lambda_-^N = (\xi_n + i|\xi'| + i)^N$ to $(\xi_n + i(|\xi'|^2 + \lambda^{\frac{2}{m}})^{\frac{1}{2}})^N$. Then any term in $R_0(\lambda)$ is homogeneous in $(\xi', \lambda^{\frac{1}{m}}, \xi_n)$ and analytic in λ, $\lambda \in C_\delta$. Therefore $R_0(\lambda)$ is a bounded operator from $\mathcal{H}_{(s)}$ to $H_s(\mathbb{R}^n_+)$, $s > m_0$, analytic in λ, $\lambda \in C_\delta$ (cf. (53.18)), and $\mathcal{A}_0(\lambda) R_0(\lambda) = I_1$, $R_0(\lambda)\mathcal{A}_0(\lambda) = I_2$. Let $|||R_0(\lambda)|||_{(s-k)}$ be the norm of $R_0(\lambda)$ as an operator from $\mathcal{H}_{(s)}$ to $H_{s-k}(\mathbb{R}^n_+)$, $0 \le k \le m$. Then

$$|||R_0(\lambda)|||_{(s-k)} \le C|\lambda|^{-\frac{k}{m}}, \quad 0 \le k \le m, \ \lambda \in C_\delta, \tag{56.4}$$

(cf. (41.1)). As in §54, we use the partition of unity. Let $T_0, T^{(0)}$ be the same as in (54.4), (54.8). Note that we replace $\Lambda^{s-m}, \Lambda_-^{s-m}$ in (54.4), (54.8) by $\left(\xi_n^2 + |\xi'|^2 + \lambda^{\frac{2}{m}}\right)^{\frac{s-m}{2}}$, $\left(\xi_n + i(|\xi'|^2 + \lambda^{\frac{2}{m}})^{\frac{1}{2}}\right)^{s-m}$, respectively. Using (56.4) we have that the norms of the operators TR_0 and $T^{(0)}R^{(0)}$, acting in $H_{s-m}(\mathbb{R}^n_+)$ and $\mathcal{H}_{(s-m)}$, respectively, satisfy

$$|||T_0 R_0||| \le \frac{C}{|\lambda|^{\frac{1}{m}}}, \qquad \left|\left|\left|T^{(0)} R^{(0)}\right|\right|\right| \le \frac{C}{|\lambda|^{\frac{1}{m}}},$$

since $\operatorname{ord} T_0 \le -1$ and $\operatorname{ord} T^{(0)} \le \operatorname{ord} \mathcal{A}_0(\lambda) - 1$. Therefore these operators have a small norm when $|\lambda|$ is large, $\lambda \in C_\delta$. Thus, if R is given by (54.11), we get $\mathcal{A}(\lambda)R = I + T$, where T has a small norm. This proves that $R(I+T)^{-1}$ is the right inverse to $\mathcal{A}(\lambda)$. Analogously, using (54.13), we can prove that $\mathcal{A}(\lambda)$ has also the left inverse, i.e., there exists $R(\lambda) = (\mathcal{A}(\lambda))^{-1}$. □

56.2. The parametrix construction.

The parametrix to (56.1), (56.2) constructed in §55 (cf. (55.37), (55.38)) undergoes some changes when the parameter λ in (56.1) is present:

a) There is no need in the cutoff functions $\chi(\xi)$ and $\chi_1(\xi')$, and the symbol $\hat{R}_0(x, \xi)$ is replaced by $(A_0(x, \xi) + \lambda)^{-1}$.

b) The symbols $A_\pm(x, \xi', \lambda^{\frac{1}{m}}, \xi_n)$ of the factorization $A_0(x, \xi', \xi_n) + \lambda = A_-(x, \xi', \lambda^{\frac{1}{m}}, \xi_n) A_+(x, \xi', \lambda^{\frac{1}{m}}, \xi_n)$ replace $A_\pm(x, \xi', \xi_n)$ (cf. (53.1)). The symbol $A_+^{-1}(x, \xi', \lambda^{\frac{1}{m}}, \xi_n)$ replaces $R_0^+(x, \xi', \xi_n)$.

c) The symbol $\left(\xi_n + i\sqrt{|\xi'|^2 + \lambda^{\frac{2}{m}}}\right)^s$ replaces the symbol $\tilde{\Lambda}_-^s = (\xi_n + i\sqrt{|\xi'|^2 + 1})^s$.

d) The symbols of all terms in (55.38) are homogeneous in $(\xi', \lambda^{\frac{1}{m}}, \xi_n)$ and analytic in λ for $\lambda \in C_\delta$, $|\lambda| \geq \lambda_0$.

e) The remainders T_{-N} have a small norm when $\lambda \in C_\delta$, $|\lambda| \geq \lambda_0$, and $\lambda_0 > 0$ is large.

Therefore we get a parametrix RF such that

$$(56.5)\qquad \begin{aligned} &(A(x,D)+\lambda I)RF = f + T_{-N-1}F, \quad x \in \Omega, \\ &B_j(x,D)RF\big|_{\partial\Omega} = g_j + T^{(j)}_{-N-1}F. \end{aligned}$$

where $F = (f, g_1, \dots, g_{m_+})$. □

In this section we consider the case where $g_j = 0, 1 \leq j \leq m_+$, in (56.2). Let D_B be the subspace of $H_s(\Omega)$, $s > m_0$, consisting of all $u(x) \in H_s(\Omega)$ satisfying $B_j u|_{\partial\Omega} = 0$, $1 \leq j \leq m_+$. Let $A_B + \lambda I$ be the operator $A(x,D) + \lambda I$ restricted to the domain D_B. We see that $(A_B + \lambda I)^{-1}$ exists for $\lambda \in C_\delta, |\lambda| > \lambda_0$, it maps $H_{s-m}(\Omega)$ onto D_B, and

$$(56.6)\qquad |||(A_B + \lambda I)^{-1}|||_{(s-m)} \leq \frac{C}{1+|\lambda|},$$

where $|||(A_B + \lambda I)^{-1}|||_{(s-m)}$ is the norm of $(A_B + \lambda I)^{-1}$ as an operator from $H_{s-m}(\Omega)$ to $H_{s-m}(\Omega)$ (cf. (56.4)).

Applying $R(\lambda)$ from the left to (56.5), we get

$$RF = R(\lambda)F + T^{(2)}_{-m_+-N-1}F,$$

where $\operatorname{ord} T^{(2)}_{-m_+-N-1} \leq -m_+ - N - 1$.

Take $F = F_1 = (f, 0, \dots, 0)$. Then $R(\lambda)F_1 = (A_B + \lambda I)^{-1} f$ by definition, and we have

$$(56.7)\qquad (A_B + \lambda I)^{-1} f = R(f, 0, \dots, 0) + T_{-m_+-N-1} f,$$

$\operatorname{ord} T_{-m_+-N-1} \leq -m_+ - N - 1$. □

Consider the initial-boundary value problem for the heat equation

$$(56.8)\qquad \frac{\partial u(x,t)}{\partial t} + A(x,D)u(x,t) = 0, \quad t > 0,\ x \in \Omega,$$

$$(56.9)\qquad \begin{aligned} &u(x,0) = u_0(x), \quad x \in \Omega, \\ &B_j(x,D)u(x,t)\ |_{\partial\Omega \times (0,+\infty)} = 0, \quad 1 \leq j \leq m_+. \end{aligned}$$

Denote by $u_+(x,t)$ the extension of $u(x,t)$ by zero for $t < 0$. Performing the Fourier-Laplace transform in t:

$$\tilde{u}_+(x, \sigma - i\tau) = \int_0^\infty u_+(x,t) e^{-it(\sigma - i\tau)} dt, \quad t \geq 0,$$

we get (cf. §47)

$$A(x,D)\tilde{u}(x,\sigma-i\tau)+i(\sigma-i\tau)\tilde{u}(x,\sigma-i\tau)=u_0(x),\quad x\in\Omega, \tag{56.10}$$

$$B_j(x,D)\tilde{u}(x,\sigma-i\tau)|_{\partial\Omega}=0,\quad 1\le j\le m_+, \tag{56.11}$$

where $\tau\ge\tau_0$ and $\tilde{u}(x,\sigma-i\tau)$ is analytic when $\tau>\tau_0$.

Since $\tilde{u}(x,\sigma-i\tau)=(A_B+i(\sigma-i\tau)I)^{-1}u_0(x)$ is the solution to (56.10), (56.11) when $\tau\ge\tau_0,\tau_0>0$ is large, and since (56.6) holds, we have that

$$u(x,t)=\frac{1}{2\pi}\int_{-\infty}^{\infty}(A_B+i(\sigma-i\tau)I)^{-1}u_0(x)e^{it(\sigma-i\tau)}d\sigma,\ \tau\ge\tau_0>0, \tag{56.12}$$

is the solution of (56.8), (56.9), $\tau\ge\tau_0,$ τ is arbitrary.

Consider first the case when $A(x,D)$ is a second order elliptic operator with zero Dirichlet boundary condition.

56.3. The heat trace for the Dirichlet Laplacian.

Consider the Laplace-Beltrami operator in $\Omega\subset\mathbb{R}^n$,

$$A(x,D)u=-\sum_{j,k=1}^{n}\frac{1}{\sqrt{g(x)}}\frac{\partial}{\partial x_j}\left(\sqrt{g(x)}g^{jk}(x)\frac{\partial u}{\partial x_k}\right), \tag{56.13}$$

where $([g^{jk}(x)]_{j,k=1}^n)^{-1}$ is the metric tensor, $g(x)=(\det[g^{jk}]_{j,k=1}^n)^{-1}$, and $[g^{jk}(x)]_{j,k=1}^n\in C^\infty(\overline{\Omega})$ is positive definite. Let $G_D(x,x^{(0)},t)$ be the Dirichlet heat kernel for the Dirichlet-Laplacian, i.e.,

(56.14)
$$\frac{\partial G_D(x,x^{(0)},t)}{\partial t}+A(x,D)G_D(x,x^{(0)},t)=0\quad\text{for}\quad t>0,\ x\in\Omega,\ x^{(0)}\in\Omega,$$

$$G_D(x,x^{(0)},0)=\delta(x-x^{(0)}), \tag{56.15}$$

$$G_D(x,x^{(0)},t)|_{\partial\Omega\times(0,+\infty)}=0. \tag{56.16}$$

Denote by $(A_D+\lambda I)^{-1}$ the operator $(A_B+\lambda I)^{-1}$ when (56.9) is the zero Dirichlet boundary condition.

As in §47, we can represent $G_D(x,x^{(0)},t)$ in the form (cf. (56.12))

$$G_D(x,x^{(0)},t)=\frac{1}{2\pi}\int_{-\infty}^{\infty}(A_D+i(\sigma-i\tau))^{-1}\delta(x-x^{(0)})e^{it(\sigma-i\tau)}d\sigma, \tag{56.17}$$

where $\tau\ge\tau_0>0$. We have, as in (56.7),

$$(A_D+\lambda I)^{-1}f=\mathcal{R}F_1+T_{-m-N-1}f, \tag{56.18}$$

where $\mathcal{R}$ is the parametrix (cf. (55.37), (55.38)), $\operatorname{ord}T_{-m-N-1}\le -m-N-1$, and $F_1=(f,0,\dots,0)$.

Analogously to §47 one can show that T_{-m-N-1} is an integral operator with continuous kernel $T_{-m-N-1}(x, x^{(0)}, (\sigma - i\tau))$, analytic in $\sigma - i\tau$ for $\tau \geq \tau_0 > 0$ and satisfying

$$(56.19) \qquad |T_{-m-N-1}(x, x^{(0)}, \sigma - i\tau)| \leq C|\sigma - i\tau|^{-N_1-2},$$

where N_1 is large if N is large.

We now compute $\int_\Omega G_D(x, x, t)dx$. Let U_j be such that $U_j \cap \partial\Omega = \emptyset$. Then analoguously to (47.20) we have

$$(56.20) \quad \int_{\mathbb{R}^n} \left(\frac{1}{2\pi} \int_{-\infty}^{\infty} \psi_j(x) R_j \varphi_j \delta(x - x^{(0)}) e^{it(\sigma - i\tau)} d\sigma \right) \Bigg|_{x = x^{(0)}} dx$$
$$= \frac{1}{t^{\frac{n}{2}}} \sum_{k=0}^{N_1'} \left(\int_{\mathbb{R}^n} \varphi_j(x) c_{2k}(x) dx \right) t^k + O\left(t^{N+1-\frac{n}{2}}\right),$$

where φ_j and ψ_j are the same as in (54.12), $c_{2k+1}(x) = 0$ (cf. §47).

It follows from (47.21) that

$$(56.21) \qquad \int_{\mathbb{R}^n} \varphi_j(x) c_0(x) dx = \frac{1}{(2\sqrt{\pi})^n} \int_{\mathbb{R}^n} \varphi_j(x) \sqrt{g(x)} dx.$$

As in §47, the contribution of $T_{-N-m-1}(x, x, \sigma - i\tau)$ to the heat trace will be

$$(56.22) \qquad \int_\Omega \int_{-\infty}^{\infty} T_{N-m-N-1}(x, x, \sigma - i\tau) e^{it(\sigma - i\tau)} d\sigma dx = O(t^{\frac{N_1}{m}}),$$

where N_1 is large, $m = 2$.

It remains to find the contribution of the neighborhoods U_j such that $U_j \cap \partial\Omega \neq \emptyset$. If U_j is such a neighborhood, we can introduce in U_j local coordinates $(y', y_n) = s_j(x)$ for $x \in U_j \cap \overline{\Omega}$.

Let $G_j(y, y^{(0)}, t)$ be the heat kernel in y-coordinates, i.e.,

$$(56.23) \qquad \frac{\partial G_j}{\partial t} + A_j(y, D_y) G_j = 0, \quad t > 0, \ y_n > 0,$$

$$(56.24) \qquad G_j(y, y^{(0)}, 0) = \delta(y - y^{(0)}), \quad G_j|_{y_n = 0} = 0,$$

where $y^{(0)} = s_j(x^{(0)})$. Note that (cf. §3)

$$\delta(x - x_0) = \delta(s_j^{-1}(y) - s_j^{-1}(y^{(0)})) = |J_j(y^{(0)})|^{-1} \delta(y - y^{(0)}),$$

$J_j(y) = \det \frac{\partial s_j^{-1}(y)}{\partial y}$. Therefore, comparing (56.23), (56.24) and (56.14), (56.15), we get

$$(56.25) \qquad G_D(s_j^{-1}(y), s^{-1}(y^{(0)}), t) |J_j(y^{(0)})| = G_j(y, y^{(0)}, t).$$

It follows from (56.25) that
(56.26)
$$\int_\Omega \psi_j(x)G_D(x,x,t)\varphi_j(x)dx = \int_{y_n>0} \psi_j(s_j^{-1}(y))G_j(y,y,t)\varphi_j(s_j^{-1}(y))dy.$$

We choose (y', y_n) such that $A(x, D)$ has the following form in (y', y_n)-coordinates:

(56.27)
$$A_j(y, D_y) = -\frac{1}{\sqrt{g_j(y)}}\frac{\partial}{\partial y_n}\left(\sqrt{g_j(y)}\frac{\partial}{\partial y_n}\right) - \sum_{k,p=1}^{n-1} \frac{1}{\sqrt{g_j(y)}}\frac{\partial}{\partial y_k}\left(\sqrt{g_j(y)}g_j^{pk}(y', y_n)\frac{\partial}{\partial y_p}\right),$$

where $g_j(y) = \det([g_j^{pk}]_{p,k=1}^n)^{-1}$, $g_j^{nn} = 1$, $g_j^{nk} = g_j^{kn} = 0$, $1 \le k \le n-1$ (cf. Remark 56.1).

Such coordinates are called semigeodesic coordinates.

Note that $([g_j^{pk}(y',0)]_{p,k=1}^{n-1})^{-1}$ is the Riemannian metric on $\partial\Omega \cap U_j$ induced by the metric $([g^{jk}(x)]_{j,k=1}^n)^{-1}$ on $\overline{\Omega}$. Consider

(56.28) $\psi_j(s_j^{-1}(y))R_{j0}\varphi_j(s_j^{-1}(y^{(0)}))\delta(y-y^{(0)}),\quad y^{(0)} = (y_0', y_{n0}) = s_j(x^{(0)}),$

where R_{j0} is the same as in (55.4).

The contribution of R_{j0} to the heat trace is the same as in (56.20):

(56.29)
$$\frac{1}{t^{\frac{n}{2}}}\sum_{k=0}^{N'}\left(\int_{\mathbb{R}_+^n} \varphi_j(s_j^{-1}(y))c_{2k}^{(j)}(y)dy\right)t^k + O(\left(t^{-\frac{n}{2}+N+1}\right);$$

in particular, the principal term is

(56.30)
$$\frac{1}{t^{\frac{n}{2}}}\frac{1}{(2\sqrt{\pi})^n}\int_{\mathbb{R}_+^n} \varphi_j(s_j^{-1}(y))\sqrt{g_j(y)}dy.$$

Note that $\sqrt{g_j(y)}$ is the density $\sqrt{g(x)}$ in y-coordinates, that is, $\sqrt{g(x)}\left|\det \frac{\partial s_j^{-1}(y)}{\partial y}\right| = \sqrt{g_j(y)}$. Therefore

(56.31)
$$\int_{\mathbb{R}_+^n} \varphi_j(s_j^{-1}(y))\sqrt{g_j(y)}dy = \int_\Omega \varphi_j(x)\sqrt{g(x)}dx.$$

Now we find the contribution of R_{j1} to the heat trace, where R_{j1} is the same as in (55.4). This contribution has the form

(56.32)
$$\psi_j(s_j^{-1}(y))R_{j1}h,$$

where $h = -p'R_{j0}\varphi_j(s_j^{-1}(y^{(0)})\delta(y-y^{(0)}))$ and R_{j1} is defined as in (55.22), (55.31) for $m_+ = 1$ and the Dirichlet boundary condition.

We see that $A_{j0}(y,\xi) = \xi_n^2 + \sigma_j^2(y,\xi',\lambda^{\frac{1}{2}})$, where

$$\sigma_j(y,\xi',\lambda^{\frac{1}{2}}) = \sqrt{\sum_{p,r=1}^{n-1} g_j^{pr}(y)\xi_p\xi_r + \lambda}.$$

Then $A_{j0} = A_j^+ A_j^-$, where $A_j^- = \xi_n + i\sigma_j$, $A_j^+ = \xi_n - i\sigma_j$. The principal part of (56.32) has the form (cf. (55.22), (55.31))

(56.33)
$$i\psi_j(s_j^{-1}(y))R_j^+(y,D_y)\left(\delta(y_n)p'A_{j0}^{(-1)}(y,D_y)\varphi_j(s^{-1}(y^{(0)}))\delta(y-y^{(0)})\right),$$

where

$$\begin{aligned} R_j^+(y,\xi) &= (\xi_n - i\sigma_j(y,\xi',\sqrt{\lambda}))^{-1}, \\ A_{j0}^{(-1)}(y,\xi) &= (A_{j0}(y,\xi)+\lambda)^{-1}, \\ y^{(0)} &= s_j(x^{(0)}), \quad x^{(0)} \in U_j \cap \Omega. \end{aligned} \tag{56.34}$$

We used in (56.33) that $p'R^+(c(y')\delta(y_n)) = ic(y')$ and therefore $c(y') = -ih(y')$ (cf. (55.28), (55.29)). Note that

$$\begin{aligned} v(y') &= ip'A_{j0}^{(-1)}\varphi_j(s_j^{-1}(y^{(0)}))\delta(y-y^{(0)}) \\ &= \frac{i}{(2\pi)^n}\int_{\mathbb{R}^n} \frac{\varphi_j(s_j^{-1}(y^{(0)}))e^{-iy_{n0}\xi_n + i(y'-y_0')\cdot\xi'}}{\xi_n^2 + \sigma_j^2(y',0,\xi',\sqrt{\lambda})}d\xi \\ &= \frac{i}{(2\pi)^{n-1}}\int_{\mathbb{R}^{n-1}} \frac{\varphi_j(s_j^{-1}(y^{(0)}))e^{-y_{n0}\sigma_j(y',0,\xi',\lambda^{\frac{1}{2}}) + i(y'-y_0')\cdot\xi'}}{2\sigma_j(y',0,\xi',\lambda^{\frac{1}{2}})}d\xi', \end{aligned} \tag{56.35}$$

where we used the Jordan lemma to compute the integral with respect to ξ_n. Again using the Jordan lemma we get

$$p_+R_j^+(v(y')\delta(y_n)) = \frac{i}{(2\pi)^{n-1}}\int_{\mathbb{R}^{n-1}} e^{-y_n\sigma_j(y,\xi',\lambda^{\frac{1}{2}}) + iy'\cdot\xi'}\tilde{v}(\xi')d\xi'. \tag{56.36}$$

Note that in (56.35) we applied the Jordan lemma to the lower half-plane, and in (56.36) to the upper half-plane. The operators (56.35), (56.36) are two ψdo's in $\mathbb{R}^{n-1}$. Taking the composition of (56.36) and (56.35), we get, modulo lower order terms,

(56.37)
$$-\frac{1}{(2\pi)^{n-1}}\int_{\mathbb{R}_+^{n-1}} \frac{\varphi_j(s_j^{-1}(y^{(0)}))e^{-y_{n0}\sigma_j(y',0,\xi',\lambda^{\frac{1}{2}}) - y_n\sigma_j(y,\xi',\lambda^{\frac{1}{2}}) + i(y'-y_0')\cdot\xi'}}{2\sigma_j(y',0,\xi',\lambda^{\frac{1}{2}})}d\xi'.$$

Taking $y = y^{(0)}$ and integrating with respect to y_n, we obtain

$$\int_0^\infty \frac{i\varphi_j(s_j^{-1}(y', y_n))e^{-y_n(\sigma_j(y',0,\xi',\lambda^{\frac{1}{2}})+\sigma_j(y',y_n,\xi',\lambda^{\frac{1}{2}}))}}{2\sigma_j(y',0,\xi',\lambda^{\frac{1}{2}})}dy_n$$
$$= \frac{\varphi_j(s^{-1}(y',0))}{4(\sum_{p,r=1}^{n-1} g_j^{pr}(y',0)\xi_p\xi_n + \lambda)} + O(\sigma_j^{-3}(y',0,\xi',\lambda^{\frac{1}{2}})). \tag{56.38}$$

Note that (cf. (47.10), (47.21))

(56.39)
$$\frac{1}{(2\pi)^{n-1}}\int_{\mathbb{R}^{n-1}} \frac{1}{2\pi}\int_{-\infty}^{\infty} \frac{1}{\sum_{p,r=1}^{n-1} g_j^{pr}(y',0,\xi') + i(\sigma - i\tau)} e^{it(\sigma - i\tau)} d\sigma d\xi'$$
$$= \frac{1}{(2\pi)^{n-1}}\int_{\mathbb{R}^{n-1}} e^{-t\sum_{p,r=0}^{n-1} g_j^{pr}(y',0)\xi_p\xi_r} d\xi' = \frac{1}{(2\sqrt{\pi})^{n-1}} t^{-\frac{n-1}{2}}\sqrt{g_j(y',0)},$$

where $g_j(y',0) = \det([g_j^{pr}(y',0)]_{p,r=1}^{n-1})^{-1}$. Therefore, performing the Fourier-Laplace transform of (56.38) in $\lambda = i(\sigma - i\tau)$ and integrating with respect to y', we show that the contribution of the principal term in (56.32) to the heat trace is

$$-t^{-\frac{n-1}{2}} \frac{1}{2^{n+1}\pi^{\frac{n-1}{2}}} \int_{\mathbb{R}^{n-1}} \varphi_j(s_j^{-1}(y',0))\sqrt{g_j(y',0)}dy' + O(t^{-\frac{n-2}{2}}). \tag{56.40}$$

Analogously we get

$$\frac{1}{2\pi}\int_{-\infty}^{\infty}\int_{\mathbb{R}_+^n} e^{it(\sigma-i\tau)}\psi_j(s_j^{-1}(y))R_{j1}h\big|_{y=y^{(0)}}\, dyd\sigma$$
$$= t^{-\frac{n-1}{2}}\left(\sum_{k=0}^{N'} c_k' t^{\frac{k}{2}} + O(t^{\frac{N'+1}{2}})\right), \tag{56.41}$$

where c_0' is the same as in (56.40).

To prove (56.41) we make the changes of variables $\lambda = \frac{\mu}{t}$, $\xi' = \frac{\eta'}{\sqrt{t}}$ in the left hand side of (56.41) and proceed with all terms of the parametrix (55.22), (55.31) as we did with the principal term.

Combining the contributions for all neighborhoods U_j and taking into account (56.20), we get the following result.

Theorem 56.2. *The Dirichlet heat trace $\int_\Omega G_D(x,x,t)dx$ has the following assymptotics expansion as $t \to +0$:*

$$\int_\Omega G_D(x,x,t)dx = t^{-\frac{n}{2}}\sum_{k=0}^{N} c_k t^{\frac{k}{2}} + O(t^{\frac{N+1-n}{2}}), \tag{56.42}$$

where

$$c_0 = \frac{1}{(2\sqrt{\pi})^n}\,\mathrm{Vol}(\Omega), \qquad c_1 = -\frac{1}{2^{n+1}\pi^{\frac{n-1}{2}}}\,\mathrm{Vol}(\partial\Omega). \tag{56.43}$$

We used the fact that $\int_\Omega \sqrt{g(x)}dx = \mathrm{Vol}(\Omega)$, and also the fact that $\sum_j' \int_{\mathbb{R}^{n-1}} \varphi_j(s_j^{-1}(y',0))\sqrt{g_j(y',0)}dy' = \mathrm{Vol}(\partial\Omega)$, where $\mathrm{Vol}(\partial\Omega)$ is the volume of the manifold $\partial\Omega$ in the metric induced by (56.1) and $\sum'$ is the summation over all j such that $U_j \cap \partial\Omega \neq \emptyset$.

56.4. The heat trace for the Neumann Laplacian.

Now consider the case of the Laplace-Beltrami operator (56.13) with Neumann boundary condition

$$\sum_{j,k=0}^n g^{jk}(x)\frac{\partial u}{\partial x_j}\nu_k(x)\left(\sum_{p,r=1}^n g^{pr}(x)\nu_p(x)\nu_r(x)\right)^{-\frac{1}{2}}\Bigg|_{\partial\Omega} = 0, \tag{56.44}$$

where $\nu(x) = (\nu_1(x),\dots,\nu_n(x))$ is the outward unit normal to $\partial\Omega$ in the Euclidean metric.

Note that in semigeodesic coordinates (y', y_n) (cf. (56.27)) the Neumann boundary condition has a simple form

$$\frac{\partial u_j(y',0)}{\partial y_n} = 0, \tag{56.45}$$

where $u_j(y', y_n) = u(s_j^{-1}(x))$.

To verify (56.45), note that $\nu^{(j)}(y) = (0,\dots,-1)$, and $g_j^{nn}(y',0) = 1$, $g_j^{nr}(y',0) = g_j^{rn}(y',0) = 0$.

The principal term in R_{j1} has the following form (cf. (56.33)):

$$-i\psi_j(s_j^{-1}(y))R_j^+\left(\delta(y_n)\sigma_j^{-1}p'\frac{\partial}{\partial y_n}A_{j0}^{(-1)}\varphi_j(s^{-1}(y^{(0)}))\delta(y-y^{(0)})\right), \tag{56.46}$$

since

$$p'\frac{\partial}{\partial y_n}R_j^+(c(y')\delta(y_n)) = -i\sigma_j(y',0,D',\lambda^{\frac{1}{2}})c(y') = h(y') \tag{56.47}$$

(cf. (55.28)) and $h(y') = -p'\frac{\partial}{\partial y_n}R_{j0}\varphi_j(s_j^{-1}(y^{(0)})\delta(y-y^{(0)}))$.

Here $R_j^+, \sigma_j^{-1}, A_{j0}^{(-1)}$ are the same as in (56.34). Analogously to (56.35) we have by Jordan lemma:

(56.48)

$$h(y') = -\frac{1}{(2\pi)^{n-1}}\int_{\mathbb{R}^{n-1}}\frac{1}{2}\varphi_j(s_j^{-1}(y^{(0)}))e^{-y_n\sigma_j(y',0,\xi',\lambda^{\frac{1}{2}})+i(y'-y_0')\cdot\xi'}d\xi'.$$

Using (56.47), (56.48) and repeating the computations of the form (56.37), (56.38), we get that the contribution of the principal term of R_{j1} in the case of the Neumann boundary condition is

$$t^{-\frac{n-1}{2}}\frac{1}{2^{n+1}\pi^{\frac{n-1}{2}}}\int_{\mathbb{R}^n}\varphi_j(s_j^{-1}(y',0))\sqrt{g_j(y',0)}dy'+O(t^{-\frac{n-2}{2}}), \tag{56.49}$$

i.e., it has an opposite sign to (56.40).

A theorem analogous to Theorem 56.2 holds.

Theorem 56.3. *Let $G_N(x,y,t)$ be the heat kernel corresponding to the operator* (56.13) *with the Neumann boundary condition* (56.44). *Then the heat trace $\int_\Omega G_N(x,x,t)dx$ has asymptotics of the form*

$$\int_\Omega G_N(x,x,t)dx=t^{-\frac{n}{2}}\left(\sum_{k=0}^{N}d_kt^{\frac{k}{2}}+O(t^{\frac{N+1}{2}})\right), \tag{56.50}$$

where $d_0=c_0,\ d_1=-c_1$ (cf. (56.43)*).*

56.5. The heat trace for the elliptic operator of an arbitrary order. Let $G_B(x,x^{(0)},t), x\in\Omega, x^{(0)}\in\Omega, t>0$, be the heat kernel corresponding to the initial-boundary value problem (56.8), (56.9). We find the asymptotics of the trace $\int_\Omega G_B(x,x,t)dx$. Using (56.12) with $u_0(x)=\delta(x-x^{(0)})$, we get

$$G_B(x,x^{(0)},t)=\frac{1}{2\pi}\int_{-\infty}^{\infty}(A_B+i(\sigma-i\tau)I)^{-1}\delta(x-x^{(0)})e^{it(\sigma-i\tau)}d\sigma.$$

We have

$$(A_B+\lambda I)^{-1}f=\sum_j\psi_jR_j\varphi F_1+T_{-m-N-1}f.$$

If U_j is such that $U_j\cap\partial\Omega=\emptyset$, then the contribution of U_j to $\int_\Omega G_B(x,x,t)dx$ is the same as in (56.20), i.e.,

$$\frac{1}{t^{\frac{n}{m}}}\sum_{k=0}^{N}t^{\frac{2k}{m}}\int_\Omega\varphi_j(x)c_{2k}(x)dx, \tag{56.51}$$

where the principal term is (cf. (47.19))

$$c_0(x)=\frac{1}{(2\pi)^n}\int_{\mathbb{R}^n}e^{-A_0(x,\xi)}d\xi. \tag{56.52}$$

We assumed in (56.51) that $A_0(x,\xi)$ is even in ξ (cf. §47). The contribution of T_{-m-N-1} is also the same as in (56.22). When $U_j\cap\partial\Omega\neq\emptyset$, we have $R_j=R_{j0}+R_{j1}$ in y-coordinates (cf. (55.4)).

The contribution of R_{j0} is similar to (56.51) (cf. (56.29)):

$$\frac{1}{t^{\frac{n}{m}}}\sum_{k=0}^{N_1}t^{\frac{k}{2m}}\int_{y_n>0}\varphi_j(s_j^{-1}(y))c_k^{(j)}(y)dy, \tag{56.53}$$

where the principal term $c_0^{(j)}(y)$ has the form

$$c_0^{(j)}(y) = \frac{1}{(2\pi)^n}\int_{\mathbb{R}^n} e^{-A_{j0}(y,\eta)}d\eta. \tag{56.54}$$

Note that $A_{j0}(y,\eta) = A_0\big(s_j^{-1}(y), \big(\big(\frac{\partial s_j^{-1}(y)}{\partial y}\big)^T\big)^{-1}\eta\big)$ (cf §45). Therefore, making the change of variables $y = s_j(x)$, $\eta = \big(\big(\frac{Ds_j(x)}{Dx}\big)^T\big)^{-1}\xi$ we get

$$\int_{y_n>0}\int_{\mathbb{R}^n}\varphi_j(s_j^{-1}(y))e^{-A_{j0}(y,\eta)}d\eta dy = \int_\Omega\int_{\mathbb{R}^n}\varphi_j(x)e^{-A_0(x,\xi)}dxd\xi. \tag{56.55}$$

Finally, consider the contribution of R_{j1} to the heat trace. Note that all symbols of R_{j1} and their compositions are rational in η_n. Take $y = y^{(0)}$ and compute the integrals in η_n using residues (cf. (56.35)). We get terms exponentially decaying in y_n. Integrating with respect to y_n (cf. (56.38)) we decrease the degree of homogeneity of symbols by one. Finally make the change of variables $\lambda = \frac{\lambda'}{t}, \eta' = \frac{\xi'}{t^{\frac{1}{m}}}$. We get that the contribution of R_{j1} is

$$t^{-\frac{n-1}{m}}\left(\sum_{k=0}^{N_1} d_k t^{\frac{k}{m}} + O(t^{\frac{N_1+1}{m}})\right). \tag{56.56}$$

Note that we have an extra factor $t^{\frac{1}{m}}$ in comparison with (56.51) because of the integration with respect to y_n. Therefore, combining (56.51), (56.53), (56.56), and taking into account (56.26), we get

$$\int_\Omega G_B(x,x,t)dx = \frac{1}{t^{\frac{n}{m}}}\sum_{k=0}^{N_1} c_k t^{\frac{2k}{m}} + \frac{1}{t^{\frac{n-1}{m}}}\sum_{k=0}^{N_1} d_k t^{\frac{k}{m}} + O(t^{\frac{N_1-n}{m}}), \tag{56.57}$$

where

$$c_0 = \frac{1}{(2\pi)^n}\int_\Omega\int_{\mathbb{R}^n} e^{-A_0(x,\xi)}dxd\xi. \tag{56.58}$$

We proved the following theorem:

Theorem 56.4. *The heat trace $\int_\Omega G_B(x,x,t)dx$ has the asymptotic expansion (56.57) as $t \to 0$ and the principal term c_0 has the form (56.58).*

Remark 56.1 (Construction of semigeodesic coordinates). Let U_0 be a small neighborhood of a point $x^{(0)} \in \partial\Omega$. Introduce in U_0 local coordinates (x', x_n) such that $x_n = 0$ is the equation of $\partial\Omega \cap U_0$.

Let

$$-\sum_{j,k=1}^{n}\frac{1}{\sqrt{g(x)}}\frac{\partial}{\partial x_j}\left(\sqrt{g(x)}\, g^{jk}(x)\frac{\partial u}{\partial x_k}\right) = 0$$

be the Laplace-Beltrami equation in these local coordinates in U_0. We construct the semigeodesic coordinates in U_0. Let $\varphi_n(x', x_n)$ be the solution of the equation

$$\sum_{j,k=1}^{n} g^{jk}(x)\frac{\partial\varphi_n}{\partial x_j}\frac{\partial\varphi_n}{\partial x_k} = 1, \quad \varphi(x', 0) = 0. \tag{56.59}$$

The solution of (56.59) exists for $|x_n| < \delta$, where $\delta > 0$ is small (cf. §64). Let $\varphi_k(x', x_n)$ be the solutions of linear first order equations

$$\sum_{p,r=1}^{n} g^{pr}(x)\frac{\partial\varphi_k}{\partial x_p}\frac{\partial\varphi_n}{\partial x_r} = 0, \quad \varphi_k(x', 0) = x_k, \ x' \in U_0 \cap \partial\Omega, \ 1 \le k \le n-1. \tag{56.60}$$

Make the change of variables

$$y = \varphi(x', x_n),$$

where $\varphi(x', x_n) = (\varphi_1(x', x_n), \dots, \varphi_n(x', x_n))$. Note that $J(x) = \det \frac{\partial\varphi(x)}{\partial x}$ $\neq 0$ when $\delta > 0$ is small, i.e., $y = (y_1, \dots, y_n)$ are coordinates in U_0. The Laplace-Beltrami equation has the following form in y-coordinates:

$$-\frac{1}{\sqrt{\hat g(y)}}\sum_{j,k=1}^{n}\frac{\partial}{\partial y_j}\left(\sqrt{\hat g(y)}\, \hat g^{jk}(y)\frac{\partial\hat u}{\partial y_k}\right) = 0, \tag{56.61}$$

where $\hat u(y) = u(x),\ y = \varphi(x),\ \hat g(y) = (\det[\hat g^{jk}(y)]_{j,k=1}^n)^{-1}$, and

$$\hat g^{jk}(y) = \sum_{p,r=1}^{n} g^{pr}(x)\frac{\partial\varphi_j}{\partial x_p}\frac{\partial\varphi_k}{\partial x_r}, \quad y = \varphi(x). \tag{56.62}$$

It follows from (56.59), (56.60) that $\hat g^{nn}(y) = 1, \hat g^{nk}(y) = \hat g^{kn}(y) = 0$, i.e., $(y_1, \dots, y_n)$ are semigeodesic coordinates.

57. Parametrix for the Dirichlet-to-Neumann operator

57.1. Construction of the parametrix.

Consider the Dirichlet problem in Ω,

$$A(x, D)u = 0, \quad x \in \Omega, \tag{57.1}$$

$$u|_{\partial\Omega} = g(x'), \tag{57.2}$$

where A is the Laplace-Beltrami operator (56.13). We assume that the Dirichlet problem (57.1), (57.2) has a unique solution.

Denote by Λ the operator (the Dirichlet-to-Neumann operator) that transforms $g(x')$ to the Neumann data on $\partial\Omega$ (cf. (56.44)):

$$\Lambda g = \sum_{j,k=1}^{n} g^{jk}(x)\frac{\partial u}{\partial x_j}\nu_k(x)\left(\sum_{p,r=1}^{n} g^{pr}(x)\nu_p\nu_r\right)^{-\frac{1}{2}}\Bigg|_{\partial\Omega}, \tag{57.3}$$

where ν is the outward normal. We show that Λ is a pseudodifferential operator on $\partial\Omega$, and we find its symbol.

Let $U_j \cap \partial\Omega \neq \emptyset$. In the semigeodesic coordinates (y', y_n) (cf. (56.27)) we have

$$A_j(y, D_y)u = 0, \tag{57.4}$$

$$-\frac{\partial u}{\partial y_n}\bigg|_{y_n=0} = g. \tag{57.5}$$

Note that the minus sign appears since we consider the outward normal to $\mathbb{R}^n_+$. As in (55.12), define

$$R_j^+ = \sum_{p-0}^{N} R_{jp}^+(y, D_y), \tag{57.6}$$

where

$$R_{j0}^+(y,\xi) = (\xi_n - i\sigma_j(y,\xi'))^{-1}(1-\chi_1(\xi')),\ \ \sigma_j(y,\xi) = \sqrt{\sum_{p,r=1}^{n-1} g_j^{pk}(y)\xi_p\xi_r},$$

and R_{jp}^+, $p \geq 1$, are defined as in (55.17) with $A_{j0} = \xi_n^2 + \sigma_j^2(y,\xi')$.

Note that $R_{jp}^+(y,\xi) = O(\frac{1}{\xi_n^2})$ for $p \geq 1$. Therefore $p'R_{jp}^+(h(y')\delta(y_n)) = 0$ when $p \geq 1$, $p'R_{j0}^+(h(y')\delta(y_n)) = ih(y')$, and we have

$$p'R_j^+(h(y')\delta(y_n)) = ih(y'). \tag{57.7}$$

It follows from (55.12) that

$$p_+A_j(y,D_y)R_j^+(g(y')\delta(y_n)) = p_+T_{-N}(g(y')\delta(y_n)), \tag{57.8}$$

where T_{-N} satisfies an estimate of the form (55.24) with $m_+ = 1$.

Let

$$\Lambda_j g = \lim_{y_n \to +0} i\frac{\partial}{\partial y_n} R_j^+(g(y')\delta(y_n)). \tag{57.9}$$

Then Λ_j is a pseudodifferential operator in $\mathbb{R}^{n-1}$ with symbol

$$\Lambda_j(y',\xi') = -\Pi'\left(\xi_n \sum_{p=0}^{N} R_{jp}^+(y',0,\xi)\right), \tag{57.10}$$

where $\Pi' D_+(y',\xi) = \frac{1}{2\pi}\int_{\gamma_+} D_+(y',\xi',\xi_n)d\xi_n$, γ_+ is a contour enclosing all the poles of $D_+(y',0,\xi',\xi_n)$ (cf. (55.27)). In (57.10) we used that $\Pi'\frac{\partial R_{jp}^+}{\partial y_n} = 0$, since $\frac{\partial R_{jp}}{\partial y_n} = O\left(\frac{1}{\xi_n^2}\right)$, $p \geq 0$. It follows from (57.10) that

$$\Lambda_j(y',\xi') = \sum_{p=0}^{N} \Lambda_{jp}(y',\xi'), \tag{57.11}$$

where $\operatorname{ord}\Lambda_{jp} = 1-p$. We compute several terms in (57.11). Note that $A_j(y,D_y)$ has the form

$$A_j(y,D_y) = -\frac{\partial^2}{\partial y_n^2} - \sum_{p,r=1}^{n-1} g_j^{pr}(y)\frac{\partial^2}{\partial y_p \partial y_r} + A_1(y,D_y), \tag{57.12}$$

where

$$A_1 = -\frac{g_{jy_n}}{2g_j}\frac{\partial}{\partial y_n} - \sum_{p,r=1}^{n-1} \frac{1}{\sqrt{g_j}}\frac{\partial}{\partial y_p}(\sqrt{g_j}g_j^{pr})\frac{\partial}{\partial y_r}. \tag{57.13}$$

We have

$$\begin{aligned}\Lambda_{j0}(y',\xi') &= -\Pi'\frac{\xi_n}{\xi_n - i\tilde{\sigma}_j}(1-\chi_1(\xi')) = \sigma_j(1-\chi_1(\xi')) \\ &= \sqrt{\sum_{p,r=1}^{n-1} g_j^{pr}(y',0)\xi_p\xi_r}(1-\chi_1(\xi')),\end{aligned} \tag{57.14}$$

$$\Lambda_{jp}(y',\xi') = \Pi'\xi_n(A_{j0}^+)^{-1}\Pi^+(A_{j0}^-)^{-1}T_p, \quad p \geq 1, \tag{57.15}$$

where T_p is given by (55.18). In particular

$$T_1 = A_1(y,\xi)R_{j0}^+ + i\sum_{k=1}^{n}\frac{\partial A_0}{\partial \xi_k}\frac{\partial R_{j0}^+}{\partial y_k}. \tag{57.16}$$

Note that $A_-^{-1}T_p$ has poles at $\xi_n = \pm i\sigma_j$, and therefore

$$\Pi^+ A_-^{-1}T_p = \sum_{k=1}^{p+1}\alpha_{pk}(y',\xi')(\xi_n - i\sigma_j)^{-k}. \tag{57.17}$$

Note that $\Pi'\xi_n(\xi_n - i\sigma_j)^{-r} = 0$ for $r \geq 3$. Since $A_+^{-1} = (\xi_n - i\sigma_j)^{-1}$, we have

$$\Lambda_{jp}(y',\xi') = \Pi'\frac{\xi_n\alpha_{p1}(y',\xi')}{(\xi_n - i\sigma_j)^2} = i\alpha_{p1}(y',\xi'); \tag{57.18}$$

i.e., to compute $\Lambda_{jp}(y',\xi')$ one needs to find the residue $\alpha_{p1}(y',\xi')$ of $A_-^{-1}T_p$ at $\xi_n = i\sigma_j$. In particular, when $p=1$, we get for $|\xi'|>1$:

$$\Lambda_{j1}(y',\xi') = i\,\mathrm{res}_{\xi_n=i\sigma_j}\left[\frac{A_1(y',0,\xi',\xi_n)}{\xi_n^2+\sigma_j^2} - \frac{\sum_{k=1}^n \frac{\partial A_0(y',0,\xi',\xi_n)}{\partial \xi_k}\sigma_{jy_k}}{(\xi_n+i\sigma_j)(\xi_n-i\sigma_j)^2}\right]$$

$$= \frac{A_1(y',0,\xi',i\sigma_j)}{2\sigma_j} - \frac{i\sum_{k=1}^n \frac{\partial A_0(y',0,\xi',-i\sigma_j)}{\partial\xi_k}\sigma_{jy_k}}{4\sigma_j^2}.$$

We used here that

$$\mathrm{res}_{\xi_n=i\sigma}\frac{\sum_{k=1}^n \frac{\partial A_0}{\partial\xi_k}\sigma_{jy_k}}{(\xi_n+i\sigma)(\xi_n-i\sigma)^2} = \frac{1}{2\pi i}\int_{-\infty}^{\infty}\frac{\sum_{k=1}^n \frac{\partial A_0}{\partial\xi_k}\sigma_{jy_k}}{(\xi_n+i\sigma)(\xi_n-i\sigma)^2}d\xi_n,$$

and we computed the integral using the residue in the lower half-plane.

Now we construct the Dirichlet-to-Neumann (D-to-N) operator in Ω. Let U_j,φ_j,ψ_j be the same as in (55.36), and let R_j^+ be the same as in (57.6) in local coordinates. Denote by Rg the following operator:

$$Rg = \sum_j{}' \psi_j(x)R_j^+\varphi_j(s_j^{-1}(y',0))g(s_j^{-1}(y',0))\delta(y_n). \tag{57.19}$$

Here $\sum_j'$ is the summation over j such that $U_j\cap\partial\Omega\neq\emptyset$ and we return to x-coordinates after applying R_j^+. We have

$$A(x,D)Rg = T_{-N}g, \tag{57.20}$$

$$Rg|_{\partial\Omega} = g + T_{-N}^{(1)}g, \tag{57.21}$$

where $\operatorname{ord}T_{-N}\le -N$, $\operatorname{ord}T_{-N}^{(1)}\le -N$ in Ω.

Denote by v the solution of the Dirichlet problem

$$A(x,D)v = -T_{-N}g, \tag{57.22}$$

$$v|_{\partial\Omega} = -T_N^{(1)}g. \tag{57.23}$$

Then (cf. §54)

$$\|v\|_{s,\Omega}\le C\|T_{-N}g\|_{s-2,\Omega} + C\|T^{(1)}g\|_{s-\frac12,\partial\Omega}\le C\|g\|_{s-\frac12-N,\partial\Omega}. \tag{57.24}$$

Therefore $Rg+v$ satisfies

$$A(Rg+v) = 0 \quad \text{in} \quad \Omega, \tag{57.25}$$

$$(Rg+v)|_{\partial\Omega} = g. \tag{57.26}$$

Substituting $u = Rf+v$ into (57.3), we get

$$\Lambda g = \sum_j{}' p'\psi_j(x)\Lambda_j p'\varphi_j g + \Lambda_{-N}g, \tag{57.27}$$

where $\operatorname{ord}\Lambda_{-N}\le -N$ and Λ_j are the same as in (57.9). □

57.2. Determination of the metric on the boundary.

We can use the parametrix of the D-to-N operator to solve the inverse problem of determining the metric tensor on the boundary.

Theorem 57.1. *Suppose we know Λ on $\partial\Omega$ for all $f \in C^\infty(\partial\Omega)$. Then we can determine the metric on $\partial\Omega$ induced by* (56.13) *and all its normal derivatives on $\partial\Omega$.*

Proof: It follows from (57.14) that we can determine

$$\sum_{p,r=1}^{n-1} g_j^{pr}(y',0)\xi_p\xi_r,$$

i.e., we know the metric on $\partial\Omega$. Therefore, we know all the derivatives of $g_j^{pr}(y',0)$ in y'. It remains to determine the derivatives $\frac{\partial^k g_j^{pr}(y',0)}{\partial y_n^k}$, $1 \leq p$, $r \leq n-1$, for all $k \geq 1$. It follows from (57.16) that the only term in T_1 that contains the derivative $\frac{\partial}{\partial y_n}$ is $T_{11} = -\frac{1}{2g_j}\frac{\partial g_j}{\partial y_n} - 2\xi_n(\xi_n - i\sigma_j)^{-2}\sigma_{jy_n}$ for $|\xi'| > 1$. Note that $\sigma_{jy_n} = \frac{1}{2\sigma_j}(\sum_{p,r=1}^{n-1} g_{jy_n}^{pr}\xi_p\xi_r)$. Computing the residue of $(A_{j0}^-)^{-1}T_{11}$, we can recover $\frac{g_{jy_n}}{g_j}$ and σ_{jy_n}. Analogously, from Λ_{jp}, $p \geq r$, we can recover $\frac{\partial^k \sigma_j}{\partial y_n^k}$ and, therefore, we can find all $\frac{\partial^k g_j^{pr}(y',0)}{\partial y_n^k}$, $k \geq 1$.

Remark 57.1. The construction of the parametrix for the Dirichlet-to-Neumann (D-to-N) operator works for arbitrary second order elliptic operators and not necessarily for operators of the form (56.13) (cf. §55). Also it is not necessary to use the semigeodesic coordinates. Consider, for example, the conductivity equation (cf. Example 34.1):

$$\sum_{j=1}^{n} \frac{\partial}{\partial x_j}\gamma(x)\frac{\partial u}{\partial x_j} = 0. \tag{57.28}$$

Here the D-to-N operator has the form

$$\Lambda f = \gamma(x)\frac{\partial u}{\partial \nu}\Big|_{\partial\Omega},$$

where $u|_{\partial\Omega} = f$, $\frac{\partial}{\partial\nu}$ is the normal derivative, and $u(x)$ satisfies (57.28).

Since the principal term in (57.28) is $\gamma(x)\Delta u$, it is natural to use the semigeodesic coordinates (y', y_n) for the Laplacian in a neighborhood of $x^{(0)} \in \partial\Omega$. In these coordinates the principal symbol of (57.28) has the form $\hat\gamma(y',y_n)(\eta_n^2 + \hat\sigma^2(y,y_n,\eta'))$, where $\hat\sigma^2(y',y_n,\eta')$ is a positive definite quadratic form in η'. We have, as in (57.6), $R_{j0}^+ = (\eta_n - i\hat\sigma(y',y_n,\eta'))^{-1}(1-\chi_1(\eta'))$. Then, as in (57.14), the principal symbol of $\hat\Lambda$ will be $\hat\gamma(y',0)\hat\sigma(y',0,\eta')$. Therefore we can recover $\gamma|_{\partial\Omega}$ from the principal symbol. Analogously to the proof of Theorem 57.1 we can recover $\frac{\partial^k\gamma}{\partial\nu^k}\big|_{\partial\Omega}$ for all $k \geq 1$.

58. Spectral theory of elliptic operators

58.1. The nonselfadjoint case.

As in §56, let D_B be the subspace of $H_s(\Omega)$ consisting of $u(x) \in H_s(\Omega)$ such that $B_j u|_{\partial\Omega} = 0$, $1 \le j \le m_+$. Denote by A_B the operator A with the domain D_B. Since $A_B + \lambda I$ is a Fredholm operator for any $\lambda \in \mathbb{C}$ and is invertible (cf. Theorem 56.1) when $\lambda \in C_\delta$, $|\lambda| \ge \lambda_0$, the operator $(A_B + \lambda I)^{-1}$ is a meromorphic function in $\mathbb{C}$ (cf. Lemma 37.5). Let $-\lambda_j$ be a pole of $(A_B + \lambda I)^{-1}$. Then λ_j is an eigenvalue of A_B and there exists a sequence of eigenvalues λ_j, $\operatorname{Re}\lambda_j \to +\infty$ and a sequence $\{\varphi_j\}$ of eigenvectors $\varphi_j \in D_B$ such that

$$(A_B - \lambda_j)\varphi_j = 0, \quad 1 \le j \le +\infty. \tag{58.1}$$

For simplicity of notation we will repeat the eigenvalues λ_j according to their multiplicity. Since A_B is not assumed to be selfadjoint, there is a finite-dimensional subspace spanned by the root vectors corresponding to φ_j:

$$\begin{gathered}\varphi_{jr},\ 1 \le r \le k_j, \quad \varphi_{jk_j} = \varphi_j, \quad \text{such that} \\ (A_B - \lambda_j)\varphi_{j1} = \varphi_{j2}, \dots, (A_B - \lambda_j)\varphi_{j,k_j-1} = \varphi_{jk_j}, \quad (A_B - \lambda_j)\varphi_{jk_j} = 0,\end{gathered} \tag{58.2}$$

i.e., we have a $k_j \times k_j$ Jordan block corresponding to λ_j. The number k_j is called an algebraic multiplicity of λ_j.

Denote by P_j the projector on the span of $(\varphi_{j1}, \dots, \varphi_{jk_j})$. Therefore $P_j^2 = P_j, P_j\varphi_{jk} = \varphi_{jk}$, $1 \le k \le k_j$. We have $P_j u_0 = \sum_{k=1}^{k_j} (u_0, \psi_{jk})\varphi_{jk}(x)$, where the ψ_{jr} belong to the dual space $H_s^*(\Omega)$, $(\varphi_{jk}, \psi_{ir}) = 0$ when $(j,k) \ne (i,r)$, and $(\varphi_{jk}, \psi_{jk}) = 1$. Since $(A_B + \lambda I)^{-1}$ is a meromorphic operator-valued function, $\lambda \in \mathbb{C}$, the singular part of $(A_B + \lambda)^{-1} u_0$ corresponding to the pole λ_j has the form

$$\sum_{k=1}^{k_j} (\lambda + \lambda_j)^{-k} P_{jk} u_0, \tag{58.3}$$

where

$$P_{jk} u_0 = \sum_{r=1}^{k_j - k + 1} (u_0, \psi_{j,r+k-1})\varphi_{jr}, \quad P_{j1} u_0 = P_j u_0.$$

Note that there is only a finite number of poles in any strip $-N < \operatorname{Re}\lambda < \lambda_0$, for all N since $|\operatorname{Im}\lambda_j| \le -C\operatorname{Re}\lambda_j \le CN$ in such a strip.

Let $G_B(x,y,t)$ be the heat kernel corresponding to the initial-boundary value problem (56.8), (56.9):

$$
\begin{aligned}
&\frac{\partial G_B(x,y,t)}{\partial t} + A(x,D)G_B(x,y,t) = 0, \quad t > 0,\ x \in \Omega,\ y \in \Omega, \\
&B_j(x,D)G_B(x,y,t)|_{\partial\Omega\times(0,+\infty)} = 0, \quad 1 \le j \le m_+, \\
&G_B(x,y,0) = \delta(x-y),
\end{aligned}
\tag{58.4}
$$

i.e., $u(x,t) = \int_\Omega G_B(x,y,t)u_0(y)dy$. It follows from the parametrix construction that $G_B(x,y,t)$ is smooth in (x,y,t) when $t > 0$.

We now find the trace $\int_\Omega G_B(x,x,t)dx$ of $G_B(x,y,t)$.

58.2. Trace class operators.

Let H be a Hilbert space and $\{e_j\}_{j=1}^\infty$ an orthonormal basis in H. An operator T is called a Hilbert-Schmidt operator (HSO) if

$$
\sum_{j,k=1}^{\infty} |(Te_j, e_k)|^2 < +\infty. \tag{58.5}
$$

By Parseval's equality, $\|Te_j\|^2 = \sum_{k=1}^\infty |(Te_j,e_k)|^2$. Therefore

$$
\sum_{j,k=1}^{\infty} |(Te_j, e_k)|^2 = \sum_{j=1}^{\infty} \|Te_j\|^2.
$$

Note that $T_N u = \sum_{j=1}^N (u,e_j)Te_j$ is a finite rank operator and thus T_N is compact. By the Cauchy-Schwartz inequality, we get

$$
\begin{aligned}
\|(T - T_N)u\| = \left\| \sum_{j=N+1}^{\infty} (u,e_j)Te_j \right\| &\le \sum_{j=N+1}^{\infty} |(u,e_j)|\ \|Te_j\| \\
&\le \|u\| \Big(\sum_{j=N+1}^{\infty} \|Te_j\|^2 \Big)^{\frac{1}{2}}.
\end{aligned}
$$

Therefore the norm of $T - T_N$ tends to 0 as $N \to \infty$. Hence HSO is a compact operator as the limit in the operator norm of compact operators. Let $H = L_2(\Omega)$. An example of HSO is an integral operator $Ku = \int_\Omega K(x,y)u(y)dy$ in $L_2(\Omega)$, where $K(x,y) \in L_2(\Omega\times\Omega)$. Indeed, if $\{e_j(x)\}_{j=1}^\infty$ is an orthonormal basis in $L_2(\Omega)$, then $\{\overline{e_j}(x)e_k(y)\}_{j,k=1}^\infty$ is an orthonormal basis in $L_2(\Omega\times\Omega)$, and

$$
\int_\Omega\int_\Omega |K(x,y)|^2 dxdy = \sum_{j,k=1}^{\infty} |(Ke_j, e_k)|^2
$$

is Parseval's equality in $L_2(\Omega\times\Omega)$. We used such HSO's in §§42 and 30.

Note that T^* is an HSO if T is an HSO, since $(Te_j, e_k) = (e_j, T^*e_k)$ and therefore (58.5) is equal to $\sum_{j,k=1}^{\infty} |(T^*e_k, e_j)|^2$.

Definition 58.1. An operator T in H is called a trace class operator (TCO) if for any orthonormal basis $\{e_j\}_{j=1}^{\infty}$ we have

$$\sum_{j=1}^{\infty} |(Te_j, e_j)| < +\infty. \tag{58.6}$$

If $T = T_1T_2$, where $T_k, k = 1, 2$, are HSO, then T is a trace class operator. We have, using the Cauchy-Schwartz inequality,

$$\sum_{j-1}^{\infty} |(T_1T_2e_j, e_j)| = \sum_{j=1}^{\infty} |(T_2e_j, T_1^*e_j)| \leq \sum_{j=1}^{\infty} \|T_2e_j\| \, \|T_1^*e_j\|$$
$$\leq \Big(\sum_{j=1}^{\infty} \|T_2e_j\|^2\Big)^{\frac{1}{2}} \Big(\sum_{j=1}^{\infty} \|T_1^*e_j\|^2\Big)^{\frac{1}{2}} < +\infty.$$

Note that T is a compact operator since $T_k, k = 1, 2$, are compact. It is also true that if T is a TCO, then T is a product of two HSO's. We skip the proof.

Now we show that for a TCO T the sum

$$\sum_{j=1}^{\infty} (Te_j, e_j) \tag{58.7}$$

is independent of the choice of an orthonormal basis in H.

Let $\{f_k\}_{k=1}^{\infty}$ be another orthonormal basis, and let $T = T_1T_2$, where T_1, T_2 are HSO's. We have

$$T_pe_j = \sum_{k=1}^{\infty} (T_pe_j, f_k)f_k, \quad p = 1, 2, \quad 1 \leq j < +\infty.$$

Therefore, by Parseval's equality,

$$(Te_j, e_j) = (T_2e_j, T_1^*e_j) = \sum_{k=1}^{\infty} (T_2e_j, f_k)\overline{(T_1^*e_j, f_k)} = \sum_{k=1}^{\infty} (e_j, T_2^*f_k)\overline{(e_j, T_1f_k)}.$$

Hence again, using Parseval's equality, we get

$$\sum_{j=1}^{\infty} (Te_j, e_j) = \sum_{j,k=1}^{\infty} (T_1f_k, e_j)\overline{(T_2^*f_k, e_j)} \tag{58.8}$$
$$= \sum_{k=1}^{\infty} \Big(\sum_{j=1}^{\infty} (T_1f_k, e_j)\overline{(T_2^*f_k, e_j)}\Big) = \sum_{k=1}^{\infty} (T_1f_k, T_2^*f_k) = \sum_{k=1}^{\infty} (T_2T_1f_k, f_k).$$

Taking $f_k = e_k, 1 \le k \le +\infty$ in (58.8), we get

$$\sum_{k=1}^{\infty}(T_1T_2f_k, f_k) = \sum_{k=1}^{\infty}(T_2T_1f_k, f_k). \tag{58.9}$$

Comparing (58.8) and (58.9), we finally obtain

$$\sum_{j=1}^{\infty}(Te_j, e_j) = \sum_{k=1}^{\infty}(Tf_k, f_k).$$

The sum (58.7) is called the trace of T and it is denoted by $\operatorname{Tr} T$. □

The main result of this subsection is the following theorem:

Theorem 58.1 (Lidskii's theorem). *Let T be a TCO and let $\lambda_j, 1 \le j < +\infty$, be the eigenvalues of T counted with their multiplicities. Let k_j be the size of the Jordan block corresponding to λ_j. Then*

$$\operatorname{Tr} T = \sum_{j=1}^{\infty} k_j\lambda_j. \tag{58.10}$$

A proof of Lidskii's theorem can be found in Simon [**Si**].

We apply Theorem 58.1 to the heat trace $\int_\Omega G_B(x,x,t)dx$. Denote by e^{-tA} the integral operator with the kernel $G_B(x,y,t)$. It follows from the construction of the parametrix that $G_B(x,y,t)$ is smooth. Therefore e^{-tA} is a Hilbert-Schmidt operator for any $t > 0$. Therefore e^{-tA} is a TCO, since $e^{-tA} = e^{-\frac{t}{2}A}e^{-\frac{t}{2}A}$, and $e^{-\frac{t}{2}A}$ is an HSO.

Applying Lidskii's theorem and combining it with Theorem 56.4, we get:

Theorem 58.2. *The following formula holds for the heat trace:*

$$\int_\Omega G_B(x,x,t)dx = \sum_{j=1}^{\infty} k_je^{-\lambda_jt} = \frac{1}{t^{\frac{n}{m}}}\sum_{k=0}^{N} c_kt^{\frac{k}{m}} + O(t^{\frac{N-n+1}{m}}), \tag{58.11}$$

where

$$c_0 = \frac{1}{(2\pi)^n}\int_\Omega\int_{R^n} e^{-A_0(x,\xi)}dxd\xi.$$

Consider the Laplace-Beltrami operators A_D and A_N with the zero Dirichlet or Neumann conditions. Since A_D and A_N are selfadjoint with respect to the scalar product $\int_\Omega u(x)\overline{v(x)}\sqrt{g(x)}dx$, we have that $k_j = 1$ in (58.10) (i.e., the Jordan blocks are 1×1).

Theorem 58.3. *Let $\int_\Omega G_D(x,x,t)dx$ and $\int_\Omega G_N(x,x,t)dx$ be the heat traces for A_D and A_N, respectively. Then*

$$\begin{aligned}(58.12)\quad \int_\Omega G_D(x,x,t)dx &= \sum_{j=1}^{\infty} e^{-\lambda_j^{(D)}t} \\ &= \frac{1}{(2\sqrt{\pi})^n}\frac{1}{t^{\frac{n}{2}}}\operatorname{Vol}(\Omega) - \frac{1}{2^{n+1}\pi^{\frac{n-1}{2}}}\frac{1}{t^{\frac{n-1}{2}}}\operatorname{Vol}(\partial\Omega) + O\Big(\frac{1}{t^{\frac{n}{2}-1}}\Big),\end{aligned}$$

$$\begin{aligned}(58.13)\quad \int_\Omega G_N(x,x,t)dx &= \sum_{j=1}^{\infty} e^{-\lambda_j^{(N)}t} \\ &= \frac{1}{(2\sqrt{\pi})^n}\frac{1}{t^{\frac{n}{2}}}\operatorname{Vol}(\Omega) + \frac{1}{2^{n+1}\pi^{\frac{n-1}{2}}}\frac{1}{t^{\frac{n-1}{2}}}\operatorname{Vol}(\partial\Omega) + O\Big(\frac{1}{t^{\frac{n}{2}-1}}\Big),\end{aligned}$$

where $\lambda_j^{(D)}$, $1 \le j < +\infty$, are the eigenvalues of A_D, and $\lambda_j^{(N)}$, $1 \le j < +\infty$, are the eigenvalues of A_N.

58.3. The selfadjoint case.

Consider the case where A_B is selfadjoint with respect to some scalar product $(u,v) = \int_\Omega u(x)\overline{v(x)}\sigma(x)dx$, and $\sigma(x) > 0$ is continuous. Suppose A_B is invertible and consider $s = m$. Then A_B^{-1} is compact in $L_2(\sigma,\Omega)$ and is also selfadjoint:

$$(58.14)\qquad (A_B^{-1}f, g) = (f, A_B^{-1}g), \quad \forall f, g \in L_2(\sigma,\Omega).$$

Therefore, there exists an orthonormal basis in $L_2(\sigma,\Omega)$ consisting of the eigenfunctions $\varphi_j(x)$ (see Rudin [**R**]):

$$A_B^{-1}\varphi_j = \mu_j\varphi_j, \quad \mu_j \to 0 \text{ and real.}$$

Then

$$A_B\varphi_j = \lambda_j\varphi_j \quad \text{when} \quad \lambda_j = \frac{1}{\mu_j}.$$

Since $k_j = 1$, we have from (58.11):

$$(58.15)\qquad \int_\Omega G_B(x,x,t)dx = \sum_{j=1}^{\infty} e^{-\lambda_j t}.$$

Denote

$$(58.16)\qquad N_B(\lambda) = \sum_{\lambda_j < \lambda} 1.$$

We find the asymptotics of $N_B(\lambda)$ as $\lambda \to +\infty$ in the selfadjoint case. Note that $N_B(\lambda)$ has jumps at $\lambda = \lambda_j$ equal to 1, and $N_B(\lambda)$ is a nondecreasing function. Therefore

$$\sum_{j=1}^{\infty} e^{-\lambda_j t} = \int_{-\lambda_0}^{\infty} e^{-\lambda t} dN_B(\lambda), \tag{58.17}$$

where the right hand side of (58.17) is the Stieltjes integral. It follows from (58.11) that

$$\int_{-\lambda_0}^{-\infty} e^{-\lambda t} dN_B(\lambda) = \frac{c_0}{t^{\frac{n}{m}}} + O\Big(\frac{1}{t^{\frac{n-1}{m}}}\Big). \tag{58.18}$$

The following Tauberian theorem holds (see Simon [**Si**]):

Theorem 58.4 (Karamata's Tauberian theorem). *If $\sigma(\lambda)$ is a nondecreasing function on $[0,+\infty)$, $\sigma(0)=0$, and if*

$$\int_0^{\infty} e^{-\lambda t} d\sigma(\lambda) = \frac{c_0}{t^{\alpha}} + o\Big(\frac{1}{t^{\alpha}}\Big),$$

then

$$\sigma(\lambda) = \frac{c_0}{\Gamma(\alpha+1)}\lambda^{\alpha} + o(\lambda^{\alpha}).$$

Applying Theorem 58.3 to $\sigma(\lambda) = N_B(\lambda)$, we get

$$N_B(\lambda) = \frac{c_0}{\Gamma(\frac{n+m}{m})}\lambda^{\frac{n}{m}} + o(\lambda^{\frac{n}{m}}), \quad \lambda \to +\infty, \tag{58.19}$$

where c_0 is the same as in (58.11). Note that $A_0(x,\xi) > 0$ for $\xi \neq 0$ in the selfadjoint case. We will rewrite c_0 in a different form. We have:

$$e^{-tA_0(x,\xi)} = \int_0^{\infty} e^{-ts}\delta(s - A_0(x,\xi))ds = \int_0^{\infty} e^{-ts} d\theta(s - A_0(x,\xi)), \tag{58.20}$$

where $\theta(s) = 1$ for $s > 0$, $\theta(s) = 0$ for $s < 0$, and $\frac{d\theta(s-A_0(x,\xi))}{ds} = \delta(s - A_0(x,\xi))$ (cf. §5). Integrating (58.20) over $\Omega \times \mathbb{R}^n$, we get:

$$\frac{c_0}{t^{\frac{n}{m}}} = \frac{1}{(2\pi)^n}\int_{\Omega}\int_{\mathbb{R}^n} e^{-tA_0(x,\xi)}dxd\xi = \int_0^{\infty} e^{-ts}dW(s),$$

where

$$W(s) = \frac{1}{(2\pi)^n}\int_{\Omega}\int_{\mathbb{R}^n} \theta(s - A_0(x,\xi))dxd\xi = \frac{1}{(2\pi)^n}\int\int_{A_0(x,\xi)<s} dxd\xi.$$

Therefore $\frac{c_0}{t^{\frac{n}{m}}}$ is the Laplace transform of $W'(s)$. Then the inverse Laplace transform of $\frac{c_0}{t^{\frac{n}{m}}}$ is $W'(s) = \frac{c_0 s^{\frac{n}{m}-1}}{\Gamma(\frac{n}{m})}$. Hence

$$W(\lambda) = \frac{c_0\lambda^{\frac{n}{m}}}{\Gamma(\frac{n}{m}+1)}. \tag{58.21}$$

Comparing (58.21) with (58.19), we get

$$N_B(\lambda) = W(\lambda) + o(\lambda^{\frac{n}{m}}) = \frac{1}{(2\pi)^n}\int_\Omega\int_{A_0(x,\xi)<\lambda} dx d\xi + o(\lambda^{\frac{n}{m}}). \tag{58.22}$$

We have proven

Theorem 58.5. *Let* $N_B(\lambda) = \sum_{\lambda_j<\lambda} 1$. *Then*

$$N_B(\lambda) = \frac{1}{(2\pi)^n\Gamma(\frac{n+m}{n})}\int_\Omega\int_{\mathbb{R}^n} e^{-A_0(x,\xi)} dx d\xi\ \lambda^{\frac{n}{m}} + o(\lambda^{\frac{n}{m}})$$
$$= \frac{1}{(2\pi)^n}\int_\Omega\int_{A_0(x,\xi)<\lambda} dx d\xi + o(\lambda^{\frac{n}{m}}).$$

□

Let A_B be selfadjoint. Replacing A_B by $A_B+\mu$ if needed, we may assume that A_B is positive definite in $L_2(\sigma,\Omega)$, i.e., all $\lambda_j > 0$, $1 \le j < +\infty$. Let $\{\varphi_j(x)\}$ be the orthonormal basis of eigenfunctions of A_B. We find conditions for the uniform convergence of the eigenfunction expansions. For any $\alpha > 0$, define

$$A_B^{-\alpha} = \frac{1}{2\pi i}\int_{\Gamma_0} (-\lambda)^{-\alpha}(A_B + \lambda I)^{-1} d\lambda, \tag{58.23}$$

where Γ_0 is the curve $\lambda = \lambda_0(\sigma)$ such that

$$\lambda(\sigma) = i\sigma - \varepsilon - \varepsilon|\sigma| \quad \text{when} \quad -\infty < \sigma \le 0,$$
$$\lambda_0(\sigma) = i\sigma - \varepsilon - \varepsilon\sigma \quad \text{when} \quad 0 \le \sigma < +\infty,$$

$\varepsilon > 0$ is small and $\lambda(\sigma) \in C_\delta$, $(-\lambda)^{-\alpha} = e^{-\alpha\ln(-\lambda)} = e^{-\alpha(\ln|\lambda| + i\arg(-\lambda))}$ and $\arg(-\lambda) = -\pi + \arg\lambda$, $0 < \arg\lambda < 2\pi$.

Let $\alpha > 0$ satisfy $m\alpha > n$. Then the explicit form of the parametrix gives that the kernel $E_\alpha(x,y)$ of $A_B^{-\alpha}$ is continuous in $\overline{\Omega}\times\overline{\Omega}$ and $\operatorname{ord} A_B^{-\alpha} \le -\alpha m$. Note that $A_B^{-\alpha}$ can be defined by (58.23) also in the nonselfadjoint case.

We have $(A_B+\lambda I)^{-1}\varphi_j = (\lambda_j+\lambda)^{-1}\varphi_j$. By the Cauchy integral formula

$$\frac{1}{2\pi i}\int_{\Gamma_0} \lambda^{-\alpha}(\lambda_j + \lambda)^{-1} d\lambda = \lambda_j^{-\alpha}.$$

Therefore

$$A_B^{-\alpha}\varphi_j = \lambda_j^{-\alpha}\varphi_j,\ 1 \le j < +\infty. \tag{58.24}$$

In the selfadjoint case it will be convenient to represent $E_\alpha(x,y)$ in the form $E_\alpha(x,y) = K_\alpha(x,y)\sigma(y)$. Then $K_\alpha(x,y) = K_\alpha(y,x)$. Since $A_B^{-\alpha}$ is positive definite, we see that the trace $K_\alpha(x,x)\sigma(x) \ge 0$. Indeed, $(A^{-\alpha}\psi,\psi) \ge 0$, $\forall\psi \in C_0^\infty$. Taking $\psi_n(x - x^{(1)})$ as a delta-like sequence $\psi_n(x - x^{(0)}) \to \delta(x - x^{(0)})$ and using that $K_\alpha(x,y)$ is continuous, we get

$$\int_\Omega K_\alpha(x,y)\psi_n(x-x^{(0)})\psi_n(y-x^{(0)})\sigma(x)\sigma(y)dxdy \to K_\alpha(x^{(0)},x^{(0)})\sigma^2(x^{(0)}) \ge 0.$$

We have

$$A_B^{-\alpha}g = \sum_{j=1}^{\infty}(A_B^{-\alpha}g, \varphi_j)\varphi_j = \sum_{j=1}^{\infty}(g, A_B^{-\alpha}\varphi_j)\varphi_j = \sum_{j=1}^{\infty}\frac{1}{\lambda_j^{\alpha}}(g, \varphi_j)\varphi_j.$$

The operator $A_B^{-\alpha}g - \sum_{j=1}^{N}\frac{1}{\lambda_j^{\alpha}}(g, \varphi_j)\varphi_j$ is also positive definite for any N. Therefore its trace is also nonnegative,

$$\int_{\Omega}\Big(K_\alpha(x,x) - \sum_{j=1}^{N}\frac{1}{\lambda_j^{\alpha}}|\varphi_j(x)|^2\Big)\sigma(x)dx \geq 0,$$

i.e.,

$$\sum_{j=1}^{N}\frac{1}{\lambda_j^{\alpha}} \leq \int_{\Omega}K_\alpha(x,x)\sigma(x)dx,$$

since $\int_{\Omega}|\varphi_j|^2\sigma(x)dx = 1$. Therefore the series $\sum_{j=1}^{\infty}\frac{1}{\lambda_j^{\alpha}}$ converges and

$$\int_{\Omega}K_\alpha(x,x)\sigma(x)dx = \sum_{j=1}^{\infty}\frac{1}{\lambda_j^{\alpha}} < +\infty \tag{58.25}$$

when $\alpha > \frac{n}{m}$.

Theorem 58.6. *Let $f = A_B^{-\alpha}g$, $g \in L_2(\sigma, \Omega)$ and let $p \in \mathbb{N}$ be such that $n + p < m\alpha$. Then the series*

$$\frac{\partial^k f}{\partial x^k} = \sum_{j=1}^{\infty}(f, \varphi_j)\frac{\partial^k \varphi_j(x)}{\partial x^k}, \tag{58.26}$$

$1 \leq |k| \leq p$, converges absolutely and uniformly in $\overline{\Omega}$.

Proof: We have (cf. §13)

$$\max_{x\in\overline{\Omega}}\left|\frac{\partial^k \varphi_j(x)}{\partial x^k}\right| \leq C\|\varphi_j\|_{\frac{n}{2}+p+\varepsilon}, \quad \forall \varepsilon > 0.$$

Since $\varphi_j = \lambda_j^{\alpha_1}A_B^{-\alpha_1}\varphi_j$ and $\operatorname{ord} A_B^{-\alpha_1} = -m\alpha_1$, we get

$$\max_{x\in\overline{\Omega}}\left|\frac{\partial^k \varphi(x)}{\partial x^k}\right| \leq \lambda_j^{\alpha_1}\|A_B^{-\alpha}\varphi_j\|_{\frac{n}{2}+p+\varepsilon}$$
$$\leq \lambda_j^{\alpha_1}\|\varphi_j\|_{\frac{n}{2}+p+\varepsilon-m\alpha_1} \leq C\lambda_j^{\alpha_1}\|\varphi_j\|_0 = C\lambda_j^{\alpha_1},$$

if we choose $m\alpha_1 > \frac{n}{2} + p$, and if ε is small. Also we have

$$(f, \varphi_j) = (A_B^{-\alpha}g, \varphi_j) = (g, A_B^{-\alpha}\varphi_j) = \frac{1}{\lambda_j^{\alpha}}(g, \varphi_j).$$

Therefore, by the Cauchy-Schwartz inequality,

$$\sum_{j=1}^{\infty} |(f,\varphi_j)| \left| \frac{\partial^k \varphi_j}{\partial x^k} \right| \leq C \sum_{j=1}^{\infty} \frac{|(g,\varphi_j)|}{\lambda_j^{(\alpha-\alpha_1)}}$$

$$\leq C\Big(\sum_{j=1}^{\infty} |(g,\varphi_j)|^2\Big)^{\frac{1}{2}} \Big(\sum_{j=1}^{\infty} \frac{1}{\lambda_j^{2(\alpha-\alpha_1)}}\Big)^{\frac{1}{2}} \leq C$$

when $2(\alpha-\alpha_1)m > n$. Thus, the series (58.26) converges uniformly and absolutely if $\alpha m > \alpha_1 m + \frac{n}{2} > n+p$.

Remark 58.1. Let $k_0 \geq \frac{n+p}{m}$ be an integer. Let $f = A_B^{-k_0} g,\ g \in L_2(\sigma,\Omega)$. We have $A_B^{k_0} f = g$, i.e., f belongs to the domain of $A_B^{k_0}$, i.e., $f \in H_{mk_0}(\Omega)$ and satisfies the boundary conditions $B_j(x,D)f|_{\partial\Omega} = 0,\ 1 \leq j \leq m_+$, $B_j A f|_{\partial\Omega} = 0,\ 1 \leq j \leq m_+$, etc., $B_j A^{k_0-1} f|_{\partial\Omega} = 0,\ 1 \leq j \leq m_+$. In particular, assuming that $\operatorname{ord} B_j < m$, we have that f belongs to the domain of $A_B^{k_0}$ if $f \in \overset{\circ}{H}_{mk_0}(\Omega)$. For any such $f(x)$ we have the uniform expansion (58.26).

58.4. The case of a compact manifold.

Consider the elliptic operator on a compact manifold M, $\partial M = \emptyset$, of the form

$$(A+\lambda I)u(x) = f(x), \quad x \in M, \tag{58.27}$$

where A is a pseudodifferential operator on M (cf. §52) with the principal symbol $A_0(x,\xi_x)$, $A_0(x,\xi_x)$ is homogeneous in ξ_x of degree $m > 0$, and $\operatorname{Re} A_0(x,\xi_x) > 0$, $\forall (x,\xi_x) \in T_0^*(M)$. As in Theorem 58.1, we can prove that $A+\lambda I$ is an invertible operator from $H_s(M)$ to $H_{s-m}(M)$ when $\lambda \in C_\delta, |\lambda| \geq \lambda_0$, $\forall s \in \mathbb{R}$. Therefore $(A+\lambda I)^{-1}$ is a meromorphic operator-valued function on C with poles at $\{-\lambda_1,\dots,-\lambda_n,\dots\}$, where $\operatorname{Re}\lambda_j \to +\infty$, $|\operatorname{Im}\lambda_j| \leq C_0 |\operatorname{Re}\lambda_j|$, λ_j are the eigenvalues of $A : A\varphi_j = \lambda_j\varphi_j$, $\varphi_j(x) \in C^\infty(M)$ are the eigenfunctions and we repeat λ_j p times if its multiplicity is p. As in (58.4), each λ_j generates a k_j-dimensional subspace of root vectors (cf. (58.2)).

Consider the parabolic equation

$$\frac{\partial u(x,t)}{\partial t} + Au(x,t) = 0, \quad t > 0,\ x \in M, \tag{58.28}$$

$$u(x,0) = u_0(x), \quad x \in M. \tag{58.29}$$

As in (47.3), for any $u_0(x) \in H_{s_0}(M)$, where s_0 is arbitrary, there exists a unique solution $u(x,t) \in C([0,+\infty), H_{s_0}(M))$ such that

$$\|e^{-\tau t} u(x,t)\|_{s_0} \leq C\|u_0\|_{s_0}, \quad \forall t \geq 0, \quad \tau \geq \lambda_0 > 0.$$

This solution has the form:

$$u(x,t) = \frac{1}{2\pi}\int_{-\infty}^{\infty}(A + i(\sigma - i\tau)I)^{-1}u_0(x)e^{it(\sigma - i\tau)}d\sigma, \tag{58.30}$$

where $\tau \geq \lambda_0$ and $|||(A + i(\sigma - i\tau)I)^{-1}|||_{s_0} \leq \frac{C}{|\sigma - i\tau|}$, $|||\ |||_{s_0}$ being the operator norm in $H_{s_0}(M)$.

We define the heat kernel $G(x,y)$ of the Cauchy problem (58.17), (58.18) as the kernel of the operator $u(x,t) = e^{-tA}u_0$, i.e.,

$$u(x,t) = \int_M G(x,y,t)u_0(y)dy, \tag{58.31}$$

where $G(x,y)$ is a scalar function with respect to $x \in M$ and a density with respect to $y \in M$; i.e., if $y^{(j)} = s_j(x)$ and $y^{(k)} = s_k(x)$ are two local system of coordinates in some neighborhood, then

$$G_j(x, y^{(j)})\left|\det \frac{\partial s_{jk}(y^{(j)})}{\partial y^{(k)}}\right| = G_k(x, y^{(k)}).$$

Here $y^{(j)} = s_{jk}(y^{(k)})$ and $G_j(x, y^{(j)})$ and $G_k(x, y^{(k)})$ are $G(x,y)$ in $y^{(j)}$- and $y^{(k)}$-coordinates, respectively. Note that the integration over M is well-defined in (58.31) when $G(x,y)$ is a density in $y \in M$.

Remark 58.2. If $\sigma(x) > 0$ is a density on M, then we can represent the heat kernel $G(x,y)$ as $G(x,y) = \hat{G}(x,y)\sigma(y)$, where $\hat{G}(x,y)$ is a scalar function on $M \times M$ (cf. (58.25)). Now (58.31) takes the form

$$u(x,t) = \int_M \hat{G}(x,y,t)u_0(y)\sigma(y)dy. \tag{58.32}$$

Such a situation occurs when $A(x,D)$ is selfadjoint with respect to a scalar product $(f,g)_\sigma = \int_M f(x)\overline{g(x)}\sigma(x)dx$, where $\sigma(x)$ is a density, $\sigma(x) > 0$. In particular, for the case of the Laplace-Beltrami operator, we have $\sigma(x) = \sqrt{g(x)}$. □

The parametrix R_j of $(A + \lambda I)^{-1}$ in a local system of coordinates in U_j is constructed as in §41. The global parametrix $R(\lambda)$ is given by the formula of the form (54.11).

Substituting the parametrix $R(\lambda)$ in (58.30), we get, as in (58.10):

$$\int_M G(x,x,t)dx = \frac{1}{t^{\frac{n}{m}}}\left(\sum_{k=0}^{N} c_k t^{\frac{k}{m}} + O(t^{\frac{N+1}{m}})\right), \tag{58.33}$$

where

$$c_0 = \frac{1}{(2\pi)^n}\int_{T_0^*(M)} e^{-A_0(x,\xi)}dxd\xi, \tag{58.34}$$

and $dx \wedge d\xi = dx_1 \wedge \cdots \wedge dx_n \wedge d\xi_1 \wedge \cdots \wedge d\xi_n$ is a differential form on $T_0^*(M)$.

As in Theorem 58.1, we have that e^{-tA} is a trace class operator, and using Lidskii's theorem, we get

$$\int_M G(x,x,t)dx = \sum_{j=1}^{\infty} k_j e^{-\lambda_j t}. \tag{58.35}$$

59. The index of elliptic operators in $\mathbb{R}^n$

59.1. Properties of Fredholm operators.

The definition of a Fredholm operator is given in §42. The number

$$\operatorname{ind} A = \dim\ker A - \dim\ker A^*$$

is called the index of the Fredholm operator A.

We need the following lemma:

Lemma 59.1. *If A is Fredholm from B_1 to B_2 and B is Fredholm from B_2 to B_3 then BA is Fredholm from B_1 to B_3 and*

$$\operatorname{ind} BA = \operatorname{ind} B + \operatorname{ind} A. \tag{59.1}$$

Proof: Let R_A, R_B be regularizers to A, B, respectively. Then $R_B R_A$ is the left and the right regularizer to AB. Therefore, AB is Fredholm (cf. §42).

Let $\operatorname{ind} A = m_1 - m_1^*$, $\operatorname{ind} B = m_2 - m_2^*$. If $u \in \ker BA$, then either $u \in \ker A$ or $Au \in \ker B$. Let $e_1, \dots, e_{m_2}$ be a basis in $\ker B$. Then $Au = \sum_{j=1}^{m_2} c_j e_j$ for some $c_j \in \mathbb{C}, 1 \le j \le m_2$.

Let $f_1^*, \dots, f_{m_1^*}^*$ be a basis in $\ker A^*$. Then

$$0 = (Au, f_k^*) = \sum_{j=1}^{m_2} c_j (e_j, f_k^*), \quad 1 \le k \le m_1^*. \tag{59.2}$$

Therefore the dimension of $\ker B \cap \operatorname{Im} A$ is equal to the dimension of the null-space of the system (59.2). This dimension is equal to $m_2 - r$, where r is the rank of the matrix (59.2). Therefore $\ker BA = m_1 + m_2 - r$. Analogously, $u^* \in \ker(BA)^* = \ker A^* B^*$ if either $u^* \in \ker B^*$ or $B^* u^* \in \ker A^*$, i.e., $B^* u^* = \sum_{j=1}^{m_1^*} d_j f_j^*$. Since $(e_k, B^* u^*) = 0, 1 \le k \le m_2$, we have that the $d_j, 1 \le j \le m_1^*$, are the solutions of

$$\sum_{j=1}^{m_1^*} d_j \overline{(e_k, f_j^*)} = 0, \quad 1 \le k \le m_2. \tag{59.3}$$

The null-space of (59.3) has dimension $m_1^* - r_1$, where r_1 is the rank of the matrix (59.3). Therefore $\dim\ker B^* A^* = m_2^* + m_1^* - r_1$. Since the matrices of the systems (59.2) and (59.3) are complex conjugate, they have the same rank. Therefore

$$\operatorname{ind} BA = \ker BA - \ker(BA)^* = (m_1 + m_2 - r) - (m_1^* + m_2^* - r)$$
$$= \operatorname{ind} B + \operatorname{ind} A. \qquad \square$$

Additional properties of Fredholm operators follow from Lemma 59.1 and Propositions 42.3 and 42.4.

1) If R is a regularizer of A, then R is Fredholm and

$$\operatorname{ind} R = -\operatorname{ind} A. \tag{59.4}$$

Proof: Since R is a regularizer, we have $AR = I + T_1,\ RA = I + T_2$, where T_1, T_2 are compact operators. Therefore A is a regularizer of R. Hence, by Proposition 42.4, R is also Fredholm. Applying Lemma 59.1 to $AR = I + T_1$, we get $\operatorname{ind} A + \operatorname{ind} R = \operatorname{ind}(I + T_1)$. By Proposition 42.3, $\operatorname{ind}(I + T_1) = 0$. Therefore, $\operatorname{ind} R = -\operatorname{ind} A$. $\square$

2) If A is Fredholm and C is compact, then $A + C$ is Fredholm and

$$\operatorname{ind}(A + C) = \operatorname{ind} A. \tag{59.5}$$

Indeed, a regularizer of A is also a regularizer of $A + C$ since $R(A + C) = I_1 + T_1 + RC$, $(A + C)R = I_2 + T_2 + CR$, and RC, CR are compact. Therefore

$$\operatorname{ind}(A + C) = -\operatorname{ind} R = \operatorname{ind} A.$$

3) Let A be a Fredholm operator and U_ε the set of all bounded operators from B_1 to B_2 such that the operator norm $\|B - A\| < \varepsilon$ for all $B \in U_\varepsilon$. Then any $B \in U_\varepsilon$ is Fredholm if $\varepsilon > 0$ is small and

$$\operatorname{ind} B = \operatorname{ind} A. \tag{59.6}$$

Proof: Let R be a regularizer for A. Then

$$RB = RA + R(B - A) = I_1 + R(B - A) + T_1.$$

If $\varepsilon < \frac{1}{\|R\|}$, then $\|R(B - A)\| \le \|R\|\ \|B - A\| < 1$. Therefore, $R_1 = (I + R(B - A))^{-1}R$ satisfies $R_1 B = I + T_2$, where T_2 is compact, i.e., R_1 is the left regularizer for B. Analogously we can show that there is R_2 such that $BR_2 = I_2 + T_3$, where T_3 is compact, i.e., R_2 is the right regularizer. Therefore any $B \in U_\varepsilon$ is Fredholm. Moreover, $\operatorname{ind} B = -\operatorname{ind} R_1 = -\operatorname{ind} R = \operatorname{ind} A$.

4) Let $A(t),\ 0 \le t \le 1$, be a family of Fredholm operators continuously dependent on t in the operator norm. Then

$$\operatorname{ind} A(t) = \operatorname{ind} A(0) \tag{59.7}$$

for all $t \in (0, 1]$.

Proof: In view of (59.6), $\operatorname{ind} A(t)$ is locally constant. Therefore, (59.7) holds. □

Property (59.7) shows that $\operatorname{ind} A$ is a topological invariant, i.e., $\operatorname{ind} A$ does not change when we continuously deform A in the operator norm.

5) Let P_1 and P_2 be the same as in (42.8), (42.10). Note that $\operatorname{Tr} P_1 = m_+, \operatorname{Tr} P_2 = m_-$. Therefore, $AR - RA$ is a trace class operator, and $\operatorname{Tr}(AR - RA) = m_+ - m_- = \operatorname{ind} A$, where R is the same as in (42.12).

59.2. Index of an elliptic ψdo.

Let M be a compact manifold without boundary and let $A(x, D)$ be an elliptic system of ψdo's on M of order α (cf. §52). Let $A_0(x, \xi_x)$ be the principal symbol of $A(x, D)$, $(x, \xi_x) \in T_0^*(M)$, $\deg A_0(x, \xi_x) = \alpha$. The ellipticity means that the matrix $A_0(x, \xi_x)$ belongs to $GL(r, \mathbb{C})$, where $GL(r, \mathbb{C})$ is the group of nonsingular $r \times r$ matrices with complex-valued entries. It was shown in §52 that $A(x, D)$ is Fredholm from $H_s(M)$ to $H_{s-\alpha}(M)$ for any s. It follows from (59.5) that $\operatorname{ind} A = \operatorname{ind}(A + A_1)$ for any ψdo A_1 of order $\alpha_1 < \alpha$. Any two ψdo's having the same principal symbol differ by an operator of a lower order (cf. §52). Therefore $\operatorname{ind} A$ is determined by $A_0(x, \xi_x)$. Note that if $A_0(x, \xi_x)$ is a real-valued function when $r = 1$ or a selfadjoint matrix when $r > 1$, then $\operatorname{ind} A = 0$. Indeed, $\operatorname{ind} A^* = \dim\ker A^* - \dim\ker A = -\operatorname{ind} A$, where A^* is the adjoint operator. When $A_0^*(x, \xi_x) = A_0(x, \xi_x)$, the operators A and A^* differ by an operator of a lower order (cf. §43). Therefore, $\operatorname{ind} A = \operatorname{ind} A^* = -\operatorname{ind} A$, i.e., $\operatorname{ind} A = 0$. We have $A_0(x, \xi_x) = |\xi_x|^\alpha A_0(x, \omega_x)$, where $\omega_x = \frac{\xi_x}{|\xi_x|}$, $|\omega_x| = 1$. The ψdo Λ^α with the symbol $|\xi_x|^\alpha I_r$ has index zero since $|\xi_x|$ is real. Moreover, the composition of Λ^α and $A^{(0)}$ differs from A by an operator of a lower order. Here $A^{(0)}$ is a ψdo with symbol $A_0(x, \frac{\xi_x}{|\xi_x|})$. Therefore, for any elliptic ψdo with principal symbol $A_0(x, \xi_x)$, $\deg A_0(x, \xi_x) = \alpha$, we have

$$\operatorname{ind} A = \operatorname{ind} A^{(0)}, \tag{59.8}$$

where $A^{(0)}$ is a ψdo of order zero with the principal symbol $A^{(0)}(x, \xi_x) = A_0(x, \frac{\xi_x}{|\xi_x|})$. Let $S^*(M) = \{(x, \xi_x) \in T_0^*(M) : |\xi_x| = 1\}$. An elliptic $A^{(0)}(x, \xi_x)$ defines a continuous map of $S^*(M)$ to $GL(r, \mathbb{C})$. If $A_t^{(0)}(x, \xi_x)$ is a continuous deformation of such a map, we can approximate it by a smooth deformation. Then (cf. Remark 40.1) we have a deformation of $A_t(x, D)$, continuous in the operator norm. Such deformations do not change the index. Therefore, $\operatorname{ind} A$ of the elliptic operator $A(x, D)$ depends only on the homotopy equivalence class of the map $S^*(M) \xrightarrow{A_0(x, \xi_x)} GL(r, \mathbb{C})$, where $A_0(x, \xi_x)$ is the principal symbol of $A(x, D)$.

Example 59.1. Let $A(x, D)$ be a scalar elliptic ψdo on M, $\deg A_0(x, \xi_x) = 0$. Suppose $S^*(M)$ is simply connected. Then $\operatorname{ind} A = 0$.

Proof: Since $S^*(M)$ is simply connected, one can define a single-valued branch of $\ln A_0(x, \omega_x) = \ln |A_0(x, \omega_x)| + i \arg A_0(x, \omega_x)$ on $S^*(M)$. Consider the family of symbols $A_t(x, \omega_x) = e^{(1-t)\ln A_0(x,\omega_x)}$, $0 \le t \le 1$. Then $A_1(x, \omega_x) = 1$, and $A_t(x, \omega_x)$ is a smooth deformation of $A_0(x, \omega_x)$ to 1. Therefore $\operatorname{ind} A_0 = 0$ by (59.7).

Example 59.2. Consider a scalar ψdo of order zero in $\mathbb{R}^1$. Its symbol has the form $A(x_1, \xi_1) = a_+(x_1)\theta^+(\xi_1) + a_1(x_1)\theta^-(\xi_1)$, where $\theta^+(\xi_1) = 1$ when $\xi_1 > 0$, $\theta^+(\xi_1) = 0$ when $\xi_1 < 0$, and $\theta^-(\xi_1) = 1 - \theta^+(\xi_1)$. The ψdo $A(x_1, D_1)$ with symbol $A(x_1, \xi_1)$ has the form (cf. Example 10.2):

$$A(x_1, D_1)u(x_1) = a_+(x_1)\Pi^+ u + a_-(x_1)\Pi^- u, \tag{59.9}$$

where

$$\Pi^\pm u = F^{-1}\theta^\pm(\xi_1)\tilde{u}(\xi_1) = \frac{\mp i}{2\pi}\int_{-\infty}^{\infty} \frac{u(y_1)}{x_1 - y_1 \mp i0} dy_1.$$

Note that $u = \Pi^+ u + \Pi^- u$.

The ellipticity of $A(x_1, D_1)$ means that $a_\pm(x_1) \ne 0$, $\forall x_1 \in \mathbb{R}^1$. We also assume that $a_\pm(x_1) = 1$ for large $|x_1|$. The operator $A(x_1, D_1)$ has obviously the same index as $A^{(1)} = g(x_1)\Pi^+ + \Pi^-$, where $g(x_1) = a_-^{-1}a_+$. Let $m = \frac{1}{2\pi}\Delta \arg g(x_1)|_{-\infty}^{\infty}$ and let

$$g^{(1)}(x_1) = g(x_1)\left(\frac{x_1 - i}{x_1 + i}\right)^{-m}. \tag{59.10}$$

Then $\Delta \arg g^{(1)}(x_1)|_{-\infty}^{\infty} = 0$ and we can represent $g^{(1)}$ as $e^{\ln g^{(1)}}$. Consider the homotopy $g_t(x) = \left(\frac{x_1-i}{x_1+i}\right)^m e^{t\ln g^{(1)}(x_1)}$, $0 \le t \le 1$. Then $g_1 = g$ and $g_0 = \left(\frac{x_1-i}{x_2+i}\right)^m$. Therefore, the operator $A^{(1)}$ has the same index as the operator $A^{(2)}$ with $g(x_1)$ replaced by $g_0(x_1)$. Note that the adjoint operator $(A^{(2)})^*$ corresponds to $\overline{g_0(x)} = \left(\frac{x_1+i}{x_1-i}\right)^m$.

If $u \in \ker A^{(2)}$, then

$$(x_1 - i)^m \Pi^+ u + (x_1 + i)^m \Pi^- u = 0. \tag{59.11}$$

Note that $\Pi^+ u$ extends analytically to the lower half-plane and $\Pi^- u$ extends analytically to the upper half-plane. Therefore, if $m \le 0$, then $\Pi^\pm u = 0$, i.e., $u = 0$, by Liouville's theorem. Hence $\ker A^{(2)} = 0$ when $m \le 0$. If $m > 0$, we again have, by the generalized Liouville theorem,

$$(x_1 - i)^m \Pi^+ u = -(x_1 + i)^m \Pi^- u = \sum_{k=0}^{m-1} c_k x_1^k. \tag{59.12}$$

Therefore

(59.13)
$$u = \Pi^+ u + \Pi^- u = \sum_{k=0}^{m-1} c_k x_1^k \left(\frac{1}{(x_1 - i)^m} - \frac{1}{(x_1 + i)^m} \right)$$

belongs to $\ker A^{(2)}$, i.e., $\dim \ker A^{(2)} = m$. Analogously $\dim \ker (A^{(2)})^* = |m|$ when $m \le 0$, and $\dim \ker (A^{(2)})^* = 0$ when $m \ge 0$. Therefore $\operatorname{ind} A = \operatorname{ind} A^{(2)} = m$. Note that

$$\frac{1}{2\pi} \Delta \arg g(x)|_{-\infty}^{\infty} = \frac{1}{2\pi i} \Delta \ln g|_{-\infty}^{\infty} = \frac{1}{2\pi i} \int_{-\infty}^{\infty} \frac{d}{dx_1} \ln g \, dx_1$$
$$= \frac{1}{2\pi i} \int_{-\infty}^{\infty} g^{-1} dg(x_1),$$

i.e.,

$$\operatorname{ind} A^{(1)} = \frac{1}{2\pi} \Delta \arg g(x)|_{-\infty}^{\infty} = \frac{1}{2\pi i} \int_{-\infty}^{\infty} g^{-1} \, dg.$$

Example 59.3. Let $A(x_1, D_1)$ be an elliptic $r \times r$ system of ψdo's of order zero in $\mathbb{R}^1$, i.e.,

$$A(x_1, D_1) u = a_+(x_1) \Pi^+ u + a_-(x_1) \Pi^- u,$$

where $\det a_\pm(x_1) \neq 0$, $\forall x_1 \in \mathbb{R}^1$, $a_\pm(x_1) = I_r$ for large $|x_1|$. Let $g(x_1) = a_-^{-1}(x_1) a_+(x_1)$. Then

(59.14)
$$\operatorname{ind} A = \frac{1}{2\pi} \Delta \arg \det g(x_1)|_{-\infty}^{\infty} = \frac{1}{2\pi i} \int_{-\infty}^{\infty} \frac{d}{dx_1} \ln \det g(x_1) \, dx_1.$$

The proof of (59.14) uses the following fact.

Any nonsingular matrix $g(x_1)$ on $\mathbb{R}^1$, $g(x_1) = I_r$ for $|x_1| > R$ (i.e., g is actually a matrix on the circle) can be continuously deformed to the diagonal matrix of the form $h = [h_j \delta_{jk}]_{j,k=1}^r$, where $h_1 = \det g(x_1)$ and $h_k = 1$ for $2 \le k \le r$. The index of the ψdo corresponding to h is obviously equal to the index of the scalar operator corresponding to h_1, and this proves (59.14). □

Suppose $c(x_1) = \frac{dg}{dx_1} g^{-1}(x_1)$, i.e., $g(x_1)$ is the fundamental solution of the system $\frac{dg}{dx_1} = c(x_1) g(x_1)$, $g(-\infty) = I_r$. It is a well-known fact of the theory of linear differential equations (the Liouville formula) that $\det g(x_1) = \exp(\int_{-\infty}^{x_1} \operatorname{Tr} c(y_1) dy_1)$. Therefore

$$\int_{-\infty}^{\infty} \frac{d}{dx_1} \ln \det g(x_1) \, dx_1 = \int_{-\infty}^{\infty} \operatorname{Tr} c(x_1) dx_1$$
$$= \int_{-\infty}^{\infty} \operatorname{Tr} \left(\frac{dg}{dx_1} g^{-1}(x_1) \right) dx_1 = \int_{-\infty}^{\infty} \operatorname{Tr} g^{-1}(x_1) dg(x_1),$$

since for any matrices B, C we have $\operatorname{Tr}(BC) = \operatorname{Tr}(CB)$.

Using (59.14), we get

$$\operatorname{ind} A = \frac{1}{2\pi i}\int_{-\infty}^{\infty} \operatorname{Tr} g^{-1}(x_1)dg(x_1). \tag{59.15}$$

Remark 59.1. It follows from (59.14) that $\operatorname{ind} A$ is equal to the winding number of $\det g(x_1)$. We evaluate this winding number counterclockwise in (59.14). If we compute it clockwise, then we get the minus sign in (59.14).

59.3. Fredholm elliptic ψdo's in $\mathbb{R}^n$.

Let $A(x, D)$ be an $r \times r$ system of ψdo's in $\mathbb{R}^n$. As in §42, we assume that the symbol $A(x,\xi) \in S^\alpha(\mathbb{R}^n \times \mathbb{R}^n)$, $A(x,\xi) = A(\infty,\xi)$ for $|x| > R$, and $|\det A(x,\xi)| \geq C(1+|\xi|)^{\alpha r}$ for $|x|^2 + |\xi|^2 \geq R^2$ for some R. It was proven in §42 that $A(x,D)$ is a Fredholm operator from $H_s(\mathbb{R}^n)$ to $H_{s-\alpha}(\mathbb{R}^n)$ for any s.

In this section we additionally assume that $A(x,\xi) = A_0(x,\xi) + A_1(x,\xi)$ for $|\xi| \geq 1$, where $A_0(x,\xi)$ is homogeneous in ξ of degree α and $|A_1(x,\xi)| \leq C|\xi|^{\alpha-1}$ when $|\xi| \geq 1$.

Let $A^{(0)}(x,\xi) = A(x,\xi)A^{-1}(\infty,\xi)$. Then $A^{(0)}(x,\xi) = I_r$ for $|x| \geq R$, and $A^{(0)}(x,D)$ is a Fredholm operator in $H_s(\mathbb{R}^n)$. Since $A(\infty, D)$ is invertible, we have that $\operatorname{ind} A = \operatorname{ind} A^{(0)}$. Therefore, without loss of generality we consider elliptic symbols $A(x,\xi)$ such that $A(x,\xi) \in S^0$, $A(x,\xi) = I_r$ for $|x| \geq R$, $|\det A(x,\xi)| \geq C$ for $|x|^2 + |\xi|^2 \geq R^2$, and

$$A(x,\xi) = A_0\Big(x, \frac{\xi}{|\xi|}\Big) + O\Big(\frac{1}{|\xi|}\Big) \quad \text{for} \quad |\xi| \geq R,\ |x| \leq R, \tag{59.16}$$

where $A_0(x,\xi) = I_r$ for $|x| \geq R$, $A_0(x,\xi)$ is homogeneous in ξ of degree 0, and $\det A_0(x,\xi) \neq 0$ for all $x \in \mathbb{R}^n$, $\xi \neq 0$.

We denote this class of symbols $A(x,\xi)$ by $\mathcal{E}ll(T^*(\mathbb{R}^n))$. Consider the following class of continuous deformation of symbols belonging to $\mathcal{E}ll(T^*(\mathbb{R}^n))$.

Let $A^{(t)}(x,\xi)$ be a family of symbols in $\mathcal{E}ll(T^*(\mathbb{R}^n))$, $t \in [0,1]$, such that $A^{(t)}(x,\xi)$ is continuous in t in the norm

$$\sum_{|k|=0}^{n+1+|s|} \max_{x,\xi} \left| \frac{\partial^k A^{(t)}(x,\xi)}{\partial x^k} \right|. \tag{59.17}$$

It follows from the Remark 40.1 that if (59.17) holds, then $A^{(t)}(x,D)$ acting in $H_s(\mathbb{R}^n)$ depends continuously on t in the operator norm. Therefore $\operatorname{ind} A^{(t)}(x,D)$ does not depend on t.

Lemma 59.2. *Let $A(x,\xi) \in \mathcal{E}ll(T^*(\mathbb{R}^n))$ and $A(x,\xi) \in GL(r,\mathbb{C})$ for all $(x,\xi) \in T^*(\mathbb{R}^n) = \mathbb{R}^n \times \mathbb{R}^n$. Then $\operatorname{ind} A = 0$.*

Proof: Let $S^{2n-1}(R)$ be the sphere $\{|x|^2+|\xi|^2 = R^2\}$. Consider the map $\alpha_R(A) : S^{2n-1}(R) \xrightarrow{A(x,\xi)} GL(r,\mathbb{C})$. For any R, the map $\alpha_R(A)$ is homotopic to a constant map $\alpha_0(A)$, i.e., when $R = 0$. It follows from (59.16) that $A(x,\xi)$ is homotopic to $A_0(x,\xi)$ for $|x|^2 + |\xi|^2 \geq R^2$ and $\alpha_R(A_0)$ is also homotopic to I_r. Therefore, we see that there exists a continuous deformation $A^{(t)}(x,\xi)$ on $|x|^2 + |\xi|^2 \geq R^2$ such that $A^{(0)}(x,\xi) = A(x,\xi)$, $A^{(1)}(x,\xi) = I_r$.

Extend $A^{(t)}(x,\xi)$ to all (x,ξ) by considering $A^{(t)}(x,\xi)(1 - \chi(\frac{|x|^2+|\xi|^2}{R^2}))$, where $\chi(s) \in C_0^\infty(\mathbb{R}^1)$, $\chi(s) = 1$ for $|s| < 1$. Then we have that $\operatorname{ind} A = \operatorname{ind} A^{(0)}(x,\xi)(1 - \chi(\frac{|x|^2+|D|^2}{R^2})) = \operatorname{ind} I_r(1 - \chi(\frac{|x|^2+|D|^2}{R^2})) = 0$. □

Let $\pi_{2n-1}(GL(r,\mathbb{C}))$ be the group of homotopy classes of maps $S^{2n-1} \rightarrow GL(x,\mathbb{C})$.

Denote by $[\alpha_R(A)] \in \pi_{2n-1}(GL(r,\mathbb{C}))$ the homotopy class of $\alpha_R(A)$. Note that $[\alpha_R(A)] = [\alpha_{R'}(A)]$ for $R' > R$ and that $[\alpha_R(A^{(t)})]$ is independent of $t \in [0,1]$, where $A^{(t)}$ is a continuous deformation.

It follows from Lemma 59.2 that if $\alpha_R(A)$ is homotopic to I_r, then $\operatorname{ind} A = 0$. Indeed, as in the proof of Lemma 59.2, we can find a deformation of $A(x,D)$ to $I_r + T_{-\infty}$, where $T_{-\infty}$ is a compact operator. Therefore $\operatorname{ind} A = \operatorname{ind}(I_r + T_{-\infty}) = 0$.

The set $\{[\alpha_R(A)] : A \in \mathcal{E}ll(T^*(\mathbb{R}^n))\}$ is a subgroup of $\pi_{2n-1}(GL(r,\mathbb{C}))$ since $A, B \in \mathcal{E}ll(T^*(\mathbb{R}^n))$ implies that $AB \in \mathcal{E}ll(T^*(\mathbb{R}^n))$, and $A(x,\xi) \in \mathcal{E}ll(T^*(\mathbb{R}^n))$ implies that $A^{-1}(x,\xi)(1 - \chi(\frac{|x|^2+|\xi|^2}{R})) \in \mathcal{E}ll(T^*(\mathbb{R}^n))$. Therefore, the map

$$[\alpha_R(A)] \xrightarrow{\operatorname{ind} A} \mathbb{Z} \tag{59.18}$$

is a group homomorphism, and hence the image of this map is a subgroup of $\mathbb{Z}$. □

In order to assign to any $A \in \mathcal{E}ll(T^*(\mathbb{R}^n))$ an integer topological invariant, we use K-theory.

59.4. Elements of K-theory.

We consider a variant of K-theory adapted for locally compact spaces (see Atiyah [**At**] and Fedosov [**F2**]). Let $T^*(\mathbb{R}^n) = \mathbb{R}^n \times \mathbb{R}^n$, and let E be a vector bundle on $T^*(\mathbb{R}^n)$. Note that any vector bundle on $T^*(\mathbb{R}^n)$ is trivial, i.e., isomorphic to $T^*(\mathbb{R}^n) \times \mathbb{C}^r$. This means that for all $(x,\xi) \in T^*(\mathbb{R}^n)$ there exists $C(x,\xi) \in GL(r,\mathbb{C})$ such that $(y, e_y) = (y, C(y)e_0)$, where $y = (x,\xi)$, e_y is a fiber of E at y, and $e_0 \in \mathbb{C}^r$. We consider only vector bundles on $T^*(\mathbb{R}^n)$ such that $C(x,\xi) \in \mathcal{E}ll(T^*(\mathbb{R}^n))$. Take a triple $(E, F, A(x,\xi))$, where E, F are vector bundles on $T^*(\mathbb{R}^n)$ and $A(x,\xi) \in \mathcal{E}ll(T^*(\mathbb{R}^n))$ is a map of E to F. In particular, we have that $A(x,\xi) \in GL(r,\mathbb{C})$ when $|x|^2 + |\xi|^2 \geq R^2$ for some $R > 0$. Therefore $A(x,\xi)$ is an isomorphism when $|x|^2 + |\xi|^2 \geq R^2$.

The triple $(E, F, A(x,\xi))$ is called trivial if $A(x,\xi) \in GL(r,\mathbb{C})$ for all $(x,\xi) \in T^*(\mathbb{R}^n)$. Two triples $(E_i, F_i, A_i(x,\xi))$, $i = 1,2$, are called isomorphic if

$$C_1(x,\xi)A_1(x,\xi) = A_2(x,\xi)C_2(x,\xi) \tag{59.19}$$

for $|x|^2 + |\xi|^2 \geq R^2$, where $C_i(x,\xi) \in GL(r,\mathbb{C})$ for all (x,ξ), $i = 1,2$. Define the sum $\xi_1 + \xi_2$ of triples $\xi_i = (E_i, F_i, A_i)$ as the triple $\xi_3 = (E_1 \oplus E_2, F_1 \oplus F_2, A_3(x,\xi))$, where $E_1 \oplus E_2$, $F_1 \oplus F_2$ are direct sums, and $A_3(x,\xi) = \left(\begin{smallmatrix} A_1(x,\xi) & 0 \\ 0 & A_2(x,\xi)\end{smallmatrix}\right)$ is an $(r_1 + r_2) \times (r_1 + r_2)$ matrix. Finally we say that triples $\xi_i = (E_i, F_i, A_i(x,\xi))$, $i = 1,2$, are equivalent if there exist trivial triples $\eta_i = (P_i, Q_i, B_i(x,\xi))$, $i = 1,2$, such that $\xi_1 + \eta_1$ is isomorphic to $\xi_2 + \eta_2$.

Denote by $[\xi] = [E, F, A(x,\xi)]$ the equivalence class containing $\xi = (E, F, A)$. The set of all equivalence classes of triples will be denoted by $K_{cpt}(T^*(\mathbb{R}^n))$.

Define addition in $K_{cpt}(T^*(\mathbb{R}^n))$ as

$$[\xi] + [\eta] = [\xi + \eta]. \tag{59.20}$$

In order to show that $K_{cpt}(T^*(\mathbb{R}^n))$ is a commutative group, we need to prove the following two properties.

1) Let $[\xi_t] = [E, F, A^{(t)}(x,\xi)]$, where $A^{(t)}(x,\xi), 0 \leq t \leq 1$, is a continuous deformation. Then

$$[E, F, A^{(0)}] = [E, F, A^{(t)}], \quad 0 \leq t \leq 1. \tag{59.21}$$

Proof: Since $C_1^{(t)} = (A^{(t)})^{-1}A^{(0)}$ is homotopic to I_r in the region $|x|^2 + |\xi|^2 \geq R^2$, we can extend the $C_1^{(t)}(x,\xi)$ as elements of $\mathcal{E}ll(T^*(\mathbb{R}^n))$ such that $C_1^{(t)} \in GL(r,\mathbb{C})$, to all (x,ξ).

Let $C_2^{(t)}(x,\xi) = I_r$. Then we have

$$A^{(t)}(x,\xi)C_1^{(t)}(x,\xi) = C_2^{(t)}(x,\xi)A^{(0)}(x,\xi)$$

for $|x|^2 + |\xi|^2 \geq R^2$, i.e., the triples $(E, F, A^{(0)})$ and $(E, F, A^{(t)})$, $0 \leq t \leq 1$, are isomorphic. Therefore (59.21) holds.

2) Consider triples $[E_1, E_2, A(x,\xi)]$ and $[E_2, E_3, B(x,\xi)]$. Then

$$[E_1, E_2, A] + [E_2, E_3, B] = [E_1, E_3, BA].$$

Proof: The matrices

$$\begin{pmatrix} B(x,\xi)A(x,\xi) & 0 \\ 0 & I_r \end{pmatrix} \quad \text{and} \quad \begin{pmatrix} B(x,\xi) & 0 \\ 0 & A(x,\xi) \end{pmatrix}$$

are homotopic. The homotopy is given by

(59.22)

$$\begin{pmatrix} B & 0 \\ 0 & I_r \end{pmatrix}\begin{pmatrix} \cos t\ I_r & \sin t\ I_r \\ -\sin t\ I_r & \cos t\ I_r \end{pmatrix}\begin{pmatrix} I_r & 0 \\ 0 & A \end{pmatrix}\begin{pmatrix} \cos t\ I_r & -\sin t\ I_r \\ \sin t\ I_r & \cos t\ I_r \end{pmatrix},$$

where $0 \le t \le \frac{\pi}{2}$. Analogously,

$$\begin{pmatrix} \cos t\ I_r & \sin t\ I_r \\ -\sin t\ I_r & \cos t\ I_r \end{pmatrix}\begin{pmatrix} A & 0 \\ 0 & B \end{pmatrix}\begin{pmatrix} \cos t\ I_r & \sin t\ I_r \\ -\sin t\ I_r & \cos t\ I_r \end{pmatrix},$$

where $0 \le t \le \frac{\pi}{2}$, gives a homotopy between $\left(\begin{smallmatrix} A & 0 \\ 0 & B \end{smallmatrix}\right)$ and $\left(\begin{smallmatrix} B & 0 \\ 0 & A \end{smallmatrix}\right)$.

Therefore $[\xi] + [\eta] = [\eta] + [\xi]$, and we get, using (59.22):

$$(59.23)\quad [E_1, E_2, A] + [E_2, E_3, B] = \left[E_2 \oplus E_1, E_3 \oplus E_2, \begin{pmatrix} B & 0 \\ 0 & A \end{pmatrix}\right]$$
$$= \left[E_1 \oplus E_2, E_3 \oplus E_2, \begin{pmatrix} BA & 0 \\ 0 & I_r \end{pmatrix}\right] = [E_1, E_3, BA] + [E_2, E_2, I_r].$$

Note that $[E_2, E_2, I_r]$ is a zero element of the group since $\xi + (E_2, E_2, I_r)$ is equivalent to ξ for any ξ. Thus (59.23) implies

$$(59.24)\qquad [E_1, E_2, A] + [E_2, E_3, B] = [E_1, E_3, BA].$$

The inverse of the element $[E, F, A]$ is $[F, E, A^{(-1)}]$, where $A^{(-1)}(x, \xi) = A^{-1}(x, \xi)(1 - \chi(\frac{|x|^2+|\xi|^2}{R^2}))$. Indeed,

$$[E, F, A] + [F, E, A^{(-1)}]$$
$$= \left[E \oplus F, F \oplus E, \begin{pmatrix} A & 0 \\ 0 & A^{(-1)} \end{pmatrix}\right] = \left[E \oplus F, E \oplus F, \begin{pmatrix} I_r & 0 \\ 0 & I_r \end{pmatrix}\right] = 0,$$

where we again used (59.22).

Therefore $K_{cpt}(T^*(\mathbb{R}^n))$ is a commutative group. □

We will need an external product of $K_{cpt}(T^*(\mathbb{R}^{n_1})$ and $K_{cpt}(T^*(\mathbb{R}^{n_2})$. First recall the definition of tensor product $V \otimes W$ of two vector bundles V and W over a manifold M: If V_y and W_y are the fiber spaces over $y \in M$, then $V_y \otimes W_y$ is spanned by the elements $v \otimes w$, where $v \in V_y$ and $w \in W_y$. If A and B are linear maps in V_y and W_y, respectively, then $A \otimes B$ acts in $V_y \otimes W_y$ as $(A \otimes B)(v \otimes w) = Av \otimes Bw$.

If $M = M_1 \times M_2$, V is a bundle over M_1, and W is a bundle over M_2, then $V \otimes W$ is a bundle over $M_1 \times M_2$ with the fiber space over $y = (y_1, y_2)$ equal to $V_{y_1} \otimes W_{y_2}$.

For every triple $(E_1, F_1, A_1(x^{(1)}, \xi^{(1)}))$ on $T^*(\mathbb{R}^{n_1})$ and every triple $(E_2, F_2, A_2(x^{(2)}, \xi^{(2)}))$ on $T^*(\mathbb{R}^{n_2})$, we define their external product as a triple $(E_3, F_3, A_3(x^{(3)}, \xi^{(3)}))$ on $T^*(\mathbb{R}^{n_1+n_2}) = T^*(\mathbb{R}^{n_1}) \times T^*(\mathbb{R}^{n_2})$, such that

$$E_3 = (E_1 \otimes E_2) \oplus (F_1 \otimes F_2), \quad F_3 = (F_1 \otimes E_2) \oplus (E_1 \otimes F_2),$$

$$A_3 = \begin{pmatrix} A_1 \otimes I & -I \otimes A_2^* \\ I \otimes A_2 & A_1^* \otimes I \end{pmatrix}. \tag{59.25}$$

This external product gives the map

$$K_{cpt}(T^*(\mathbb{R}^{n_1})) \otimes K_{cpt}(T^*(\mathbb{R}^{n_2})) \to K_{cpt}(T^*(\mathbb{R}^{n_1}) \times T^*(\mathbb{R}^{n_2})). \tag{59.26}$$

The following theorem holds (see Atiyah [**At**]):

Theorem 59.3 (Bott theorem). *For any $n \geq 1$, the map*

$$K_{cpt}(T^*(\mathbb{R}^1)) \otimes K_{cpt}(T^*(\mathbb{R}^n)) \to K_{cpt}(T^*(\mathbb{R}^{n+1}))$$

is onto, i.e.,

$$K_{cpt}(T^*(\mathbb{R}^1)) \otimes K_{cpt}(T^*(\mathbb{R}^n)) = K_{cpt}(T^*(\mathbb{R}^{n+1})). \tag{59.27}$$

The next lemma follows from Bott theorem.

Lemma 59.4. *For any $n \geq 1$, the group $K_{cpt}(\mathbb{R}^n)$ is isomorphic to $\mathbb{Z}$.*

Proof: We prove Lemma 59.4 by induction on n. It follows from Example 59.3 that $K_{cpt}(T^*(\mathbb{R}^1))$ is isomorphic to $\mathbb{Z}$. Therefore $K_{cpt}(T^*(\mathbb{R}^1)) \otimes K_{cpt}(T^*(\mathbb{R}^1))$ is isomorphic to the tensor product $\mathbb{Z} \otimes \mathbb{Z}$. By Theorem 59.3, $K_{cpt}(T^*(\mathbb{R}^1)) \otimes K_{cpt}(T^*(\mathbb{R}^1)) = K_{cpt}(T^*(\mathbb{R}^2))$. Since $\mathbb{Z} \otimes \mathbb{Z} = \mathbb{Z}$, we see that $K_{cpt}(T^*(\mathbb{R}^2))$ is isomorphic to Z.

Analogously, for $n \geq 2$ we have, by induction,

$$K_{cpt}(T^*(\mathbb{R}^{n+1})) = K_{cpt}(T^*(\mathbb{R}^1)) \otimes K_{cpt}(T^*(\mathbb{R}^n)) \approx \mathbb{Z} \otimes \mathbb{Z} = \mathbb{Z}, \tag{59.28}$$

where $\approx$ means isomorphic. Therefore $K_{cpt}(T^*(\mathbb{R}^n)) = \mathbb{Z}$. For any triple $[E, F, A(x, \xi)] \in K_{cpt}(T^*(\mathbb{R}^n))$, denote by $\text{t-ind}\, A$ the image of $[E, F, A]$ under the map $K_{cpt}(T^*(\mathbb{R}^n)) \to \mathbb{Z}$. We call $\text{t-ind}\, A$ the topological index of A.

It follows from (59.28) that

$$(\text{t-ind}\, A_1) \cdot (\text{t-ind}\, A_2) = \text{t-ind}\, A_3, \tag{59.29}$$

where A_1, A_2, A_3 are the same as in (59.25).

Let $[E, F, A_1(x_1, \xi_1)$ be an arbitrary triple in $K_{cpt}(T^*(\mathbb{R}^1))$, where $A_1(x_1, \xi_1)$ is the same as in Example 59.3. We have defined the isomorphism $K_{cpt}(T^*(\mathbb{R}^1)) \to \mathbb{Z}$ by formula (59.14), i.e., $\text{t-ind}\, A = \frac{1}{2\pi}\Delta \arg \det a_-^{-1} a_+|_{-\infty}^{\infty}$. With this choice of the isomorphism we have $\text{t-ind}\, A = \text{ind}\, A$ for $n = 1$. An alternative choice of the isomorphism $K_{cpt}(T^*(\mathbb{R}^1)) \to \mathbb{Z}$ would lead to $\text{t-ind}\, A = -\,\text{ind}\, A$.

Denote by β the element of $K_{cpt}(T^*(\mathbb{R}^1))$ corresponding to the triple $[E, F, A(x, \xi)]$, where $E = \mathbb{C}$, $F = \mathbb{C}$, and $A_1(x_1, \xi_1) = \frac{x_1 - i}{x_1 + i}\theta(\xi_1) + \theta(-\xi_1)$

(cf. Example 59.2). Then t-ind $A_1 = 1$ and ind $A_1(x_1, D_1) = 1$. It follows from Theorem 59.3 and Lemma 59.4 that the map

$$\beta \otimes K_{cpt}(T^*(\mathbb{R}^n)) \to K_{cpt}(T^*(\mathbb{R}^{n+1})) \tag{59.30}$$

gives an isomorphism of $K_{cpt}(T^*(\mathbb{R}^n))$ and $K_{cpt}(T^*(\mathbb{R}^{n+1})$, $\forall n \geq 1$. The isomorphism (59.30) is called the Bott isomorphism, and the statement (59.28) is called the Bott periodicity theorem (see Atiyah [**At**]).

Sometimes the index of a Fredholm operator A is called the analytical index of A.

59.5. Proof of the index theorem.

Theorem 59.5 (Atiyah-Singer index theorem in $\mathbb{R}^n$). *Suppose* $A(x, \xi) \in \mathcal{E}ll(T^*(\mathbb{R}^n))$. *Then* ind A = t-ind A.

Proof: The image of the map $K_{cpt}(T^*(\mathbb{R}^n)) \xrightarrow{\text{ind}\,A} \mathbb{Z}$ is a subgroup of $\mathbb{Z}$. If t-ind $A = 0$, then $A(x, \xi)$ is homotopic to the identity, and therefore ind $A = 0$ (cf. Lemma 59.2). Hence

$$\text{ind}\, A = C_n\, \text{t-ind}\, A \tag{59.31}$$

for all $A \in \mathcal{E}ll(T^*(\mathbb{R}^n))$, where C_n is independent of A. We show that there exists $A \in \mathcal{E}ll(T^*(\mathbb{R}^n))$ such that ind $A = 1$ and t-ind $A = 1$. Substituting such an A into (59.31), we get $C_n = 1$, and this will prove Theorem 59.5. □

First we compute ind A_3, where A_3 is the same as in (59.25).

Lemma 59.6. *Let A_i be Fredholm operators from H_i^1 to H_i^2, where H_i^p, $i = 1, 2$, $p = 1, 2$, are Hilbert spaces. Denote by A_3 the following operator acting from $(H_1^1 \otimes H_2^1) \oplus (H_1^2 \otimes H_2^2)$ to $(H_1^2 \otimes H_2^1) \oplus (H_1^1 \otimes H_2^2)$:*

$$A_3 = \begin{pmatrix} A_1 \otimes I & -I \otimes A_2^* \\ I \otimes A_2 & A_1^* \otimes I \end{pmatrix}. \tag{59.32}$$

Then A_3 is Fredholm and ind A_3 = ind $A_1 \cdot$ ind A_2.

Proof: We have

$$A_3^* A_3 = \begin{pmatrix} A_1^* A_1 \otimes I + I \otimes A_2^* A_2 & 0 \\ 0 & A_1 A_1^* \otimes I + I \otimes A_2 A_2^* \end{pmatrix}. \tag{59.33}$$

Note that $\ker(A_1^* A_1 \otimes I + I \otimes A_2^* A_2)$ consists of $u \otimes v$, where $A_1 u = 0$, $A_2 v = 0$. Therefore, $\dim\ker(A_1^* A_1 \otimes I + I \otimes A_2^* A_2) = p_1^+ p_2^+$, where $p_i^+ = \dim\ker A_i$, $i = 1, 2$. Analogously, $\dim\ker(A_1 A_2^* \otimes I + I \otimes A_2 A_2^*) = p_1^- p_2^-$, where $p_i^- = \dim\ker A_i^*$, $i = 1, 2$. Therefore $\dim\ker A_3 = p_1^+ p_2^+ + p_1^- p_2^-$. Consider

$$A_3 A_3^* = \begin{pmatrix} A_1 A_1^* \otimes I + I \otimes A_2^* A_2 & 0 \\ 0 & A_1^* A_1 \otimes I + I \otimes A_2 A_2^* \end{pmatrix}. \tag{59.34}$$

Analogously to (59.33), we get from (59.34) that $\dim\ker A_3^* = p_1^- p_2^+ + p_1^+ p_2^-$. Therefore

$$\operatorname{ind} A_3 = (p_1^+ p_2^+ + p_1^- p_2^-) - (p_1^- p_2^+ + p_1^+ p_2^-) = (p_1^+ - p_1^1)(p_2^+ - p_2^-)$$
$$= \operatorname{ind} A_1 \ \operatorname{ind} A_2. \qquad \square$$

Remark 59.2. Let the operators A_1 and A_2 in (59.32) be ψdo's in $\mathbb{R}^{n_1}$ and $\mathbb{R}^{n_2}$, respectively, with elliptic symbols

$$A_1(x^{(1)}, \xi^{(1)}) \in \mathcal{E}ll(T^*(\mathbb{R}^{n_1})), \qquad A_2(x^{(2)}, \xi^{(2)}) \in \mathcal{E}ll(T^*(\mathbb{R}^{n_2})),$$

$x^{(1)} \in \mathbb{R}^{n_1}$, $x^{(2)} \in \mathbb{R}^{n_2}$. Then the operator A_3 in (59.32) will be a ψdo in $\mathbb{R}^{n_1+n_2} = \mathbb{R}^{n_1} \times \mathbb{R}^{n_2}$ with symbol

$$A_3(x^{(3)}, \xi^{(3)}) = \begin{pmatrix} A_1(x^{(1)}, \xi^{(1)}) & -A_2^*(x^{(2)}, \xi^{(2)}) \\ A_2(x^{(2)}, \xi^{(2)}) & A_1^*(x^{(1)}, \xi^{(1)}) \end{pmatrix}, \tag{59.35}$$

where $x^{(3)} = (x^{(1)}, x^{(2)})$, $\xi^{(3)} = (\xi_1^{(1)}, \xi^{(2)})$.

The operator $A_3(x^{(3)}, D^{(3)})$ is Fredholm and $|\det A_3(x^{(3)}, \xi^{(3)})| \geq C$ for $|x^{(3)}|^2 + |\xi^{(3)}|^2 \geq R^2$. Therefore the analytic and the topological indices of A_3 are well defined. However, $A_3(x^{(3)}, \xi^{(3)})$ is not an elliptic symbol of class $\mathcal{E}ll(T^*(\mathbb{R}^{n_1+n_2}))$. We replace $A_3(x^{(3)}, \xi^{(3)})$ by an elliptic symbol belonging to $\mathcal{E}ll(T^*(\mathbb{R}^{n_1+n_2}))$ without changing its analytical and topological indices.

Suppose $A_i'(x^{(i)}, \xi^{(i)}) = A_i(x^{(i)}, \xi^{(i)})\Lambda_i(\xi^{(i)})$, $i = 1, 2$, where $\Lambda_i(\xi^{(i)}) = (1 + |\xi^{(i)}|^2)^{\frac{1}{2}}$. The operators $A_1'(x^{(1)}, D^{(1)})$ and $A_2'(x^{(1)}, D^{(1)})$ have the same analytical and topological indices as A_1 and A_2, respectively.

First we apply Lemma 59.6 to $A_i'(x^{(i)}, D^{(i)})$, $i = 1, 2$, and then apply the operator $I\Lambda_3^{-1}(D^{(3)})$ from the right, where $\Lambda_3(\xi^{(3)}) = (1 + |\xi^{(1)}|^2 + |\xi^{(2)}|^2)^{\frac{1}{2}}$. We get an operator $A_4(x^{(3)}, D^{(3)})$ with the symbol

$$A_4(x^{(3)}, \xi^{(3)}) = \Lambda_3^{-1}(\xi^{(3)}) \begin{pmatrix} A_1'(x^{(1)}, \xi^{(1)}) & -(A_2'(x^{(2)}, \xi^{(2)}))^* \\ A_2'(x^{(2)}, \xi^{(2)}) & (A_1'(x^{(1)}, \xi^{(1)}))^* \end{pmatrix}. \tag{59.36}$$

Note that the analytical and the topological indices of A_4 are the same as for $A_3(x^{(3)}, D^{(3)})$ and (cf. (59.16))

$$A_4(x^{(3)}, \xi^{(3)}) = A_4^{(0)}(x^{(3)}, \xi^{(3)}) + O\left(\frac{1}{|\xi^{(3)}|}\right) \quad \text{for} \quad |\xi^{(3)}| \geq R,$$

where $A_4^{(0)}(x^{(3)}, \xi^{(3)})$ is C^∞ in $x^{(3)}$, homogeneous in $\xi^{(3)}$ of degree zero, continuous on the sphere $|\xi^{(3)}| = 1$, and has bounded first derivatives with respect to $\xi^{(3)}$ when $|\xi^{(3)}| = 1$.

Let C_i be ψdo's of order zero, $i = 1, 2$, having the same properties as $A_4(x^{(3)}, \xi^{(3)})$; i.e., they are C^∞ in x, continuous in (x, ξ), and the first derivatives of $C_i(x^{(3)}, \xi^{(3)})$ with respect to $\xi^{(3)}$ are $O(\frac{1}{|\xi^{(3)}|})$ for $|\xi^{(3)}| \geq R$. Then the composition of C_1 and C_2 is a ψdo with the symbol $C_1(x^{(3)},\xi^{(3)})C_2(x^{(3)},\xi^{(3)})$ plus an operator of order -1 (cf. Theorem 40.2 with $N=0$). Therefore, $\operatorname{ind} C_1C_2 = \operatorname{ind} C_1 + \operatorname{ind} C_2$ if C_1, C_2 are Fredholm.

It remains to replace A_4 by a ψdo with the symbol equal to I when $|x^{(3)}| > R$ without changing the analytical and the topological indices. Note that $A_1(x^{(1)},\xi^{(1)}) = I$ if $|x^{(1)}| \geq R$, and $A_2(x^{(2)},\xi^{(2)}) = I$ if $|x^{(2)}| \geq R$. For $|x^{(2)}| \geq R$ we have

$$A_4(x^{(1)},\infty,\xi^{(3)}) = \Lambda_3^{-1}(\xi^{(3)})\begin{pmatrix} A_1'(x^{(1)},\xi^{(1)}) & -\Lambda_2 I \\ \Lambda_2 I & (A_1'(x^{(1)},\xi^{(1)}))^* \end{pmatrix}.$$

Since $A_4(x^{(1)},\infty,\xi^{(1)})$ is obtained from $A_4(x^{(3)},\xi^{(3)})$ by replacing A_2' with $\Lambda_2 I$ and since $\operatorname{ind}\Lambda_2 I = \text{t-ind}\,\Lambda_2 I = 0$, we have that the analytic and the topological indices of $A_4(x^{(1)},\infty,\xi^{(3)})$ are zero. Denote $A_5(x^{(3)},\xi^{(3)}) = A_4(x^{(3)},\xi^{(3)})(A_4(x^{(1)},\infty,\xi^{(3)}))^{-1}$. Then $\operatorname{ind} A_5 = \operatorname{ind} A_4$ and $\text{t-ind}\, A_5 = \text{t-ind}\, A_4$ (since the t-index of product is a sum of t-indices), and $A_5 = I$ when $|x^{(2)}| \geq R$. Finally, define

$$A_6(x^{(3)},\xi^{(3)}) = A_5(x^{(3)},\xi^{(3)})A_5^{-1}(\infty,x^{(2)},\xi^{(3)}).$$

Then $A_6 = I$ when $|x^{(3)}|^2 = |x^{(1)}|^2 + |x^{(2)}|^2 \geq R^2$, and $\operatorname{ind} A_6 = \operatorname{ind} A_5 = \operatorname{ind} A_3$, $\text{t-ind}\, A_6 = \text{t-ind}\, A_5 = \text{t-ind}\, A_3$, since the analytical and the topological indices of $A_5(\infty,x^{(2)},D^{(2)})$ are zero. Approximating $A_6(x^{(3)},\xi^{(3)})$ by C^∞ symbols, we get a symbol belonging to the class $\mathcal{E}ll(T^*(\mathbb{R}^n))$. □

Now we complete the proof of Theorem 59.5 by constructing $A(x,\xi) \in \mathcal{E}ll(T^*(\mathbb{R}^n))$ such that $\operatorname{ind} A = \text{t-ind}\, A = 1$. As in Lemma 59.4, we proceed by induction on n. When $n=1$, Example 59.2 shows that there exists $a(x_1,D_1)$ in $\mathbb{R}^1$ such that $\operatorname{ind} a = \text{t-ind}\, a = 1$. Let $a_2(x_1,x_2,D_1,D_2)$ be an operator with symbol (59.32), where $A_1 = a(x_1,D_1)$ and $A_2 = a(x_2,D_2)$. Then $\operatorname{ind} a_2 = 1$ by Lemma 59.6 and $\text{t-ind}\, a_2 = 1$ by (59.29). As in Remark 59.2, we can replace a_2 by an operator $A_2(x_1,x_2,D_1,D_2)$ with $A_2 \in \mathcal{E}ll(T^*(\mathbb{R}^2))$ such that $\operatorname{ind} A_2 = \text{t-ind}\, A_2 = 1$.

Now define $a_3(x_1,x_2,x_3,D_1,D_2,D_3)$ by (59.32) using $a(x_1,D_1)$ and A_2. We obtain that $\operatorname{ind} a_3 = 1$ and $\text{t-ind}\, a_3 = 1$, etc. Finally, we construct $A_n(x,D)$ in $\mathbb{R}^n$ such that $\operatorname{ind} A_n = \text{t-ind}\, A_n = 1$. Substituting this A_n into (59.31), we get $C_n = 1$.

Remark 59.3 (See Fedosov [**F1**]). There is an explicit formula for t-ind A. It follows from (59.16) that $A(x,\xi)$ on $S^{2n-1}(R)$ is homotopic to $A_0(x,\omega)$, $|\omega| = 1$. We have

$$\begin{aligned}\operatorname{ind} A &= \text{t-ind}\, A \\ &= -\frac{1}{(-2\pi i)^n}\frac{(n-1)!}{(2n-1)!}\int_{S^{2n-1}(R)} \operatorname{Tr}(A_0^{-1}(x,\xi)dA_0(x,\xi))^{2n-1},\end{aligned}$$

where $(A_0^{-1}dA_0)^{2n-1} = A_0^{-1}dA_1 \wedge A_0^{-1}dA_0 \wedge \cdots \wedge A_0^{-1}dA_0$ ($(2n-1)$ times). When $n = 1$, we get (59.15).

60. Problems

1. Consider a boundary value problem

(60.1)
$$-\Delta u + \lambda u = f \quad \text{in} \quad \Omega,$$

(60.2)
$$\sum_{k=1}^{n} a_k(x)\frac{\partial u}{\partial x_k} = 0 \quad \text{on} \quad \partial\Omega,$$

where $n \geq 3$ and $(a_1(x), \ldots, a_n(x))$ is not tangent to $\partial\Omega$ at any $x \in \partial\Omega$.

Find the second term of the heat trace asymptotics for (60.1), (60.2).

2. (See Seeley [**Se2**].) Let A_B be the same as in §58, and let $k > \frac{n}{m}$.

a) Prove that $\operatorname{Tr}(A_B + \lambda I)^{-k} = \sum_{j=1}^{\infty} \frac{k_j}{(\lambda_j+\lambda)^k}$.

b) Find the asymptotics of $\operatorname{Tr}(A_B + \lambda I)^{-k}$ as $\lambda \to +\infty$ and compare the coefficients of this asymptotics with the coefficients of the heat trace expansion $\operatorname{Tr} e^{-tA_B}$, $t \to 0$.

3. Consider the initial-boundary value problem for the parabolic equation

(60.3)
$$\frac{\partial u(x,t)}{\partial t} + A(x,t,D)u(x,t) = f(x,t), \quad (x,t) \in \Omega \times (0,\infty),$$

(60.4)
$$u(x,0) = 0,$$

(60.5)
$$B_j(x,t,D)u|_{\partial\Omega\times(0,\infty)} = g_j(x',t), \quad 1 \leq j \leq m_+ = \frac{m}{2}.$$

Assume that $A(x,t,D)$ is elliptic in $\overline{\Omega}\times\mathbb{R}$, $\operatorname{Re} A_0(x,t,\xi) > 0$, $\xi \neq 0$. Assume that for each (x,t), $x \in \partial\Omega$, $t \geq 0$, condition (56.3) is satisfied. Let $u_+(x,t), f_+(x,t), g_j^+(x,t)$ be the extensions of u, f, g_j by zero to $t < 0$. Prove that for any $e^{-t\tau} f_+ \in \overset{\circ}{\Pi}_{\frac{s-m}{m},s-m}(\Omega \times \mathbb{R})$, $e^{-t\tau} g_j^+ \in \overset{\circ}{\Pi}_{\frac{s-m_j-\frac{1}{2}}{m},s-m_j-\frac{1}{2}}(\partial\Omega \times \mathbb{R})$, $1 \leq j \leq m_+$, there exists a unique solution $e^{-t\tau}u_+(x,t) \in \overset{\circ}{\Pi}_{\frac{s}{m},s}(\Omega \times \mathbb{R})$.

Here $s > \max_{1\leq j\leq m_+}(m_j + \frac{1}{2})$, and the spaces $\overset{\circ}{\Pi}_{\frac{s}{m},s}(\Omega \times \mathbb{R})$, $\overset{\circ}{\Pi}_{\frac{s-m_j-\frac{1}{2}}{m},s-m_j-\frac{1}{2}}(\partial\Omega \times \mathbb{R})$ are defined as in §46, $\tau \geq \tau_0$, τ_0 is large.

(Hint: Consider first the initial boundary value problem in the half-space $x_n > 0$:

$$\frac{\partial u_+}{\partial t} + A_0(D)u_+ = f_+, \quad x_n > 0, \quad (x', t) \in \mathbb{R}^n, \tag{60.6}$$

$$B_{j0}(D)u_+|_{x_n=0} = g_j^+(x', t), \quad 1 \le j \le m_+, \tag{60.7}$$

with constant coefficients and homogeneous principal parts (cf. §53). Then use the solution of (60.6), (60.7) and the partition of unity to find the left and the right regularizers for (60.3), (60.4), (60.5) (cf. §54 and §56.1).)

4. Let $A(x, D)$ be an elliptic operator of order m, $\operatorname{Re} A_0(x, \xi) > 0$ for all $x \in \mathbb{R}^n$, $\xi \neq 0$, and $A(x, \xi) = A(\infty, \xi)$ for $|x| > R$. Let Σ_δ be the exterior of a parabolic region in $\mathbb{C}$: $\{\lambda \in \mathbb{C} : \operatorname{Re}\lambda < \lambda_0, \ |\operatorname{Im}\lambda| \le C|\operatorname{Re}\lambda - \lambda_0|^{1-\delta}\}$.

a) Prove that $R_0(x, \xi, \lambda) = (A_0(x, \xi) + \lambda)^{-1}$ satisfies the following estimates in Σ_δ:

$$\left| \frac{\partial^{p+k} R_0(x, \xi, \lambda)}{\partial x^p \partial \xi^k} \right| \le C_{pk} |R_0| (|\xi| + |\operatorname{Re}\lambda|^{\frac{1}{m}})^{|p|m\delta - |k|(1-m\delta)}. \tag{60.8}$$

b) Use (60.8) to construct the parametrix of $(A(x, D) + \lambda I)u(x) = f(x)$ in $\mathbb{R}^n$ and to prove that $(A(x, D) + \lambda I)^{-1}$ exists when $\lambda \in \Sigma_\delta$, $|\lambda_0|$ is large, and $0 < \delta < \frac{1}{2m}$.

5. Let $A(x, D)$ be an elliptic operator of order m on a manifold M. Suppose $A_0(x, \xi_x) > 0$, $\forall x \in M$, $\forall \xi_x \neq 0$. Let $\sigma(x)$ be a density on M, $\sigma(x) > 0$, $\forall x \in M$. Suppose that $A(x, D)$ is selfadjoint in $L_2(\sigma, M)$, where $(u, v)_\sigma = \int_M u(x)\overline{v(x)}\sigma(x)dx$.

Let $\{\varphi_j(x)\}$ be the orthonormal basis in $L_2(\sigma, M)$ consisting of the eigenfunctions of $A(x, D)$, i.e., $A(x, D)\varphi_j = \lambda_j \varphi(x)$, $1 \le j < +\infty$, $\lambda_j \to +\infty$. Denote by $E(x, y, \lambda)$ the kernel of the spectral projection $E(\lambda)$, i.e., $E(x, y, \lambda) = \sum_{\lambda_j < \lambda} \varphi_j(x)\overline{\varphi_j(y)}\sigma(y)$.

Suppose that $R(x, y, \lambda)\sigma(y)$ is the kernel of the resolvent $(A(x, D) + \lambda I)^{-1}$, i.e.,

$$(A(x, D) + \lambda I)^{-1} f(x) = \int_M R(x, y, \lambda) f(y)\sigma(y)dy.$$

a) Prove that $R(x, y, \lambda)$ is continuous in $M \times M$ when $m > n$.

b) Prove that

$$\int_M R(x, x, \lambda)\sigma(x)dx = \sum_{j=1}^{\infty} \frac{1}{\lambda + \lambda_j},$$

where $m > n$.

c) Let $N(\lambda) = \sum_{\lambda_j < \lambda} 1$, and thus

$$N(\lambda) = \int_M E(x, x, \lambda)\sigma(x)dx.$$

The following Pleijel theorem holds (see Agmon-Kannai [**AK**]):

$$\left| E(x, x, t) - \frac{1}{2\pi i} \int_{L_\zeta} R(x, x, \lambda)d\lambda \right| \le \frac{\tau}{1 + \pi^2} |R(x, x, \zeta)|, \ \sigma(x) = 1, \tag{60.9}$$

where $m > n_1$, $\zeta = -t + i\tau$, and L_ζ is a contour connecting ζ and $\overline{\zeta}$ and intersecting the real axis at $\lambda = \lambda_0$. Using (60.9), prove that

$$N(\lambda) = \frac{1}{(2\pi)^n} \int_{A_0(x,\xi_x)<\lambda} dxd\xi + O\left(\lambda^{\frac{n-\delta_1}{m}}\right), \quad 0 \le \delta_1 < \frac{1}{2}, \ \lambda > 0. \tag{60.10}$$

Note that an analogous result holds for $E(x, y, \lambda)$ (see Agmon-Kannai [**AK**]). For a generalization to the case of domains with boundaries see Eskin [**E2**].

6. Consider the following boundary value problem in $\mathbb{R}^n_+$ with a small parameter $\varepsilon > 0$:

$$\begin{aligned} -\Delta u_\varepsilon(x', x_n) + u_\varepsilon + \varepsilon^2 \Delta^2 u_\varepsilon &= f(x', x_n), \quad x_n > 0, \ x' \in \mathbb{R}^{n-1}, \\ u_\varepsilon(x', 0) = 0, \qquad \frac{\partial u_\varepsilon(x', 0)}{\partial x_n} &= 0. \end{aligned} \tag{60.11}$$

Let $1 + \varepsilon^2|\xi|^2 = (\varepsilon\xi_n - i\sqrt{\varepsilon^2|\xi'|^2 + 1})(\varepsilon\xi_n + i\sqrt{\varepsilon^2|\xi'|^2 + 1})$ be the factorization of $1 + \varepsilon^2|\xi|^2$ (cf. §53). Denote by $R^\pm$ the ψdo with symbol $(\varepsilon\xi_n \mp i\sqrt{\varepsilon^2|\xi'|^2 + 1})^{-1}$. Let $f(x)$ be smooth, let lf be an extension of $f(x)$ to $\mathbb{R}^n$, and let p_+ be the restriction operator to $\mathbb{R}^n_+$. Substituting $u_\varepsilon = R^+ v_\varepsilon^+$ and applying R^- to (60.11), we get:

$$-\Delta v_\varepsilon^+ + (1 - \varepsilon^2\Delta)^{-1} v_\varepsilon^+ = p_+ R_- lf, \quad x_n > 0, \tag{60.12}$$

where $v_\varepsilon^+ = 0$ for $x_n < 0$.

a) Prove that v_ε^+ converges to u_0^+ in the $H_1(\mathbb{R}^n_+)$ norm, where

$$\begin{aligned} -\Delta u_0^+ + u_0^+ &= f(x), \quad x_n > 0, \\ u_0^+(x', 0) &= 0. \end{aligned} \tag{60.13}$$

b) Prove that

$$u_\varepsilon = (1 - e^{-\frac{x_n}{\varepsilon}}) u_0^+(x', x_n) + O(\varepsilon).$$

7. **a)** Using the operators of the form $R^{\pm}$, find the asymptotics as $\varepsilon \to 0$ of the solution u_ε to the boundary value problem (see Eskin [**E1**], §27):

$$\begin{aligned} &A_0(x,D)u_\varepsilon + \varepsilon^2 A_1(x,D)u_\varepsilon = f(x), \quad x_n > 0, \\ &u_\varepsilon\big|_{x_n=0} = \frac{\partial u_\varepsilon}{\partial x_n}\bigg|_{x_n=0} = 0, \end{aligned} \tag{60.14}$$

where A_0, A_1 are elliptic operators of order 2 and 4, respectively. Assume that the Dirichlet problem

$$\begin{aligned} &A_0(x,D)u_0 = f(x), \quad x_n > 0, \\ &u\big|_{x_n=0} = 0, \end{aligned} \tag{60.15}$$

has a unique solution and

$$1 + \varepsilon^2 \frac{A_{10}(x,\xi)}{A_{00}(x,\xi)} \neq 0 \quad \forall x,\ \forall \xi,\ \varepsilon > 0, \tag{60.16}$$

where A_{00}, A_{10} are the principal symbols of $A_0(x,D), A_1(x,D)$.

b) Consider the case of nonhomogeneous boundary conditions (see Frank-Wendt [**FrW**]):

$$\begin{aligned} &A_0(x,D)u_\varepsilon + \varepsilon^2 A_1(x,D)u_\varepsilon = 0, \quad x_n > 0, \\ &u_\varepsilon\big|_{x_n=0} = g(x'), \qquad \frac{\partial u(x',0)}{\partial x_n} = 0. \end{aligned} \tag{60.17}$$

8. Let $A(x,D)$ be an elliptic operator on a manifold M, and $A^*(x,D)$ the adjoint operator with respect to some scalar product $(u,v)_\sigma = \int_M u(x)\overline{v(x)}\sigma(x)dx$. Prove that

$$\operatorname{ind} A = \operatorname{Tr} e^{-tA^*A} - \operatorname{Tr} e^{-tAA^*}. \tag{60.18}$$

9. Let $A(x,D)$ be a Fredholm elliptic operator in $\mathbb{R}^n$ (cf. §42). Let $R(x,D)$ be a parametrix of $A(x,D)$ such that $AR = I + T_{-N_1}$ and $RA = I + T_{-N_2}$, where $\operatorname{ord} T_{N_i} < -n$, $i = 1,2$. Prove that

$$\operatorname{ind} A = \operatorname{Tr}(AR - RA). \tag{60.19}$$

10. Let Γ be a Jordan curve in $\mathbb{C}$ and let $Sf = \frac{1}{2\pi i}\,\text{p.v.}\int_\Gamma \frac{f(w)dw}{w-z}$ be the Cauchy integral. Consider a singular integral equation on Γ,

$$a(z)f(z) + b(z)Sf = g(z),\ z \in \Gamma. \tag{60.20}$$

Prove that $A = af + bSf$ is a Fredholm operator in $L_2(\Gamma)$ provided that $a(z) \pm b(z) \neq 0$ on Γ and compute its index.

11. Consider the Laplace equation in $\Omega \subset \mathbb{R}^2$ with oblique derivative boundary condition:

$$\Delta u = 0 \quad x \in \Omega, \tag{60.21}$$

$$a(x_1, x_2)\frac{\partial u}{\partial x_1} + b(x_1, x_2)\frac{\partial u}{\partial x_2}\bigg|_{\partial\Omega} = g, \tag{60.22}$$

where $a^2 + b^2 > 0$ on $\partial\Omega$.

Prove that the operator defined by (60.21), (60.22) is Fredholm and compute its index.

(Hint: Use the reduction to the singular integral equation on $\partial\Omega$.)

Chapter VIII

Fourier Integral Operators

Introduction to Chapter VIII

This chapter is an introduction to the theory of Fourier integral operators (FIO's). We start with the simplest class of FIO's defined by a generating function $S(x, \eta)$ of the canonical transformation $(y, \eta) \to (x, \xi)$. In §61 we prove the boundedness theorem for such FIO's (see Eskin [**E3**]). In §62 we consider the products $A\Phi$ and ΦA, where A is a ψdo and Φ is a FIO. At the end of this section we study the conjugation $\Phi^* A \Phi$ of a ψdo by the FIO's (Egorov's theorem). The wave front sets of FIO's are considered in §63. In §64 the geometric optics construction of the parametrix of the Cauchy problem for a second order hyperbolic equation is studied on a small time interval $0 \leq t \leq t_1$. More advanced topics for FIO's are discussed in §65. We study FIO's with nondegenerate phase functions and associate a Lagrangian manifold with a FIO. The general theory of such FIO's was developed by Hörmander [**H1**], [**H2**]. In this book we study only a restricted class of FIO's where the associated Lagrangian manifolds correspond to the graphs of canonical transformations. This restriction makes the results and the proofs much simpler. We also make an extensive use of phase functions determined by the generating functions of canonical transformations analogously to Maslov [**M1**], [**MF**] and Eskin [**E4**], Appendix. In particular, we construct a global FIO given by a global canonical transformation and prove a theorem on the composition of global FIO's.

In §66, following Maslov [**M1**], [**M2**], [**MF**], we extend the geometric optics construction of §64 to the case of an arbitrary time interval $[0, T]$.

In §67 we study the oblique derivative problem as an example of non-elliptic boundary value problems for elliptic equations. The lack of ellipticity of the boundary conditions leads to the loss of smoothness of the solutions in comparison with the elliptic case. Moreover, it may lead to a completely new boundary condition on a submanifold of the boundary or an appearance of a "coboundary" condition (see Eskin [**E3**]). This section is included in the chapter on FIO's because the use of FIO's is an important part of the solution of such problems.

Similar results were obtained independently by Sjöstrand [**Sj**] and Duistermaat and Sjöstrand [**DSj**]. The chapter ends with the problem section (§68).

61. Boundedness of Fourier integral operators (FIO's)

61.1. The definition of a FIO.

Let $S(x,\eta)\in C^\infty(\mathbb{R}^n\times(\mathbb{R}^n\setminus\{0\}))$, $\deg S(x,\eta)=1$, and

$$\det S_{x\eta}\neq 0,\quad \text{where}\quad S_{x\eta}=\left[\frac{\partial^2 S}{\partial x_i\partial\eta_j}\right]_{i,j=1}^n. \tag{61.1}$$

Denote $\xi=S_x(x,\eta)$, $y=S_\eta(x,\eta)$. It follows from (61.1) that

$$y=S_\eta(x,\eta),\quad \xi=S_x(x,\eta) \tag{61.2}$$

define a diffeomorphism $(y,\eta)\to(x,\xi)$ in a neighborhood of every point $(x,\eta)\in\mathbb{R}^n\times(\mathbb{R}^n\setminus\{0\})$. We assume for simplicity that $S(x,\eta)=x\cdot\eta+s(\eta)$ for $|x|>R$ and that (61.2) is a diffeomorphism of $\mathbb{R}^n\times(\mathbb{R}^n\setminus\{0\})\to\mathbb{R}^n\times(\mathbb{R}^n\setminus\{0\})$.

Let $a(x,\eta)\in S^m$,

$$a(x,\eta)=\sum_{k=0}^N a_k(x,\eta)(1-\chi(\eta))+a^{(N+1)}(x,\eta),\ \forall N, \tag{61.3}$$

where $a_k(x,\eta)$ are homogeneous in η of degree $m-k$, $a^{(N+1)}(x,\eta)\in S^{m-N-1}$.

An operator of the form

$$\begin{aligned}\Phi u&=\frac{1}{(2\pi)^n}\int_{\mathbb{R}^n}a(x,\eta)e^{iS(x,\eta)}\tilde{u}(\eta)d\eta\\&=\frac{1}{(2\pi)^n}\int_{\mathbb{R}^n}\int_{\mathbb{R}^n}a(x,\eta)e^{iS(x,\eta)-iy\cdot\eta}u(y)dyd\eta,\quad u\in C_0^\infty(\mathbb{R}^n),\end{aligned} \tag{61.4}$$

is called the Fourier integral operator with phase function $S(x,\eta)-y\cdot\eta$ and amplitude (symbol) $a(x,\eta)$. Represent $a(x,\eta)=a(x,\eta)(1-\chi(\eta))+a(x,\eta)\chi(\eta)$. The operator $T_{-\infty}$ with amplitude $a(x,\eta)\chi(\eta)$ is an integral operator with a $C^\infty(\mathbb{R}^n\times\mathbb{R}^n)$ kernel and $\operatorname{ord}T_{-\infty}=-\infty$. Therefore, modulo such an operator, we can assume that $a(x,\eta)=0$ when $|\eta|<1$. This will be convenient since $S(x,\eta)$ is not smooth for $\eta = 0$. Note that (61.4) is a ψdo when $S(x,\eta) = x \cdot \eta$. Denote by Φ^* the formally adjoint operator to Φ (cf. (43.26)):

$$(\Phi u, v) = (u, \Phi^* v), \quad \forall u, v \in C_0^\infty(\mathbb{R}^n),$$

where (u,v) is the L_2-scalar product. We have

$$\Phi^* v = \frac{1}{(2\pi)^n} \int_{\mathbb{R}^n} \int_{\mathbb{R}^n} \overline{a(y,\xi)} e^{ix\cdot\xi - iS(y,\xi)} v(y) dy d\xi, \quad v \in C_0^\infty(\mathbb{R}^n). \tag{61.5}$$

As in §43, we treat the integral in (61.5) as a repeated integral, i.e., first integrate with respect to y and then with respect to ξ. We call (61.5) also a Fourier integral operator. The operators (61.4) and (61.5) are particular cases of a more general class of Fourier integral operators of the form

$$\Psi u - \int_{\mathbb{R}^n} \int_{\mathbb{R}^r} a(x,y,\xi) e^{i\psi(x,y,\xi)} u(y) dy d\xi. \tag{61.6}$$

61.2. The boundedness of FIO's.

Theorem 61.1. *The operators Φ and Φ^* are bounded from $H_s(\mathbb{R}^n)$ to $H_{s-m}(\mathbb{R}^n)$ for all s.*

Proof: Consider the product $\Phi \Lambda^{2s} \Phi^* u$, where Λ^{2s} is a ψdo with the symbol $\Lambda^{2s}(\xi) = (1+|\xi|^2)^s$. We have

$$\Lambda^{2s} \Phi^* u = \frac{1}{(2\pi)^n} \int_{\mathbb{R}^n} \int_{\mathbb{R}^n} (1+|\xi|^2)^s \overline{a(y,\xi)} e^{ix\cdot\xi - iS(y,\xi)} u(y) dy d\xi$$

and
(61.7)

$$\Phi \Lambda^{2s} \Phi^* u = \frac{1}{(2\pi)^n} \int_{\mathbb{R}^n} \int_{\mathbb{R}^n} a(x,\xi) \overline{a(y,\xi)} (1+|\xi|^2)^s e^{iS(x,\xi) - iS(y,\xi)} u(y) dy d\xi.$$

As in (45.6), we have

$$S(x,\xi) - S(y,\xi) = \Sigma(x,y,\xi) \cdot (x-y), \tag{61.8}$$

where

$$\Sigma(x,y,\xi) = \int_0^1 S_x(y + t(x-y), \xi) dt. \tag{61.9}$$

Let

(61.10) $K_1 u =$

$$\frac{1}{(2\pi)^n} \int_{\mathbb{R}^n} \int_{\mathbb{R}^n} a(x,\xi) \overline{a(y,\xi)} (1+|\xi|^2)^s \chi\Big(\frac{x-y}{\delta}\Big) e^{iS(x,\xi) - iS(y,\xi)} u(y) dy d\xi,$$

where $\chi(x) \in C_0^\infty(\mathbb{R}^n)$, $\chi(x) = 1$ for $|x| \le 1$, $\chi(x) = 0$ for $|x| > 2$, and δ is small. Let

$$K_2 u = \Phi \Lambda^{2s} \Phi^* u - K_1 u,$$

i.e., K_2 has the form (61.10) with $\chi(\frac{x-y}{\delta})$ replaced by $1-\chi(\frac{x-y}{\delta})$. Since $S_\xi(x,\xi)-S_\xi(y,\xi)\neq 0$ when $x-y\neq 0$, we have

$$|S_\xi(x,\xi)-S_\xi(y,\xi)|\geq C$$

when $|x-y|\geq\delta>0$. The following identity holds for any N:

(61.11)

$$e^{iS(x,\xi)-iS(y,\xi)}=\left[\frac{S_\xi(x,\xi)-S_\xi(y,\xi)}{|S_\xi(x,\xi)-S_\xi(y,\xi)|^2}\cdot\left(-i\frac{\partial}{\partial\xi}\right)\right]^N e^{iS(x,\xi)-iS(y,\xi)}.$$

Note that we assumed that $a(x,\xi)=0$ when $|\xi|\leq 1$. Substituting the identity (61.11) in K_2u and integrating by parts N times with respect to ξ we conclude (cf. (44.5), (44.6)) that K_2u is an integral operator with a $C^\infty(\mathbb{R}^n\times\mathbb{R}^n)$ kernel, since N is arbitrary.

In (61.10) make the change of variables

$$\eta=\Sigma(x,y,\xi). \tag{61.12}$$

Since $\Sigma(x,x,\xi)=S_x(x,\xi)$, the inverse map $\xi=\Sigma^{-1}(x,y,\eta)$ exists when $|x-y|<2\delta$. To justify the change of variables, we introduce the cutoff function $\chi(\varepsilon\xi)$, make the change of variables, and then let $\varepsilon\to 0$. We get

$$K_1u=\frac{1}{(2\pi)^n}\int_{\mathbb{R}^n}\int_{\mathbb{R}^n}a(x,\Sigma^{-1}(x,y,\eta))\overline{a(y,\Sigma^{-1}(x,y,\eta)}$$
$$\times(1+|\Sigma^{-1}(x,y,\eta)|^2)^s\chi\left(\frac{x-y}{\delta}\right)e^{i(x-y)\cdot\eta}u(y)\left|\frac{\partial\Sigma^{-1}}{\partial\eta}\right|dyd\eta,$$

where $\xi=\Sigma^{-1}(x,y,\eta)$ is the inverse to (61.12), $|\frac{\partial\Sigma^{-1}}{\partial\eta}|=|\frac{\partial\eta}{\partial\xi}|^{-1}$, and $\frac{\partial\eta}{\partial\xi}$ is the Jacobian matrix of the map (61.12), $\left|\frac{\partial\eta}{\partial\xi}\right|$ being the absolute value of $\det\frac{\partial\eta}{\partial\xi}$. Note that $\left|\frac{\partial\eta}{\partial\xi}\right|=|S_{x\xi}(x,\xi)|$ for $x=y$, where $|S_{x\xi}(x,\xi)|$ is the absolute value of the determinant of $S_{x\xi}(x,\xi)$. Note also that K_1 is a pseudodifferential operator of the form (43.2) and has order $2m+2s$. The principal symbol of K_1 is

$$|a(x,\Sigma^{-1}(x,x,\eta))|^2(1+|\Sigma^{-1}(x,x,\eta)|^2)^s\,|S_{x\xi}(x,\Sigma^{-1}(x,x,\eta)|^{-1}. \tag{61.13}$$

Since $\operatorname{ord}K_2=-\infty$, we get

$$\|\Phi\Lambda^{2s}\Phi^*u\|^2_{-m-s}=\|K_1u+K_2u\|^2_{-m-s}\leq C\|u\|^2_{m+s}. \tag{61.14}$$

Therefore

$$\begin{aligned}\|\Phi^*u\|_s^2=(\Lambda^s\Phi^*u,\Lambda^s\Phi^*u)&=((K_1+K_2)u,u)\\&\leq\|(K_1+K_2)u\|_{-m-s}\|u\|_{m+s}\leq C\|u\|^2_{m+s},\end{aligned} \tag{61.15}$$

i.e., $\operatorname{ord}\Phi^*=m$. It follows from (61.15) that

$$|(\Phi u,v)|=|(u,\Phi^*v)|\leq C\|u\|_s\|\Phi^*v\|_{-s}\leq C\|u\|_s\|v\|_{m-s}$$

for all $v \in H_{m-s}(\mathbb{R}^n)$. Therefore (cf. §13)

$$\|\Phi u\|_{s-m} \leq C\|u\|_s. \tag{61.16}$$

Since s is arbitrary, Theorem 61.1 is proven. □

61.3. Canonical transformations.

Let

$$\hat{x} = \hat{x}(x,\xi), \quad \hat{\xi} = \hat{\xi}(x,\xi) \tag{61.17}$$

be a map of $\mathbb{R}^n \times (\mathbb{R}^n \setminus \{0\})$ to $\mathbb{R}^n \times (\mathbb{R}^n \setminus \{0\})$, where $\hat{x}(x,\xi)$ is homogeneous in ξ of degree zero, $\hat{\xi}(x,\xi)$ is homogeneous in ξ of degree one: $\deg_\xi \hat{x}(x,\xi) = 0$, $\deg_\xi \hat{\xi}(x,\xi) = 1$, and $\hat{x}(x,\xi), \hat{\xi}(x,\xi)$ are $C^\infty(\mathbb{R}^n \times (\mathbb{R}^n \setminus \{0\}))$ functions. The transformation (61.17) is called canonical if it preserves the differential form $\sum_{j=1}^n dx_j \wedge d\xi_j$, i.e.,

$$\sum_{k=1}^n d\hat{x}_k \wedge d\hat{\xi}_k = \sum_{k=1}^n dx_k \wedge d\xi_k, \tag{61.18}$$

where $\hat{x}, \hat{\xi}$ are given by (61.17). We present some examples of canonical transformations.

Let $S(x,\hat{\xi}) \in C^\infty(\mathbb{R}^n \times (\mathbb{R}^n \setminus \{0\}))$, $S(x,\hat{\xi})$ is homogeneous in $\hat{\xi}$ of degree one, and $\det \frac{\partial^2 S(x,\hat{\xi})}{\partial x \partial \hat{\xi}} \neq 0$, $\forall x \in \mathbb{R}^n$, $\hat{\xi} \in \mathbb{R}^n \setminus \{0\}$. Define the map $(x,\xi) \to (\hat{x},\hat{\xi})$ by

$$\xi = \frac{\partial S(x,\hat{\xi})}{\partial x}, \quad \hat{x} = \frac{\partial S(x,\hat{\xi})}{\partial \hat{\xi}}. \tag{61.19}$$

Assuming that the map $(x,\xi) \to (\hat{x},\hat{\xi})$ is a diffeomorphism, we show that it is a canonical transformation. The function $S(x,\hat{\xi})$ is called the generating function of this canonical transformation.

We have

$$d\xi \wedge dx = \left(\frac{\partial^2 S}{\partial x^2} dx + \frac{\partial^2 S}{\partial \hat{\xi} \partial x} d\hat{\xi}\right) \wedge dx$$

and

$$d\hat{\xi} \wedge d\hat{x} = d\hat{\xi} \wedge \left(\frac{\partial^2 S(x,\hat{\xi})}{\partial x \partial \hat{\xi}} dx + \frac{\partial^2 S(x,\hat{\xi})}{\partial \hat{\xi}^2} d\hat{\xi}\right).$$

Note that $\frac{\partial^2 S(x,\hat{\xi})}{\partial x^2} dx \wedge dx = 0$ and $d\hat{\xi} \wedge \frac{\partial^2 S(x,\hat{\xi})}{\partial \hat{\xi}^2} d\hat{\xi} = 0$. Therefore $d\hat{\xi} \wedge d\hat{x} = d\xi \wedge dx$. □

Remark 61.1. Suppose $S(x,\hat{\xi}) = s(x)\cdot\hat{\xi}$, where $\hat{x} = s(x)$ is a diffeomorphism of $\mathbb{R}^n$. The canonical transformation $(x,\xi)\to(\hat{x},\hat{\xi})$ with the generating function $S(x,\hat{\xi})$ has the form:

$$\hat{x} = S_{\hat{\xi}}(x,\hat{\xi}) = s(x), \quad \xi = \hat{S}_x(x,\hat{\xi}) = \left(\frac{\partial s(x)}{\partial x}\right)^T \hat{\xi}, \tag{61.20}$$

i.e., $\hat{\xi} = \left(\left(\frac{\partial s(x)}{\partial x}\right)^T\right)^{-1}\xi$.

Therefore the canonical transformation (61.20) in $T_0^*(\mathbb{R}^n)$ corresponds to the change of variables $\hat{x} = s(x)$ in $\mathbb{R}^n$ (cf. §45). □

Let $\lambda(x,t,\xi)$ be a real-valued function $\lambda(x,t,\xi) \in C^\infty(\mathbb{R}^n\times\mathbb{R}\times(\mathbb{R}^n\setminus\{0\}))$, and let $\lambda(x,t,\xi)$ be homogeneous in ξ of degree one.

Consider the Hamiltonian system (cf. (50.11)):

$$\begin{aligned}\frac{dx(t)}{dt} &= \frac{\partial\lambda(x(t),t,\xi(t))}{\partial\xi}, \quad x(0) = y,\\ \frac{d\xi(t)}{dt} &= -\frac{\lambda(x(t),t,\xi(t))}{\partial x}, \quad \xi(0) = \eta.\end{aligned} \tag{61.21}$$

Let

$$x = x(t,y,\xi), \quad \xi = \xi(t,y,\xi) \tag{61.22}$$

be the solution of (61.21). It was shown in Remark 50.1 that the solution (61.22) exists for all $t\in\mathbb{R}$. We now show that (61.22) is a canonical transformation of $\mathbb{R}^n\times(\mathbb{R}^n\setminus\{0\})$ to $\mathbb{R}^n\times(\mathbb{R}^n\setminus\{0\})$ for each $t\in\mathbb{R}$. We have

$$\begin{aligned}\frac{d}{dt}(dx\wedge d\xi) &= d\frac{dx}{dt}\wedge d\xi + dx\wedge d\frac{d\xi}{dt} = d\frac{\partial\lambda}{\partial\xi}\wedge d\xi - dx\wedge d\frac{\partial\lambda}{\partial x}\\ &= \left(\frac{\partial^2\lambda}{\partial\xi\partial x}dx + \frac{\partial^2\lambda}{\partial\xi^2}d\xi\right)\wedge d\xi - dx\wedge\left(\frac{\partial^2\lambda}{\partial x\partial\xi}d\xi + \frac{\partial^2\lambda}{\partial x^2}dx\right) = 0\end{aligned}$$

since $\frac{\partial^2\lambda}{\partial\xi^2}d\xi\wedge d\xi = 0$, $\frac{\partial^2\lambda}{\partial x^2}dx\wedge dx = 0$. Therefore $dx(t,y,\eta)\wedge d\xi(t,y,\eta)$ is independent of t, and so

$$dx\wedge d\xi = dx(0,y,\eta)\wedge d\xi(0,y,\eta) = dy\wedge d\eta. \tag{61.23}$$

Note that in (61.23) we consider t to be a parameter, and $dx, d\xi$ are the differentials of x and ξ as functions of (y,η) only.

62. Operations with Fourier integral operators

62.1. The stationary phase lemma.

We need the following stationary phase lemma (cf. Lemma 19.4).

Lemma 62.1. *Let $\psi(x,\xi,y,\eta)$ be a C^∞ function for $\xi \neq 0, \eta \neq 0$, homogeneous in (ξ,η) of degree one, $y \in \mathbb{R}^{n_1}$, $\eta \in \mathbb{R}^{n_2}$. Suppose $y_0 = y(x,\xi)$, $\eta_0 = \eta(x,\xi)$, $\xi \neq 0$, are the unique solutions of the equations*

$$\psi_y(x,\xi,y,\eta) = 0, \quad \psi_\eta(x,\xi,y,\eta) = 0. \tag{62.1}$$

Suppose $|H| \neq 0$, where $H = \begin{bmatrix} \psi_{yy} & \psi_{y\eta} \\ \psi_{\eta y} & \psi_{\eta\eta} \end{bmatrix}$, $|H| = |\det H|$. Assume that $C(x,\xi,y,\eta) \in C^\infty$ for $\xi \neq 0$, $\eta \neq 0$, is homogeneous in (ξ,η), and that the support of $C(x,\xi,y,\eta)$ is contained in a conic neighborhood of (y_0,η_0). Then

$$\begin{aligned} &\frac{1}{(2\pi)^{\frac{n_1+n_2}{2}}} \int_{\mathbb{R}^{n_1}} \int_{\mathbb{R}^{n_2}} C(x,\xi,y,\eta) e^{i\psi(x,\xi,y,\eta)} dy d\eta \\ &\quad = \frac{C(x,\xi,y_0(x,\xi),\eta_0(x,\xi)) e^{i\psi(x,\xi,y_0,\eta_0) + i\frac{\pi}{4}\operatorname{sgn} H}}{\sqrt{|H(x,\xi,y_0,\eta_0)|}} \\ &\qquad\qquad \times \Big(1 + \sum_{k=1}^{N} C_k(x,\xi) + O\Big(\frac{1}{|\xi|^{N+1}}\Big)\Big), \end{aligned} \tag{62.2}$$

where $\operatorname{sgn} H(x,\xi,y_0,\eta_0)$ *is the signature of the matrix H and* $\operatorname{ord}_\xi C_k(x,\xi) = -k$.

Note that (62.2) is understood as an oscillatory integral, i.e., we take a cutoff $\chi(\varepsilon\eta)$ and consider the limit in the distribution sense as $\varepsilon \to 0$.

The proof of Lemma 62.1 follows from the proof of Lemma 19.4.

Note that if the symbol $C(x,\xi,y,\eta)$ is such that there is no solution to (62.1) on $\operatorname{supp} C$ and

$$|\psi_y| + |\psi_\eta|(|\xi| + |\eta|) \geq C(|\xi| + |\eta|) \quad \text{on} \quad \operatorname{supp} C, \tag{62.3}$$

then using an identity of the form (61.11) and integrating by parts with respect to (y,η), we see that the integral (62.2) is $O(\frac{1}{|\xi|^N})$, $\forall N$.

62.2. Composition of a ψdo and a FIO.

Lemma 62.2. *Let $A(x,\xi) \in S^\alpha$ and let Φ be an operator of the form (61.4). Then*

$$A(x,D)\Phi u = \frac{1}{(2\pi)^n} \int_{\mathbb{R}^{3n}} C_N(x,\eta) e^{iS(x,\eta)} \tilde{u}(\eta) d\eta + T_{\alpha+m-N-1} u, \tag{62.4}$$

where $C_N(x,\eta) \in S^{\alpha+m}$, $\operatorname{ord} T_{\alpha+m-N-1} \leq \alpha + m - N - 1$ *and the principal part $C^{(0)}$ of $C_N(x,\eta)$ has the form*

$$C_0(x,\eta) = A_0(x, S_x(x,\eta)) a(x,\eta), \tag{62.5}$$

where A_0 is the principal part of $A(x,\xi)$.

Proof: We have

$$A(x,D)\Phi u = \frac{1}{(2\pi)^{2n}} \int_{\mathbb{R}^n} A(x,\xi)a(y,\eta)e^{i(x-y)\cdot\xi + iS(y,\eta)}\tilde{u}(\eta)d\eta dy d\xi.$$

Compute the integral with respect to (y,ξ) using Lemma 62.1. We have

$$\psi = (x-y)\cdot\xi + S(y,\eta), \quad \psi_y = -\xi + S_y(y,\eta) = 0, \quad \psi_\xi = x - y = 0,$$

$$\psi_{\xi y} = -I, \quad \psi_{yy} = S_{yy}, \quad \psi_{\xi\xi} = 0.$$

Therefore, at the critical point, $\psi = S(x,\eta)$, $H = \left[\begin{smallmatrix} S_{yy} & -I \\ -I & 0 \end{smallmatrix}\right]$, $\det H = 1$, and the signature of H is zero. Thus Lemma 62.1 implies (62.4), (62.5).

Remark 62.1. We prove that $\operatorname{sgn}\left[\begin{smallmatrix} S_{yy} & -I \\ I & 0 \end{smallmatrix}\right] = 0$. Consider a more general case of the matrix $\left[\begin{smallmatrix} B & A \\ A & 0 \end{smallmatrix}\right] = 0$, where A, B are symmetric $n\times n$ matrices and A is nonsingular. Note that $\det\left[\begin{smallmatrix} B & A \\ A & 0 \end{smallmatrix}\right] = (\det A)^2 \neq 0$. Consider the homotopy $\left[\begin{smallmatrix} \varepsilon B & A \\ A & 0 \end{smallmatrix}\right]$, $0 \le \varepsilon \le 1$. Since $\det\left[\begin{smallmatrix} \varepsilon B & A \\ A & 0 \end{smallmatrix}\right] \neq 0$ for all $\varepsilon > 0$, the signature does not change. Therefore, $\operatorname{sgn}\left[\begin{smallmatrix} B & A \\ A & 0 \end{smallmatrix}\right] = \operatorname{sgn}\left[\begin{smallmatrix} 0 & A \\ A & 0 \end{smallmatrix}\right]$. Write any vector in $\mathbb{R}^{2n}$ as (x,y), where $x \in \mathbb{R}^n$, $y \in \mathbb{R}^n$. The quadratic form $\left(\left[\begin{smallmatrix} 0 & A \\ A & 0 \end{smallmatrix}\right](x,y),(x,y)\right)$ can be written as $2(Ax,y) = \frac{1}{2}(Ax',x') - \frac{1}{2}A(y',y')$, where $x' = x+y, y' = x-y$. Since $x', y' \in \mathbb{R}^n$ are arbitrary, the quadratic form (Ax,y) has zero signature. Therefore $\operatorname{sgn}\left[\begin{smallmatrix} B & A \\ A & 0 \end{smallmatrix}\right] = 0$.

Analogously we can prove

Lemma 62.3. *Let $B(x,\xi) \in S^\alpha$ and let Φ be the same as in (61.4). Then*

$$\Phi B u = \frac{1}{(2\pi)^n} \int_{\mathbb{R}^n} C_N^{(1)}(x,\eta)e^{iS(x,\eta)}\tilde{u}(\eta)d\eta + T_{\alpha+m-N-1}u, \tag{62.6}$$

where $C_N^{(1)} \in S^{m+\alpha}$, $\operatorname{ord} T_{\alpha+m-N-1} \le \alpha + m - N - 1$, *and the principal part $C^{(1)}$ of $C_N^{(1)}(x,\eta)$ is*

$$C^{(1)}(x,\eta) = B_0(S_\eta(x,\eta),\eta)a(x,\eta), \tag{62.7}$$

where B_0 is the principal part of $B(x,\xi)$.

Proof: We have

$$\Phi B u = \frac{1}{(2\pi)^{2n}} \int_{\mathbb{R}^n}\int_{\mathbb{R}^n}\int_{\mathbb{R}^n} e^{iS(x,\xi)-iy\cdot\xi}a(x,\xi)B(y,\eta)e^{iy\cdot\eta}\tilde{u}(\eta)d\xi dy d\eta.$$

Compute the integral with respect to (y,ξ) using Lemma 62.1. We have

$$\psi = S(x,\xi) - y\cdot\xi + y\cdot\eta, \quad \psi_y = -\xi + \eta = 0,$$

$$\psi_\xi = S_\xi(x,\xi) - y = 0, \quad \psi_{y\xi} = -I, \quad \psi_{yy} = 0, \quad \psi_{\xi\xi} = S_{\xi\xi}(x,\xi).$$

Therefore, at the critical point,

$$\psi = S(x,\eta), \quad H = \begin{bmatrix} 0 & -I \\ -I & S_{\xi\xi} \end{bmatrix}, \quad |H| = 1, \quad \operatorname{sgn} H = 0$$

(cf. Remark 62.1), and we see from (62.2) that (62.6) and (62.7) hold.

Lemma 62.4. *Let Φ and Φ^* have the form* (61.4) *and* (61.5), *respectively. Then for arbitrary N,*

$$\Phi^*\Phi u = C_N u + T_{2m-N-1}, \tag{62.8}$$

where C_N is a ψdo, $C_N(x,\xi) \in S^{2m}$.

Proof: We have

$$(\Phi^*\Phi)u(y) = \frac{1}{(2\pi)^{2n}} \int_{\mathbb{R}^{3n}} \overline{a(x,\xi)}a(x,\eta)e^{iy\cdot\xi - iS(x,\xi)+iS(x,\eta)}\tilde{u}(\eta)d\eta dx d\xi.$$

Apply Lemma 62.1 to the integral with respect to (x,ξ). Let $\psi = y\cdot\xi - S(x,\xi) + S(x,\eta)$. Then

$$\psi_x = -S_x(x,\xi) + S_x(x,\eta) = 0, \quad \psi_\xi = y - S_\xi(x,\xi) = 0, \quad \psi_{x\xi} = -S_{x\xi}(x,\xi),$$

$$\psi_{xx} = -S_{xx}(x,\xi) + S_{xx}(x,\eta), \quad \psi_{\xi\xi} = -S_{\xi\xi}(x,\xi).$$

Since we assume that the map $\xi = S_x(x,\eta)$ is one-to-one for each x, we obtain that $\psi_x = 0$ implies that $\xi = \eta$ at the critical point. Therefore $S(x,\eta) = S(x,\xi)$, $\psi_{xx} = 0$, and

$$H = \begin{bmatrix} 0 & -S_{x\xi}(x,\xi) \\ -S_{x\xi}(x,\xi) & -S_{\xi\xi}(x,\xi) \end{bmatrix}.$$

Here $|H| = |S_{x\xi}(x,\xi)|^2$, sgn $H = 0$ (cf. Remark 62.1), and we get

$$(\Phi^*\Phi u)(y) = \frac{1}{(2\pi)^n} \int_{\mathbb{R}^n} C_N(y,\eta)e^{iy\cdot\eta}\tilde{u}(\eta)d\eta + T_{2m-N}u,$$

where the principal symbol of $C_N(y,\eta)$ is

$$|a_0(x_0,\eta)|^2|S_{x\eta}(x_0,\eta)|^{-1}, \tag{62.9}$$

and where $x_0(y,\eta)$ is such that $S_\eta(x_0,\eta) = y$.

62.3. Elliptic FIO's.

Denote by Φ_0 the Fourier integral operator of the form

$$\Phi_0 u = \frac{1}{(2\pi)^n} \int_{\mathbb{R}^n} |S_{x\eta}(x,\eta)|^{\frac{1}{2}}(1-\chi(\eta))e^{iS(x,\eta)}\tilde{u}(\eta)d\eta, \tag{62.10}$$

where $|S_{x\eta}(x,\eta)|$ is the absolute value of $\det S_{x\eta}$. Then

$$\Phi_0^* u = \frac{1}{(2\pi)^n} \int_{\mathbb{R}^n}\int_{\mathbb{R}^n} |S_{y\xi}(y,\xi)|^{\frac{1}{2}}(1-\chi(\xi))e^{ix\cdot\xi - iS(y,\xi)}u(y)dyd\xi. \tag{62.11}$$

The next lemma follows from (61.13) with $s = 0$, and from (62.9).

Lemma 62.5. *Let* Φ_0, Φ_0^* *be as in* (62.10), (62.11). *Then*

$$\begin{aligned}\Phi_0^*\Phi_0 &= I + T_{-1}^{(1)},\\ \Phi_0\Phi_0^* &= I + T_{-1}^{(2)},\end{aligned} \tag{62.12}$$

where $\operatorname{ord} T_{-1}^{(p)} \le -1,\ p = 1, 2.$ □

We call the Fourier integral operator (61.4) an elliptic Fourier integral operator (FIO) if the principal part $a_0(x,\eta)$ of $a(x,\eta) \in S^m$ is elliptic, i.e., $a_0(x,\eta) \neq 0,\ \forall x \in \mathbb{R}^n,\ \forall \eta \neq 0.$

Lemma 62.6. *Suppose* $a(x,\xi) \in S^m$, $a(x,\xi)$ *is elliptic and* $|a(x,\xi)| \geq C(1+|\xi|)^m$ *when* $|x|^2 + |\xi|^2 \geq R^2$ *for some* $R > 0$. *Then the FIO* (61.4) *is Fredholm from* $H_s(\mathbb{R}^n)$ *to* $H_{s-m}(\mathbb{R}^n)$ *for every* $s \in \mathbb{R}$.

Proof: It follows from (62.8), (62.9) that $\Phi^*\Phi = C + T_{-1}$, where C is an elliptic ψdo in $\mathbb{R}^n$. Therefore there exists an operator R_1, $\operatorname{ord} R_1 = -2m$, such that $R_1 C = I + T_{-1}^{(1)}$, where $\operatorname{ord} T_{-1}^{(1)} \le -1$ (cf. §42). Note that $T_{-1}^{(1)}$ is compact (cf. Remark 40.2). Therefore $R\Phi^*$ is a left regularizer for Φ. Analogously it follows from (61.7) with $s = 0$ and from (61.13) that

$$\Phi\Phi^* = C_1 + T_{-1}^{(2)},$$

where C_1 is an elliptic ψdo, $\operatorname{ord} T_{-1}^{(2)} \le -1$. Therefore Φ has also a right regularizer, and subsequently Φ is Fredholm (cf. §42).

62.4. Egorov's theorem.

Theorem 62.7. *Let* $A_0(x,\xi)$ *be the principal part of* $A(x,\xi) \in S^m$. *Then*

$$\Phi_0^* A(x,D)\Phi_0 = B^{(N)}(x,D) + T_{m-N-1}, \quad \forall N, \tag{62.13}$$

where $B^{(N)}(x,\xi) \in S^m$ *with the principal part*

$$B_0(y,\eta) = A_0(x,\xi), \tag{62.14}$$

and where

$$y = S_\eta(x,\eta), \quad \xi = S_x(x,\eta), \tag{62.15}$$

i.e., (y,η) *and* (x,ξ) *are related by the canonical transformation* (62.15) *(cf.* (61.19)*).*

This is Egorov's theorem.

Proof: As in Lemmas 62.2 and 62.4, we have that $B^{(N)}$ is a ψdo. Since $a_0(x,\eta) = |S_{x\eta}(x,\eta)|^{\frac{1}{2}}$, we get from (62.5) that the principal symbol of $A\Phi$ is

$$A_0(x, S_x(x,\eta))|S_{x\eta}(x,\eta)|^{\frac{1}{2}}.$$

Analogously to the proof of Lemma 62.4 we have (cf. (62.9)):

$$B_0(y,\eta) = A_0(x_0(y,\eta), S_x(x_0(y,\eta),\eta)), \tag{62.16}$$

where $y = S_\eta(x_0,\eta)$. Therefore

$$B_0(S_\eta(x,\eta),\eta) = A_0(x, S_x(x,\eta)), \tag{62.17}$$

i.e., (62.14) holds. □

Remark 62.2. Theorem 62.7 follows also from Lemmas 62.2, 62.3. If $A\Phi u = \Phi B u + T_{-N}$, then, comparing (62.5) and (62.7), we conclude that $A_0(x, S_x(x,\eta)) = B_0(S_\eta(x,\eta),\eta)$. Replacing Φ by Φ_0 and multiplying $A\Phi_0$ by Φ_0^*, we get $\Phi_0^* A\Phi = \Phi_0^*\Phi_0 B + T_{-1} = B + T_{-1}^{(1)}$.

Example 62.1. Consider the symbol $\xi_0 - \lambda(x,t,\xi)$, where $\xi \in \mathbb{R}^n$, $\lambda(x,t,\xi) \in C^\infty$ when $\xi \neq 0$, $\deg_\xi \lambda(x,t,\xi) = 1$, $\lambda(x,t,\xi) = \lambda(\infty,\xi)$ when $|x| > R$, and $\lambda(x,t,\xi)$ is real-valued. We find a phase function $S(x,t,\eta,\eta_0) - y\cdot\eta - y_0\eta_0$ such that

$$\Phi_0^* A \Phi_0 = B + T_{-1}, \tag{62.18}$$

where $A(x,t,D_x,D_t) = D_t - \lambda(x,t,D_x)(1 - \chi(D_x))$, B is a ψdo with the symbol $B(y,y_0,\eta,\eta_0) = \eta_0$, and Φ_0 has the form (62.10). Let $\varphi(x,t,\eta)$ be the solution of the equation

$$\begin{aligned} &\varphi_t - \lambda(x,t,\varphi_x) = 0, \\ &\varphi(x,0,\eta) = x\cdot\eta. \end{aligned} \tag{62.19}$$

Denote $S(x,t,\eta,\eta_0) = \varphi(x,t,\eta) + t\eta_0$. We have

$$\begin{aligned} \xi &= S_x = \varphi_x(x,t,\eta), \\ \xi_0 &= S_t = \eta_0 + \varphi_t(x,t,\xi) = \eta_0 + \lambda(x,t,\varphi_x(x,t,\eta)), \\ y &= S_\eta = \varphi_\eta(x,t,\eta), \quad y_0 = S_{\eta_0} = t. \end{aligned} \tag{62.20}$$

Note that $y = \varphi_\eta(x,t,\eta) = x$ when $t = 0$. Therefore $S(x,t,\eta,\eta_0)$ defines a canonical transformation

$$(y,y_0,\eta,\eta_0) \to (x,t,\xi,\xi_0)$$

equal to the identity for $t = 0$. It is a diffeomorphism when $|t|$ is small. The solution of (62.19) also exists when $|t|$ is small. Note that (62.20) implies that $\eta_0 = \xi_0 - \lambda(x,t,\xi)$, and therefore (62.18) holds. We assumed in this section that the solution of (62.19) exists for all $t \in \mathbb{R}$ and that the canonical transformation $(y,y_0,\eta,\eta_0) \to (x,t,\xi,\xi_0)$ defined by (62.20) is a diffeomorphism on $\mathbb{R}^{n+1} \times (\mathbb{R}^{n+1} \setminus \{0\})$. We consider the general case in §65.

63. The wave front set of Fourier integral operators

Consider FIO of the form (61.4). Denote by $\sigma \circ \Sigma$ the image of the set $\Sigma \subset \mathbb{R}^n \times (\mathbb{R}^n \setminus \{0\})$ under the canonical transformation $\sigma : (y, \eta) \to (x, \xi)$ given by (61.2).

Theorem 63.1. *Suppose $u \in H_s(\mathbb{R}^n)$ for some $s \in \mathbb{R}$. The wave front set $WF(\Phi u)$ of Φu is contained in $\sigma \circ WF(u)$, where $WF(u)$ is the wave front set of u.*

Proof: We have

$$\Phi u = \frac{1}{(2\pi)^n} \int_{\mathbb{R}^n} \int_{\mathbb{R}^n} e^{iS(x,\eta) - iy\cdot\eta} a(x, \eta) u(y) dy d\eta.$$

We represent Φ as a sum $\Phi = \Phi_1 + \Phi_2$, where

$$\Phi_1 u = \frac{1}{(2\pi)^n} \int_{\mathbb{R}^n} \int_{\mathbb{R}^n} e^{iS(x,\eta) - iy\cdot\eta} \chi\Big(\frac{S_\eta(x, \eta) - y}{\delta}\Big) a(x, \eta) u(y) dy d\eta$$

and $\Phi_2 = \Phi - \Phi_1$. The following identity holds (cf. (61.11)):

$$\Big[\frac{(S_\eta(x, \eta) - y)}{|S_\eta - y|^2} \cdot \Big(-i\frac{\partial}{\partial \eta}\Big)\Big]^N e^{iS(x,\eta) - iy\cdot\eta} = e^{iS(x,\eta) - iy\cdot\eta}, \quad \forall N. \tag{63.1}$$

Substituting (63.1) into $\Phi_2 u$ and integrating by parts with respect to η, we see that $\Phi_2 u \in C^\infty$.

Suppose $(x_0, \xi_0) \notin \sigma \circ WF(u)$. Then there exist $\varphi_0(x) \in C_0^\infty(\mathbb{R}^n)$, $\varphi_0(x_0) \neq 0$, and $\alpha_0(\xi) \in C^\infty$ for $\xi \neq 0$, $\deg_\xi \alpha_0(\xi) = 0$, $\alpha_0(\xi_0) \neq 0$, such that $(\operatorname{supp} \varphi_0(x) \alpha_0(\xi)) \cap \sigma \circ WF(u) = \emptyset$. Consider the Fourier transform of $\varphi_0 \Phi_1 u$. We get

$$I(\xi) = \int_{\mathbb{R}^n} \int_{\mathbb{R}^n} \int_{\mathbb{R}^n} \alpha_0(\xi) \varphi_0(x) e^{-ix\cdot\xi + iS(x,\eta) - iy\cdot\eta} a(x, \eta) \chi\Big(\frac{S_\eta - y}{\delta}\Big) \times u(y) dy d\eta dx. \tag{63.2}$$

Let $I = I_1 + I_2$, where

$$I_1 = \int_{\mathbb{R}^n} \int_{\mathbb{R}^n} \int_{\mathbb{R}^n} \alpha_0(\xi) \varphi_0(x) \chi\Big(\frac{\xi - S_x(x, \eta)}{\delta\sqrt{|\xi|^2 + |\eta|^2}}\Big) \chi\Big(\frac{S_\eta - y}{\delta}\Big) a(x, \eta) \times e^{-ix\cdot\xi + iS(x,\eta) - iy\cdot\eta} u(y) dy d\eta dx, \tag{63.3}$$

and $I_2 = I - I_1$. Using the identity

$$\Big[\frac{(\xi - S_x(x, \eta))}{|\xi - S_x(x, \eta)|^2} \cdot \Big(i\frac{\partial}{\partial x}\Big)\Big]^N e^{iS(x,\eta) - ix\cdot\xi} = e^{iS(x,\eta) - ix\cdot\xi}, \quad \forall N, \tag{63.4}$$

and integrating by parts in I_2, we obtain that

$$|I_2(\xi)| \leq C_N (1 + |\xi|)^{-N}, \quad \forall N. \tag{63.5}$$

Let (y_0, η_0) be such that $(x_0, \xi_0) = \sigma \circ (y_0, \eta_0)$. By assumption, $(y_0, \eta_0) \notin WF(u)$. Let $\psi_0(y) \in C_0^\infty(\mathbb{R}^n)$, $\psi_0(y) = 1$ in a neighborhood of y_0, and let $\beta_0(\eta) \in C^\infty(\mathbb{R}^n \setminus \{0\})$, $\deg_\eta \beta_0(\eta) = 0$, $\beta_0(\eta) = 1$ in a conic neighborhood of η_0. We assume that $(\operatorname{supp} \psi_0(y)\beta_0(\eta)) \cap WF(u) = \emptyset$. We take $\delta > 0$ so small that $\psi_0(y)\beta_0(\eta) = 1$ on $\operatorname{supp} \alpha_0(\xi)\varphi_0(x)\chi(\frac{\xi - S_x}{\delta(\sqrt{|\xi|^2+|\eta|^2})})\chi(\frac{y-S_\eta}{\delta})$. Then

$$(63.6) \quad I_1 = \int_{\mathbb{R}^n}\int_{\mathbb{R}^n}\int_{\mathbb{R}^n} \alpha_0(\xi)\varphi_0(x)\chi\Big(\frac{\xi - S_x(x,\eta)}{\delta\sqrt{|\xi|^2 + |\eta|^2}}\Big)\chi\Big(\frac{y - S_\eta}{\delta}\Big)a(x,\eta) \times e^{-ix\cdot\xi + iS(x,\eta) - iy\cdot\eta}\beta_0(\eta)\psi_0(y)u(y)dyd\eta dx.$$

Let

$$I_3(\eta) = \int_{\mathbb{R}^n} e^{-iy\cdot\eta}\chi\Big(\frac{y - S_\eta}{\delta}\Big)\beta_0(\eta)\psi_0(y)u(y)dy = \beta_0(\eta)\frac{1}{(2\pi)^n}\int_{\mathbb{R}^n}\tilde{\chi}(\eta - \xi)(\widetilde{\psi_0 u})(\xi)d\xi,$$

where $\widetilde{\psi_0 u}(\xi), \tilde{\chi}(\xi)$ are the Fourier transforms of $\psi_0(y)u(y)$ and $\chi(\frac{y-S_\eta}{\delta})$ respectively. Let $\beta_1(\eta)$ be such that $(\operatorname{supp} \beta_1(\eta)\psi_0(y)) \cap WF(u) = \emptyset$ and $\beta_1(\eta) = 1$ in a small conic neighborhood of $\beta_0(\eta)$. Note that $|\beta_1(\xi)\widetilde{\psi u}(\xi)| \le C_N(1 + |\xi|)^{-N}$, $\forall N$ and $|\tilde{\chi}(\eta - \xi)| \le C_N(1 + |\eta - \xi|)^{-N}$, $\forall N$. We have (cf. the proof of Proposition 14.1):

$$I_3(\eta) = I_{31}(\eta) + I_{32}(\eta),$$

where

$$I_{31}(\eta) = \frac{1}{(2\pi)^n}\int_{\mathbb{R}^n} \beta_0(\eta)\tilde{\chi}(\eta - \xi)\beta_1(\xi)(\widetilde{\psi_0 u})(\xi)d\xi,$$

and $I_{32} = I_3 - I_{31}$. It is obvious that $|I_{31}(\eta)| \le C_N(1 + |\eta|)^{-N}$, $\forall N$. When $\xi \in \operatorname{supp}(1 - \beta_1(\xi))$ and $\eta \in \operatorname{supp} \beta_0(\eta)$, we have $|\xi - \eta| \ge C(|\xi| + |\eta|)$. Therefore $|I_{32}(\eta)| \le C_N(1 + |\eta|)^{-N}$, $\forall N$. Hence

$$(63.7) \qquad |I_3(\eta)| \le C_N(1 + |\eta|)^{-N}.$$

We represent $e^{-ix\cdot\xi} = \frac{(-\Delta_x + 1)^N e^{-ix\cdot\xi}}{(1+|\xi|^2)^N}$, $\forall N$, in (63.3) and integrate by parts with respect to x. Using (63.7), we get

$$|I_1(\xi)| \le C_N(1 + |\xi|)^{-N}, \quad \forall N,$$

i.e., $(x_0, \xi_0) \notin WF(\Phi u)$. □

Denote by $E(x, y)$ the Schwartz's kernel of the operator (61.4) (cf. §44), i.e.,

$$(63.8) \qquad E(x,y) = \frac{1}{(2\pi)^n}\int_{\mathbb{R}^n}\int_{\mathbb{R}^n} a(x,\theta)e^{iS(x,\theta) - iy\cdot\theta}d\theta,$$

where the integral is understood as a distribution limit when $a(x, \theta)$ is replaced by $a(x, \theta)\chi(\varepsilon\theta)$.

Lemma 63.2. *The wave front set of $E(x,y)$ is contained in the set*

$$(x, S_x(x,\eta), S_\eta(x,\eta), -\eta) \in T_0^*(\mathbb{R}^n) \times T_0^*(\mathbb{R}^n).$$

The proof of Lemma 63.2 is similar to the proof of Theorem 63.1.

64. Parametrix for the hyperbolic Cauchy problem

64.1. Asymptotic expansion.

Let $H(x,t,D_x,D_t)$ be a second order hyperbolic operator of the form:

$$\begin{aligned}(64.1)\quad H(x,t,D_x,D_t) = &-\frac{\partial^2}{\partial t^2} + \sum_{j,k=1}^n g^{jk}(x,t)\frac{\partial^2}{\partial x_j \partial x_k} \\ &+ \sum_{j=1}^n b_j(x,t)\frac{\partial}{\partial x_j} + b_0(x,t)\frac{\partial}{\partial t} + c(x,t).\end{aligned}$$

We assume that $\sum_{j,k=1}^n g^{jk}(x,t)\xi_j\xi_k \geq C|\xi|^2$.

Let H_0 be the principal symbol of (64.1). Then

$$(64.2)\qquad H_0(x,t,\xi,\xi_0) = (\xi_0 - \lambda_1(x,t,\xi))(\xi_0 - \lambda_2(x,t,\xi)),$$

where

$$(64.3)\qquad \lambda_1 = \sqrt{\sum_{j,k=1}^n g^{jk}(x,t)\xi_j\xi_k}, \qquad \lambda_2 = -\sqrt{\sum_{j,k=1}^n g^{jk}(x,t)\xi_j\xi_k}.$$

Consider the Cauchy problem

$$(64.4)\qquad Hu(x,t) = 0, \quad t > 0, \quad x \in \mathbb{R}^n,$$

$$(64.5)\qquad u(x,0) = g_1(x), \quad \frac{\partial u}{\partial t}(x,0) = g_2(x), \quad x \in \mathbb{R}^n,$$

where $g_k(x)$ have compact support, $k = 1,2$. We construct a parametrix for (64.4), (64.5). Denote by $\varphi_j(x,t,\eta)$, $j = 1,2$, the solution of the equation

$$(64.6)\qquad \frac{\partial \varphi_j(x,t,\eta)}{\partial t} - \lambda_j\Big(x,t,\frac{\partial \varphi_j}{\partial x}\Big) = 0, \quad t > 0,$$

$$(64.7)\qquad \varphi_j(x,0,\eta) = x\cdot\eta, \quad j = 1,2.$$

The solution of (64.6), (64.7) exists for $|t| \leq t_0$ if t_0 is small.

We look for the solution of (64.4) in the form

$$(64.8)\qquad u_N(x,t) = \frac{1}{(2\pi)^n}\int_{\mathbb{R}^n}\sum_{j=1}^2\sum_{k=0}^N a_{jk}(x,t,\eta)e^{i\varphi_j(x,t,\eta)}(1-\chi(\eta))d\eta,$$

where $a_{jk}(x,t,\eta) \in C^\infty$ when $\eta \neq 0$. Substituting (64.8) into (64.4), we get the following equations:

$$H_0(x,t,\varphi_{jx},\varphi_{jt}) = 0, \quad j = 1,2, \tag{64.9}$$

$$\frac{\partial H_0(x,t,\varphi_{jx},\varphi_{jt})}{\partial \xi_0}\frac{\partial a_{j0}}{\partial t} + \sum_{k=1}^{n} \frac{\partial H_0(x,t,\varphi_{jx},\varphi_{jt})}{\partial \xi_k}\frac{\partial a_{j0}(x,t,\eta)}{\partial x_k} + \left(H_0\left(x,t,\frac{\partial}{\partial x},\frac{\partial}{\partial t}\right)\varphi_j\right)a_{j0} - \left(\sum_{k=1}^{n} b_k(x,t)\varphi_{jx_k} + b_0(x,t)\varphi_{jt}\right)a_{j0}(x,t,\eta) = 0, \tag{64.10}$$

$$L_j a_{jk} = -iH(x,t,D_x,D_t)a_{j,k-1}, \quad k > 1, \ j = 1,2, \tag{64.11}$$

where L_j is the left hand side of (63.10). Since we want to have $u_N(x,0) = g_1(x)$, $\frac{\partial u_N(x,0)}{\partial t} = g_2(x)$, we impose the following initial conditions on a_{jk}, $j = 1,2$, $0 \leq k \leq N$:

$$\sum_{j=1}^{2}\sum_{k=0}^{N} a_{jk}(x,0,\eta) = \tilde{g}_1(\eta), \tag{64.12}$$

$$i\varphi_{1t}(x,0,\eta)\sum_{k=0}^{N} a_{1k}(x,0,\eta) + i\varphi_{2t}(x,0,\eta)\sum_{k=0}^{N} a_{2k}(x,0,\eta) + \sum_{j=1}^{2}\sum_{k=0}^{N}\frac{\partial a_{jk}(x,0,\eta)}{\partial t} = \tilde{g}_2(\eta). \tag{64.13}$$

Since $\varphi_{jt}(x,0,\eta) = \lambda_j(x,0,\eta)$, $j = 1,2$, we can find $a_{10}(x,0,\eta)$, $a_{20}(x,0,\eta)$ as the unique solution of the algebraic 2×2 system

$$\begin{aligned} a_{10}(x,0,\eta) + a_{20}(x,0,\eta) &= \tilde{g}_1(\eta), \\ i\lambda_1(x,0,\eta)a_{10} + i\lambda_2(x,0,\eta)a_{20} &= \tilde{g}_2(\eta). \end{aligned} \tag{64.14}$$

Note that $\det \left[\begin{smallmatrix} 1 & 1 \\ \lambda_1 & \lambda_2 \end{smallmatrix}\right] = \lambda_2 - \lambda_1 \neq 0$, $\forall x,t,\eta \neq 0$. We determine $a_{jk}(x,0,\eta)$ from the equations

(64.15)

$$\begin{aligned} a_{1k}(x,0,\eta) + a_{2k}(x,0,\eta) &= 0, \\ i\lambda_1(x,0,\eta)a_{1k} + i\lambda_2(x,0,\eta)a_{2k} &= -\frac{\partial a_{1,k-1}(x,0,\eta)}{\partial t} - \frac{\partial a_{2,k-1}(x,0,\eta)}{\partial t}, \end{aligned}$$
$$1 \leq k \leq N.$$

Let $w_N(x,t,\eta) = \sum_{j=1}^{2}\sum_{k=0}^{N} a_{jk}(x,t,\eta)e^{i\varphi_j(x,t,\eta)}(1-\chi(\eta))$. We have

$$
\begin{aligned}
Hw_N(x,t,\eta) &= \sum_{j,p=1}^{2} t_{pj}(x,t,\eta)e^{i\varphi_p}\tilde{g}_j(\eta), \\
w_N(x,0,\eta) &= \tilde{g}_1(\eta)e^{ix\cdot\eta}, \\
\frac{\partial w_N(x,0,\eta)}{\partial t} &= \tilde{g}_2(\eta)e^{ix\cdot\eta} + \sum_{j=1}^{2} t_{0j}(x,\eta)\tilde{g}_j(\eta)e^{ix\cdot\eta},
\end{aligned}
\tag{64.16}
$$

where $t_{pj} = O\big(\frac{1}{(|\eta|+1)^{N-1}}\big)$, $j=1,2,\ 0\le p\le 2$.

Let $w^{(N+1)}(x,t,\eta) = \sum_{j=1}^{2} w_j^{(N+1)}(x,t,\eta)\tilde{g}_j(\eta)$, where

$$
\begin{gathered}
Hw_j^{(N+1)}(x,\eta) = -\sum_{p=1}^{2} t_{pj}(x,t,\eta)e^{i\varphi_p}, \\
w_j^{(N+1)}(x,0,\eta) = 0, \qquad \frac{\partial w_j^{(N)}(x,0,\eta)}{\partial t} = -t_{0j}(x,\eta)e^{ix\cdot\eta}.
\end{gathered}
\tag{64.17}
$$

Assume that $H(x,t,\xi,\sigma) = H(\infty,t,\xi,\sigma)$ for $|x|>R$; then $w_j^{(N+1)}(x,t,\eta) = 0$ for $|x|>R$.

It follows from Theorems 48.6 and 13.4 with $s > \frac{n}{2}+|k|+r$ that

$$
\frac{\partial^{k+r} w_j^{(N+1)}(x,t,\eta)}{\partial x^k \partial t^r} = O\Big(\frac{1}{(1+|\eta|)^{N-1-\frac{n}{2}-|k|-r}}\Big), \quad \forall k,\ \forall r.
\tag{64.18}
$$

Therefore

$$
u(x,t) = \frac{1}{(2\pi)^n}\int_{\mathbb{R}^n} (w_N(x,t,\eta) + w^{(N+1)}(x,t,\eta))d\eta
\tag{64.19}
$$

is the solution of (63.4), (63.5) for $0<t\le t_0$. □

64.2. Solution of the eikonal equation.

For simplicity of notation we temporarily drop the subscript $j=1,2$ in (64.2), (64.3) and consider the bicharacteristic system for the symbol $\xi_0 - \lambda(x,t,\xi)$:

$$
\begin{aligned}
\frac{dx}{dt} &= -\frac{\partial\lambda(x(t),t,\xi(t))}{\partial\xi}, \quad x(0)=y, \\
\frac{d\xi}{dt} &= \frac{\partial\lambda(x(t),t,\xi(t))}{\partial x}, \quad \xi(0)=\eta.
\end{aligned}
\tag{64.20}
$$

It follows from Remark 50.1 that the solution of (64.20) exists for all $-\infty < t < \infty$. It was shown at the end of §61 that the map

$$
x = x(t,y,\eta), \quad \xi = \xi(t,y,\eta)
\tag{64.21}
$$

is a canonical transformation of $\mathbb{R}^n \times (\mathbb{R}^n \setminus \{0\})$ to $\mathbb{R}^n \times (\mathbb{R}^n \setminus \{0\})$ for any $t \in \mathbb{R}$, i.e.,

$$\hat{d}x(t,y,\eta) \wedge \hat{d}\xi(t,y,\eta) = dy \wedge d\eta, \quad \forall t \in \mathbb{R}. \tag{64.22}$$

Here the notation $\hat{d}$ means the differential of a function depending on (y,η) for a fixed t.

Let $dx(t,y,\eta),\ d\xi(t,y,\eta)$ be the differentials of functions depending on (t,y,η). Then $dx = \hat{d}x + \frac{\partial x}{\partial t}dt,\ d\xi = \hat{d}\xi + \frac{\partial \xi}{\partial t}dt$.

It follows from (64.20) that $\frac{\partial x}{\partial t} = -\frac{\partial \lambda}{\partial \xi},\ \frac{\partial \xi}{\partial t} = \frac{\partial \lambda}{\partial x}$. Therefore

$$dx \wedge d\xi = \left(\hat{d}x + \frac{\partial x}{\partial t}dt\right) \wedge \left(\hat{d}\xi + \frac{\partial \xi}{\partial t}dt\right) = \hat{d}x \wedge \hat{d}\xi + \left(\frac{\partial \lambda}{\partial \xi}\hat{d}\xi + \frac{\partial \lambda}{\partial x}\hat{d}x\right) \wedge dt.$$

Since $\frac{\partial \lambda}{\partial t}dt \wedge dt = 0$ and since $\hat{d}x \wedge dt = dx \wedge dt,\ \hat{d}\xi \wedge dt = d\xi \wedge dt$, we have

$$\hat{d}x \wedge \hat{d}\xi = dx \wedge d\xi - d\lambda \wedge dt.$$

Therefore

$$dx \wedge d\xi - d\lambda \wedge dt = dy \wedge d\eta, \tag{64.23}$$

where $x = x(t,y,\eta)$ and $\xi = \xi(t,y,\eta)$ are the same as in (64.21). Note that $x(t,y,\eta),\ \xi(t,y,\eta)$ are homogeneous in η of degree 0 and 1, respectively. It will be proven in Lemma 65.1 that in this case (61.22) is equivalent to

$$\xi(t,x,y) \cdot \hat{d}x = \eta \cdot dy, \tag{64.24}$$

where $\xi \cdot \hat{d}x = \sum_{j=1}^n \xi_j \hat{d}x_j,\ \eta \cdot dy = \sum_{j=1}^n \eta_j dy_j$.

We often write ξdx and ηdy instead of $\xi \cdot dx$ and $\eta \cdot dy$.

Using that $dx = \hat{d}x + \frac{\partial x}{\partial t}dt$ and that $\xi \cdot \frac{\partial \lambda}{\partial \xi} = \lambda$, we get from (64.24) analogously to (64.23):

$$\xi dx + \lambda dt = \eta dy. \tag{64.25}$$

□

Let t_0 be such that the determinant $\det \frac{\partial x}{\partial y} \neq 0,\ \forall y, \forall \eta \neq 0$ and $0 \leq t \leq t_0$ and there is an inverse map

$$y = y(t,x,\eta),\ \forall x,\ \eta \neq 0,\ 0 \leq t \leq t_0.$$

Then we can express ξ as a function of (x,t,η):

$$\xi = p(t,x,\eta) = \xi(t,y(t,x,\eta),\eta). \tag{64.26}$$

We rewrite (64.25) in the form

$$d(\eta \cdot y(t,x,\eta)) = p(t,x,\eta)dx + \lambda(x,t,p(t,x,\eta))dt + yd\eta. \tag{64.27}$$

Let

$$S_0(x,t,\eta) = y(t,x,\eta) \cdot \eta. \tag{64.28}$$

It follows from (64.27) that

$$\frac{\partial S_0(x,t,\eta)}{\partial x} = \xi, \qquad \frac{\partial S_0(x,t,\eta)}{\partial \eta} = y, \tag{64.29}$$

$$\frac{\partial S_0(x,t,\eta)}{\partial t} = \lambda(x,t,\xi), \tag{64.30}$$

where $\xi = p(t,x,\eta)$, $y = y(t,x,\eta)$. Note that (64.29) means that $S_0(x,t,\eta)$ is a generating function of the canonical transformation (64.21) for $0 \le t \le t_0$. Since $\xi = \frac{\partial S_0}{\partial x}$, we see from (64.30) that

$$S_{0t}(x,t,\eta) = \lambda(x,t,S_{0x}(x,t,\eta)), \tag{64.31}$$

i.e., $S_0(x,t,\eta)$ is the solution of the eikonal equation and

$$S_0(x,0,\eta) = x \cdot \eta, \tag{64.32}$$

since $x(0,y,\eta) = y$.

Reinstating the subscript $j = 1,2$, and denoting by $\varphi_j(x,t,\eta)$ the function $S_0(x,t,\eta)$ for $\lambda(x,t,\xi)$ replaced by $\lambda_j(x,t,\xi)$, $j = 1,2$, we get (64.6), (64.7). $\square$

64.3. Solution of the transport equation.

Now we find an explicit expression for the solution of equation (64.10), called the transport equation.

Let $x = x(t,y,\eta)$, $\xi = \xi(t,y,\eta)$ be the solutions of (64.20), where we dropped the subscript $j = 1,2$ and write $\varphi(x,t,\eta)$ instead of $\varphi_j, j = 1,2$, in (64.6), (64.7), and (64.10). It follows from (64.31), (64.32) and (64.28) that $\varphi(x,t,\eta) = S_0(x,t,\eta) = y(t,x,\eta)\cdot\eta$. Substitute $x = x(t,y,\eta)$ into (64.10) taking into account that $\varphi_x(x(t,y,\eta),t,\eta) = \xi(t,y,\eta)$ and that $\varphi_t(x(t,y,\eta),t,\eta) = \xi_0(t,y,\eta)$, where $\xi_0(t,y,\eta) = \lambda(x(t,y,\eta),t,\xi(t,y,\eta))$ (cf. (64.29), (64.30)). Since $\xi_0(t,y,\eta) - \lambda(x(t,y,\eta),t,\xi(t,y,\eta)) = 0$, we obtain from (64.2) that

$$H_{0\xi_0} = 2\xi_0(t,y,\eta) \neq 0, \quad H_{0\xi_k} = H_{0\xi_0}(-\lambda_{\xi_k}(x(t,y,\eta),t,\xi(t,y,\eta))).$$

Therefore

$$\begin{aligned} H_{0\xi_0}\frac{d}{dt}a_0(x(t,y,\eta),t,\eta) &= H_{0\xi_0}\left(\frac{\partial a_0}{\partial t} + \sum_{k=1}^n \frac{\partial a_0}{\partial x_k}\frac{\partial x_k}{\partial t}\right) \\ &= H_{0\xi_0}\left(\frac{\partial a_0}{\partial t} - \sum_{k=1}^n \frac{\partial a_0}{\partial x_k}\lambda_{\xi_k}\right). \end{aligned} \tag{64.33}$$

We used the equation (64.20) in (64.33).

Denote $\hat{a}_0(t,y,\eta) = a_0(x(t,y,\eta),t,\eta)$. Then we can rewrite (64.10) in the following form:

$$\frac{d\hat{a}_0(t,y,\eta)}{dt} - M(x(t,y,\eta),t,\eta)\hat{a}_0(t,y,\eta) = 0, \tag{64.34}$$

where $M(x,t,\eta) = -H_{0\xi_0}^{-1}\big(H_0(x,t,\frac{\partial}{\partial x},\frac{\partial}{\partial t})\varphi(x,t,\eta) - \sum_{j=1}^n b_j\varphi_{x_j} - b_0\varphi_t\big)$. Therefore

$$\hat{a}_0(t,y,\eta) = \hat{a}_0(0,y,\eta)\exp\int_0^t M(x(t',y,\eta),t',\eta)dt'. \tag{64.35}$$

Differentiating $\frac{dx}{dt} = -\lambda_\xi(x,t,\varphi_x) = \frac{H_{0\xi}(x,t,\varphi_x,\varphi_t)}{H_{0\xi_0}(x,t,\varphi_x,\varphi_t)}$ with respect to y, we have

$$\frac{d}{dt}\frac{\partial x(t,y,\eta)}{\partial y} = \left(\frac{H_{0\xi}}{H_{0\xi_0}}\right)_x \frac{\partial x(t,y,\eta)}{\partial y}, \tag{64.36}$$

where $\frac{\partial x(t,y,\eta)}{\partial y}$ is the Jacobian matrix. Let $\det\frac{\partial x(t,y,\eta)}{\partial y}$ be the Jacobian. Denote by $\left|\frac{\partial x}{\partial y}\right|$ the absolute value of the Jacobian. Since $\det\frac{\partial x(t,y,\eta)}{\partial y} > 0$, we have $\left|\frac{\partial x(t,y,\eta)}{\partial y}\right| = \det\frac{\partial x(t,y,\eta)}{\partial y}$. Using Liouville's formula, we get from (64.36):

$$\frac{d}{dt}\left|\frac{\partial x}{\partial y}\right| = \operatorname{Tr}\left(\frac{H_{0\xi}}{H_{0\xi_0}}\right)_x\left|\frac{\partial x}{\partial y}\right|. \tag{64.37}$$

We have $H_{0\xi_0} = 2\varphi_t$ and

$$\operatorname{Tr}\left(\frac{H_{0\xi}}{H_{0\xi_0}}\right)_x = \operatorname{Tr}\left[\frac{H_{0\xi_k x_r}}{2\varphi_t}\right]_{k,r=1}^n - \operatorname{Tr}\left[\frac{H_{0\xi_k}\varphi_{tx_r}}{2\varphi_t^2}\right]_{k,r=1}^n. \tag{64.38}$$

Note that

$$\operatorname{Tr}\left[\frac{H_{0\xi_k x_r}}{2\varphi_t}\right]_{k,r=1}^n = -\varphi_t^{-1}\left(\sum_{p,k=1}^n g^{pk}\varphi_{x_p x_k} + \sum_{p,k=1}^n g^{pk}_{x_k}\varphi_{x_p}\right) \tag{64.39}$$

and

$$\operatorname{Tr}\left[\frac{H_{0\xi_k}\varphi_{tx_r}}{2\varphi_t^2}\right]_{k,r=1}^n = -\frac{1}{\varphi_t^2}\sum_{p,k=1}^n g^{pk}(x,t)\varphi_{x_p}\varphi_{tx_k}, \tag{64.40}$$

where $H_0(x,t,\varphi_x,\varphi_t) = \varphi_t^2 - \sum_{p,k=1}^n g^{pk}(x,t)\varphi_{x_p}\varphi_{x_k}$.

We have, differentiating $H_0(x,t,\varphi_x,\varphi_t) = 0$ with respect to t:

$$2\varphi_t\varphi_{tt} - 2\sum_{p,k=1}^n g^{pk}(x,t)\varphi_{x_p}\varphi_{tx_k} - \sum_{p,k=1}^n g_t^{pk}(x,t)\varphi_{x_p}\varphi_{x_k} = 0.$$

Therefore

$$\operatorname{Tr}\left[\frac{H_{0\xi_k}\varphi_{tx_r}}{2\varphi_t^2}\right]_{k,r=1}^n = -\frac{\varphi_{tt}}{\varphi_t} + \frac{1}{2\varphi_t^2}\sum_{p,k=1}^n g_t^{pk}\varphi_{x_p}\varphi_{x_k}. \tag{64.41}$$

Note that $H_0(x,t,\frac{\partial}{\partial x},\frac{\partial}{\partial t})\varphi = \varphi_{tt} - \sum_{j,k=1}^{n} g^{pk}\varphi_{x_px_k}$. Combining (64.38), (64.39), and (64.41), we get

(64.42)

$$\frac{H_0(x,t,\frac{\partial}{\partial x},\frac{\partial}{\partial t})\varphi}{H_{0\xi_0}} = \frac{1}{2}\operatorname{Tr}\left(\frac{H_{0\xi}}{H_{0\xi_0}}\right)_x + \frac{1}{2\varphi_t}\sum_{p.k=1}^{n} g^{pk}_{x_k}\varphi_{x_p} + \frac{1}{4\varphi_t^2}\sum_{p,k=1}^{n} g^{pk}_t\varphi_{x_p}\varphi_{x_k}.$$

Using (64.37) and substituting $\frac{H_0(x,t,\frac{\partial}{\partial x},\frac{\partial}{\partial t})\varphi}{H_{0\xi_0}}$ into (64.35) give

$$\hat{a}_0(t,y,\eta) = \frac{\hat{a}_0(0,y,\eta)}{\sqrt{|\frac{\partial x}{\partial y}|}}\exp\int_0^t M_0(t',y,\eta)dt', \tag{64.43}$$

where $M_0(t,y,\eta) = M_1(x(t,y,\eta),t,\eta)$,

$$\begin{aligned} M_1(x,t,\eta) = &-\frac{1}{2\varphi_t(x,t)}\sum_{p,k=1}^{n} g^{pk}_{x_k}(x,t)\varphi_{x_p} - \frac{1}{4\varphi_t^2}\sum_{p,k=1}^{n} g^{pk}_t\varphi_{x_p}\varphi_{x_k} \\ &+ \frac{1}{2\varphi_t}\left(\sum_{k=1}^{n} b_k(x,t)\varphi_{x_k} + b_0(x,t)\varphi_k\right). \end{aligned} \tag{64.44}$$

Note that $\varphi_x(x(t,y,\eta),t,\eta) = p(x(t,y,\eta),t,\eta) = \xi(t,y,\eta)$ and that $\varphi_t(x(t,y,\eta),t,\eta) = \lambda(x(t,y,\eta),t,\xi(t,y,\eta))$. Therefore $M_1(x(t,y,\eta),t)$ is a smooth function of (t,y,η) for all $t\in\mathbb{R}$, $\eta\neq 0$. □

64.4. Propagation of singularities.

Now we find the wave front set of $u(x,t)$ knowing $WF(g_j)$, $j=1,2$. Note that (64.19) is a sum of FIO's with phase functions $\varphi_j(x,t,\eta)-y\cdot\eta$, $j=1,2$. Denote by $\varphi_j\circ WF(g)$ the image of $WF(g)$ under the canonical transformation with the generating function $\varphi_j(x,t,\eta)$. We get, using Theorem 63.1, that

$$WF(u(\cdot,t)) \subset \bigcup_{j=1}^{2}(\varphi_j\circ(WF(g_1)\cup WF(g_2))),$$

where $WF(u(\cdot,t))$ is the wave front set of $u(x,t)$ in $\mathbb{R}^n\times\mathbb{R}^n\setminus\{0\}$ for a fixed t. This result coincides with Theorem 50.5 when $m=2, f=0$. Indeed, the canonical transformation defined by the generating function $\varphi_j(x,t,\eta)$ for t fixed is

$$\xi = \varphi_{jx}(x,t,\eta),\quad y = \varphi_{j\eta}(x,t,\eta). \tag{64.45}$$

Therefore the canonical transformation (64.45) coincides with the map $(y,\eta)\to(x_j(t,y,\eta),\xi_j(t,y,\eta))$, where $x=x_j(t,y,\eta)$, $\xi=\xi_j(t,y,\eta)$ is the solution of (64.21), $j=1,2$.

Using Remark 50.2, we can also describe $WF(u)$ in $\mathbb{R}^{n+1}\times(\mathbb{R}^{n+1}\setminus\{0\})$.

65. Global Fourier integral operators

65.1. Lagrangian manifolds.

A manifold $\Lambda \subset T_0^*(\mathbb{R}^N) = \mathbb{R}^N \times \mathbb{R}^N \setminus \{0\}$ is called Lagrangian if $\dim \Lambda = N$ and the symplectic form

$$d\xi \wedge dx = \sum_{j=1}^{N} d\xi_j \wedge dx_j \tag{65.1}$$

vanishes on Λ.

A Lagrangian manifold (65.1) is called conic if $(x, \xi) \in \Lambda$ implies that $(x, \tau\xi) \in \Lambda$ for any $\tau > 0$.

Lemma 65.1. *If Λ is a conic Lagrangian manifold, then*

$$\sum_{j=1}^{N} \xi_j dx_j = 0 \quad \textit{on} \quad \Lambda. \tag{65.2}$$

Proof: Let (x, ξ) be an arbitrary point of Λ, and let $(\tau_1, \tau_2, \dots, \tau_N)$ be local coordinates in a neighborhood $U_0 \subset \Lambda$ of (x, ξ), i.e., $x = x(\tau_1, \dots, \tau_N)$, $\xi = \xi(\tau_1, \dots, \tau_N)$ are the parametric equations of Λ in U_0. We have

$$\begin{aligned} d\xi \wedge dx &= \left(\sum_{j=1}^{N} \frac{\partial \xi}{\partial \tau_j} d\tau_j \right) \wedge \left(\sum_{k=1}^{N} \frac{\partial x}{\partial \tau_k} d\tau_k \right) \\ &= \sum_{1 \le j < k \le N} \left(\frac{\partial \xi}{\partial \tau_j} \cdot \frac{\partial x}{\partial \tau_k} - \frac{\partial \xi}{\partial \tau_k} \cdot \frac{\partial x}{\partial \tau_j} \right) d\tau_j \wedge d\tau_k. \end{aligned}$$

Therefore (65.1) implies

$$\frac{\partial \xi}{\partial \tau_j} \cdot \frac{\partial x}{\partial \tau_k} - \frac{\partial \xi}{\partial \tau_k} \cdot \frac{\partial x}{\partial \tau_j} = 0, \quad 1 \le j, k \le N. \tag{65.3}$$

Note that $\left(\frac{\partial x}{\partial \tau_j}, \frac{\partial \xi}{\partial \tau_j} \right)$, $1 \le j \le N$, form a basis of the tangent space to Λ at (x, ξ). Since Λ is conic, the curve $x(\tau) = x$, $\xi(\tau) = \tau\xi$ lies on Λ. Therefore $(0, \xi)$ is the tangent vector to Λ. Representing $(0, \xi)$ as a linear combination of $\left(\frac{\partial x}{\partial \tau_k}, \frac{\partial \xi}{\partial \tau_k} \right)$, $1 \le k \le N$, and using (65.3), we get

$$\frac{\partial x}{\partial \xi_j} \cdot \xi - 0 \cdot \frac{\partial \xi}{\partial \tau_j} = 0, \ 1 \le j \le N. \tag{65.4}$$

Hence, $\xi \cdot dx = \sum_{j=1}^{N} \xi \cdot \frac{\partial x}{\partial \tau_j} d\tau_j = 0$ on Λ. $\square$

In the case $T^*(\mathbb{R}^N) = T^*(\mathbb{R}^n) \times T^*(\mathbb{R}^n)$, Λ is a Lagrangian manifold if

$$dx \wedge d\xi + dy \wedge d\eta = 0 \quad \text{on} \quad \Lambda. \tag{65.5}$$

We say that the points $(x, \xi, y, \eta) \in \Lambda$ form a canonical relation between (x, ξ) and (y, η).

Let $x = x(y, \eta)$, $\xi = \xi(x, \eta)$ be a canonical transformation of $T_0^*(\mathbb{R}^n)$ onto $T_0^*(\mathbb{R}^n)$, i.e.,

$$dx \wedge d\xi - dy \wedge d\eta = 0 \quad \forall (y, \eta). \tag{65.6}$$

Therefore, if the points (x, ξ, y, η) belong to the graph of the canonical transformation, then $(x, \xi, y, -\eta)$ belong to the Lagrangian manifold Λ, i.e., $(x, \xi, y, -\eta)$ form a canonical relation.

We denote the graph of the canonical transformation by Λ'. It follows from Lemma 65.1 that if $x = x(y, \eta)$ and $\xi = \xi(y, \eta)$ are homogeneous in ξ of degree 0 and 1, respectively, then (65.6) implies

$$\xi dx - \eta dy = 0, \quad \forall (y, \eta). \tag{65.7}$$

□

An example of such a canonical transformation is the map

$$(y, \eta) \to (x(t), \xi(t)),$$

where $(x(t), \xi(t))$, $t \in \mathbb{R}$, is the solution of the system (64.20).

A generating function $\varphi(x, t, \eta)$ of the form (64.28) for this canonical transformation exists when $|t| \leq t_0$ and may not exist for large t. Therefore the FIO of the form (61.4) may not exist globally.

We shall define the Fourier integral operator corresponding to the global canonical transformation.

65.2. FIO's with nondegenerate phase functions.

Consider a Fourier integral operator of the form

$$\Phi u = \int_{\mathbb{R}^n} \int_{\mathbb{R}^r} a(x, y, \theta) e^{i\psi(x,y,\theta)} u(y) dy d\theta, \tag{65.8}$$

where $\psi(x, y, \theta) \in C^\infty$ for all (x, y, θ), $\theta \neq 0$ and $\psi(x, y, \theta)$ is homogeneous in θ of degree one, $a(x, y, \theta) \in S^m(\mathbb{R}^n \times \mathbb{R}^n \times \mathbb{R}^r)$ (cf. §40). Let $C_0 = \{(x, y, \theta) : \psi_{\theta_j} = 0, 1 \leq j \leq r\}$.

a) Suppose that the $r \times (2n+r)$ matrix $[\psi_{\theta x}, \psi_{\theta y}, \psi_{\theta\theta}]$ has rank r on C_0. Then C_0 is a smooth $2n$-manifold in $\mathbb{R}^n \times \mathbb{R}^n \times (\mathbb{R}^r \setminus \{0\})$.

b) Assume that $\psi_x(x, y, \theta)$ and $\psi_y(x, y, \theta)$ do not vanish on C_0.

A function $\psi(x, y, \theta)$ having properties a) and b) is called a nondegenerate phase function.

Consider the subset C of $T_0^*(\mathbb{R}^n) \times T_0^*(\mathbb{R}^n)$ of the form (x, ψ_x, y, ψ_y), where $(x, y, \theta) \in C_0$.

Lemma 65.2. *The map $\sigma : C_0 \to C$ is a local diffeomorphism.*

Proof: Let $(dx, dy, d\theta)$ be tangent to C_0 at $(x_0, y_0, \theta_0) \in C_0$. Then $d\psi_\theta = \psi_{\theta x}dx + \psi_{\theta y}dy + \psi_{\theta\theta}d\theta = 0$, and $d\sigma = (dx, d\psi_x, dy, d\psi_y)$ is tangent to C. We shall show that the linear map: $(dx, dy, d\theta) \to d\sigma$ is injective at (x_0, y_0, θ_0). Therefore, the implicit function theorem implies that σ is a local diffeomorphism. We see that $d\sigma = 0$ implies $dx = 0$, $dy = 0$, $d\psi_x = 0$, $d\psi_y = 0$. When $dx = dy = 0$, we get $d\psi_x = \psi_{x\theta}d\theta = 0$, $d\psi_y = \psi_{y\theta}d\theta = 0$. Taking into account that $dx = dy = 0$, we get $d\psi_0 = \psi_{\theta\theta}d\theta = 0$. Then $d\theta$ satisfies $\psi_{x\theta}d\theta = 0$, $\psi_{y\theta}d\theta = 0$, $\psi_{\theta\theta}d\theta = 0$, and since the rank of the matrix $[\psi_{x\theta}, \psi_{y\theta}, \psi_{\theta,\theta}]$ is equal to r, we see that $d\theta = 0$, i.e., $d\sigma$ is injective.

We show that C is a Lagrangian manifold. We have $\xi dx + \eta dy = \psi_x dx + \psi_y dy = d\psi - \psi_\theta d\theta = d\psi$, since $\psi_\theta = 0$ on C_0. Also, $\psi = \theta \cdot \psi_\theta = 0$, since $\psi(x, y, \theta)$ is homogeneous of degree 1. Therefore $\xi dx + \eta dy = d\psi = 0$ when $(x, y, \theta) \in C_0$. By Lemma 65.1, C is a Lagrangian manifold. Note that $\xi dx + \eta dy = 0$ obviously implies that $d\xi \wedge dx + d\eta \wedge dy = 0$ on C since $d^2 = 0$. □

Suppose that $a(x, y, \theta) = \sum_{k=0}^{N} a_k(x, y, \theta)(1 - \chi(\theta)) + a^{(N+1)}(x, y, \theta)$, where $a_k(x, y, \theta)$ are homogeneous in ξ of degree $m - k$, $\chi(\theta) \in C_0^\infty(\mathbb{R}^r)$, $\chi(\theta) = 1$ for $|\theta| < 1$, $a^{(N+1)}(x, y, \theta) \in S^{m-N-1}$. Assume that the Lagrangian manifold C corresponds to the graph Λ' of the canonical transformation $\alpha : (y, \eta) \to (x, \xi)$, where $(x, \xi, y, -\eta) \in C$ in a conic neighborhood of $(x_0, \xi_0, y_0, -\eta_0)$ and where $\xi_0 = \psi_x(x_0, y_0, \theta_0), \eta_0 = -\psi_y(x_0, y_0, \theta_0)$.

We find the principal part of the amplitude $a(x, y, \theta)$ of $\Phi(a)u$, where $\Phi(a)u$ has the form (65.8). Let $\varphi(x, y, \theta) \in C^\infty(\mathbb{R}^n \times \mathbb{R}^n \times \mathbb{R}^r \setminus \{0\})$, $\deg_\theta \varphi = 0$. If $\operatorname{supp} \varphi(x, y, \theta) \cap C_0 = \emptyset$, then using the identity

$$\Big(- i \frac{\psi_\theta}{|\psi_\theta|^2} \cdot \frac{\partial}{\partial \theta} \Big)^N e^{i\psi} = e^{i\psi}, \quad \forall N,$$

and integrating by parts with respect to θ, we get

$$\Phi(a\varphi)u = \int_{\mathbb{R}^n} \int_{\mathbb{R}^r} T_N(x, y, \theta) e^{i\psi} u(y) dy d\theta,$$

where $T_{-N} \in S_{-N}$, $\forall N$.

Suppose $\operatorname{supp} \varphi \cap C_0 \neq \emptyset$. Let $(z_1, \dots, z_{2n+r})$ be local coordinates on $\operatorname{supp} \varphi(x, y, \theta)$ such that $z_{2n+j} = \psi_{\theta_j}, 1 \le j \le r$, i.e., $z_{2n+1} = \dots = z_{2n+r} = 0$ is the equation of C_0. Let $\hat{a}_0(z', z'')$ be $a_0(x, y, \theta)$ in the new coordinates $z' = (z_1, \dots, z_{2n})$, $z'' = (z_{2n+1}, \dots, z_{2n+r})$. By the Taylor formula we have $\hat{a}_0(z', z'') = \hat{a}_0(z', 0) + \sum_{j=1}^{r} \hat{a}_{j0}(z', z'') z_{2n+j}$.

Let $b(x, y, \theta), b_j(x, y, \theta)$ be $\hat{a}_0(z', 0), \hat{a}_{j0}(z', z'')$ in the (x, y, θ)-coordinates,

$1 \le j \le r$. We have $a_0(x,y,\theta) = b + \sum_{k=1}^{n} b_k(x,y,\theta)\psi_{\theta_k}$, where $(x,y,\theta) \in \operatorname{supp}\varphi$. Since $\psi_{\theta_k} e^{i\psi} = -i\frac{\partial}{\partial\theta_k} e^{i\psi}$, integration by parts gives

$$\Phi(a_0\varphi)u = \int_{\mathbb{R}^n}\int_{\mathbb{R}^r} b(x,y,\theta)\varphi(x,y,\theta)e^{i\psi(x,y,\theta)}u(y)dyd\theta + \int_{\mathbb{R}^n}\int_{\mathbb{R}^r} b^{(1)}(x,y,\theta)e^{i\psi(x,y,\theta)}u(y)dyd\theta,$$

where $b^{(1)} \in S^{m-1}$. Since C_0 and C are locally diffeomorphic, the push-forward of $b|_{C_0}$ under the map $C_0 \to C$ will be a function $\hat{b}$ on the graph Λ' of the canonical transformation $\alpha : (y,\eta) \to (x,\xi)$. Since any point of Λ' is determined by (y,η), we can view $\hat{b}$ as a function of (y,η). We call $\hat{b}(y,\eta)$ the principal part of the amplitude of a FIO of the form (64.8) near (x_0,y_0,θ_0).

Example 65.1. In the case of the FIO (61.4) we have
(65.9)
$$\psi(x,y,\eta) = S(x,\eta) - y\cdot\eta,\quad \psi_x = S_x(x,\eta),\quad \psi_y = -\eta,\quad \psi_\eta = S_\eta(x,\eta) - y = 0.$$

Therefore, the canonical relation has the form $(x, S_x, y, -\eta)$, where $y - S_\eta = 0$. Since $S(x,\eta)$ is the generating function of the canonical transformation $\alpha : (y,\eta) \to (x,\xi)$ (cf. (61.19)), we have $d\xi \wedge dx - d\eta \wedge dy = 0$ if $y = S_\eta(x,\eta)$, $\xi = S_x(x,\eta)$. Note that the graph Λ' of the canonical transformation is $(x,\xi,y,\eta) = (x, S_x(x,\eta), S_\eta(x,\eta), \eta)$, and it differs from the canonical relation (65.9) corresponding to (61.4) by the change of the sign of η.

Example 65.2. Let $\alpha : (y,\eta) \to (x,\xi)$ be a homogeneous canonical transformation. Suppose there exists a conic neighborhood of (y_0,η_0) in $T_0^*(\mathbb{R}^n)$ with the following properties.

There is a splitting $x = (x^{(1)},\ x_{(1)})$, where $x_{(1)} \in \mathbb{R}^{n_1}$, $x^{(1)} \in \mathbb{R}^{n_2}$, $n_1 + n_2 = n$, and analogously the splittings $y = (y^{(1)}, y_{(1)})$, $\xi = (\xi^{(1)}, \xi_{(1)})$, $\eta = (\eta^{(1)}, \eta_{(1)})$ such that the $(x^{(1)}, y_{(1)}, \eta^{(1)}, \xi_{(1)})$ are coordinates on the graph of α near (y_0,η_0,x_0,ξ_0). Then there exists a generating function $L(x^{(1)}, \eta^{(1)}, y_{(1)}, \xi_{(1)})$ such that

$$\xi^{(1)} = L_{x^{(1)}},\quad \eta_{(1)} = -L_{y_{(1)}},\quad x_{(1)} = -L_{\xi_{(1)}},\quad y^{(1)} = L_{\eta^{(1)}}. \tag{65.10}$$

Proof: Since $x = x(y,\eta)$, $\xi = \xi(y,\eta)$ is a homogeneous canonical transformation, we have (cf. Lemma 65.1):

$$\xi^{(1)}\cdot dx^{(1)} + \xi_{(1)}\cdot dx_{(1)} - \eta^{(1)}\cdot dy^{(1)} - \eta_{(1)}\cdot dy_{(1)} = 0. \tag{65.11}$$

We can rewrite (65.11) in the form

$$\xi^{(1)} \cdot dx^{(1)} + d(x_{(1)} \cdot \xi_{(1)}) - x_{(1)} \cdot d\xi_{(1)} - d(y^{(1)} \cdot \eta^{(1)}) + y^{(1)} \cdot d\eta^{(1)} - \eta_{(1)} \cdot dy_{(1)} = 0. \tag{65.12}$$

Set

$$L(x^{(1)}, \eta^{(1)}, y_{(1)}, \xi_{(1)}) = y^{(1)} \cdot \eta^{(1)} - x_{(1)} \cdot \xi_{(1)}, \tag{65.13}$$

where we expressed $x_{(1)}$ and $y^{(1)}$ in the right hand side of (65.13) as functions of $(x^{(1)}, \eta^{(1)}, y_{(1)}, \xi_{(1)})$. Then (65.12) has the form

$$dL(x^{(1)}, \eta^{(1)}, y_{(1)}, \xi_{(1)}) = \xi^{(1)} \cdot dx^{(1)} - x_{(1)} d\xi_{(1)} + y^{(1)} d\eta^{(1)} - \eta_{(1)} \cdot dy_{(1)}. \tag{65.14}$$

Therefore (65.10) follows from (65.14). □

Given a generating function $L(x^{(1)}, \eta^{(1)}, y_{(1)}, \xi_{(1)})$, we define the phase function

$$\psi_{(1)}(x, y, \theta) = x_{(1)} \cdot \xi_{(1)} + L(x^{(1)}, \eta^{(1)}, y_{(1)}, \xi_{(1)}) - y^{(1)} \cdot \eta^{(1)}, \tag{65.15}$$

where $\theta = (\eta^{(1)}, \xi_{(1)})$. We define a FIO corresponding to the generating function L with symbol $a(y, \eta) \in S^m(\mathbb{R}^n \times \mathbb{R}^n)$ as

$$\begin{aligned} \Phi_{(1)} u = \frac{1}{(2\pi)^n} \int_{\mathbb{R}^n} \int_{\mathbb{R}^n} & a(L_{\eta^{(1)}}, y_{(1)}, \eta^{(1)}, -L_{y_{(1)}}) \\ & \times \exp\left(ix_{(1)} \cdot \xi_{(1)} + iL(x^{(1)}, \eta^{(1)}, y_{(1)}, \xi_{(1)}) - iy^{(1)} \cdot \eta^{(1)}\right) \\ & \times |L_{x^{(1)}\eta^{(1)}y_{(1)}\xi_{(1)}}|^{\frac{1}{2}} u(y) dy d\eta^{(1)} d\xi_{(1)}, \end{aligned} \tag{65.16}$$

where $|L_{x^{(1)}\eta^{(1)}y_{(1)}\xi_{(1)}}|$ denotes the absolute value of the determinant of $L_{x^{(1)}\eta^{(1)}y_{(1)}\xi_{(1)}}$ and $\operatorname{supp} a(y, \eta)$ is contained in a small conic neighborhood of (y_0, η_0). Note that $\psi_{(1)}(x, y, \theta)$ is a nondegenerate phase function since $|L_{x^{(1)}\eta^{(1)}y_{(1)}\xi_{(1)}}| \neq 0$.

Note also that when $x^{(1)} = (x_1, \dots, x_n)$ and $x_{(1)}$ is empty, we get a FIO of the form (61.4), and when $x_{(1)} = (x_1, \dots, x_n)$ and $x^{(1)}$ is empty, we get a FIO of the form (61.5). □

65.3. Local coordinates for a graph of a canonical transformation.

If $\alpha(y, \eta) = (x, \xi)$ is an arbitrary canonical transformation, then the coordinates of the form $(x^{(1)}, \eta^{(1)}, y_{(1)}, \xi_{(1)})$ may not exist in a neighborhood of some point $(x_0, \xi_0, y_0, \eta_0)$ (cf. Example 65.3). Therefore we prove the following lemma.

Lemma 65.3. *Let $\alpha(y, \eta) = (x, \xi)$ be a homogeneous canonical transformation of $T_0^*(\mathbb{R}^n)$ onto $T_0^*(\mathbb{R}^n)$. For any $(y_0, \eta_0) \in T_0^*(\mathbb{R}^n)$ there exists a* *splitting* $x = (x^{(1)}, x_{(1)})$, $\xi = (\xi^{(1)}, \xi_{(1)})$ *such that* $(x^{(1)}, \xi_{(1)}, \eta)$ *are the coordinates on the graph* Λ' *of* α *in a small conic neighborhood of* $(y_0, \eta_0, x_0, \xi_0)$.

There exists a generating function $S(x^{(1)}, \xi_{(1)}, \eta)$ *such that*

(65.17)
$$\xi^{(1)} = S_{x^{(1)}}(x^{(1)}, \xi_{(1)}, \eta),\ x_{(1)} = -S_{\xi_{(1)}}(x^{(1)}, \xi_{(1)}, \eta),\ y = S_\eta(x^{(1)}, \xi_{(1)}, \eta),$$

and

$$(x^{(1)}, -S_{\xi_{(1)}}, S_{x^{(1)}}, \xi_{(1)}, S_\eta, \eta) \in \Lambda' \quad near \quad (x_0, \xi_0, y_0, \eta_0), \tag{65.18}$$

where Λ' *is the graph of* α.

Proof: Let $(y_0, \eta_0) \in T_0^*(\mathbb{R}^n)$. Consider $x = x(y, \eta_0)$, $\xi = \xi(y, \eta_0)$, where η_0 is fixed and y belongs to a small neighborhood of y_0. Since $\alpha(y, \eta) = (x, \xi)$ is a canonical transformation and since $d\eta_0 = 0$, we have: $dx \wedge d\xi = dy \wedge d\eta_0 = 0$. Consider the $n \times 2n$ matrix $\frac{\partial(x,\xi)}{\partial y}$ at the point (y_0, η_0). We show that there exists a splitting $(x^{(1)}, x_{(1)})$, $(\xi^{(1)}, \xi_{(1)})$ such that $\det \frac{\partial(x^{(1)},\xi_{(1)})}{\partial y} \neq 0$ at (y_0, η_0). Let r be the rank of the matrix $\frac{\partial x}{\partial y}$. If $r = n$, then we take $x^{(1)} = x, \xi_{(1)} = \emptyset$. Suppose $r < n$. Denote by V the linear span of vectors $\frac{\partial x}{\partial y_1}, \dots, \frac{\partial x}{\partial y_n}$. Consider n vectors $e_j \in \mathbb{R}^n$, $j = 1, \dots, n$, of the form $e_j = (a_{j1}, \dots, a_{jn})$, where $a_{jj} = 1$, $a_{jk} = 0$ for $k \neq j$. Since $\dim V = r$, there exist $(n-r)$ vectors $e_{j_1}, e_{j_2}, \dots, e_{j_{n-r}}$ such that the vector space W spanned by these vectors is transversal to V, i.e., $W \cap V = 0$. Let $x_{(1)} = (x_{j_1}, \dots, x_{j_{n-r}})$. Then $x = (x^{(1)}, x_{(1)})$ belongs to W if $x^{(1)} = 0$ and $x_{(1)} \in \mathbb{R}^{n-r}$ is arbitrary.

Let $\xi = (\xi^{(1)}, \xi_{(1)})$ be the same splitting. We prove that $\det \frac{\partial(x^{(1)},\xi_{(1)})}{\partial y} \neq 0$. Consider the $n \times n$ system $\sum_{k=1}^n c_k \frac{\partial(x^{(1)},\xi_{(1)})}{\partial y_k} = 0$. It is enough to prove that $c_k = 0$, $1 \le k \le n$. Define

$$(0, x_{(1)0}, \xi_0^{(1)}, 0) = \sum_{k=1}^n c_k \frac{\partial(x,\xi)}{\partial y_k}, \tag{65.19}$$

where $c_1, \dots, c_n$ are the same as above. Since $(0, x_{(1)0}) = \sum_{k=1}^n c_k \frac{\partial x}{\partial y_k}$, we have that $(0, x_{(1)0}) \in V \cap W = 0$. Therefore $x_{(1)0} = 0$. Denote by $[(\hat{x}, \hat{\xi}), (x, \xi)] = x \cdot \hat{\xi} - \xi \cdot \hat{x}$ the symplectic scalar product of (x, ξ) and $(\hat{x}, \hat{\xi})$. Note that $dx \wedge d\xi = 0$ is equivalent to $\left[\frac{\partial(x,\xi)}{\partial y_j}, \frac{\partial(x,\xi)}{\partial y_k}\right] = 0$, $1 \le j, k \le n$. Taking the symplectic scalar product of (65.19) with $\frac{\partial(x,\xi)}{\partial y_j}$, $j = 1, \dots, n$, we get

$$\left[(0, 0, \xi_0^{(1)}, 0), \frac{\partial(x,\xi)}{\partial y_j}\right] = -(\xi_0^{(1)}, 0) \cdot \frac{\partial x}{\partial y_j} = 0, \quad 1 \le j \le n.$$

Since $(\xi_0^{(1)}, 0) \cdot \frac{\partial x}{\partial y_k} = 0$, we have that $(\xi_0^{(1)}, 0)$ is orthogonal to V. Also $(\xi_0^{(1)}, 0)$ is orthogonal to W since W consists of vectors of the form $(0, x_{(1)})$.

Therefore $\xi_0^{(1)} = 0$ because $\dim V + \dim W = n$. Having the left hand side of (65.19) equal to zero, we get that $c_k = 0,\ 1 \le k \le n$, since $\frac{\partial(x,\xi)}{\partial y_k},\ 1 \le n$, are linearly independent. Therefore $\sum_{k=1}^n c_k \frac{\partial(x^{(1)},\xi_{(1)})}{\partial y_k} = 0$ has only a trivial solution and hence $|\frac{\partial(x^{(1)},\xi_{(1)})}{\partial y}| = |\det \frac{\partial(x^{(1)},y_{(1)})}{\partial y}| \neq 0$ at (y_0, η_0). □

Since $|\frac{\partial(x^{(1)},\xi_{(1)})}{\partial y}| \neq 0$, we can use the implicit function theorem and obtain the expression $y = y(x^{(1)}, \xi_{(1)}, \eta)$. Note that

$$\left|\frac{\partial(x^{(1)}, \xi_{(1)}, \eta)}{\partial(y, \eta)}\right| = \left|\frac{\partial(x^{(1)}, \xi_{(1)})}{\partial y}\right| \neq 0.$$

Therefore we can use $(x^{(1)}, \xi_{(1)}, \eta)$ as coordinates on Λ' instead of (y, η) near $(x_0, \xi_0, y_0, \eta_0)$. Note also that $\eta \neq 0,\ \xi_{(1)} \neq 0$ in a neighborhood of $(x_0, \xi_0, y_0, \eta_0)$.

As in (65.12), we have:

$$\xi^{(1)} \cdot dx^{(1)} - x_{(1)} \cdot d\xi_{(1)} + y \cdot d\eta = d(y \cdot \eta - x_{(1)} \cdot \xi_{(1)}).$$

Let

$$S(x^{(1)}, \xi_{(1)}, \eta) = y \cdot \eta - x_{(1)} \cdot \xi_{(1)}, \tag{65.20}$$

where we express y and $x_{(1)}$ as functions of $(x^{(1)}, \xi_{(1)}, \eta)$. Then
(65.21)

$$\xi^{(1)} = S_{x^{(1)}}(x^{(1)}, \xi_{(1)}, \eta), \quad x_{(1)} = -S_{\xi_{(1)}}(x^{(1)}, \xi_{(1)}, \eta), \quad y = S_\eta(x^{(1)}, \xi_{(1)}, \eta).$$

Remark 65.1. Analogously to the proof of Lemma 65.3 one can find a splitting $(x_{(1)}, \xi^{(1)}),\ \xi^{(1)} \in \mathbb{R}^{n_1},\ n_1 \ge 1$, such that $|\frac{\partial(x_{(1)},\xi^{(1)})}{\partial \eta}| \neq 0$. Then, by the implicit function theorem, $\eta = \eta(x_{(1)}, \xi^{(1)}, y)$. Therefore $(x_{(1)}, \xi^{(1)}, y)$ are coordinates on Λ' near $(x_0, \xi_0, y_0, \eta_0)$. Analogously to (65.19) we can rewrite $\xi \cdot dx - \eta \cdot dy = 0$ in the form

$$-x^{(1)} \cdot d\xi^{(1)} + \xi_{(1)} \cdot dx_{(1)} - \eta \cdot dy = d(-x^{(1)} \cdot \xi^{(1)}). \tag{65.22}$$

Therefore

$$S_1(x_{(1)}, \xi^{(1)}, y) = -x^{(1)} \cdot \xi^{(1)} \tag{65.23}$$

is also a generating function for the canonical transformation α near $(x_0, \xi_0, y_0, \eta_0)$. In (65.23), $x^{(1)} = x^{(1)}(x_{(1)}, \xi^{(1)}, y)$. Therefore

$$x^{(1)} = -S_{1\xi^{(1)}}, \quad \xi_{(1)} = S_{1x_{(1)}}, \quad \eta = -S_{1y}.$$

Corollary 65.4. *Let*

$$\psi(x, y, \theta) = x_{(1)} \cdot \xi_{(1)} + S(x^{(1)}, \xi_{(1)}, \eta) - y \cdot \eta, \tag{65.24}$$

where $\theta = (\xi_{(1)}, \eta),\ \eta \neq 0,\ \xi_{(1)} \neq 0$. *Then* $\psi(x, y, \theta)$ *is a nondegenerate phase function.*

Proof: Note that $\psi_\theta = 0$ means that $x_{(1)} + S_{\xi_{(1)}}(x^{(1)}, \xi_{(1)}, \eta) = 0$ and $S_\eta(x^{(1)}, \xi_{(1)}, \eta) - y = 0$. The rank of the $(n_1+n)\times(n_1+3n)$ matrix $\psi_{\theta x}\psi_{\theta y}\psi_{\theta\theta}$ is equal to $n + n_1$ since $\psi_{y\eta} = -I_n$, $\psi_{y\xi_{(1)}} = 0$, $\psi_{\xi_{(1)}x_{(1)}} = I_{n_{(1)}}$. Also

$$\psi_x = (S_{x^{(1)}}(x^{(1)}, \xi_{(1)}, \eta), \xi_{(1)}) \neq 0, \quad \psi_y = -\eta \neq 0,$$

since $\xi_{(1)} \neq 0, \eta \neq 0$. Therefore $\psi(x, y, \xi_{(1)}, \eta)$ is a nondegenerate phase function. □

Define the FIO with the phase function $\psi(x, y, \xi_{(1)}, \eta)$:

$$(65.25) \quad \Phi u = \frac{1}{(2\pi)^{n+\frac{n_1}{2}}} \int_{\mathbb{R}^n}\int_{\mathbb{R}^{n+n_1}} a(S_\eta(x^{(1)}, \xi_{(1)}, \eta), \eta) \times e^{i\psi(x,y,\xi_{(1)},\eta)} u(y)|J|^{\frac{1}{2}} dy d\eta d\xi_{(1)},$$

where n_1 is the dimension of $\xi_{(1)}$, $\psi(x, y, \xi_{(1)}, \eta)$ is the same as in (65.24), $\xi^{(1)} \in \mathbb{R}^{n_2}$, $x^{(1)} \in \mathbb{R}^{n_2}$, $n_1 + n_2 = n$, $a(y, \eta) \in S^m$ and has the support in a conic neighborhood of (y_0, η_0), $|S_{x^{(1)},\xi_{(1)},\eta}|$ is the absolute value of $\det S_{x^{(1)},\xi_{(1)},\eta}$, and

$$(65.26) \qquad |J(x^{(1)}, \xi_{(1)}, \eta)| = |S_{x^{(1)},\xi_{(1)},\eta}| = \left|\frac{\partial(y, \eta)}{\partial(x^{(1)}, \xi_{(1)}, \eta)}\right|.$$

Remark 65.2. Given a canonical transformation α near $(y_0, \eta_0, x_0, \xi_0)$, there are many choices to select coordinates on the graph Λ'. Therefore there are many ways to construct FIO's with nondegenerate phase functions such that their Lagrangian manifold Λ corresponds to the graph of α. In addition to (65.25) we can take the nondegenerate phase function (cf. (65.23))

$$(65.27) \qquad \psi_1(x, y, \xi^{(1)}) = x^{(1)} \cdot \xi^{(1)} + S_1(x_{(1)}, \xi^{(1)}, y),$$

and construct a FIO with the phase function ψ_1 of the form

$$\Phi_1 u = \frac{1}{(2\pi)^{\frac{n_2}{2}}} \int_{\mathbb{R}^n}\int_{\mathbb{R}^{n_2}} a_1(y, -S_{1y}) e^{i\psi_1(x,y,\xi^{(1)})} u(y)|J_1|^{\frac{1}{2}} dy d\xi^{(1)},$$

where n_2 is the dimension of $\xi^{(1)}$, $a_1(y, \eta) \in S^m$, and $|J_1(x_{(1)}, \xi^{(1)}, y)| = |S_{1x_{(1)}\xi^{(1)}y}| = \left|\frac{\partial(y,\eta)}{\partial(x_{(1)},\xi^{(1)},y)}\right|$. □

Another type of FIO is the following. Repeat the proof of Lemma 65.3 reversing the roles of (y, η) and $(x, \xi) = \alpha(y, \eta)$ and choose the coordinates in the form $(y^{(2)}, \eta_{(2)}, \xi)$, where $y = (y^{(2)}, y_{(2)})$, $\eta = (\eta^{(2)}, \eta_{(2)})$ are some splittings. Then, analogously to (65.20), we get a generating function

$$S_2(y^{(2)}, \eta_{(2)}, \xi) = -x \cdot \xi + y_{(2)} \cdot \eta_{(2)},$$

where $x = x(y^{(2)}, \eta_{(2)}, \xi)$, and $y_{(2)} = y_{(2)}(y^{(2)}, \eta_{(2)}, \xi)$, since $(y^{(2)}, \eta_{(2)}, \xi)$ are the coordinates on Λ'. We have (cf. (65.21)):

$$\eta^{(2)} = S_{2y^{(2)}}(y^{(2)}, \eta_{(2)}, \xi),\ y_{(2)} = -S_{2\eta_{(2)}}(y^{(2)}, \eta_{(2)}, \xi),\ x = S_{2\xi}(y^{(2)}, \eta_{(2)}, \xi).$$

Let

(65.28)

$$\psi_2(x,y,\theta) = x\cdot\xi + S_2(y^{(2)}, \eta_{(2)}, \xi) - y_{(2)}\cdot\eta_{(2)}, \quad \theta = (\eta_{(2)}, \xi),\ \eta_{(2)} \neq 0,\ \xi \neq 0.$$

As in Corollary 65.4, we see that $\psi_2(x,y,\theta)$ is a nondegenerate phase function. Let Φ_2 be the following FIO:

(65.29)

$$\Phi_2 u = \frac{1}{(2\pi)^{n+\frac{m_2}{2}}} \int_{\mathbb{R}^n} \int_{\mathbb{R}^{n+m_2}} a_2(S_{2\xi}, \xi) e^{i\psi_2(x,y,\eta_{(2)},\xi)} |J_2|^{\frac{1}{2}} u(y)\, dy\, d\xi\, d\eta_{(2)},$$

where $a_2(x,\xi) \in S^m$, $\eta_{(2)} \in \mathbb{R}^{m_2}$, and

$$|J_2| = |S_{2y^{(2)}\eta_{(2)}\xi}| = \left| \frac{\partial(x,\xi)}{\partial(y^{(2)}, \eta_{(2)}, \xi)} \right|.$$

Remark 65.3. Let Φ^* be the operator formally adjoint to (65.25). Then analogously to (65.5) we have

$$\Phi^* u = \frac{1}{(2\pi)^{n+\frac{n_1}{2}}} \int_{\mathbb{R}^n} \int_{\mathbb{R}^{n+n_1}} \overline{a_1(S_\eta(y^{(1)}, \xi_{(1)}, \eta), \eta)} \times e^{i\psi^{(1)}(x,y,\xi_{(1)},\eta)} u(y) |J|^{\frac{1}{2}}\, dy\, d\eta\, d\xi_{(1)},$$

where

$$\psi^{(1)}(x,y,\xi_{(1)},\eta) = -\psi(y,x,\xi_{(1)},\eta) = x\cdot\eta - S(y^{(1)}, \xi_{(1)}, \eta) - y_{(1)}\cdot\xi_{(1)}.$$

Note that Φ^* corresponds to the canonical transformation α^{-1} and has a phase functions of the form (65.28).

Example 65.3. Let $S(x', \xi_n, \eta) = x'\cdot\eta' + \frac{\xi_n\eta_n}{|\eta|}$, where $x' = (x_1, \dots, x_{n-1})$, $\xi' = (\xi_1, \dots, \xi_{n-1})$. We consider a small conic neighborhood of $y_0 = 0$, $\eta_0 = \xi_0 = (\eta_0', 0) \neq 0$, $x_0 = 0$. Note that $|S_{x'\eta'\xi_n\eta_n}|$ is different from zero at $(y_0, \eta_0, x_0, \xi_0)$. Therefore we can define a canonical transformation with generating function $S(x', \xi_n, \eta)$ (cf. (65.21)):

$$\xi' = S_{x'} = \eta',\ y' = S_{\eta'} = x' - \frac{\xi_n\eta_n\eta'}{|\eta|^3},\ x_n = -S_{\xi_n} = \frac{-\eta_n}{|\eta|},\ y_n = S_{\eta_n} = \frac{\xi_n|\eta'|^2}{|\eta|^3}.$$

Therefore the canonical transformation $\alpha(y,\eta) = (x,\xi)$ has the following form near $(0, \eta_0', 0, 0, \eta_0', 0)$:

$$x_n = -\frac{\eta_n}{|\eta|},\ \xi' = \eta',\ \xi_n = y_n\frac{|\eta|^3}{|\eta'|^2},\ x' = y' + \frac{y_n\eta_n\eta'}{|\eta'|^2}. \qquad \square$$

65.4. Definition of a global FIO.

Lemma 65.5. *Let* Φu *be a FIO of the form* (65.8) *with nondegenerate phase function* $\psi(x, y, \theta)$*. Suppose* $\operatorname{supp} a(x, y, \theta)$ *is contained in a small neighborhood of* $(x_0, y_0, \theta_0) \in C_0$ *and let* $(x, \xi, y, -\eta) = \sigma(x, y, \theta)$*, where* σ *is the map of* C_0 *onto* C *(cf. Lemma 65.2). We assume that* (x, ξ, y, η) *belongs to the graph* Λ' *of the canonical transformation* $\alpha(y, \eta) = (x, \xi)$ *near* (x_0, y_0, θ_0)*. Let* $(x^{(1)}, \xi_{(1)}, \eta)$ *be local coordinates on* Λ'*. Then*

$$\Phi u = \Phi_1 u + T_{-N-1} u, \tag{65.30}$$

where $\operatorname{ord} T_{-N-1} \leq -N-1$*, and* Φ_1 *is an operator of the form* (65.25).

Proof: Using the identity

$$f(x) = \frac{1}{(2\pi)^n} \int_{\mathbb{R}^n} \int_{\mathbb{R}^n} e^{i(x-z)\cdot\zeta} f(z) dz d\zeta, \tag{65.31}$$

we can rewrite (65.8) in the form:

(65.32)

$$\Phi u = \frac{1}{(2\pi)^{n+n_1}} \int_{\mathbb{R}^{2n+2n_1}} \int_{\mathbb{R}^n} \int_{\mathbb{R}^r} a(x^{(1)}, \hat{z}_{(1)}, z, \theta) \exp\{i(x_{(1)} - \hat{z}_{(1)}) \cdot \xi_{(1)} + i\psi(x^{(1)}, \hat{z}_{(1)}, z, \theta) - i(y-z)\cdot\eta\} u(y) dz d\theta dy d\hat{z}_{(1)} d\xi_{(1)} d\eta.$$

We used (65.31) twice in (65.32): in the variables $(\hat{z}_{(1)}, \xi_{(1)})$ and in the variables (z, η).

As is usually done for justification, we introduce a cutoff function $\chi(\varepsilon(|\theta| + |\eta| + |\xi_{(1)}|))$ inside the integral (65.32) and take the distribution limit as $\varepsilon \to 0$.

We apply the stationary phase Lemma 62.1 to the integral with respect to $(z, \hat{z}_{(1)}, \theta)$ in (65.32). The equations for the critical point are:

$$\begin{aligned} \psi_\theta(x^{(1)}, \hat{z}_{(1)}, z, \theta) &= 0, \\ -\xi_{(1)} + \psi_{\hat{z}_{(1)}}(x^{(1)}, \hat{z}_{(1)}, z, \theta) &= 0, \\ \eta + \psi_z(x^{(1)}, \hat{z}_{(1)}, z, \theta) &= 0. \end{aligned} \tag{65.33}$$

When $\psi_\theta = 0$, the points

$$(x^{(1)}, \hat{z}_{(1)}, \psi_{x^{(1)}}, \psi_{\hat{z}_{(1)}}, z, -\psi_z) = (x^{(1)}, \hat{z}_{(1)}, \psi_{x^{(1)}}, \xi_{(1)}, z, \eta)$$

belong to the graph Λ' of the canonical transformation α. By assumption, $(x^{(1)}, \xi_{(1)}, \eta)$ are local coordinates on Λ'. Therefore $(x^{(1)}, \xi_{(1)}, \eta)$ uniquely determine the point $(x^{(1)}, \hat{z}_{(1)}, \xi^{(1)}, \xi_{(1)}, z, \eta)$ on Λ'. By Lemma 65.2, C and C_0 are diffeomorphic, i.e., the coordinates $(x^{(1)}, \xi_{(1)}, \eta)$ uniquely determine the point $(x^{(1)}, \hat{z}_{(1)0}, z_0, \theta_0)$ as a function of $(x^{(1)}, \xi_{(1)}, \eta)$.

Thus (65.33) has a unique solution $(\hat{z}_{(1)0}, \hat{z}_0, \theta_0) \in C_0$ for any given $(x^{(1)}, \xi_{(1)}, \eta)$. We shall prove that $(z_{(1)0}, z_0, \theta_0)$ is a nondegenerate critical point, i.e., the matrix

$$H = \begin{bmatrix} \psi_{\theta \hat{z}_{(1)}} & \psi_{\theta z} & \psi_{\theta\theta} \\ \psi_{\hat{z}_{(1)}\hat{z}_{(1)}} & \psi_{\hat{z}^{(1)} z} & \psi_{\hat{z}^{(1)}\theta} \\ \psi_{z\hat{z}_{(1)}} & \psi_{zz} & \psi_{z\theta} \end{bmatrix}$$

is nonsingular at $(\hat{z}_{(1)0}, z_0, \theta_0)$. Let $x^{(1)}$ be fixed. Considering $(\hat{z}_{(1)0}, z_0, \zeta_0)$ as functions of $(\xi_{(1)}, \eta)$ and taking the differentials of (65.63), we get

$$\begin{aligned} \psi_{\theta \hat{z}_{(1)}} d\hat{z}_{(1)} + \psi_{\theta z} dz + \psi_{\theta\theta} d\theta &= 0, \\ \psi_{\hat{z}_{(1)}\hat{z}_{(1)}} d\hat{z}_{(1)} + \psi_{\hat{z}_{(1)} z} dz + \psi_{\hat{z}^{(1)}\theta} d\theta &= d\xi^{(1)}, \\ \psi_{z\hat{z}_{(1)}} d\hat{z}_{(1)} + \psi_{zz} dz + \psi_{z\theta} d\theta &= -d\eta. \end{aligned}$$

Note that $(d\hat{z}_{(1)}, dz, d\zeta)$ are the vectors tangent to C_0 at $(x^{(1)}, \hat{z}_{(1)0}, z_0, \zeta_0)$. It follows from Lemma 65.2 that there is one-to-one correspondence between the vectors of the form $(d\hat{z}_{(1)}, dz, d\zeta)$ and the vectors $(d\xi_{(1)}, d\eta)$. If $(dz_{(1)}, dz, d\theta)$ is a solution of the homogeneous equation with matrix H, then $d\xi^{(1)} = 0, d\eta = 0$. Also $dx^{(1)} = 0$ since we do not change $x^{(1)}$. Therefore $d\hat{z}_{(1)} = dz = d\theta = 0$. Hence $\det H \neq 0$.

The application of the stationary phase lemma gives

$$\begin{aligned} \text{(65.34)} \quad \Phi u = \frac{1}{(2\pi)^{\frac{n+n_1-r}{2}}} \int_{\mathbb{R}^n} \int_{\mathbb{R}^{n+n_1}} \sum_{p=0}^{N} e^{i\frac{\pi}{4} \operatorname{sgn} H} b_p(x, \xi_{(1)}, \eta) \\ \times \exp\{i(x_{(1)} - \hat{z}_{(1)0}) \cdot \xi_{(1)} - i(y - z_0) \cdot \eta\} \\ \times u(y) dy d\xi_{(1)} d\eta + T_{m-N-1} u, \end{aligned}$$

where $\operatorname{ord} T_{m-N-1} \le m - N - 1,\ \operatorname{ord} b_p \le m - p$, and

$$b_0(x^{(1)}, \xi_{(1)}, \eta) = \frac{a(x^{(1)}, \hat{z}_{(1)0}, z_0, \theta_0)}{|H|^{\frac{1}{2}}}.$$

Note that $(x^{(1)}, \hat{z}_{(1)0}, \psi_{x^{(1)}}(x^{(1)}, \hat{z}_{(1)0}, z_0, \theta_0), \xi_{(1)}, z_0, \eta)$ belongs to the graph of α, and therefore $\psi(x^{(1)}, \hat{z}_{(1)0}, z_0, \theta_0) = \theta \cdot \psi_\theta = 0$.

At the critical point $(\hat{z}_{(1)0}, z_0, \theta_0)$, the phase function of (65.34) has the form

$$\text{(65.35)} \qquad \psi_1(x, y, \xi_{(1)}, \eta) = x_{(1)} \cdot \xi_{(1)} + S(x^{(1)}, \xi_{(1)}, \eta) - y \cdot \eta,$$

where

$$\text{(65.36)} \qquad S(x^{(1)}, \xi_{(1)}, \eta) = -\hat{z}_{(1)0} \cdot \xi_{(1)} + z_0 \cdot \eta.$$

Comparing (65.36) with (65.20), we can see that (65.36) is the generating function of α. Therefore (65.34) has the same phase function as in (65.25).

We can rewrite the right hand side of (65.34) in the form (65.25) with the principal symbol having the form

$$e^{\frac{\pi}{4}\operatorname{sgn} H} a_0(S_\eta(x^{(1)}, \xi_{(1)}, \eta), \eta),$$

where

$$a_0(S_\eta(x^{(1)}, \xi_{(1)}, \eta), \eta) = b_0(x^{(1)}, \xi_{(1)}, \eta)|J(x^{(1)}, \xi_{(1)}, \eta)|^{-\frac{1}{2}}. \qquad \square$$

Remark 65.4. Let $\Psi^{(k)}(a_k), k = 1, 2$, be FIO's of the form (65.8) with nondegenerate phase functions $\psi_k(x, y, \theta^{(k)}), \theta^{(k)} \in \mathbb{R}^{n_k} \setminus \{0\}$ and suppose that $\operatorname{supp} a_k(x, y, \theta^{(k)})$ is contained in a small neighborhood of $(x_0, y_0, \theta_0^{(k)})$, where $\psi_{k\theta^{(k)}}(x_0, y_0, \theta_0^{(k)}) = 0,\ k = 1, 2$. We assume that the image $(x_0, \xi_0, y_0, -\eta_0)$ of $(x_0, y_0, \theta_0^{(1)})$ and $(x_0, y_0, \theta_0^{(2)})$ under the maps σ_1 and σ_2, respectively, is the same. Here σ_k is the map defined in Lemma 65.2. We also assume that the canonical transformations α_1 and α_2 corresponding to $\Psi^{(1)}$ and $\Psi^{(2)}$ coincide in a neighborhood of (y_0, η_0). Finally, assume that $\Psi^{(1)} = \Psi^{(2)} + T$, where T is a lower order operator. Then, applying Lemma 65.5 separately to Ψ_1 and Ψ_2, we get

$$e^{i\frac{\pi}{4}H_1} a_0^{(1)}(S_\eta, \eta) = e^{i\frac{\pi}{4}H_2} a_0^{(2)}(S_\eta, \eta),$$

where $H_k, a_0^{(k)}, k = 1, 2$, are the same as in Lemma 65.5. $\square$

We define a global FIO as a locally finite sum $\Phi = \sum_k \Phi^{(k)}$ of FIO's of the form (65.8) with nondegenerate phase functions. We assume that all $\Phi^{(k)}$ correspond to the same canonical transformation $\alpha(y, \eta) = (x, \xi)$ of $T_0^*(\mathbb{R}^n)$ onto $T_0^*(\mathbb{R}^n)$. Using the partition of unity to localize each $\Phi^{(k)}$, we can apply Lemma 65.5 and represent $\Phi^{(k)}$ as a sum of FIO's of the form (65.25). Therefore, any global FIO can be written as a sum

$$A = \sum_j \Phi_j(a_j \varphi_j),$$

where $\Phi_j(a_j\varphi_j)$ have the form (65.25).

65.5. Construction of a global FIO given a global canonical transformation.

Consider a global canonical transformation α of $T_0^*(\mathbb{R}^n)$ onto $T_0^*(\mathbb{R}^n)$. We construct a global FIO corresponding to this canonical transformation.

We can find a cover $\{U_j\}$ of $\mathbb{R}^n \times (\mathbb{R}^n \setminus \{0\})$ by conic neighborhoods U_j such that the generating function $S_j(x^{(j)}, \xi_{(j)}, \eta)$ exists in $\overline{U_j}$. Let $\{\varphi_j(y, \eta)\}$ be a partition of unity corresponding to $\{U_j\}$. We assume that $\alpha(y, \eta) = (y + \alpha_1(\eta), \eta)$ when $|y| > R$. Therefore there is a finite cover $\{U_j\}$ such that $S_j(x^{(j)}, \xi_{(j)}, \eta)$ exists in $\overline{U_j}$. Let $(x^{(j)}, \xi_{(j)}, \eta)$ and $(x^{(k)}, \xi_{(k)}, \eta)$ be the coordinates of the graph of α in U_j and U_k, where $U_j \cap U_k \neq \emptyset$. Let

$\Phi_j(\varphi_j a), \Phi_k(\varphi_k b)$ be the FIO's of the form (65.25) with phase functions $\psi_j(x, y, \xi_{(j)}, \eta), \psi_k(x, y, \xi_{(k)}, \eta)$ and symbols $\varphi_j a, \varphi_k b$, respectively.

The following lemma is a particular case of Lemma 65.5. We give an independent proof to have an explicit expression of the principal symbol of a FIO.

Lemma 65.6. *Let $a(y,\eta) \in S^m$ and let $a_0(y,\eta)$ be the principal part of a. Then*

$$\Phi_j(\varphi_j\varphi_k a)u = e^{i\frac{\pi}{4}\sigma_{jk}}\Phi_k(\varphi_j\varphi_k a_0 + b_N)u + T_{m-N-1}u, \tag{65.37}$$

where N is arbitrary, $\operatorname{ord} T_{m-N-1} \le m - N - 1$, $b_N \in S^{m-1}$, *and σ_{jk} is the signature of the matrix* (65.41).

Proof: It will be convenient to change notation for the proof of Lemma 65.6. We denote $x^{(k)}, x_{(k)}$ by $x_{\overline{k}}, x_{\underline{k}}$, where $\underline{k} \cup \overline{k} = (1, 2, \dots, n)$, $\underline{k} \cap \overline{k} = \emptyset$. Analogously, $x^{(j)} = x_{\overline{j}}$, $x_{(j)} = x_{\underline{j}}$, etc. Let $\overline{jk} = \overline{k} \cap \overline{j}$, $\underline{jk} = \underline{k} \cap \underline{j}$. Define $\overline{k}_1 = \overline{k} \setminus \overline{jk}$, $\overline{j}_1 = \overline{j} \setminus \overline{jk}$, $\underline{k}_1 = \underline{k} \setminus \underline{jk}$, etc.

We have, in the new notation:

$$\begin{aligned}(x^{(j)}, \xi_{(j)}) &= (x_{\overline{jk}}, x_{\overline{j}_1}, \xi_{\underline{jk}}, \xi_{\underline{j}_1}),\\ (x^{(k)}, \xi_{(k)}) &= (x_{\overline{jk}}, x_{\overline{k}_1}, \xi_{\underline{jk}}, \xi_{\underline{k}_1}).\end{aligned} \tag{65.38}$$

Note that $\overline{j}_1 = \underline{k}_1$, $\overline{k}_1 = \underline{j}_1$. Using the identity of the form (65.31) in variables $(z_{\underline{k}_1}, \zeta_{\underline{k}_1})$, we get

$$\begin{aligned}\Phi_j(\varphi_j\varphi_k a)u = &\frac{1}{(2\pi)^{n+\frac{1}{2}|\underline{j}|+|\underline{k}_1|}} \int_{\mathbb{R}^{2n+|\underline{j}|+2|\underline{k}_1|}} (\varphi_j\varphi_k a)(S_{j\eta}, \eta)\\ &\times \exp\{i(x_{\underline{k}_1} - z_{\underline{k}_1}) \cdot \zeta_{\underline{k}_1} + ix_{\underline{jk}} \cdot \xi_{\underline{jk}} + ix_{\underline{j}_1} \cdot \xi_{\underline{j}_1}\\ &+ iS_j(x_{\overline{jk}}, z_{\overline{j}_1}, \xi_{\underline{jk}}, \xi_{\underline{j}_1}, \eta) - i(y \cdot \eta)\}\\ &\times |J_j|^{\frac{1}{2}} u(y) dy d\eta d\xi_{\underline{j}} dz_{\underline{k}_1} d\zeta_{\underline{k}_1},\end{aligned} \tag{65.39}$$

where $J_j = \left|S_{jx_{\overline{jk}}z_{\overline{j}_1}\xi_{\underline{jk}}\xi_{\underline{j}_1}\eta}\right|^{\frac{1}{2}} = \left|\frac{\partial(y,\eta)}{\partial(x_{\overline{jk}}, z_{\overline{j}_1}, \xi_{\underline{jk}}, \xi_{\underline{j}_1})}\right|^{\frac{1}{2}}$, and $|\underline{j}|, |\underline{k}_1|$ are the cardinalities of $\underline{j}, \underline{k}_1$. Note that $z_{\overline{j}_1} = z_{\underline{k}_1}$.

We now apply the stationary phase lemma to the integral with respect to $z_{\underline{k}_1}, \xi_{\underline{j}_1}$. The equations for the critical point are:

$$-\zeta_{\underline{k}_1} + S_{jz_{\overline{j}_1}} = 0, \quad x_{\underline{j}_1} + S_{j\xi_{\underline{j}_1}} = 0. \tag{65.40}$$

Since $x_{\underline{j}_1} = x_{\overline{k}_1}$, the Hessian of (65.40) is

$$H_{jk} = \begin{bmatrix} S_{j\xi_{\underline{j}_1}z_{\overline{j}_1}} & S_{j\xi_{\underline{j}_1}\xi_{\overline{j}_1}} \\ S_{jz_{\overline{j}_1}z_{\overline{j}_1}} & S_{jz_{\overline{j}_1}\xi_{\overline{j}_1}} \end{bmatrix}. \tag{65.41}$$

If we add to (65.40) the equations

$$\psi_{j\eta} = 0, \quad \psi_{j\xi_{\underline{jk}}} = 0,$$

where ψ_j is the same as in (65.24), then $(x_{\overline{jk}}, z^{(0)}_{\overline{j}_1}, \xi_{\underline{jk}}, \xi^{(0)}_{\underline{j}_1}, \eta)$ are coordinates on Λ' when the $(z^{(0)}_{\overline{j}_1}, \xi^{(0)}_{\underline{j}_1})$ satisfy (65.40). Since $(x_{\overline{jk}}, x_{\overline{k}_1}, \xi_{\underline{jk}}, \zeta_{\underline{k}_1}, \eta)$ are also coordinates of Λ' on $U_j \cap U_k$, we have, when $z_{\overline{j}_1} = z^{(0)}_{\overline{j}_1}$, $\xi_{\underline{j}_1} = \xi^{(0)}_{\underline{j}_1}$:

$$(65.42) \qquad |\det H_{jk}| = \left|\frac{\partial(x_{\overline{k}_1}, \zeta_{\underline{k}_1})}{\partial(z_{\overline{j}_1}, \xi_{\underline{j}_1})}\right| = \left|\frac{\partial(x_{\overline{jk}}, x_{\overline{k}_1}, \xi_{\underline{jk}}, \zeta_{\underline{k}_1}, \eta)}{\partial(x_{\overline{jk}}, z_{\overline{j}_1}, \xi_{\underline{jk}}, \xi_{\underline{j}_1}, \eta)}\right| = \frac{|J_j|}{|J_k|} \neq 0,$$

where $|J_k| = \left|\frac{\partial(y,\eta)}{\partial(x_{\overline{jk}}, x_{\overline{k}_1}, \xi_{\underline{jk}}, \zeta_{\underline{k}_1}, \eta)}\right|$ and $|J_j| = \left|\frac{\partial(y,\eta)}{\partial(x_{\overline{jk}}, z_{\overline{j}_1}, \xi_{\underline{jk}}, \zeta_{\underline{j}_1}, \eta)}\right|$. Therefore (65.40) has a unique solution, and we can apply the stationary phase method.

Now, for convenience, denote $\zeta_{\underline{k}_1}$ by $\xi_{\underline{k}_1}$ and $z^{(0)}_{\underline{k}_1}$ by $x^{(0)}_{\underline{k}_1}$.

It follows from (65.20) that

$$(65.43) \qquad \begin{aligned} &- x^{(0)}_{\underline{k}_1} \cdot \xi_{\underline{k}_1} + S_j(x_{\overline{jk}}, x^{(0)}_{\underline{k}_1}, \xi_{\underline{jk}}, \xi^{(0)}_{\underline{j}_1}, \eta) + x_{\underline{j}_1} \cdot \xi^{(0)}_{\underline{j}_1} \\ &\qquad = -x^{(0)}_{\underline{k}_1} \cdot \xi_{\underline{k}_1} + y^{(0)} \cdot \eta - x^{(0)}_{\underline{jk}} \cdot \xi_{\underline{jk}} - x^{(0)}_{\underline{j}_1} \cdot \xi^{(0)}_{\underline{j}_1} + x_{\underline{j}_1} \cdot \xi^{(0)}_{\underline{j}_1}, \end{aligned}$$

where $x^{(0)}_{\underline{k}_1}, \xi^{(0)}_{\underline{j}_1}, y^{(0)}, x^{(0)}_{\underline{j}_1}, x^{(0)}_{\underline{jk}}$ are functions of $(x_{\overline{k}}, \xi_{\underline{k}}, \eta)$. Note that $\underline{j}_1 = \overline{k}_1$. In the variables $(x_{\overline{k}}, \xi_{\underline{k}}, \eta)$ we have $x^{(0)}_{\underline{j}_1} = x_{\overline{k}_1}$. As a result, the last two terms in (65.43) are cancelled out. Therefore, (65.43) has the following form (cf. (65.20)):

$$y^{(0)} \cdot \eta - x^{(0)}_{\underline{jk}} \cdot \xi_{\underline{jk}} - x^{(0)}_{\underline{k}_1} \cdot \xi_{\underline{k}_1} = S_k(x_{\overline{k}}, \xi_{\underline{k}}, \eta),$$

where $y^{(0)}, x^{(0)}_{\underline{jk}}, x^{(0)}_{\underline{k}_1}$ are considered as functions of $(x_{\overline{k}}, \xi_{\underline{k}}, \eta)$. □

The application of the stationary phase lemma in $z_{\underline{k}_1}, \xi_{\underline{j}_1}$ gives a factor $(2\pi)^{\frac{1}{2}|\underline{k}_1| + \frac{1}{2}|\underline{j}_1|} e^{i\frac{\pi}{4}\sigma_{jk}}$, where σ_{jk} is the signature of the matrix (65.41). Note that

$$\frac{1}{2}|\underline{j}| + |\underline{k}_1| - \frac{1}{2}(|\underline{k}_1| + |\underline{j}_1|) = \frac{1}{2}(|\underline{jk}| + |\underline{j}_1|) + \frac{1}{2}|\underline{k}_1| - \frac{1}{2}|\underline{j}_1| = \frac{1}{2}|\underline{k}|.$$

Therefore, using (65.42) and (65.43), we get

$$(65.44) \quad \begin{aligned} \Phi_j(\varphi_j\varphi_k a)u = \frac{e^{i\frac{\pi}{4}\sigma_{jk}}}{(2\pi)^{n+\frac{1}{2}|\underline{k}|}} &\int_{\mathbb{R}^{2n+|\underline{k}|}} \sum_{p=0}^{n} b_p(x^{(k)}, \xi_{(k)}, \eta) \\ &\times \exp\{ix_{(k)} \cdot \xi_{(k)} + iS_k(x^{(k)}, \xi_{(k)}, \eta) - y \cdot \eta\} \\ &\qquad \times |J_k|^{\frac{1}{2}} u(y) dy d\eta d\xi_{(k)} + T_{m-N-1}u, \end{aligned}$$

where $\operatorname{ord} T_{m-N-1} \leq m - N - 1$, $\operatorname{ord} b_p = m - p$, and

$$b_0(x^{(k)}, \xi_{(k)}, \eta) = \varphi_j(S_{k\eta}, \eta)\varphi_k(S_{k\eta}, \eta)a_0(S_{k\eta}, \eta). \qquad \square$$

The complex vector bundle on a manifold M with fibers of dimension one is called a line bundle. It follows from Lemma 65.6 that the integers $\sigma_{jk} = \operatorname{sgn} H_{jk}$ are defined in $U_j \cap U_k$ and have the following properties:

$$e^{i\frac{\pi}{4}\sigma_{jk}} = e^{-i\frac{\pi}{4}\sigma_{kj}}, \quad e^{i\frac{\pi}{4}\sigma_{jk}}e^{i\frac{\pi}{4}\sigma_{kr}}e^{i\frac{\pi}{4}\sigma_{rj}} = 1 \quad \text{if} \quad U_j \cap U_k \cap U_r \neq \emptyset.$$

Therefore $e^{i\frac{\pi}{4}\sigma_{jk}}$ are the transition functions of the line bundle on $T_0^*(\mathbb{R}^n)$ called the Maslov bundle $M(\alpha)$. More precisely, $M(\alpha)$ is a line bundle on the graph Λ' of a canonical transformation α, but it will be more convenient to regard it as a bundle on $T_0^*(\mathbb{R}^n)$ since Λ' and $T_0^*(\mathbb{R}^n)$ are diffeomorphic. A section of $M(\alpha)$ consists of nonvanishing functions $c_j(y, \eta)$ on U_j, $1 \leq j \leq N$, such that $c_j c_k^{-1} = e^{-i\frac{\pi}{4}\sigma_{jk}}$ in $U_j \cap U_k$.

Let S_0^m be the class of $C^\infty(T_0^*(\mathbb{R}^n))$ functions $a(x, \eta)$ homogeneous in η of degree m. Lemma 65.6 implies that the principal symbol of a global FIO corresponding to the canonical transformation α is a section of the tensor product $S_0^m \otimes M(\alpha)$; i.e., there exist a section

$$\{e^{i\frac{\pi}{4}\sigma_j(y,\eta)}, (y, \eta) \in U_j\}_{j=1}^N$$

of $M(\alpha)$ and $a(y, \eta) \in S_0^m$ such that the principal symbol is the section $e^{i\frac{\pi}{4}\sigma_j}a(y, \eta)$, $1 \leq j \leq N$, of $S_0^m \otimes M(\alpha)$. We assume that the σ_j are homogeneous in η of degree 0. Note that (65.37) implies

$$\Phi_j(e^{i\frac{\pi}{4}\sigma_j}a\varphi_j\varphi_k)u = \Phi_k(e^{i\frac{\pi}{4}\sigma_k}a\varphi_j\varphi_k)u + T_{jk},$$

where T_{jk} is a lower order operator. Therefore, given a global canonical transformation $\alpha(y, \eta) = (x, \xi)$, the symbol $a_0(y, \eta) \in S_0^m$, and the nonvanishing section $\{e^{i\frac{\pi}{4}\sigma_j(y,\eta)}, U_j\}_{j=1}^N$ of the Maslov bundle $M(\alpha)$, we can construct a global FIO in the form

$$\Phi u = \sum_j \Phi_j(e^{i\frac{\pi}{4}\sigma_j}a_0\varphi_j)u \tag{65.45}$$

with the principal symbol $\{e^{i\frac{\pi}{4}\sigma_j}a_0, U_j\}_{j=1}^N$ belonging to $S_0^m \otimes M(\alpha)$. Since the $\sigma_j(y, \eta)$ are homogeneous in η of degree zero, we can view $M(\alpha)$ as a line bundle on $\mathbb{R}^n \times S^{n-1}$. $\square$

Vice versa, given an arbitrary global FIO A corresponding to the canonical transformation α, we represent A in the form (65.45) and we define the principal symbol of A.

We have already shown that any A can be written in the form

$$Au = \sum_j \Phi_j(a_j\varphi_j)u,$$

where $\Phi_j(a_j\varphi_j)u$ have the form (65.25). Applying Lemma 65.6, we have

$$Au = \sum_k \sum_j \Phi_j(a_j\varphi_j\varphi_k)u$$
$$= \sum_k \sum_j e^{i\frac{\pi}{4}\sigma_{jk}}\Phi_k(a_j\varphi_j\varphi_k)u + Tu = \sum_k \Phi_k(b_k\varphi_k)u + Tu,$$

where

$$b_k\varphi_k = \Big(\sum_j e^{i\frac{\pi}{4}\sigma_{jk}} a_j\varphi_j\Big)\varphi_k,$$

and T is a lower order operator.

Suppose $U_k \cap U_p \neq \emptyset$. By the definition of b_k, we have

$$\Phi_k(b_k\varphi_k\varphi_p)u = \sum_j e^{i\frac{\pi}{4}\sigma_{jk}}\Phi_j(a_j\varphi_j\varphi_k\varphi_p)u + T_k u,$$

where T_k is a lower order operator.

Analogously,

$$\Phi_p(b_p\varphi_k\varphi_p)u = \sum_j e^{i\frac{\pi}{4}\sigma_{jp}}\Phi_j(a_j\varphi_j\varphi_k\varphi_p)u + T_p u,$$

Since $e^{i\frac{\pi}{4}\sigma_{jp}}e^{i\frac{\pi}{4}\sigma_{pk}} = e^{i\frac{\pi}{4}\sigma_{jk}}$, we see that

$$\Phi_k(b_k\varphi_k\varphi_p)u = e^{i\frac{\pi}{4}\sigma_{pk}}\Phi_p(b_p\varphi_p\varphi_k)u + T_2 u.$$

Therefore,

$$e^{i\frac{\pi}{4}\sigma_{kp}} b_k = b_p,$$

i.e., $\{b_k, U_k\}_{k=1}^N$ is a section of $S_0^m \otimes M(\alpha)$.

Let $\{e^{i\frac{\pi}{4}\sigma_k(y,\eta)}, U_k\}_{k=1}^N$ be an arbitrary nonvanishing section of $M(\alpha)$. Then $b_k e^{-i\frac{\pi}{4}\sigma_k} = b \in S_0^m$, $\forall k$, and $b_k = e^{i\frac{\pi}{4}\sigma_k} b \in S_0^m \otimes M(\alpha)$.

Therefore we determined the principal symbol $\{(b_k, U_k)\}_{k=1}^N$ of an arbitrary FIO A and represented A in the form (65.45). □

Note that the existence of a nonvanishing section of a line bundle is equivalent to the triviality of the bundle. When $n \geq 4$, any line bundle on $\mathbb{R}^n \times S^{n-1}$ is trivial. When $n = 2$ or 3, the size of the matrices H_{jk} (cf. (65.41)) implies that $|\sigma_{jk} + \sigma_{kp} + \sigma_{pj}| < 8$ if $U_j \cap U_k \cap U_p \neq \emptyset$. Therefore $e^{i\frac{\pi}{4}(\sigma_{jk}+\sigma_{kp}+\sigma_{pj})} = 1$ is equivalent to $\sigma_{jk} + \sigma_{kp} + \sigma_{pj} = 0$. Using the partition of unity, one can find real-valued $\sigma_j(y,\eta)$ in U_j, homogeneous in η of degree zero, such that $\sigma_{ij} = \sigma_i - \sigma_j$ in $U_j \cap U_k$ (see Hörmander [**H2**]). Therefore we obtain a section $\{e^{i\frac{\pi}{4}\sigma_j}, U_j\}$ such that $e^{i\frac{\pi}{4}\sigma_j}e^{-i\frac{\pi}{4}\sigma_k} = e^{-i\frac{\pi}{4}\sigma_{jk}}$ in $U_j \cap U_k$. Thus, $M(\alpha)$ is trivial for all n.

65.6. Composition of global FIO's.

Theorem 65.7. *Let $\alpha_k, k = 1, 2$, be canonical transformations of $T_0^*(\mathbb{R}^n)$ onto $T_0^*(\mathbb{R}^n)$ and let $M(\alpha_k), k = 1, 2$, be the corresponding Maslov bundles. Let $\Phi_k, k = 1, 2$, be global FIO's corresponding to α_k with the principal symbols $\{e^{i\frac{\pi}{4}\sigma_j^{(k)}} a_k(y,\eta), U_j\}_{j=1}^N \in S_0^{m_k} \otimes M(\alpha_k), k = 1, 2$. Consider the composition $\alpha_3 = \alpha_2 \circ \alpha_1$ of α_1 and α_2. Let $M(\alpha_3)$ be the Maslov bundle for α_3. Then $\Phi_2\Phi_1 = \Phi_3 + T_{m_1+m_2-N-1}$, where Φ_3 is a global FIO corresponding to α_3 with the principal symbol $\{c_j^{(3)} a_3(y,\eta), U_j\}_{j=1}^N \in S_0^{m_1+m_2} \otimes M(\alpha_3)$, where*

$$a_3(y,\eta) = a_1(y,\eta) a_2(z,\zeta). \tag{65.46}$$

Here $(z,\zeta) = \alpha_1(y,\eta)$, $\operatorname{ord} T_{m_1+m_2-N-1} \le m_1+m_2-N-1$, and $\{c_j^{(3)}, U_j\}_{j=1}^N$ is a nonvanishing section of $M(\alpha_3)$. For simplicity of notation we assumed that the same cover $\{U_j\}$ works for α_1, α_2, and α_3.

Proof: Since Φ_1 and Φ_2 can be represented in the form (65.45), it is enough to consider the composition of $\Phi_1(e^{i\frac{\pi}{4}\sigma^{(1)}}\varphi_1 a_1)$ and $\Phi_2(e^{i\frac{\pi}{4}\sigma^{(2)}}\varphi_2 a_2)$.

We have

$$\begin{aligned} &\Phi_2(e^{i\frac{\pi}{4}\sigma^{(2)}}\varphi_2 a_2)\Phi_1(e^{i\frac{\pi}{4}\sigma^{(1)}}\varphi_1 a_1)u \\ &= \frac{1}{(2\pi)^{2n+\frac{n_1+n_2}{2}}} \int_{\mathbb{R}^{2n}} \int_{\mathbb{R}^{n_1+n_2}} (e^{i\frac{\pi}{4}\sigma^{(2)}}\varphi_2 a_2)(S_{2\zeta},\zeta) \\ &\times e^{i\psi_2(x,z,\xi_{(2)},\zeta)+i\psi_1(z,y,\hat{\zeta}_{(1)},\eta)} \cdot (e^{i\frac{\pi}{4}\sigma^{(1)}}\varphi_1 a_1)(S_{1\eta},\eta) \\ &\times |J_1|^{\frac{1}{2}} |J_2|^{\frac{1}{2}} u(y)\,dy\,d\eta\,d\hat{\zeta}_{(1)}\,dz\,d\zeta\,d\xi_{(2)}, \end{aligned} \tag{65.47}$$

where

$$\begin{aligned} \psi_1(z,y,\hat{\zeta}_{(1)},\eta) &= z_{(1)}\cdot\hat{\zeta}_{(1)} + S_1(z^{(1)},\hat{\zeta}_{(1)},\eta) - y\cdot\eta, \\ \psi_2(x,z,\xi_{(2)},\zeta) &= x_{(2)}\cdot\xi_{(2)} + S_2(x^{(2)},\xi_{(2)},\zeta) - z\cdot\zeta. \end{aligned} \tag{65.48}$$

Here $(z^{(1)},\zeta_{(1)},\eta)$ are the coordinates on $\Lambda_1'\cap\operatorname{supp}\varphi_1(y,\eta)$ and $(x^{(2)},\xi_{(2)},\zeta)$ are the coordinates on $\Lambda_2'\cap\operatorname{supp}\varphi_2(z,\zeta)$, and Λ_1' and Λ_2' are the graphs of $\alpha_1(y,\eta) = (z,\zeta)$ and $\alpha_2(z,\zeta) = (x,\xi)$, respectively. We assume that $\operatorname{supp}\varphi_2(z,\zeta)\cap\operatorname{supp}\varphi_1(y,\eta) \neq 0$ when $(z,\zeta) = \alpha_1(y,\eta)$. Also S_1 and S_2 are the generating functions of α_1 and α_2 as in (65.20), (65.21), $\xi_{(2)} \in \mathbb{R}^{n_2}$, $\hat{\zeta}_{(1)} \in \mathbb{R}^{n_1}$, and we have

$$|J_1(z^{(1)},\hat{\zeta}_{(1)},\eta)| = |S_{1\eta z^{(1)}\hat{\zeta}_{(1)}}| = \left|\frac{\partial(y,\eta)}{\partial(z^{(1)},\hat{\zeta}_{(1)},\eta)}\right|,$$

$$|J_2(x^{(2)},\xi_{(2)},\zeta)| = |S_{2\zeta x^{(2)}\xi_{(2)}}| = \left|\frac{\partial(z,\zeta)}{\partial(x^{(2)},\xi_{(2)},\zeta)}\right|.$$

Consider $\alpha_3(y,\eta) = \alpha_2\circ\alpha_1(y,\eta) = (x,\xi)$. Assume, for simplicity of notation, that $(x^{(2)},\xi_{(2)},\eta)$ are the coordinates on $\Lambda_3'\cap\operatorname{supp}\varphi_1(y,\eta)$, where Λ_3' is the graph of α_3.

In the notation of (65.8) we have $\theta = (\eta,\hat{\zeta}_{(1)},\zeta,\xi_{(2)})$. Therefore the set $\psi_\theta = (\psi_1+\psi_2)_\theta = 0$ has the form

$$S_{1\eta}(z^{(1)},\hat{\zeta}_{(1)},\eta) - y = 0, \quad S_{2\xi_{(2)}}(x^{(2)},\xi_{(2)},\zeta) + x_{(2)} = 0, \tag{65.49}$$

$$S_{1\hat{\zeta}_{(1)}}(z^{(1)},\hat{\zeta}_{(1)},\eta) + z_{(1)} = 0, \quad S_{2\zeta}(x^{(2)},\xi_{(2)},\zeta) - z = 0. \tag{65.50}$$

We apply the stationary phase lemma to (65.47) in $(z,\zeta,\hat{\zeta}_{(1)})$. The equations for the critical points are (65.50) and

$$S_{1z^{(1)}}(z^{(1)},\hat{\zeta}_{(1)},\eta) - \zeta^{(1)} = 0, \quad \hat{\zeta}_{(1)} - \zeta_{(1)} = 0. \tag{65.51}$$

Note that we can perform integration with respect to $z_{(1)},\hat{\zeta}_{(1)}$ in (65.46), but it will be more convenient to apply the stationary phase lemma to all variables $z^{(1)},z_{(1)},\zeta^{(1)},\zeta_{(1)},\hat{\zeta}_{(1)}$. We prove that there is a unique nondegenerate critical point $z^{(1)} = z_0^{(1)}, z_{(1)} = z_{(1)0}, \zeta^{(1)} = \zeta_0^{(1)}, \zeta_{(1)} = \hat{\zeta}_{(1)} = \zeta_{(1)0}$ smoothly dependent of $(x^{(2)},\xi_{(2)},\eta)$.

Note that when equations (65.49), (65.50), (65.51) are satisfied, the corresponding points

$$(z_0^{(0)},z_{(1)0},\zeta_0^{(1)},\zeta_{(1)0},y,\eta) \quad\text{and}\quad (x,S_{2x^{(2)}}(x^{(2)},\xi_{(2)},\zeta_0),\xi_{(2)},z_0,\zeta_0)$$

belong to the graphs Λ_1',Λ_2' of the canonical transformations α_1,α_2. By our assumption, $(x^{(2)},\xi_{(2)},\eta)$ are the coordinates on Λ_3' where $\alpha_3(y,\eta) = (x,\xi)$. Therefore we can express $(x_{(2)},\xi^{(2)},y)$ as functions of $(x^{(2)},\xi_{(2)},\eta)$. Knowing (y,η), we can find $(\hat{z},\hat{\zeta}) = \alpha_1(y,\eta)$ and $(z,\zeta) = \alpha_2^{-1}(x,\xi)$. Since $\hat{z} = z$, $\hat{\zeta} = \zeta$, we see that there exists a unique solution $(z_0,\zeta_0^{(1)},\zeta_{(1)0},\hat{\zeta}_{(1)0})$ of (65.49), (65.50), (65.51) for a given $(x^{(2)},\xi_{(2)},\eta)$. It remains to show that this critical point is nondegenerate. Take the differentials of (65.50), (65.51) regarding all variables as functions of $(x^{(2)},\xi_{(2)},\eta)$. We get

$$\begin{gathered} S_{2\zeta\zeta}d\zeta - dz = -S_{2\zeta x^{(2)}}dx^{(2)} - S_{2\zeta\xi_{(2)}}d\xi_{(2)}, \\ S_{1z^{(1)}z^{(1)}}dz^{(1)} + S_{1z^{(1)}\zeta_{(1)}}d\zeta_{(1)} - d\zeta^{(1)} = -S_{1z^{(1)}\eta}d\eta, \\ S_{1\zeta_{(1)}z^{(1)}}dz^{(1)} + S_{1\zeta_{(1)}\zeta_{(1)}}d\zeta_{(1)} + dz_{(1)} = -S_{1\zeta_{(1)}\eta}d\eta, \\ d\zeta_{(1)} = d\hat{\zeta}_{(1)}. \end{gathered} \tag{65.52}$$

We shall rewrite (65.52) in a matrix form. Let

$$
\begin{aligned}
S_{2\zeta x^{(2)}\xi_{(2)}} &= \begin{bmatrix} S_{2\zeta x^{(2)}} \\ S_{2\zeta\xi_{(2)}} \end{bmatrix}, \\
S_{1\zeta_{(1)}z^{(1)}\eta} &= \begin{bmatrix} S_{1z^{(1)}\eta}, S_{1\zeta_{(1)}\eta} \end{bmatrix}
\end{aligned}
\tag{65.53}
$$

be two $n \times n$ matrices. Let H be the following $2n \times 2n$ matrix:

$$
\left[\begin{array}{ccccc}
 & -I_n & & S_{2\zeta\zeta} & \\
S_{1z^{(1)}z^{(1)}} & 0 & -I_{n-n_1} & & S_{1z^{(1)}\zeta_{(1)}} \\
S_{1\zeta_{(1)}z^{(1)}} & I_{n-n_1} & 0 & & S_{1\zeta_{(1)}\zeta_{(1)}}
\end{array}\right].
\tag{65.54}
$$

Therefore, (65.52) can be written as

$$
\begin{aligned}
H\begin{pmatrix} dz \\ d\zeta \end{pmatrix} &= -\begin{bmatrix} S_{2\zeta x^{(2)}\xi_{(2)}} & 0 \\ 0 & S_{1\eta\zeta_{(1)}z^{(1)}} \end{bmatrix}\begin{pmatrix} dx^{(2)} \\ d\xi_{(2)} \\ d\eta \end{pmatrix}, \\
[I_{n_1}, -I_{n_1}]\begin{pmatrix} d\zeta_{(1)} \\ d\hat{\zeta}_{(1)} \end{pmatrix} &= \begin{pmatrix} 0 \\ 0 \end{pmatrix}.
\end{aligned}
\tag{65.55}
$$

Note that the Hessian of the critical point is

$$
H_1 = \left[\begin{array}{ccc} \multicolumn{2}{c}{H} & 0 \\ 0 & I_{n_1} & I_{n_1} \end{array}\right],
$$

where we added the equations $d\zeta_{(1)} - d\hat{\zeta}_{(1)} = 0$. Therefore $\det H_1 = \det H$, and it is enough to prove that $\det H \neq 0$ at the critical point. For this we must show that the homogeneous system $H\binom{dz}{d\zeta} = \binom{0}{0}$ has only a trivial solution. Note that $H\binom{dz}{d\zeta} = \binom{0}{0}$ means that the right hand sides in (65.55) are equal to zero, i.e., $dx^{(2)} = d\xi_{(2)} = d\eta = 0$. Note that there is a diffeomorphism between (z, ζ) and $(x^{(2)}, \xi_{(2)}, \eta)$, and the Jacobian $\left|\frac{\partial(x^{(2)},\xi_{(2)},\eta)}{\partial(z,\zeta)}\right| \neq 0$. Therefore $(dx^{(2)}, d\xi_{(2)}, d\eta) = (0, 0, 0)$ implies that $(dz, d\zeta) = 0$. So $H\binom{dz}{d\zeta} = \binom{0}{0}$ if and only if $(dz, d\zeta) = (0, 0)$. Therefore $\det H \neq 0$, and we can apply the stationary phase lemma. It follows from (65.55) that

$$
\begin{pmatrix} dx^{(2)} \\ d\xi_{(2)} \\ d\eta \end{pmatrix} = \begin{bmatrix} S_{2\zeta x^{(2)}\xi_{(2)}} & 0 \\ 0 & S_{1\eta z^{(1)}\zeta_{(1)}} \end{bmatrix}^{-1} H\begin{pmatrix} dz \\ d\zeta \end{pmatrix}.
\tag{65.56}
$$

Therefore

$$
\left|\frac{\partial(x^{(2)}, \xi_{(2)}, \eta)}{\partial(z, \zeta)}\right| = |J_1|^{-1}|J_2|^{-1}|H|,
\tag{65.57}
$$

where $|H| = |\det H|$ and $|J_1|$, $|J_2|$ are the same as in (65.47). Note that

$$\left|\frac{\partial(x^{(2)},\xi_{(2)},\eta)}{\partial(z,\zeta)}\right| = \left|\frac{\partial(x^{(2)},\xi_{(2)},\eta)}{\partial(y,\eta)}\right|\left|\frac{\partial(y,\eta)}{\partial(z,\zeta)}\right| = \left|\frac{\partial(y,\eta)}{\partial(x^{(2)},\xi_{(2)},\eta)}\right|^{-1} \tag{65.58}$$

because $\alpha_1(y,\eta) = (z,\zeta)$ is a canonical transformation, and therefore $\left|\frac{\partial(y,\eta)}{\partial(z,\zeta)}\right| = 1$. Hence $\left|\frac{\partial(y,\eta)}{\partial(x^{(2)},\xi_{(2)},\eta)}\right| = |J_1|\,|J_2|\,|H|^{-1}$.

Let

$$\psi_3(x,y,\xi_{(2)},\eta) = \psi_1(z_0,y,\hat{\zeta}_{(1)0},\eta) + \psi_2(x,z_0,\xi_{(2)},\zeta_0),$$

where $(z_0,\zeta_0,\hat{\zeta}_{(1)0})$ is the critical point. Using (65.20), we get

$$\psi_3(x,y,\xi_{(2)},\eta) = x_{(2)}\cdot\xi_{(2)} + S_3(x^{(2)},\xi_{(2)},\eta) - y\cdot\eta, \tag{65.59}$$

where

(65.60)

$$\begin{aligned} S_3(x^{(2)},\xi_{(2)},\eta) &= S_2(x^{(2)},\xi_{(2)},\zeta_0) - z_0\cdot\zeta_0 + z_{(1)0}\cdot\zeta_{(1)0} + S_1(z_0^{(1)},\zeta_{(1)0},\eta) \\ &= z_0\cdot\zeta_0 - x_{(2)}\cdot\xi_{(2)} - z_0\cdot\zeta_0 + z_{(1)0}\cdot\zeta_{(1)0} + y\cdot\eta - z_{(1)0}\cdot\zeta_{(1)0} \\ &= y\cdot\eta - x_{(2)}\cdot\xi_{(2)}. \end{aligned}$$

Hence $S_3(x^{(2)},\xi_{(2)},\eta)$ is a generating function for α_3 (cf. (65.20)).

The application of the stationary phase lemma gives

$$\begin{aligned} &\Phi_2(e^{i\frac{\pi}{4}\sigma^{(2)}}\varphi_2 a_2)\Phi_1(e^{i\frac{\pi}{4}\sigma^{(1)}}\varphi_1 a_1)u \\ &= \frac{1}{(2\pi)^{n+\frac{n_2}{2}}}\int_{\mathbb{R}^n}\int_{\mathbb{R}^{n+n_2}} e^{i\frac{\pi}{4}(\sigma^{(1)}+\sigma^{(2)}+\operatorname{sgn} H_1)}\sum_{p=0}^{N} b_p(x^{(2)},\xi_{(2)},\eta) \\ &\quad\times e^{i\psi_3(x,y,\xi_{(2)},\eta)}|J_3(x^{(2)},\xi_{(2)},\eta)|^{\frac{1}{2}}u(y)dyd\eta d\xi_{(2)} + T_{m_1+m_2-N-1}u, \end{aligned} \tag{65.61}$$

where $\xi_{(2)}\in\mathbb{R}^{n_2}$, $b_p\in S^{m_1+m_2-p}$, $\operatorname{ord} T_{m_1+m_2-N-1}\le m_1+m_2-N-1$, $b_0 = (\varphi_2 a_2)(z_0,\zeta_0)(\varphi_1 a_1)(S_{3\eta},\eta)$, $(z_0,\zeta_0) = \alpha_1(S_{3\eta},\eta)$, and

$$|J_3| = |S_{3x^{(2)}\xi_{(2)},\eta}| = \left|\frac{\partial(y,\eta)}{\partial(x^{(2)},\xi_{(2)},\eta)}\right| \tag{65.62}$$

(cf. (65.57), (65.58)), i.e., the right hand side of (65.61) is an operator of the form (65.25).

Therefore we proved that Φ_3 is a global FIO corresponding to the canonical transformation $\alpha_3 = \alpha_2\circ\alpha_1$ with the principal symbol of the form $\{c_j^{(3)}(y,\eta)a_3(y,\eta),U_j\}_{j=1}^N \in S_0^{m_1+m_2}\otimes M(\alpha_3)$, where $a_3(y,\eta)$ has the form (65.46) and $\{c_j^{(3)},U_j\}$ is a nonvanishing section of $M(\alpha_3)$. $\square$

Remark 65.5. Suppose $(x^{(3)}, \xi_{(3)}, \eta)$, $\xi_{(3)} \in \mathbb{R}^{n_3}$, are coordinates on Λ_3', where $\alpha_3 = \alpha_2 \circ \alpha_1$ and $(x^{(3)}, \xi^{(3)}) \neq (x^{(2)}, \xi^{(2)})$. We indicate the changes in the proof of Theorem 65.7. For convenience, we change the notation (x, ξ) to $(\hat{x}, \hat{\xi})$, i.e., we have $\alpha_2(z, \zeta) = (\hat{x}, \hat{\xi})$. Using the identity (cf. (65.31))

$$f(x_{(3)}) = \frac{1}{(2\pi)^{n_3}} \int_{\mathbb{R}^{n_3}} \int_{\mathbb{R}^{n_3}} e^{i(x_{(3)} - \hat{x}_{(3)}) \cdot \xi_{(3)}} f(\hat{x}_{(3)}) d\hat{x}_{(3)} d\xi_{(3)},$$

we get (cf. (65.47))

$$\begin{aligned}
&\Phi_2(e^{i\frac{\pi}{4}\sigma^{(2)}}\varphi_2 a_2)\Phi_1(e^{i\frac{\pi}{4}\sigma^{(1)}}\varphi_1 a_1) \\
&\quad = \frac{1}{(2\pi)^{2n + \frac{n_1+n_2}{2} + n_3}} \int_{\mathbb{R}^{2n+n_1+n_2+2n_3}} e^{i\frac{\pi}{4}(\sigma^{(2)}+\sigma^{(1)})}(\varphi_2 a_2)(\varphi_1 a_1) \\
&\qquad \times e^{i(x_{(3)} - \hat{x}_{(3)}) \cdot \xi_{(3)} + i\psi_2(x^{(3)}, \hat{x}_{(3)}, z, \hat{\xi}_{(2)}, \zeta)} e^{i\psi_1(z, y, \hat{\zeta}_{(1)}, \eta)} \\
&\qquad\qquad \times |J_1|^{\frac{1}{2}} |J_2|^{\frac{1}{2}} u(y) dy d\eta d\hat{\zeta}_{(1)} dz d\zeta d\hat{\xi}_{(2)} d\hat{x}_{(3)} d\xi_{(3)}.
\end{aligned}$$

After applying the stationary phase lemma in variables $(\hat{\zeta}_{(1)}, z, \zeta, \hat{\xi}_{(2)}, \hat{x}_{(3)})$, we proceed as in the proof of Theorem 65.7. □

65.7. Conjugation by a global FIO and the boundedness theorem.

Denote by Φ_0 the global FIO (cf. (65.45)) with $a(y, \eta) = 1$ corresponding to the canonical transformation α. Analogously to the proof of Lemma 62.4 and Theorem 65.7 we have

$$\Phi_0 \Phi_0^* = I + C_N^{(1)} + T_{-N-1}^{(1)}, \quad \Phi_0^* \Phi_0 = I + C_N^{(2)} + T_{-N-1}^{(2)}, \tag{65.63}$$

where $\operatorname{ord} T_{-N-1}^{(k)} \leq -N - 1$, $C_N^{(k)}(x, \xi) \in S^{-1}$, $k = 1, 2$, and N is arbitrary. Note that Φ_0^* corresponds to the canonical transformation α^{-1}. Using Theorem 65.7, we can prove the global version of Theorem 62.7.

Theorem 65.8. *Let $A(x, \xi) \in S^m$. Then*

$$\Phi_0^* A(x, D) \Phi_0 = B_N(x, D) + T_{m-N-1}, \tag{65.64}$$

where $B_N(x, \xi) \in S^m$, $\operatorname{ord} T \leq m - N - 1$, and the principal symbol of $B_N(x, D)$ is the image of the principal symbol of $A(x, \xi)$ under the canonical transformation α.

Obviously the canonical transformation corresponding to a ψdo is the identity $x = y$, $\xi = \eta$ and we can choose $c(y, \eta) = 1$ for all (y, η) as a section of the Maslov bundle. Since $\Phi^* \Phi$ is a ψdo for any global Φ, we can prove, as in §61, the boundedness theorem for global FIO's.

Theorem 65.9. *Let Φ be a global FIO of the form* (65.45), $a(x,\eta) \in S^m$. *Then*

$$\|\Phi u\|_s \le C\|u\|_{s+m}, \quad \forall s. \tag{65.65}$$ □

Example 65.4 (See Eskin [**E4**], Appendix). Consider the canonical transformation defined by the Hamiltonian system (cf. (64.20))

$$\begin{aligned} \frac{dx}{dt} &= -\frac{\partial\lambda(x,t,\xi)}{\partial\xi}, \quad x(0) = y, \\ \frac{d\xi}{dt} &= \frac{\partial\lambda(x,t,\xi)}{\partial x}, \quad \xi(0) = \eta. \end{aligned} \tag{65.66}$$

Then, as in (64.23), we have:

$$dx \wedge d\xi - d\lambda \wedge dt = dy \wedge d\eta, \tag{65.67}$$

where $x = x(t,y,\eta)$ and $\xi = \xi(t,y,\eta)$ are solutions of (65.66). Consider the following map of $T_0^*(\mathbb{R}^{n+1})$ to $T_0^*(\mathbb{R}^{n+1})$:

$$x = x(t,y,\eta), \quad \xi = \xi(t,y,\eta), \tag{65.68}$$

$$t = y_0, \quad \xi_0 - \lambda(x,t,\xi) = \eta_0. \tag{65.69}$$

Then, using (65.68) and (65.69), we get:

$$dx\wedge d\xi + dt\wedge d\xi_0 = dx\wedge d\xi - d\lambda\wedge dt + dt\wedge d(\xi_0 - \lambda) = dy\wedge d\eta + dy_0\wedge d\eta_0, \tag{65.70}$$

i.e., the map (65.68), (65.69) is a canonical transformation of $T_0^*(\mathbb{R}^{n+1})$ onto $T_0^*(\mathbb{R}^{n+1})$.

Let $A(x,t,\xi,\xi_0) = \xi_0 - \lambda(x,t,\xi) + i\mu(x,t,\xi)$, where $\mu(x,t,\xi) \in C^\infty$ when $\xi \ne 0$, $\deg_\xi \mu(x,t,\xi) = 1$ and $\mu(x,t,\xi)$ is real-valued.

Let Φ_0^*, Φ_0 be global FIO's as in (65.63) corresponding to the canonical transformations (65.68), (65.69). Applying (65.64), we get

$$\Phi_0^* A(x,t,D,D_t)\Phi_0 = B_N(x,t,D,D_t) + T_{-N-1},$$

where $\operatorname{ord} T_{-N-1} \le -N-1$ and $B_N(x,t,D,D_t)$ is a ψdo with principal symbol

$$\eta_0 + i\mu_1(y,y_0,\eta). \tag{65.71}$$

66. Geometric optics at large

66.1. Generating functions and the Legendre transforms.

In §64 we constructed an asymptotic solution to the equation (64.4) of the form (cf. (64.8))

$$a_j(x,t,\eta)e^{i\varphi_j(x,t,\eta)}, \quad j = 1,2, \ |\eta| \to +\infty. \tag{66.1}$$

These solutions hold for $0 \le t \le t_0$, where in general t_0 is small. In this section we shall construct asymptotic solutions of (64.4) for an arbitrary time interval $[0, T]$.

Again, as in (64.20), we drop the subscript j, $j = 1, 2$, and study asymptotic solutions associated with the Hamiltonian $\xi_0 - \lambda(x, t, \xi)$. We look for the solutions $u(x, t, \tau)$ with initial values

$$u(x, 0, \tau) = c(x)e^{i\tau x \cdot \eta}, \tag{66.2}$$

where τ is a large parameter, $c(x) \in C_0^\infty(U_0)$, U_0 is a neighborhood of some $y_0 \in \mathbb{R}^n$, and $\eta \neq 0$ is fixed. Note that (64.21) exists for all $t \in \mathbb{R}$ and (64.23) holds. Since η is fixed, we have $dy \wedge d\eta = 0$ and (64.23) has the form:

$$d\xi(y, t) \wedge dx(y, t) + d\lambda(x(y, t), t, \xi(y, t)) \wedge dl = 0. \tag{66.3}$$

Since η is fixed, we do not indicate the dependence on η in (66.3). Denote by Λ the union of all curves $x = x(t, y)$, $\xi = \xi(t, y)$, $\xi_0 = \xi_0(t, y) = \lambda(x(t, y), t, \xi(t, y))$, $t = t$, $0 \le t < +\infty$, where $x(t, y), \xi(t, y)$ satisfy (64.20) and $y \in \overline{U}_0$. It follows from (66.3) that Λ is an $(n + 1)$-dimensional Lagrangian manifold.

Denote by γ_y the curve $x = x(t, y)$, $t = t$ in $\mathbb{R}^{n+1}$, $t \ge 0$. A point $(x^{(0)}, t^{(0)})$ on γ_y is called a focal point (caustics) if

$$\det \left[\frac{\partial x(t, y)}{\partial y}\right] = 0 \quad \text{at} \quad (x^{(0)}, t^{(0)}). \tag{66.4}$$

Generically there are finitely many focal points $(x(t_k, y_0), t_k)$, $k = 1, \dots, m$, on a curve γ_{y_0} for $0 \le t \le T$, where T is fixed.

For simplicity we assume that U_0 is a small neighborhood of y_0, and therefore there is a finite number of intervals (t_{k1}, t_{k2}) surrounding t_k, $1 \le k \le m$, such that there is no focal points in $\Lambda - \bigcup_{k=1}^m (\Lambda \bigcap (t_{k_1}, t_{k_2}))$.

Therefore, since $\det \left[\frac{\partial x(t,y)}{\partial y}\right] \neq 0$ for $0 \le t \le t_{11}$, we have a generating function $S_0(x, t)$ as in (64.28). Then (cf. (64.27)):

$$dS_0(x, t) = p(x, t)dx + \lambda(x, t, p(x, t))dt, \quad S_0(x, 0) = x \cdot \eta, \tag{66.5}$$

and

$$\frac{\partial S_0(x, t)}{\partial x} = p(x, t), \quad \frac{\partial S_0(x, t)}{\partial t} = \lambda(x, t, p(x, t)). \tag{66.6}$$

Our goal in this section is the construction of an asymptotic solution for $0 \le t \le T$ that is supported modulo $(\frac{1}{\tau^N})$ on a small neighborhood of γ_{y_0}.

When $0 \le t \le t_{11}$, we constructed in §64 an asymptotic solution $u_0(x,t)$ of the form:

$$u_0(x,t) = e^{i\tau S_0(x,t)} \sum_{p=0}^{N} \frac{a_p(x,t)}{\tau^p} + O\Big(\frac{1}{\tau^{N+1}}\Big), \tag{66.7}$$

where $S_0(x,t)$ is the same as in (66.6).

Analogous solutions can be constructed on each interval $[t_{k2}, t_{k+1,1}]$, $1 \le k \le m$, since $\det\left[\frac{\partial x(t,y)}{\partial y}\right] \ne 0$ on these intervals.

Consider the $n \times 2n$ matrix

$$\left[\frac{\partial x(t,y)}{\partial y} \quad \frac{\partial \xi(t,y)}{\partial y}\right]. \tag{66.8}$$

It has rank n for all (y,t). When $t \in (t_{k1}, t_{k2})$, the determinant (66.4) can be zero. However (cf. the proof of Lemma 65.3), we can find a subset $(k) = \{i_1, \dots, i_{n_1}\} \subset \{1,2,\dots,n\}$ such that

$$\det\left[\frac{\partial x^{(k)}(t,y)}{\partial y} \quad \frac{\partial \xi_{(k)}(t,y)}{\partial y}\right] \ne 0 \quad \text{in} \quad \Lambda \cap [t_{k1}, t_{k2}], \ k = 1,\dots,m, \tag{66.9}$$

where $x^{(k)} = (x_{i_1}, x_{i_2}, \dots, x_{i_{n_1}})$, $\xi_{(k)} = (\xi_{j_1}, \dots, \xi_{j_{n_2}})$, $(j_1, \dots, j_{n_2}) \cap (k) = \emptyset$, $n_1 + n_2 = n$, $x^{(k)} \in \mathbb{R}^{n_1}$, $\xi_{(k)} \in \mathbb{R}^{n_2}$.

We have taken advantage of the fact that U_0 is small. Therefore, by the implicit function theorem,

$$y = y(x^{(k)}, \xi_{(k)}, t), \tag{66.10}$$

i.e., $(x^{(k)}, \xi_{(k)}, t)$ are coordinates in $\Lambda_k = \Lambda \cap [t_{k1} - \varepsilon, t_{k2} + \varepsilon]$, $\varepsilon > 0$.

Let

$$\begin{aligned} p^{(k)}(x^{(k)}, \xi_{(k)}, t) &= \xi^{(k)}(y(x^{(k)}, \xi_{(k)}, t), t), \\ q_{(k)}(x^{(k)}, \xi_{(k)}, t) &= x_{(k)}(y(x^{(k)}, \xi_{(k)}, t), t). \end{aligned} \tag{66.11}$$

Consider the differential form

$$p^{(k)} dx^{(k)} - q_{(k)} d\xi_{(k)} + \lambda(x^{(k)}, q_{(k)}, t, p^{(k)}, \xi_{(k)}) dt. \tag{66.12}$$

It follows from (66.3) that (66.12) is a closed form in Λ_k, and since Λ_k is simply connected, there exists $S^{(k)}(x^{(k)}, \xi_{(k)}, t)$ such that

$$dS^{(k)} = p^{(k)} dx^{(k)} - q_{(k)} d\xi_{(k)} + \lambda(x^{(k)}, q_{(k)}, t, p^{(k)}, \xi_{(k)}) dt. \tag{66.13}$$

Therefore

$$p^{(k)} = S^{(k)}_{x^{(k)}}(x^{(k)}, \xi_{(k)}, t), \quad q_{(k)} = -S^{(k)}_{\xi_{(k)}}(x^{(k)}, \xi_{(k)}, t), \tag{66.14}$$

$$S^{(k)}_t(x^{(k)}, \xi_{(k)}, t) = \lambda(x^{(k)}, q_{(k)}, t, p^{(k)}, \xi_{(k)}). \tag{66.15}$$

It follows from (66.14), (66.15) that $S^{(k)}(x^{(k)}, \xi_{(k)}, t)$ satisfies the eikonal equation in the $(x^{(k)}, \xi_{(k)})$-coordinates,

(66.16)
$$S_t^{(k)}(x^{(k)}, \xi_{(k)}, t) - \lambda(x^{(k)}, -S_{\xi_{(k)}}^{(k)}, t, S_{x^{(k)}}^{(k)}, \xi_{(k)}) = 0 \quad \text{on} \quad [t_{k1} - \varepsilon, t_{k2} + \varepsilon].$$

In the domain $\Lambda \cap [t_{k2} - \varepsilon, t_{k+1,1} + \varepsilon]$, where (66.4) holds, i.e., where (x, t) are coordinates, we have as in (66.5), (66.6):

$$dS^{(0)}(x, t) = p(x, t)dx + \lambda(x, t, p(x, t))dt, \tag{66.17}$$

$$\frac{\partial S^{(0)}(x, t)}{\partial x} = p(x, t), \quad \frac{\partial S^{(0)}(x, t)}{\partial t} = \lambda(x, t, p(x, t)), \tag{66.18}$$

where $p(x, t) = \xi(y(x, t), t)$.

Consider a small neighborhood $\Lambda \cap (t_{k1} - \varepsilon, t_{k1} + \varepsilon)$, where both (x, t) and $(x^{(k)}, \xi_{(k)}, t)$ are coordinates. We find a relation between $S^{(0)}(x, t)$ and $S^{(k)}(x^{(k)}, \xi_{(k)}, t)$. We rewrite (66.17) in the form:

$$d(S^{(0)} - p_{(k)} \cdot x_{(k)}) = p^{(k)}dx^{(k)} - x_{(k)}dp_{(k)} + \lambda(x^{(k)}, x_{(k)}, t, p^{(k)}, p_{(k)})dt. \tag{66.19}$$

In the $(x^{(k)}, \xi_{(k)}, t)$-coordinates we have

$$p_{(k)} = \xi_{(k)}, \quad x_{(k)} = q_{(k)}.$$

Therefore the right hand side of (66.19) has the form (cf. (66.13)):

(66.20)
$$dS^{(k)}(x^{(k)}, \xi_{(k)}, t) = p^{(k)}dx^{(k)} + q_{(k)}d\xi_{(k)} - \lambda(x^{(k)}, q_{(k)}, t, p^{(k)}, \xi_{(k)})dt.$$

Consequently

$$S^{(0)} - p_{(k)} \cdot x_{(k)} = S^{(k)}(x^{(k)}, \xi_{(k)}, t). \tag{66.21}$$

Note the relation between the coordinates $(x^{(k)}, x_{(k)}, t)$ and $(x^{(k)}, \xi_{(k)}, t)$:

$$\xi_{(k)} = S_{x_{(k)}}^{(0)}(x^{(k)}, x_{(k)}, t), \tag{66.22}$$

since $p_{(k)} = \frac{\partial S^{(0)}(x,t)}{\partial x_{(k)}}$ (cf. (66.18)). The change of variables $x^{(k)} \to \xi_{(k)}$ given by (66.22) is called the Legendre transform in $x_{(k)}$, and $S^{(k)}(x^{(k)}, \xi_{(k)}, t)$ is called the Legendre transform of $S^{(0)}(x^{(k)}, x_{(k)}, t)$. □

It follows from (66.14) that

$$x_{(k)} = -S_{\xi_{(k)}}^{(k)}(x^{(k)}, \xi_{(k)}, t). \tag{66.22$'$}$$

Therefore, the change of variables given by (66.22′) is also the Legendre transform of $\xi_{(k)}$, and $S^{(0)}(x^{(k)}, x_{(k)}, t)$ is the Legendre transform of $S^{(k)}(x^{(k)}, \xi_{(k)}, t)$.

66.2. Asymptotic solutions.

Now we construct an asymptotic solution $u_k(x,t,\tau)$ in a neighborhood of the interval $[t_{k1}, t_{k2}]$.

We are looking for $u_k(x,t,\tau)$ in the form

(66.23)
$$u_k(x,t,\tau) = \Big(\frac{\tau}{2\pi}\Big)^{\frac{n_2}{2}} \int_{\mathbb{R}^{n_2}} a_k(x^{(k)}, \xi_{(k)}, t, \tau) e^{i\tau S^{(k)}(x^{(k)},\xi_{(k)},t) + i\tau x_{(k)}\cdot\xi_{(k)}} d\xi_{(k)},$$

where $S^{(k)}$ is the same as in (66.13), $t_{k1} - \varepsilon \le t \le t_{k2} + \varepsilon$, and

$$a_k(x^{(k)}, \xi_{(k)}, t, \tau) = \sum_{p=0}^{N} a_{kp}(x^{(k)}, \xi_{(k)}, t)\tau^{-p} + O\Big(\frac{1}{\tau^{N+1}}\Big). \tag{66.23$'$}$$

Substituting (66.23) into (64.4), we derive equations for a_{kp} similar to (64.9), (64.10), (64.11).

The following remark is important.

Consider an integral of the form

$$\int_{\mathbb{R}^{n_2}} b(x^{(k)}, x_{(k)}, \xi_{(k)}, t) e^{i\tau S^{(k)}(x^{(k)},\xi_{(k)},t) + i\tau x_{(k)}\cdot\xi_{(k)}} d\xi_{(k)}. \tag{66.24}$$

From the Taylor expansion of $b(x^{(k)}, x_{(k)}, \xi_{(k)}, t)$ at the point $x_{(k)} = -S^{(k)}_{\xi_{(k)}}(x^{(k)}, \xi_{(k)}, t)$we obtain

$$b(x^{(k)}, x_{(k)}, \xi_{(k)}, t) = \sum_{|p|=0}^{N} \frac{1}{p!} \frac{\partial^p b(x^{(k)}, -S^{(k)}_{\xi_{(k)}}, \xi_{(k)}, t)}{\partial x^p_{(k)}} (x_{(k)} + S^{(k)}_{\xi_{(k)}})^p + O\left((x_{(k)} + S^{(k)}_{\xi_{(k)}})^{N+1}\right). \tag{66.25}$$

Using the identity

$$(S^{(k)}_{\xi_{(k)}} + x_{(k)}) e^{i\tau(S^{(k)} + x_{(k)}\cdot\xi_{(k)})} = \frac{1}{i\tau}\frac{\partial}{\partial\xi_{(k)}} e^{i\tau(S^{(k)} + x^{(k)}\cdot\xi_{(k)})}$$

and integrating by parts with respect to $\xi_{(k)}$, we get

$$\int_{\mathbb{R}^{n_2}} b(x^{(k)}, x_{(k)}, \xi_{(k)}, t) e^{i\tau(S^{(k)} + x^{(k)}\cdot\xi_{(k)})} d\xi_{(k)} = \int_{\mathbb{R}^{n_2}} \sum_{r=0}^{N} \frac{1}{\tau^r} b_r(x^{(k)}, \xi_{(k)}, t) e^{i\tau(S^{(k)} + x_{(k)}\cdot\xi_{(k)})} d\xi_{(k)} + O\Big(\frac{1}{\tau^{N+1}}\Big), \tag{66.26}$$

where

$$b_0(x^{(k)}, \xi_{(k)}, t) = b(x^{(k)}, -S^{(k)}_{\xi_{(k)}}(x^{(k)}, \xi_{(k)}, t), \xi_{(k)}, t). \tag{66.27}$$

Note that the contribution of the set $\{\xi_{(k)} : |S^{(k)}_{\xi_{(k)}} - x_{(k)}| > \delta\}$ is $O(\frac{1}{\tau^N})$, $\forall N$.

Therefore, using the expansion (66.26), we see that after substituting (66.23) into (64.4), the term of the highest order with respect to τ has the form:

$$\tau^2 \int_{\mathbb{R}^{n_2}} H_0(x^{(k)}, -S^{(k)}_{\xi_{(k)}}(x^{(k)},\xi_{(k)},t),t,S^{(k)}_{x^{(k)}}(x^{(k)},\xi_{(k)},t),\xi_{(k)},S^{(k)}_t)$$
$$\times\, a_0(x^{(k)},\xi_{(k)},t)e^{i\tau(S^{(k)}+x_{(k)}\cdot\xi_{(k)})}d\xi_{(k)}.$$

Since $S^{(k)}(x^{(k)},\xi_{(k)},t)$ satisfies the eikonal equation (66.16) and $\lambda(x,t,\xi)$ is either $\lambda_1(x,t,\xi)$ or $\lambda_2(x,t,\xi)$ (cf. (64.2)), we get

$$H_0(x^{(k)}, -S^{(k)}_{\xi_{(k)}}, t, S^{(k)}_{x^{(k)}}, \xi_{(k)}, S^{(k)}_t) = 0. \tag{66.28}$$

Now we derive the equation for $a_{k0}(x^{(k)},\xi_{(k)},t)$ on $(t_{k1}-\varepsilon, t_{k2}+\varepsilon)$. Collecting all terms of order τ and using (66.25), we get (cf. (64.10)):

$$\begin{aligned}(66.29)\quad & -iH_{0\xi_0}(x^{(k)},-S^{(k)}_{\xi_{(k)}},t,S^{(k)}_{x^{(k)}},\xi_{(k)},S^{(k)}_t)\frac{\partial a_{k0}(x^{(k)},\xi_{(k)},t)}{\partial t} \\ & -iH_{0\xi^{(k)}}(x^{(k)},-S^{(k)}_{\xi_{(k)}},t,S^{(k)}_{x^{(k)}},\xi_{(k)},S^{(k)}_t)\cdot\frac{\partial a_{k0}(x^{(k)},\xi_{(k)},t)}{\partial x^{(k)}} \\ & -i\left(H_0(x^{(k)},-S^{(k)}_{\xi_{(k)}},t,-i\frac{\partial}{\partial x^{(k)}},0,-i\frac{\partial}{\partial t})S^{(k)}\right)a_{k0} \\ & +(ib_{(k)}\cdot\xi_{(k)}+ib^{(k)}(x^{(k)},-S^{(k)}_{\xi_{(k)}},t)\cdot S^{(k)}_{x^{(k)}})a_{k0}+ib_0S^{(k)}_t a_{k0} \\ & +i\frac{\partial}{\partial\xi_{(k)}}\left(H_{0x_{(k)}}(x^{(k)},-S^{(k)}_{\xi_{(k)}},t,S^{(k)}_{x^{(k)}},\xi_{(k)},S^{(k)}_t)a_{k0}\right)=0.\end{aligned}$$

Note that the last term in (66.29) comes from the Taylor expansion of $H_0(x^{(k)},x_{(k)},S^{(k)}_{x^{(k)}},\xi_{(k)},S^{(k)}_t)$ at $x_{(k)}=-S^{(k)}_{\xi_{(k)}}$ and integration by parts with respect to $\xi_{(k)}$.

Substitute $x^{(k)}=x^{(k)}(t,y)$, $\xi_{(k)}=\xi_{(k)}(t,y)$ into (66.29). As in (64.20), we have

$$\frac{dx^{(k)}}{dt}=-\lambda_{\xi^{(k)}}=H_{0\xi^{(k)}}H_{0\xi_0}^{-1},\qquad \frac{d\xi_{(k)}}{dt}=\lambda_{x_{(k)}}=-H_{0x_{(k)}}H_{0\xi_0}^{-1}. \tag{66.30}$$

Note that $H_{0\xi_0}=2S^{(k)}_t\neq 0$. Therefore (cf. (64.33)):

(66.31)
$$H_{0\xi_0}\frac{d}{dt}a_{k0}(x^{(k)}(t,y),\xi_{(k)}(t,y),t)=H_{0\xi_0}\frac{\partial a_{k0}}{\partial t}+H_{0\xi^{(k)}}\cdot\frac{\partial a_{k0}}{\partial x^{(k)}}-H_{0x_{(k)}}\cdot\frac{\partial a_{k0}}{\partial\xi_{(k)}}.$$

Hence the equation (66.29) in the (y,t)-coordinates has the form:

$$\frac{\partial\tilde a_{k0}(y,t)}{\partial t}+M_k(y,t)\tilde a_{k0}(t,y)=0, \tag{66.32}$$

where $\tilde a_{k0}(y,t)=a_{k0}(x^{(k)}(t,y),\xi_{(k)}(t,y),t)$, and $M_k(y,t)$ are similar to (64.34).

Analogously we obtain that the equations for

$$\tilde{a}_{kp}(y,t) = a_{kp}(x^{(k)}(t,y), \xi_{(k)}(t,y), t)$$

have the form

$$\frac{d\tilde{a}_{kp}(y,t)}{dt} + M_k(y,t)\tilde{a}_{kp}(y,t) = f_p(t,y), \quad p \geq 1, \tag{66.33}$$

where $f_p(y,t)$ depends on $a_{kr}, \quad 0 \leq r \leq p-1$.

Returning to the $(x^{(k)}, \xi_{(k)}, t)$-coordinates and using (66.28), (66.32), and (66.33), we get:

(66.34)

$$H\left(x, t, -i\frac{\partial}{\partial x}, -i\frac{\partial}{\partial t}\right)u_k = O\left(\frac{1}{\tau^N}\right) \text{ for } t_{k1} - \varepsilon \leq t \leq t_{k2} + \varepsilon, \ 1 \leq k \leq m.$$

Let u_0 be the solution obtained in §64 on $[0, t_{11} + \varepsilon]$ and u_1 the solution of the form (66.23) on $[t_{11} - \varepsilon, t_{12} + \varepsilon]$, where $\varepsilon > 0$ is small.

We will match these solutions at $t = t_{11}$. When $t \in [t_{11} - \varepsilon, t_{11} + \varepsilon]$, both (x,t) and $(x^{(1)}, \xi_{(1)}, t)$ are coordinates on Λ.

In particular, $|S_{\xi_{(1)}\xi_{(1)}}| \neq 0$ since $S_{\xi_{(1)}\xi_{(1)}} = -\frac{\partial x_{(1)}}{\partial \xi_{(1)}}$. Apply the stationary phase lemma to (66.23). The critical point $\xi_{(1)} = p_{(1)}(x,t)$ satisfies

$$S^{(1)}_{\xi_{(1)}}(x^{(1)}, p_{(1)}(x,t), t) + x_{(1)} = 0, \tag{66.35}$$

i.e., $x^{(1)}$ is the Legendre transform of $\xi_{(1)}$. At the critical point we have

$$S^{(1)}(x^{(1)}, p_{(1)}(x,t), t) + x_{(1)} \cdot p_{(1)}(x,t) = S_0(x,t)$$

(cf. (66.21)). The Hessian is

$$-S^{(1)}_{\xi_{(1)}\xi_{(1)}}(x^{(1)}, p_{(1)}(x,t), t) = \frac{\partial x_{(1)}}{\partial \xi_{(1)}} = \left(\frac{\partial \xi_{(1)}}{\partial x_{(1)}}\right)^{-1} = \left(\frac{\partial^2 S_0}{\partial x_{(1)}^2}\right)^{-1}. \tag{66.36}$$

Therefore $u_1(x,t,\tau)$ has the following form on $(t_{11} - \varepsilon, t_{11} + \varepsilon)$, after applying the stationary phase lemma:

$$u_1(x,t,\tau) = e^{i\tau S_0(x,t)} \sum_{p=0}^{N} \frac{c_{1p}(x,t)}{\tau^p} + O\left(\frac{1}{\tau^{N+1}}\right). \tag{66.37}$$

In particular, the principal term in (66.37) has the form:

$$u_1(x,t,\tau) = \frac{e^{i\tau S_0(x,t)}\, e^{i\frac{\pi}{4}\operatorname{sgn} S^{(1)}_{\xi_{(1)}\xi_{(1)}}}}{|S^{(1)}_{\xi_{(1)}\xi_{(1)}}(x^{(1)}, p_{(1)}, t|^{\frac{1}{2}}} a_{10}(x^{(1)}, p_{(1)}, t) + O\left(\frac{1}{\tau}\right). \tag{66.38}$$

On the other hand, we have the asymptotic solution $u_0(x,t,\tau)$ of (64.4) on the interval $[0, t_{11}+\varepsilon]$ having the form:

$$u_0(x,t,\tau) = e^{i\tau S_0(x)} \sum_{p=0}^{N} \frac{a_{0p}(x,t)}{\tau^p} + O\left(\frac{1}{\tau^{N+1}}\right). \tag{66.39}$$

Note that c_{1p} and a_{0p} satisfy the same first order differential equations on $[t_{11}-\varepsilon, t_{11}+\varepsilon]$. Therefore a_{0p} and c_{1p} will be equal if we match c_{1p} with a_{0p} at $t = t_{11}$. In particular, we have

$$a_{00}(x,t) = a_{10}\frac{e^{i\frac{\pi}{4}\sigma(y,t_{11})}}{|S^{(1)}_{\xi_{(1)}\xi_{(1)}}|^{\frac{1}{2}}} \quad \text{on} \quad [t_{11}-\varepsilon, t_{11}+\varepsilon], \tag{66.40}$$

where $\sigma = \operatorname{sgn} S^{(1)}_{\xi_{(1)}\xi_{(1)}}$.

Knowing $c_{1p}(x,t)$ on $[t_{11}-\varepsilon, t_{11}+\varepsilon]$, we can recover $a_{1p}(x^{(1)}, \xi_{(1)}, t)$ on $[t_{11}-\varepsilon, t_{11}+\varepsilon]$. Since $a_{1p}(x^{(1)}, \xi_{(1)}, t)$ satisfies differential equations of the form (66.32), (66.33) on $[t_{11}-\varepsilon, t_{12}+\varepsilon]$, the solution $a_{1p}(x^{(1)}, \xi_{(1)}, t)$ on a larger interval $[t_{11}-\varepsilon, t_{12}+\varepsilon]$ is determined by its values on a smaller interval $[t_{11}-\varepsilon, t_{11}+\varepsilon]$. Therefore $u(x,t,\tau) = u_0(x,t,\tau)$ on $[0,t_{11}]$, and $u(x,t) = u_1(x,t,\tau)$ on $[t_{11}, t_{12}]$ is smooth on $[0,t_{12}]$ and satisfies $Hu = O(\frac{1}{\tau^{N-1}})$ on $[0, t_{12}]$.

Analogously we match asymptotic solutions u_k of the form (66.23) and asymptotic solutions u_0 of the form (66.7) on each pair of intervals $[t_{k1}, t_{k2}]$ and $[t_{k2}, t_{k+1,1}]$ respectively, $k \geq 1$. We obtain a smooth asymptotic solution $u(x,t,\tau)$ on $[0,T]$ such that

$$Hu = h,$$

where $h = O(\frac{1}{\tau^{N-1}})$. As in §64, let $v(x,t)$ be the solution of the Cauchy problem

$$Hv = -h, \quad v|_{t=0} = \frac{\partial v}{\partial t}\Big|_{t=0} = 0.$$

Then $v = O(\frac{1}{\tau^{N-\frac{n}{2}-2}})$ and $u+v$ will be an exact solution of (64.4). □

66.3. The Maslov index.

We find an explicit form of the principal term of $u(x,t,\tau)$ on $[t_{k2}, t_{k+1,1}]$, where $t_{k1} < t_k < t_{k2}$, the t_k being the focal points on γ_{y_0}, $1 \leq k \leq m$.

The principal term of $u(x,t,\tau)$ (cf. (66.39) and (64.43)) when $t \in [t_{11}-\varepsilon, t_{11}+\varepsilon]$ can be written as

$$\hat{a}_{00}(y,t) = \frac{\hat{a}_{00}(y,0)}{\left|\frac{\partial x(t,y)}{\partial y}\right|^{\frac{1}{2}}} \exp \int_0^t M_0(y,t')dt'. \tag{66.41}$$

Equating (66.40) and (66.41), we get

$$(66.42)\qquad \tilde{a}_{10}(y,t) = \frac{\hat{a}_{00}(0,y)\exp\int_0^t M_0(t',y)dt'}{\left|\frac{\partial x(t,y)}{\partial y}\right|^{\frac{1}{2}}} \left|\frac{\partial x_{(1)}}{\partial \xi_{(1)}}\right|^{\frac{1}{2}} e^{-i\frac{\pi}{4}\sigma(y,t_{11})},$$

where $\sigma(y,t_{11}) = \operatorname{sgn} S^{(1)}_{\xi_{(1)}\xi_{(1)}}$, and $\tilde{a}_{10}(y,t) = a_{10}(x^{(k)}(y,t),\xi^{(k)}(y,t),t)$.

Note that

$$(66.43)\quad \left|\frac{\partial x(t,y)}{\partial y}\right| = \left|\frac{\partial(x^{(1)},x_{(1)})}{\partial(x^{(1)},\xi_{(1)})}\right| \left|\frac{\partial(x_{(1)},\xi_{(1)})}{\partial y}\right| = \left|\frac{\partial x_{(1)}}{\partial \xi_{(1)}}\right| \left|\frac{\partial(x^{(1)},\xi_{(1)})}{\partial y}\right|.$$

Therefore

$$(66.44)\qquad \tilde{a}_{10}(y,t) = \frac{\hat{a}_{00}(0,y)\exp\int_0^t M_0(t',y)dt'}{\left|\frac{\partial(x^{(1)},\xi_{(1)})}{\partial y}\right|^{\frac{1}{2}}} e^{-i\frac{\pi}{4}\sigma(y,t_{11})}.$$

Note that (66.44) holds on $[t_{11}-\varepsilon, t_{12}+\varepsilon]$.

Consider now the interval $[t_{12}-\varepsilon, t_{12}+\varepsilon]$. We have two solutions on this interval: u_1 and the second solution

$$u_0^{(1)} = e^{i\tau S^{(0)}(x,t)} a_0^{(1)}(x,t) + O\left(\frac{1}{\tau}\right),$$

where

$$(66.45)\quad \hat{a}_0^{(1)} = \frac{c(y)\hat{a}_{00}(0,y)\exp\int_0^t M_0(t',y)dt'}{\left|\frac{\partial x(t,y)}{\partial y}\right|^{\frac{1}{2}}}, \qquad t_{12}-\varepsilon < t < t_{21}+\varepsilon,$$

and $c(y)$ is unknown. Note that $M_0(y,t)$ is defined and smooth for all $t \in \mathbb{R}$ and $\hat{a}_0^{(1)}(t,y)$ satisfies the equation (64.34). Therefore $\hat{a}_0^{(1)}(t,y)$ can be represented in the form (66.45) on any interval that does not contain focal points.

After applying the stationary phase method to u_1 on $(t_{12}-\varepsilon, t_{12}+\varepsilon)$, we get (66.38). Matching u_1 and $u_0^{(1)}$ at $t = t_{12}$, we obtain, as in (66.42):

$$(66.46)\qquad \frac{\tilde{a}_{10}(t,y)e^{i\frac{\pi}{4}\sigma(y,t_{12})}}{\left|\frac{\partial x_{(1)}}{\partial \xi_{(1)}}\right|^{\frac{1}{2}}} = \frac{c(y)\hat{a}_{00}(0,y)\exp\int_0^t M_0(t',y)dt'}{\left|\frac{\partial x(t,y)}{\partial y}\right|^{\frac{1}{2}}}.$$

Substituting (66.44) into (66.46) and taking into account (66.43), we get

$$(66.47)\qquad c(y) = e^{i\frac{\pi}{4}(\sigma(y,t_{12})-\sigma(y,t_{11}))}.$$

Therefore the asymptotic solution $u_0^{(1)}$ has the form

$$(66.48)\quad u_0^{(1)}(x(t,y),y) = \frac{e^{i\tau S_0(x(t,y),t)}\hat{a}_{00}(0,y)\exp\int_0^t M_0(t',y)dt' e^{i\frac{\pi}{4}(\sigma(y,t_{12})-\sigma(y,t_{11}))}}{\left|\frac{\partial x(t,y)}{\partial y}\right|^{\frac{1}{2}}} + O\left(\frac{1}{\tau}\right)$$

on $(t_{12}-\varepsilon, t_{21}+\varepsilon)$. Analogously, for $t \in [t_{m2}, t_{m+1,1}]$ we have

$$(66.49)\quad u(x(t,y),t) = \frac{e^{i\tau S_0(x(t,y),t)}\hat{a}_{00}(0,y)\exp\int_0^t M_0(t',y)dt' e^{i\frac{\pi}{4}\sum_{k=1}^m(\sigma(y,t_{k2})-\sigma(y,t_{k1}))}}{\left|\frac{\partial x(t,y)}{\partial y}\right|^{\frac{1}{2}}} + O\left(\frac{1}{\tau}\right),$$

where m is the number of focal points on $\gamma_{y0} \cap (0,t)$.

The sum $\sum_{k=1}^m(\sigma(y_0,t_{k2}) - \sigma(y_0,t_{k1}))$ over the focal points on γ_{y0} is called the Maslov index of γ_{y0}.

Example 66.1. Consider the solution $u(x,t,\tau)$ of (64.4) of the form (66.23). At the focal point $x^{(1)} = x(t,y_0)$, $t = t_1$ we have $\det\left[\frac{\partial x(t,y)}{\partial y}\right] = 0$. Let $x' = (x_1,\dots,x_{n-1})$. We assume that $\det\left[\frac{\partial(x',\xi_n)}{\partial y}\right] \neq 0$, i.e., (x',ξ_n,t) are the coordinates on Λ near $t = t_1$. Let $S(x',\xi_n,t)$ be the generating function corresponding to the coordinates (x',ξ_n,t). Analogously to (66.43) we have:

$$(66.50)\qquad \left|\frac{\partial x(t,y)}{\partial y}\right| = \left|\frac{\partial(x',x_n)}{\partial(x',\xi_n)}\right|\ \left|\frac{\partial(x',\xi_n)}{\partial y}\right| = \left|\frac{\partial x_n}{\partial \xi_n}\right|\ \left|\frac{\partial(x',\xi_n)}{\partial y}\right|.$$

Therefore

$$(66.51)\qquad -S_{\xi_n^2} = \frac{\partial x_n}{\partial \xi_n} = 0 \quad \text{at the focal point}\ \ (x^{(1)},t_1).$$

We assume that

$$(66.52)\qquad S_{\xi_n^3}(x',\xi_n,t) \neq 0 \quad \text{at} \quad (x^{(1)},t_1),$$

i.e., when $x' = x'(t,y_0)$, $\xi_n = \xi_n(t,y_0)$, $t = t_1$.

The asymptotic solution (66.23) has the following form when S is the generating function:

$$(66.53)\qquad u(x,t,\tau) = \left(\frac{\tau}{2\pi}\right)^{\frac{1}{2}} \int_{-\infty}^{\infty} a(x',\xi_n,t,\tau)e^{i\tau S(x',\xi_n,t)+i\tau x_n\xi_n}d\xi_n,$$

where $a(x',\xi_n,t)=\sum_{p=0}^N a_p(x',\xi_n,t)\tau^{-p}$, and $\operatorname{supp} a_0(x',\xi_n,0)$ is contained in a small neighborhood of the focal point.

It follows from (66.51), (66.52) that

$$S_{\xi_n^2}(x',\xi_n,t)=(\xi_n-\alpha(x',t))c_1(x',\xi_n,t), \tag{66.54}$$

where $\xi_n-\alpha(x',t)=0$ is the equation for the focal points (caustics) near $(x^{(1)},t_1)$ and $c_1(x',\xi_n,t)\neq 0$ near $(x^{(1)},t_1)$.

By the Taylor formula we have

$$\begin{aligned} S(x',\xi_n,t) &= S(x',\alpha(x',t),t)+S_{\xi_n}(x',\alpha(x',t),t)(\xi_n-\alpha(x',t)) \\ &\quad+\frac{1}{3!}S_{\xi_n^3}(x',\alpha,t)(\xi_n-\alpha(x',t))^3+O\left((\xi_n-\alpha)^4\right). \end{aligned} \tag{66.55}$$

Make the change of variables in (66.53),

$$\eta_n=\tau^{\frac13}\left(\frac12 S_{\xi_n^3}(x',\alpha,t)\right)^{\frac13}(\xi_n-\alpha). \tag{66.56}$$

Then the exponential in (66.53) has the form:

$$\exp i\tau(S(x',\alpha,t)+x_n\alpha)\exp\left(i\frac{\eta_n^3}{3}+i\tau^{\frac23}\eta_n c_2(x,t)+O\Big(\frac{\eta_n^4}{\tau^{\frac13}}\Big)\right), \tag{66.57}$$

where

$$c_2(x,t)=\left(\frac12 S_{\xi_n^3}(x',\alpha,t)\right)^{-\frac13}(x_n+S_{\xi_n}(x',\alpha,t)). \tag{66.58}$$

The principal term in (66.53) has the form:
(66.59)

$$\frac{(2\pi\tau)^{\frac12}}{\tau^{\frac13}\left(\frac12 S_{\xi_n^3}(x',\alpha,t)\right)^{\frac13}}a_0(x',\alpha(x',t),t)e^{i\tau(S(x',\alpha,t)+x_n\alpha(x',t))}\operatorname{Ai}(\tau^{\frac23}c_2(x,t)),$$

where $\operatorname{Ai}(z)$ is the Airy function:

$$\operatorname{Ai}(z)=\frac{1}{2\pi}\int_{-\infty}^{\infty}\exp\left(i\frac13\eta_n^3+iz\eta_n\right)d\eta_n.$$

It follows from the asymptotics of the Airy function that

$$|\operatorname{Ai}(z)|\le\frac{C}{(1+|z|^{\frac14})},\quad z=\tau^{\frac23}c_2(x,t). \tag{66.60}$$

Therefore we get the following estimate of $u(x,t,\tau)$ near the focal set (caustics):

$$|u(x,t,\tau)|\le\frac{C\tau^{\frac16}}{1+\tau^{\frac16}|x_n+S_{\xi_n}(x',\alpha(x',t),t)|^{\frac14}}. \tag{66.61}$$

67. Oblique derivative problem

67.1. Reduction to the boundary.

Let Ω be a smooth bounded domain in $\mathbb{R}^n, n \geq 3$. Consider a second order elliptic equation in Ω:

$$(67.1)\quad A(x,D)u \stackrel{\text{def}}{=} -\sum_{j,k=1}^{n} g^{jk}(x)u_{x_jx_k} + \sum_{j=1}^{n} b_j(x)u_{x_j} + c(x)u = 0, \quad x \in \Omega,$$

where $[g^{jk}(x)]_{jk=1}^n$ is a positive definite matrix. In §57 we constructed a parametrix for the Dirichlet problem, i.e., an operator Rv acting from $H_s(\partial\Omega)$ to $H_{s+\frac{1}{2}}(\Omega)$ such that (cf. (57.19))

$$A(x,D)Rv = T_{-N-1}v \quad \text{in} \quad \Omega,$$

$$Rv|_{\partial\Omega} = v + T_{-N-1}^{(1)}v,$$

where $\operatorname{ord} T_{-N-1} \leq -N-1$, $\operatorname{ord} T_{-N-1}^{(1)} \leq -N-1$.

For simplicity we assume that the Dirichlet problem for (67.1) has a unique solution. Then, as in §57, we can add to R an operator of order $-N$ such that the new operator R_1 satisfies

$$(67.2)\qquad A(x,D)R_1v = 0, \quad x \in \Omega,$$

$$(67.3)\qquad R_1v|_{\partial\Omega} = v.$$

Consider a boundary value problem

$$(67.4)\qquad B(x,D)u|_{\partial\Omega} = g$$

for the equation (67.1) , where $\operatorname{ord} B = r$. The substitution $u = R_1v$ reduces the boundary value problem (67.1), (67.4) to the solution of a pseudodifferential equation on $\partial\Omega$:

$$(67.5)\qquad BR_1v|_{\partial\Omega} = g, \quad x \in \partial\Omega.$$

We shall find the principal symbol of this operator. Let $x_0 \in \partial\Omega$ and let U_0 be a neighborhood of x_0. As in §57, we can find a local system of coordinates (x', x_n) in U_0 such that $x_n = 0$ is the equation of $\partial\Omega \cap U_0, x_n > 0$ in $U_0 \cap \Omega$ and $A(x,D)$ has the following form in $U_0 \cap \Omega$:

$$(67.6)\qquad \frac{\partial^2 u}{\partial x_n^2} + L_2u + L_1u = 0,$$

where $L_2 = \sum_{j,k=1}^{n-1} a_{jk}(x',x_n)u_{x_jx_k}$, $[a_{jk}(x',x_n)]_{j,k=1}^{n-1}$ is a positive definite matrix for $x_n \geq 0$, $\operatorname{ord} L_1 \leq 1$, and $x' = (x_1,\ldots,x_{n-1})$. The principal part

R_{10} of the operator R_1 has the following form in the (x', x_n)-coordinates (cf. §57):

$$R_{10}v = \frac{1}{(2\pi)^{(n-1)}} \int_{\mathbb{R}^{n-1}} e^{-x_n\sigma(x',x_n,\xi')}\tilde{g}(\xi')e^{ix'\cdot\xi'}\,d\xi', \tag{67.7}$$

where $\sigma(x,\xi') = \sqrt{\sum_{j,k=1}^{n-1} a_{jk}(x)\xi_j\xi_k}$.

Let $B_0(x', 0, D', D_n)$ be the principal part of $B(x, D)$ in the coordinates (x', x_n). The principal symbol $b_0(x', \xi,)$ of $BRg|_{x_n=0}$ has the following form in $U_0 \cap \partial\Omega$:

$$b_0(x', \xi') = B_0(x', 0, \xi', i\sigma(x', 0, \xi')), \ \operatorname{ord} b_0 = r. \tag{67.8}$$

Globally on $\partial\Omega$, the equation (67.5) is a ψdo

$$b(x', D')v = g \tag{67.9}$$

with the principal symbol $b_0(x', \xi'_{x'})$, where $(x', \xi'_{x'}) \in T_0^*(\partial\Omega)$. The ellipticity of the boundary operator $B(x, D)$ means that $b(x', D')$ is an elliptic operator on $\partial\Omega$, i.e.,

$$b_0(x', \xi'_{x'}) \neq 0 \quad \text{on} \quad T_0^*(\partial\Omega). \tag{67.10}$$

Let γ be a closed $(n-2)$-dimensional surface on $\partial\Omega$. We consider the case where $b_0(x', \xi'_{x'}) \neq 0$ in $\partial\Omega \setminus \gamma$ and

$$b_0(x'_0, \xi'_0) = 0 \quad \text{for all} \quad x'_0 \in \gamma \quad \text{and some} \quad \xi'_0 \neq 0. \tag{67.11}$$

Moreover, we assume that the gradient

$$\frac{\partial b_0(x'_0, \xi'_0)}{\partial \xi'} \quad \text{is not tangent to} \quad \gamma. \tag{67.12}$$

67.2. Formulation of the oblique derivative problem.

Our main example of a nonelliptic boundary value problem is the oblique derivative problem for the Laplace equation, i.e.,

$$\Delta u = 0 \quad \text{in} \quad \Omega, \tag{67.13}$$

$$\sum_{k=1}^{n} l_k(x)\frac{\partial u}{\partial x_k} = g \quad \text{on} \quad \partial\Omega, \tag{67.14}$$

where $l(x) = (l_1(x), \ldots, l_n(x)) \neq (0, \ldots, 0)$ for all $x \in \partial\Omega$. If $l(x)$ is not tangent to $\partial\Omega$, then the boundary value problem (67.13), (67.14) is elliptic. We shall assume that $l(x)$ is tangent to $\partial\Omega$ when $x \in \gamma \subset \partial\Omega$, where γ is a closed smooth $(n-2)$-dimensional surface. The condition (65.12) means that $l(x)$ is not tangent to γ when $x \in \gamma$. □

In a neighborhood $U_0 \cap \partial\Omega$ of $x_0 \in \gamma$ we choose coordinates $x' = (x_1, \ldots, x_{n-1})$ in which $x_{n-1} = 0$ is the equation of γ. Suppose that

$b_0(x'', x_{n-1}, D'', D_{n-1})$ is the principal symbol b_0 in these coordinates. The conditions (67.11), (67.12) imply that

$$(67.15) \qquad b_0(x'', 0, \xi'', \xi_{n-1}) = 0, \qquad \frac{\partial b_0(x'', 0, \xi'', \xi_{n-1})}{\partial \xi_{n-1}} \neq 0 \text{ in } U_0 \cap \partial\Omega,$$

where $x'' = (x_1, \ldots, x_{n-2})$, $\xi'' = (\xi_1, \ldots, \xi_{n-2})$. In the case of the oblique derivative problem we have

$$(67.16) \qquad b_0(x'', x_{n-1}, \xi'', \xi_{n-1}) = -l_{n0}(x')|\xi'| + i\sum_{k=1}^{n-1} l_{k0}(x')\xi_k,$$

where $(l_{10}, \ldots, l_{n0})$ is the vector $(l_1, \ldots, l_n)$ in local coordinates in U_0. Therefore the condition (67.15) means

$$(67.17) \qquad l_{n0}(x'', 0) = 0, \quad l_{n-1,0}(x'', 0) \neq 0.$$

Note that $b_0(x', 0, \xi'', \xi_{n-1})$ is real analytic in (ξ'', ξ_{n-1}) when $(\xi'', \xi_{n-1}) \neq (0, 0)$. Therefore, applying the implicit function theorem, we get from (67.15) that

$$(67.18) \qquad b_0(x'', x_{n-1}, \xi', \xi_{n-1}) = (\xi_{n-1} - \lambda(x', \xi''))b_1(x', \xi'),$$

where $b_1(x', \xi') \neq 0$ in $U_0 \cap \partial\Omega$, $\operatorname{ord} b_1 = r - 1$, and $\lambda(x', \xi'') = \lambda_1(x', \xi'') + i\lambda_2(x', \xi'')$ is homogeneous of degree one in ξ'', $\lambda \in C^\infty$ when $\xi'' \neq 0$. Note that $\lambda_2 = \operatorname{Im} \lambda(x', \xi'') \neq 0$ when $x_{n-1} \neq 0$, and $\lambda_2(x', \xi'') = 0$ when $x_{n-1} = 0$ (cf. (67.15)). For simplicity we assume that

$$(67.19) \qquad \lambda_2(x', \xi'') = x_{n-1}^k c(x', \xi''),$$

where $k > 0$ is an integer and $c(x', \xi'') \neq 0$ in $U_0 \cap \partial\Omega$.

For the case of oblique derivative problem, condition (67.19) is equivalent to the condition that $l_{n0}(x') = x_{n-1}^k c(x')$, $c(x') \neq 0$ in $U_0 \cap \partial\Omega$. Condition (67.19) contains three cases:

a) $k = 2k'$ is even;

b) $k = 2k' + 1$ is odd and $c(x', \xi'') > 0$;

c) $k = 2k' + 1$ is odd and $c(x', \xi'') < 0$.

For the oblique derivative problem this means the following. Denote by γ_+ neighborhood of γ where the vector

$$l'(x') = (l_{10}(x'), \ldots, l_{n-1,0}(x')), \; x' \in \gamma,$$

points in, and by γ_- neighborhood of γ where $l(x')$ points out. Then

in case **a)**, $l_{n0}(x')$ has the same sign in γ_+ and γ_-;

in case **b)**, $l_{n0}(x') > 0$ in γ_+ and $l_{n0}(x') < 0$ in γ_-;

in case **c)**, $l_{n0}(x') < 0$ in γ_+ and is positive in γ_-.

Note that $l'(x')$ is the tangential component of the vector $l(x')$, $x' \in \partial\Omega$, and $l_{n0} > 0$ ($l_{n0} < 0$) means that $l(x')$ is pointed inward (outward) with respect to Ω.

67.3. Model problem.

We start with studying the following equation in $\mathbb{R}^{n-1}$:

$$\frac{\partial u(x'', x_{n-1})}{\partial x_n} + x_{n-1}^k \hat{c}(D'')u = f(x'', x_{n-1}), \tag{67.20}$$

where $c(\xi'')$ is a real-valued function of ξ'', homogeneous of degree one, $c(\xi'') \neq 0$ when $\xi'' \neq 0$, and $\hat{c}(\xi'') = c\big((1+\xi''|)\frac{\xi''}{|\xi''|}\big)$. Making the Fourier transform in x'', we get an ordinary differential equation in $\mathbb{R}^1$:

$$\frac{\partial \tilde{u}(\xi'', x_{n-1})}{\partial x_{n-1}} + x_{n-1}^k \hat{c}(\xi'')\tilde{u}(\xi'', x_{n-1}) = \tilde{f}(\xi'', x_{n-1}). \tag{67.21}$$

a) The case where k is even.

Proposition 67.1. *Let k be even. For any $f(x'', x_{n-1}) \in H_{0,\frac{k}{k+1}}(\mathbb{R}^{n-1})$, there exists a unique solution $u \in H_1(\mathbb{R}^{n-1})$ of the equation (67.20) and*

$$\|u\|_1 \leq C\|f\|_{0,\frac{k}{k+1}}. \tag{67.22}$$

Here $H_{0,\frac{k}{k+1}}(\mathbb{R}^n)$ is the Sobolev space with the norm

$$\|f\|_{0,\frac{k}{k+1}}^2 = \int_{\mathbb{R}^{n-1}} (1+|\xi''|)^{\frac{2k}{k+1}} |\tilde{f}(\xi'', \xi_{n-1})|^2 d\xi'' d\xi_{n-1}.$$

Proof: Suppose for definiteness that $\hat{c}(\xi'') > 0$. Then the unique solution of (67.21) has the form

$$\tilde{u}(\xi'', x_{n-1}) = \int_{-\infty}^{x_{n-1}} e^{-\frac{x_{n-1}^{k+1} - y_{n-1}^{k+1}}{k+1}\hat{c}(\xi'')} \tilde{f}(\xi'', y_{n-1}) dy_{n-1}.$$

Using the Cauchy-Schwartz inequality, we have

$$|\tilde{u}(\xi'', x_{n-1})|^2 \leq \int_{-\infty}^{x_{n-1}} e^{-\frac{x_{n-1}^{k+1} - y_{n-1}^{k+1}}{k+1}\hat{c}(\xi'')} dy_{n-1} \times \int_{-\infty}^{x_{n-1}} e^{-\frac{x_{n-1}^{k+1} - y_{n-1}^{k+1}}{k+1}\hat{c}(\xi'')} |\tilde{f}(\xi'', y_{n-1})|^2 dy_{n-1}. \tag{67.23}$$

Making the change of variables $\hat{x}_{n-1} = \hat{c}^{\frac{1}{k+1}} x_{n-1}$, $\hat{y}_{n-1} = \hat{c}^{\frac{1}{k+1}} y_{n-1}$, we get

$$\int_{-\infty}^{x_{n-1}} e^{-\frac{x_{n-1}^{k+1} - y_{n-1}^{k+1}}{k+1}\hat{c}(\xi'')} dy_n = \hat{c}^{-\frac{1}{k+1}} \int_{-\infty}^{\hat{x}_{n-1}} e^{-\frac{\hat{x}_{n-1}^{k+1} - \hat{y}_{n-1}^{k+1}}{k+1}} d\hat{y}_{n-1} \leq C\hat{c}^{-\frac{1}{k+1}}. \tag{67.24}$$

Analogously,

$$\int_{y_n}^{\infty} e^{-\frac{x_{n-1}^{k+1}-y_{n-1}^{k+1}}{k+1}\hat{c}(\xi'')} dx_{n-1} \le C\hat{c}^{-\frac{1}{k+1}}. \tag{67.25}$$

Therefore, integrating (67.23) with respect to x_{n-1} and using (67.24) and (67.25), we get:

(67.26)

$$\int_{-\infty}^{\infty} |\tilde{u}(\xi'', x_{n-1})|^2 dx_{n-1} \le C \int_{-\infty}^{\infty} (|1+|\xi''|)^{-\frac{2}{k+1}} |\tilde{f}(\xi'', y_{n-1})|^2 dy_{n-1},$$

since $\hat{c} \ge C(1+|\xi''|)$.

The estimate (67.26) implies

$$\|u\|_{0,1} \le C\|f\|_{0,\frac{k}{k+1}}. \tag{67.27}$$

Analogously to (67.24)–(67.26) we get

$$\|x_{n-1}^k u\|_{0,1} \le C\|f\|_0. \tag{67.28}$$

Therefore using (67.21) we obtain:

$$\left\| \frac{\partial \tilde{u}}{\partial x_{n-1}} \right\|_0 \le \|x_{n-1}^k \hat{c}\tilde{u}\|_0 + \|\tilde{f}\|_0 \le C\|\tilde{f}\|_0. \tag{67.29}$$

Combining (67.29) and (67.27), we get (67.22).

Remark 67.1. In the case $\hat{c} < 0$ the unique solution of (67.21) has the form

$$\tilde{u}(\xi'', x_{n-1}) = \int_{x_{n-1}}^{\infty} e^{-\frac{x_{n-1}^{k+1}-y_{n-1}^{k+1}}{k+1}\hat{c}(\xi'')} \tilde{f}(\xi'', y_{n-1}) dy_{n-1},$$

and the proof of the estimate (67.22) is similar. Note also that multiplying (67.21) by $(1+|\xi''|)^{2s}$, we can prove that for any s,

$$\|u\|_{1,s} \le C\|f\|_{0,\frac{k}{k+1}+s}. \tag{67.30}$$

b) The case where k is odd and $c > 0$.

Proposition 67.2. *Consider the equation (67.20) in $\mathbb{R}^n$ assuming that k is odd and $\hat{c}(\xi'') > 0$. Then for any $g(x'') \in H_{s+1-\frac{1}{2(k+1)}}(\mathbb{R}^{n-2})$ and any $f \in H_{0,s+\frac{k}{k+1}}(\mathbb{R}^{n-1})$, there exists a unique solution $u(x') \in H_{1,s}(\mathbb{R}^{n-1})$ such that $u(x')$ satisfies (67.20) and*

$$u(x'', 0) = g(x'').$$

The following estimate holds:

$$\|u\|_{1,s} \le C\|f\|_{0,s+\frac{k}{k+1}} + C[g]_{s+1-\frac{1}{2(k+1)}}, \tag{67.31}$$

where $[g]_s$ is the norm in $H_s(\mathbb{R}^{n-2})$.

Proof: Let

(67.32)
$$\tilde K_2\tilde f(\xi'',x_{n-1})=\int_0^{x_{n-1}} e^{-\frac{x_{n-1}^{k+1}-y_{n-1}^{k+1}}{k+1}\hat c(\xi'')}\tilde f(\xi'',y_{n-1})dy_{n-1},\quad x_{n-1}\in\mathbb{R},$$

(67.33)
$$\tilde K_{20}\tilde g=e^{-\frac{x_{n-1}^{k+1}}{k+1}\hat c(\xi'')}\tilde g(\xi'').$$

Then $\tilde K_2\tilde f+\tilde K_2\tilde g$ is the general solution of (67.21) in $\mathbb{R}^1$ and

$$(\tilde K_2\tilde f+K_{20}\tilde g)|_{x_{n-1}=0}=\tilde g(\xi'').$$

We have, making the change of variables $x_{n-1}=|\hat c|^{-\frac{1}{k+1}}y_n$

$$\|K_{20}g\|_{0,s}^2\le C\int_{-\infty}^{\infty} e^{-\frac{2x_{n-1}^{k+1}\hat c}{k+1}}(1+|\xi''|)^{2s}|\tilde g(\xi'')|^2dx_{n-1}d\xi''$$

(67.34)
$$\le C[g]^2_{s-\frac{1}{2(k+1)}},$$

$$\left\|\frac{\partial}{\partial x_{n-1}}K_{20}g\right\|^2_{0,s}\le C[g]_{\frac{1}{2(k+1)}+s}.$$

Analogously to (67.30),

(67.35)
$$\|K_2f\|_{1,s}\le C\|f\|_{0,s+\frac{k}{k+1}}.$$

Here

(67.36)
$$K_2f+K_{20}f=\frac{1}{(2\pi)^{n-2}}\int_{\mathbb{R}^{n-2}}(\tilde K_2\tilde g+\tilde K_{20}\tilde g)e^{ix''\cdot\xi''}d\xi''.$$

Therefore we have (67.31). □

c) The case where k is odd and $c<0$. Let

(67.37)
$$(\tilde K_3\tilde f)(\xi',x_n)=\theta(-x_{n-1})\int_{-\infty}^{x_{n-1}} e^{-\frac{x_{n-1}^{k+1}-y_{n-1}^{k+1}}{k+1}\hat c(\xi'')}\tilde f(\xi'',y_{n-1})dy_{n-1}$$
$$-\theta(x_{n-1})\int_{x_{n-1}}^{\infty} e^{-\frac{x_{n-1}^{k+1}-y_{n-1}^{k+1}}{k+1}\hat c(\xi'')}\tilde f(\xi'',y_{n-1})dy_{n-1}.$$

Substituting into (67.21) we get:

(67.38)
$$\left(\frac{\partial u}{\partial x_{n-1}}+x_{n-1}^k\hat c(\xi'')\right)\tilde K_3\tilde f$$
$$=\tilde f(\xi'',x_{n-1})-\delta(x_{n-1})\int_{-\infty}^{\infty} e^{\frac{y_{n-1}^{k+1}}{k+1}\hat c}\tilde f(\xi'',y_{n-1})dy_{n-1}.$$

Therefore (67.21) has a solution iff

(67.39)
$$\int_{-\infty}^{\infty} e^{\frac{y_{n-1}^{k+1}}{k+1}\hat c}\tilde f(\xi'',y_{n-1})dy_{n-1}=0.$$

Instead of imposing the orthogonality condition (67.39) on f, we add "a coboundary" term to (67.21), i.e., we replace (67.20) (or (67.21)) by the equation:

$$\frac{\partial u}{\partial x_{n-1}} + x_n^k \hat{c}(D'')u(x') + \rho(x'')\delta(x_{n-1}) = f(x'), \quad x' \in \mathbb{R}^{n-1}. \tag{67.40}$$

In (67.40) the unknowns are u and ρ. It follows from (67.38) that

$$\tilde{\rho}(\xi'') = \int_{-\infty}^{\infty} e^{\frac{y_{n-1}^{k+1}}{k+1}\hat{c}(\xi'')} \tilde{f}(\xi'', y_{n-1})dy_{n-1}, \tag{67.41}$$

$$\tilde{u} = \tilde{K}_3\tilde{f}. \tag{67.42}$$

The advantage of changing the equation (67.20) to (67.40) in comparison with subjecting f to the orthogonality condition (67.39) is that "the coboundary" problem (67.40) can be generalized to the case of pseudodifferential operators with variable coefficients in a bounded domain.

Using the Cauchy-Schwartz inequality, we get from (67.41):

$$[\rho]_{s+\frac{1}{2(k+1)}} \le C\|f\|_{0,s}^2. \tag{67.43}$$

Estimating $\tilde{K}_3\tilde{f}$ as in (67.30), we get $\|K_3 f\|_{0,1} \le C\|f\|_{0,\frac{k}{k+1}}$.

We have proven the following proposition.

Proposition 67.3. *For any $f \in H_{0,s+\frac{k}{k+1}}(\mathbb{R}^{n-1})$, there exist a unique $\rho \in H_{s+1-\frac{1}{2(k+1)}}(\mathbb{R}^{n-2})$ and a unique $u \in H_{0,s+1}(\mathbb{R}^{n-1})$ that solve (67.40).*

67.4. First order differential equations with symbols depending on x'.

We generalize Propositions 67.1–67.3 to the case of the equation of the form

$$Lu \overset{\text{def}}{=} \frac{\partial u}{\partial x_{n-1}} + x_{n-1}^k c(x', D'')u + c_1(x', D'')u = f, \quad x' \in \mathbb{R}^{n-1}, \tag{67.44}$$

where $c(x', \xi'') \in S^1(\mathbb{R}^{n-1} \times \mathbb{R}^{n-2})$, $c_1(x', \xi'') \in S^0(\mathbb{R}^{n-1} \times \mathbb{R}^{n-2})$, and

$$|c(x', \xi'')| \ge c_0(1 + |\xi''|^2)^{\frac{1}{2}}. \tag{67.45}$$

a) The case where k is even and $c(x', \xi'') \neq 0$. Suppose, for definiteness, that $c > 0$. Denote

$$E(x_{n-1}, y_{n-1}, x'', \xi'') = \exp\left(-\int_{y_{n-1}}^{x_{n-1}} t^k c(x'', t, \xi'')dt\right). \tag{67.46}$$

Lemma 67.4. *Let $a(x', y_{n-1}, \xi'') \in S^m$ and let K be the operator*

$$(67.47)\quad Kf = \frac{1}{(2\pi)^{n-2}} \int_{\mathbb{R}^{n-2}} \int_{-\infty}^{x_{n-1}} a(x', y_{n-1}, \xi'') E(x_{n-1}, y_{n-1}, x'', \xi'') \times e^{ix''\cdot\xi''} \tilde{f}(\xi'', y_{n-1}) d\xi'' dy_{n-1}.$$

Then

$$(67.48)\qquad \|Kf\|_{1,s-1} \leq C\|f\|_{0,s+m-\frac{1}{k+1}}, \quad \forall s.$$

Proof: Denote

$$E_1 = \exp\left(-\int_{y_{n-1}}^{x_{n-1}} t^k\left(c(x'', t, \xi'') - \frac{c_0}{2}\Lambda''\right)dt\right) = E(x_{n-1}, y_{n-1}, x'', \xi'') \exp\left(\frac{x_{n-1}^{k+1} - y_{n-1}^{k+1}}{k+1}\frac{c_0}{2}\Lambda''\right),$$

where $\Lambda'' = (1 + |\xi''|^2)^{\frac{1}{2}}$ and c_0 is the same as in (67.45). Note that

$$(67.49)\qquad \left|\frac{\partial^{p+k} E_1}{\partial x''^p \partial \xi''^k}\right| \leq C_{pk} e^{-\frac{x_{n-1}^{k+1} - y_{n-1}^{k+1}}{k+1}\frac{c_0}{4}\Lambda''} \leq C_{pk} e^{-\frac{x_{n-1}^{k+1} - y_{n-1}^{k+1}}{k+1}\frac{c_0}{4}}.$$

Therefore $a(x', y_{n-1}, \xi'')E_1 \in S^m$ in (x'', ξ'') uniformly in (x_{n-1}, y_{n-1}). We have

$$Kf = \int_{-\infty}^{x_{n-1}} \frac{1}{(2\pi)^{n-2}} \int_{\mathbb{R}^{n-2}} a(x', y_{n-1}, \xi'') E_1(x_{n-1}, y_{n-1}, x'', \xi'') \times \tilde{f}_1(\xi'', y_{n-1}) e^{ix''\cdot\xi''} d\xi'' dy_{n-1},$$

where $\tilde{f}_1 = e^{(-\frac{x_{n-1}^{k+1} - y_{n-1}^{k+1}}{k+1}\frac{c_0}{2}\Lambda'')}\tilde{f}(\xi'', y_{n-1})$. Applying the theorem on the boundedness of ψdo's (cf. §40), we have:

$$[Kf(\cdot, x_{n-1})]_s \leq C \int_{-\infty}^{x_{n-1}} e^{(-\frac{x_{n-1}^{k+1} - y_{n-1}^{k+1}}{k+1}\frac{c_0}{4})} [f_1]_{s+m} dy_{n-1}.$$

By the Cauchy-Schwartz inequality,

$$\left(\int_{-\infty}^{x_{n-1}} e^{(-\frac{x_{n-1}^{k+1} - y_{n-1}^{k+1}}{k+1}\frac{c_0}{4})} [f_1]_{s+m} dy_{n-1}\right)^2 \leq C \int_{-\infty}^{x_n} [f_1]_{s+m}^2 dy_{n-1},$$

since $\int_{-\infty}^{x_{n-1}} e^{(-\frac{x_{n-1}^{k+1}-y_{n-1}^{k+1}}{k+1}\frac{c_0}{2})}dy_{n-1} \le C$. Therefore

$$\begin{aligned}\int_{-\infty}^{\infty} & [Kf(\cdot, x_{n-1})]_s^2 dx_{n-1} \\ & \le C \int_{-\infty}^{\infty}\int_{-\infty}^{x_{n-1}}\int_{\mathbb{R}^{n-2}} (1+|\xi''|)^{2s+2m} e^{(-\frac{x_{n-1}^{k+1}-y_{n-1}^{k+1}}{k+1}c_0\Lambda'')} \\ & \qquad \times |\tilde{f}(\xi'', y_{n-1})|^2 dy_{n-1} d\xi'' dx_{n-1} \\ & = C\int_{\mathbb{R}^{n-2}} (1+|\xi''|)^{2s+2m} \int_{-\infty}^{\infty}\int_{-\infty}^{x_{n-1}} e^{(-\frac{x_{n-1}^{k+1}-y_{n-1}^{k+1}}{k+1}c_0\Lambda'')} \\ & \qquad \times |\tilde{f}(\xi'', y_{n-1})|^2 dy_{n-1} dx_{n-1} d\xi''.\end{aligned}$$

Integrating with respect to x_{n-1}, and then with rspect to y_{n-1} (cf. (67.26)), we get:

$$\|Kf\|_{0,s}^2 \le C\|f\|_{0,s+m-\frac{1}{k+1}}^2, \quad \forall s.$$

Analogously to (67.29) we have

$$\left\|\frac{\partial}{\partial x_{n-1}}Kf\right\|_{0,s-1} \le C\|f\|_{0,s+m-\frac{1}{k+1}}.$$

Therefore we get (67.48). □

Lemma 67.5. *Suppose k is even and $c > 0$. Let $K_1 f$ be the operator* (67.47) *with $a = 1$. Then*

$$LK_1 = I + T^{(1)}_{-\frac{1}{k+1}}, \quad K_1 L = I + T^{(2)}_{-\frac{1}{k+1}}, \tag{67.50}$$

where $L = \frac{\partial}{\partial x_{n-1}} + x_{n-1}^k c(x', D'') + c_1(x', D'')$ and $T^{(j)}_{-\frac{1}{k+1}}$ are operators of order $-\frac{1}{k+1}$, $j = 1, 2$.

We call an operator of negative order a smoothing operator. When (67.50) holds, we call K_1 the left and right regularizer for L.

Proof: Applying L to K_1 we get

$$LK_1 = I + T^{(1)}_{-\frac{1}{k+1}},$$

where $T^{(1)}_{-\frac{1}{k+1}}$ is an operator of the form (67.47) with a having order zero. Therefore, by (67.48), $T^{(1)}_{-\frac{1}{k+1}}$ is a smoothing operator.

Analogously,

$$K_1 L = I + T^{(2)}_{-\frac{1}{k+1}},$$

where $T^{(2)}_{-\frac{1}{k+1}}$ has the form $K_1 a_1(x', D'')$, $\operatorname{ord} a_1 = 0$. Therefore we have that $T^{(2)}_{-\frac{1}{k+1}}$ is smoothing of order $-\frac{1}{k+1}$.

b) The case where k is odd and $c(x', \xi'') > 0$. We consider the solution $u(x')$ of (67.44) subject to the boundary condition:

$$P(x', D')u|_{x_{n-1}=0} = g, \tag{67.51}$$

where $P(x', D')$ is a ψdo of order $\alpha < \frac{1}{2}$. We assume that

$$P_0(x'', 0, \xi'', 0) \neq 0, \tag{67.52}$$

where $P_0(x', \xi')$ is the principal part of $P(x', \xi')$, $\deg_{\xi'} P_0(x', \xi') = \alpha$.

We construct a regularizer of the boundary value problem (67.44), (67.51). Denote by $K_2 f$ and $K_{20} g$ the operators

$$K_2 f = \frac{1}{(2\pi)^{n-2}} \int_0^{x_{n-1}} \int_{\mathbb{R}^{n-2}} E(x_{n-1}, y_{n-1}, x'', \xi'') \tilde{f}(\xi'', y_{n-1}) \times e^{ix''\cdot\xi''} d\xi'' dy_{n-1}, \tag{67.53}$$

$$K_{20} g = \frac{1}{(2\pi)^{n-2}} \int_{\mathbb{R}^{n-2}} E(x_{n-1}, 0, x'', \xi'') \tilde{g}(\xi'') e^{ix''\cdot\xi''} d\xi'', \tag{67.54}$$

where $E(x_{n-1}, y_{n-1}, x'', \xi'')$ is the same as in (67.46). Analogously to Lemma 67.4 we have:

$$\begin{aligned} \|K_2 f\|_{1,s-1} &\leq C\|f\|_{0,s-\frac{1}{k+1}}, \\ \|K_{20} g\|_{1,s-1} &\leq C[g]_{s-\frac{1}{2(k+1)}}. \end{aligned} \tag{67.55}$$

Applying L to $K_2 f + K_{20} g$, we get, analogously to (67.50):

$$L(K_2 f + K_{20} g) = f + T^{(1)}(f, g),$$

where $T^{(1)}(f, g)$ is a smoothing operator in f and g.

Now consider

$$P(K_2 f + K_{20} g)|_{x_{n-1}=0}.$$

We assume, for simplicity, that $P_0(x', \xi'', \xi_{n-1})$ is real analytic in ξ'. Then the Taylor formula gives

$$P_0(x', \xi'', \xi_{n-1}) = P_0(x', \xi'', ix_{n-1}^k c(x', \xi'')) + (\xi_{n-1} - ix_{n-1}^k c(x', \xi''))P_1(x', \xi'),$$

where $\operatorname{ord} P_1 = \alpha - 1$. Therefore,

$$P(x', D') = \hat{P}_0(x', D'', ix_{n-1}^k c) + \hat{P}_1(x', D')\Big(-i\frac{\partial}{\partial x_{n-1}} - ix_{n-1}c(x', D'')\Big) + P_2, \tag{67.56}$$

where $\operatorname{ord} P_2 \le \alpha - 1$, $\hat{P}_1, \hat{P}_0$ are ψdo's with the principal symbols $P_1(x',\xi')$, $P_0(x',\xi'', ix_{n-1}^k c(x',\xi''))$.

We have:

$$PK_2 f = \hat{P}_0(x', D'', -ix_{n-1}^k c(x', D''))K_2 f \\ + \hat{P}_1\left(-i\frac{\partial}{\partial x_{n-1}} - ix_{n-1}^k c(x', D'')\right) K_2 f + P_2 K_2 f.$$

Note that $\left(\frac{\partial}{\partial x_{n-1}} + x_{n-1}^k c(x', D'')\right)K_2 f = T_{-\frac{1}{k+1}} f$ and $K_2 f|_{x_{n-1}=0} = 0$. Therefore

$$\hat{P}_0(x'', D'', -ix_{n-1}c(x', D''))K_2 f|_{x_{n-1}=0} = 0.$$

Since $\operatorname{ord} P_1 = \alpha - 1$, $\operatorname{ord} P_2 = \alpha - 1$, we have

$$PK_2 f|_{x_{n-1}=0} = T^{(1)}_{\alpha-1-\frac{1}{k+1}} f|_{x_{n-1}=0}, \tag{67.57}$$

where $\operatorname{ord} T_{\alpha-1-\frac{1}{k+1}} \le \alpha - 1 - \frac{1}{k+1}$.

Using (67.56) again, we get:

$$PK_{20} g|_{x_{n-1}=0} = \hat{P}_0(x'', 0, D'', 0)g + T_{\alpha-\frac{1}{2}-\frac{1}{2(k+1)}} g. \tag{67.58}$$

Therefore, if we replace g by Qg, where Q is a ψdo in $\mathbb{R}^{n-2}$ with the principal symbol $P_0^{-1}(x'', 0, \xi'', 0)$ (cf. (67.52)), we get:

$$\begin{aligned} L(K_2 f + K_{20}Qg) &= f + T_1(f, g), \\ P(K_2 f + K_{20}Qg)|_{x_{n-1}=0} &= g + T_2(f, g), \end{aligned} \tag{67.59}$$

where T_1, T_2 are operators of negative order. Therefore $K_2 f + K_{20}Qg$ is the right regularizer of the boundary value problem (67.44), (67.51).

The following estimate holds:

$$\|u\|_{1,s-1} \le C\|f\|_{0,s-\frac{1}{k+1}} + [g]_{s-\alpha-\frac{1}{2(k+1)}}. \tag{67.60}$$

Analogously, one can check that $K_2\left(\frac{\partial}{\partial x_{n-1}} + x_{n-1}^k c + c_1\right)u + K_{20}QPu|_{x_{n-1}=0}$ $= u - K_{20}u|_{x_{n-1}=0} + T_1 u + K_{20}Q(Pu)|_{x_{n-1}=0} = u + Tu$, i.e., $K_2 f + K_{20}Qg$ is the left regularizer. We have used that $K_{20}QPu|_{x_{n-1}=0} = K_{20}u|_{x_{n-1}=0} + T_2 u$. Here T_1, T_2, T are smoothing operators and Q is the ψdo with the principal symbol $P_0^{-1}(x'', 0, \xi'', 0)$.

c) The case where k is odd and $c(x', \xi'') < 0$. Then analogously to (67.40) we consider the equation

$$\frac{\partial u}{\partial x_{n-1}} + x_{n-1}^k c(x', D'')u \\ + c_1(x', D')u + G(x', D')(\rho(x'')\delta(x_{n-1})) = f(x'), \quad x' \in \mathbb{R}^{n-1}, \tag{67.61}$$

where $G(x', D')$ is a ψdo of order β. We assume that $\beta < -\frac{1}{2}$ and

(67.62) $$G_0(x'', 0, \xi'', 0) \neq 0,$$

where G_0 is the principal symbol of the ψdo $G(x', D')$. As in (67.37), denote by $K_3 f$ the operator

(67.63) $$K_3 f = \theta(-x_{n-1}) \int_{-\infty}^{x_{n-1}} \frac{1}{(2\pi)^{n-2}} \int_{\mathbb{R}^{n-2}} E(x_{n-1}, y_{n-1}, x'', \xi'') \times \tilde{f}(\xi'', y_{n-1}) e^{ix''\cdot\xi''} d\xi'' dy_{n-1} - \theta(x_{n-1}) \int_{x_{n-1}}^{\infty} \frac{1}{(2\pi)^{n-2}} \int_{\mathbb{R}^{n-2}} E(x_{n-1}, y_{n-1}, x'', \xi'') \times \tilde{f}(\xi'', y_n) e^{ix''\cdot\xi''} d\xi'' dy_{n-1},$$

where $E(x_{n-1}, y_{n-1}, x'', \xi'')$ is the same as in (67.46). As in (67.38), we have:

(67.64) $$LK_3 f = f + T_{-\frac{1}{k+1}} f - \delta(x_{n-1}) \frac{1}{(2\pi)^{n-2}} \int_{\mathbb{R}^{n-2}} E(0, y_{n-1}, x'', \xi'') \times \tilde{f}(\xi'', y_{n-1}) e^{ix''\cdot\xi''} d\xi'' dy_{n-1}.$$

Denote

(67.65) $$e(f) = \frac{1}{(2\pi)^{n-2}} \int_{-\infty}^{\infty} E(0, y_{n-1}, x'', \xi'') \tilde{f}(\xi'', y_{n-1}) e^{ix''\cdot\xi''} d\xi'' dy_{n-1}.$$

Suppose, for simplicity, that $G(x', \xi')$ is real analytic in ξ' when $\xi' \neq 0$. Then by the Taylor formula we have, as in (67.56):

(67.66)
$$G_0(x', \xi'', \xi_{n-1}) = G_0(x', \xi'', ix_{n-1}^k c) + G_1(x', \xi')(\xi_{n-1} - ix_{n-1}^k c(x', \xi'')),$$

where $\operatorname{ord} G_0 = \beta$, $\operatorname{ord} G_1 = \beta - 1$, and G_0 is the principal symbol of G. Note that $\hat{G}_0(x', D'', ix_{n-1}^k c(x', D''))(\rho(x'')\delta(x_{n-1})) = (G_0(x'', 0, D'', 0)\rho)\delta(x_{n-1})$. Since $E(0, 0, x'', \xi'') = 1$, we have, using (67.66):

(67.67) $$e(G(\rho\delta)) = \hat{G}_0 \rho + T_{\beta-1} \rho,$$

where $\operatorname{ord} T_{\beta-1} \leq \beta - 1$ and $\hat{G}_0$ is a ψdo with the principal symbol $G_0(x'', 0, \xi'', 0)$. Denote by $H(x'', D'')$ the ψdo with the principal symbol $G_0^{-1}(x'', 0, \xi'', 0)$. Let

(67.68) $$\rho(x'') = H(x'', D'') e(f).$$

We have

$$[\rho]_s \leq C \|f\|_{0, s-\beta-\frac{1}{2(k+1)}}.$$

We construct a regularizer for the equation (67.61). For this we need to modify $K_3 f$ (cf. (67.64)) to make $K_{\text{mod}} f \in H_1(\mathbb{R}^{n-1})$. We define

(67.69) $$K_{\text{mod}} f = K_3 f + \theta(x_{n-1}) e^{-x_{n-1}\Lambda''(D'')} e(f),$$

where $e(f)$ is defined in (67.65).

Now we have

$$LK_{\text{mod}}f = f + T_{-\frac{1}{k+1}}f + \theta(x_{n-1})(-\Lambda'' + x_{n-1}^k c(x', D'') + c_1(x', D''))e^{-x_{n-1}\Lambda''}e(f). \tag{67.70}$$

We show that

$$u = K_{\text{mod}}(f - G(\rho\delta)), \tag{67.71}$$

and $\rho = H(x'', D'')e(f)$ is a regularizer of (67.61). If ρ has the form (67.68), then (cf. (67.66), (67.67))

$$e(f - G(\rho\delta)) = e(f) - (\hat{G}_0(x'', 0, D'', 0) + T_{\beta-1})H(x'', D'')e(f) = T_{-1}^{(1)}e(f),$$

where $\operatorname{ord} T_{-1} = -1$. Therefore

$$[e(f - G(\rho\delta))]_s \le C[e(f)]_{s-1} \le C\|f\|_{s-1-\frac{1}{2(k+1)}}.$$

We have, as in Lemma 67.4:

$$\|K_3 f\|_{0,1} \le C\|f\|_{0,\frac{k}{k+1}}.$$

Using (67.66), we get

$$K_3 G(\rho\delta) = K_3\bigg(\hat{G}_0(x'', 0, D'', 0)\rho\delta(y_{n-1}) + \Big(-i\frac{\partial}{\partial y_{n-1}} - iy_{n-1}^k c\Big)\, G_1(x', D')(\rho\delta) + T_{\beta-1}(\rho\delta)\bigg).$$

We note that $K_3(\hat{G}_0(x'', 0, D'', 0)\rho\delta(y_{n-1})) = 0$ when $|x_{n-1}| > 0$ and $K_3(\frac{\partial}{\partial y_{n-1}} + y_{n-1}^k c)v = v + T_{-\frac{1}{k+1}}v$. Therefore

$$\begin{aligned}\|K_3 G(\rho\delta)\|_{0,1} &\le C\|G_1(\rho\delta)\|_{0,1} + C\|T_{\beta-1}(\rho\delta)\|_{0,1-\frac{1}{k+1}}\\ &\le C[\rho]_{\beta+\frac{1}{2}} + C[\rho]_{\frac{k}{k+1}+\beta-\frac{1}{2}} \le C[\rho]_{\beta+\frac{1}{2}}.\end{aligned}$$

It follows from (67.68) that $\|K_3 G(\rho\delta)\|_{0,1} \le C[\rho]_{\beta+\frac{1}{2}} \le C\|f\|_{0,\frac{k}{k+1}}$, and therefore $\|K_3(f - G(\rho\delta))\|_{0,1} \le C\|f\|_{0,\frac{k}{k+1}}$. Thus (cf. (67.69)):

$$\|K_{\text{mod}}(f - G(\rho\delta))\|_{0,1} \le C\|f\|_{0,\frac{k}{k+1}} + C[e(\tilde{f} - G(\rho\delta))]_{\frac{1}{2}} \le C\|f\|_{0,\frac{k}{k+1}}.$$

Substituting $K_{\text{mod}}(f - G(\rho\delta))$ into (67.62), we get:

$$\begin{aligned}LK_{\text{mod}}(f - G(\rho\delta)) &= f - G(\rho\delta) + T_{-\frac{1}{k+1}}f + T_{\beta+\frac{1}{2}}\rho + T_{\frac{1}{2}}e(f - G(\rho\delta))\\ &= f - G(\rho\delta) + T_{-\frac{1}{k+1}}^{(1)}f,\end{aligned}$$

assuming that (67.68) holds. Therefore

$$LK_{\text{mod}}(f - G(\rho\delta) + G(\rho\delta)) = f + T_{-\frac{1}{k+1}}f. \tag{67.72}$$

The estimate

$$\left\| \frac{\partial}{\partial x_{n-1}} K_{\text{mod}}(f - G(\rho\delta)) \right\|_0 \leq C\|f\|_{0,\frac{k}{k+1}}$$

follows from equation (67.72). Therefore $\|K_{\text{mod}}(f - G(\rho\delta))\|_1 \leq C\|f\|_{0,\frac{k}{k+1}}$. Moreover, (67.72) shows that $u = K_{\text{mod}}(f - G(\rho\delta))$, $\rho = H(x', D')e(f)$ is a right regularizer for (67.62). Analogously, one can show that (67.71), (67.68) is also a left regularizer.

67.5. The boundary value problem on $\partial\Omega$.

Let γ_δ be the δ-neighborhood of γ. We introduce local coordinates $(x'', x_{n-1}, \xi''_{x''}, \xi_{n-1})$ in $T^*(\gamma_\delta)$, where $(x'', \xi''_{x''}) \in T^*(\gamma)$, $|x_{n-1}| < \delta$, and $\xi_{n-1} \in \mathbb{R}$.

We will study the equation

$$b(x', D')v = f, \quad x' \in \partial\Omega.$$

Let $b_0(x', \xi'_{x'})$ be the principal symbol of b. We assume that $b_0(x', \xi'_{x'}) \neq 0$ when $x' \notin \gamma$, $\xi'_{x'} \neq 0$, and

(67.73) $$b_0(x', \xi'_{x'}) = b_1(x', \xi'_{x'})(\xi_{n-1} - \lambda_1(x', \xi''_{x'}) - ix^k_{n-1}c(x', \xi''_{x'}))$$

in $T^*(\gamma_\delta)$, where $b_1 \neq 0, \lambda_1, c$ are real-valued, $\operatorname{ord}_{\xi''} \lambda_1 = \operatorname{ord}_{\xi''} c = 1$, and $|c| \neq 0$ when $\xi''_{x''} \neq 0$. As above, we consider separately three cases:

a) k is even;

b) k is odd, $c > 0$;

c) k is odd, $c < 0$.

a) The case where k is even.

Theorem 67.6. *Let D_b be the domain of $b(x', D')$ consisting of $v \in H_1(\partial\Omega)$ such that $bv \in H_{1-r+\frac{k}{k+1}}(\partial\Omega)$. Suppose that k is even. Then $b(x', D')$ is a Fredholm operator from D_b to $H_{1-r+\frac{k}{k+1}}(\partial\Omega)$.*

Remark 67.2. Note the loss of smoothness of order $\frac{k}{k+1}$ in comparison with the case where $b(x', \xi')$ is elliptic everywhere in $\partial\Omega$. In fact, the loss occurs only near γ and only in the variables x'' tangential to γ. However, we use the space $H_s(\partial\Omega)$ for simplicity. □

Theorem 67.6 can be reformulated for the original boundary value problem

(67.74) $$Au = 0 \quad \text{in} \quad \partial\Omega, \qquad Bu|_{\partial\Omega} = f.$$

Denote by D_A the subspace of all $u \in H_{\frac{3}{2}}(\Omega)$, $Au = 0$, such that $Bu|_{\partial\Omega} \in H_{1-r+\frac{k}{k+1}}(\partial\Omega)$. Then the boundary value problem (67.74) defines a Fredholm operator from D_A to $H_{\frac{k}{k+1}}(\partial\Omega)$.

Note that we consider here unbounded Fredholm operators whose domains are not closed.

Proof of Theorem 67.6. We shall construct an operator R bounded from $H_{1-r+\frac{k}{k+1}}(\partial\Omega)$ to $H_1(\partial\Omega)$ such that $bR = I + T$ when T is a compact operator in $H_{1-r+\frac{k}{k+1}}(\partial\Omega)$, i.e., R is a right regularizer of b. Let $\{\varphi_j, U_j\}$ be a partition of unity in $\partial\Omega$, $\varphi_j \in C_0^\infty(U_j)$, $\sum_{j=1}^N \varphi_j = 1$. Let $\psi_j(x) \in C_0^\infty(U_j)$, and $\psi_j = 1$ in a neighborhood of $\operatorname{supp}\varphi_j$. We look for R in the form $R = \sum_{j=1}^N \psi_j R_j \varphi_j$. When $U_j \cap \gamma = \emptyset$, we take $R_j = b_0^{(-1)}(x', D')$, where $b_0^{(-1)}$ is a ψdo with the principal symbol $b_0^{-1}(x', \xi')$. Let $U_j \cap \gamma \neq \emptyset$ and let (x'', x_{n-1}) be local coordinates in U_j. We have

$$b_{0j}(x', \xi') = b_{1j}(x', \xi')(\xi_{n-1} - \lambda_{j1}(x', \xi'') - ix_{n-1}^k c_j(x', \xi'')),$$

where b_{0j} is the symbol b_0 in these local coordinates in U_j. We have

$$b_j(x', D') = \left(-i\frac{\partial}{\partial x_{n-1}} - \lambda_{j1}(x', D'') - ix_{n-1}^k c(x', D'')\right)\hat{b}_{1j} + b_{2j}, \tag{67.75}$$

where $\operatorname{ord} b_{2j} \leq r - 1$, and $\hat{b}_{1j}$ is a ψdo with the principal symbol $b_{1j}(x', \xi')$.

Let $l_j(x', \eta'')$ be the solution of the eikonal equation

$$l_{jx_{n-1}} - \lambda_{j1}(x', l_{jx''}) = 0, \quad l_j|_{x_{n-1}=0} = x'' \cdot \eta''. \tag{67.76}$$

Denote

$$L_j(x', \eta') = x_{n-1}\eta_{n-1} + l_j(x', \eta''). \tag{67.77}$$

Consider the canonical transformation $(y', \eta') \to (x', \xi')$ with the generating function $L_j(x', \eta')$ (cf. Example 62.1):

$$\begin{aligned} \xi'' &= L_{jx''} = l_{jx''}, \quad \xi_{n-1} = L_{jx_{n-1}} = \eta_{n-1} + l_{jx_{n-1}}, \\ y'' &= L_{j\eta''} = l_{j\eta''}, \quad y_{n-1} = L_{j\eta_n} = x_{n-1}. \end{aligned} \tag{67.78}$$

It follows from (67.78) that

$$\xi_{n-1} - \lambda_{j1}(x', \xi'') = \eta_{n-1}. \tag{67.79}$$

Denote by Φ_j the FIO of the form

$$\Phi_j u = \frac{1}{(2\pi)^{n-1}} \int_{\mathbb{R}^{n-1}} e^{iL_j(x',\eta')} |L_{jx'\eta'}|^{\frac{1}{2}} \tilde{u}(\eta') d\eta'' d\eta_{n-1}. \tag{67.80}$$

Integrating with respect to η_{n-1}, we get:

$$\Phi_j u = \frac{1}{(2\pi)^{n-2}} \int_{\mathbb{R}^{n-2}} e^{il_j(x',\eta'')} \tilde{u}(\eta'', x_{n-1}) |l_{jx''\eta''}|^{\frac{1}{2}} d\eta''. \tag{67.81}$$

We look for R_j in the form:

(67.82) $$\psi_j R_j = \psi_j b_{1j}^{(-1)} \Phi_j R_{j1},$$

where R_{j1} will be chosen later, and $b_{1j}^{(-1)}$ is a ψdo with the principal symbol $b_{1j}^{-1}(x', \xi')$. We have

$$b_j \psi_j R_j = \psi_j \left(-i \frac{\partial}{\partial x_{n-1}} - \lambda_{j1} - i x_{n-1}^k c_j \right) \Phi_j R_{j1} + T_0 \Phi_j R_{j1},$$

where T_p means an operator of order p.

It follows from Theorem 62.7 and Remark 62.1 that

(67.83) $$\left(-i \frac{\partial}{\partial x_{n-1}} - \lambda_{j1} - i x_{n-1}^k c_j \right) \Phi_j = \Phi_j \left(-i \frac{\partial}{\partial y_{n-1}} - i y_{n-1}^k c_{j1}(y', D_{y''}) \right) + T_0^{(1)},$$

where $\xi_{n-1} - \lambda_{j1}(x', \xi'') - i x_{n-1}^k c(x', \xi') = \eta_{n-1} - i y_{n-1}^k c_{j1}(y', \eta'')$, $|c_{j1}| \neq 0$, (x', ξ') and (y', η') being related by the canonical transformation (67.78). Note that the operator $\frac{\partial}{\partial y_{n-1}} + y_{n-1}^k c_{j1}(y', D_{y''})$ has the form (67.44), and therefore by Lemma 67.5 there exists an operator K_{j1} bounded from $H_{0, \frac{k}{k+1}}(\mathbb{R}^{n-1})$ to $H_1(\mathbb{R}^{n-1})$ and such that

$$\left(\frac{\partial}{\partial y_{n-1}} + y_{n-1}^k c_{j1} \right) K_{j1} = I + T_{-\frac{k}{k+1}}.$$

Finally, note that the FIO

(67.84) $$\Phi_j^* u = \frac{1}{(2\pi)^{n-1}} \int_{\mathbb{R}^{n-1}} \int_{\mathbb{R}^{n-1}} e^{i x' \cdot \xi' - i L_j(y', \xi')} |L_{j y' \xi'}|^{\frac{1}{2}} u(y') dy' d\xi'$$

satisfies (cf. (62.12))

$$\Phi_j \Phi_j^* = I + T_{-1}.$$

Therefore $\psi_j R_j \varphi_j = -i \psi_j b_{1j}^{(-1)} \Phi_j K_{j1} \Phi_j^* \varphi_j$ satisfies

(67.85) $$b_j \psi_j R_j \varphi_j = \varphi_j + T_{-\frac{1}{k+1}}^{(j)}.$$

We proved that $R = \sum_{j=1}^N \psi_j R_j \varphi_j$ is a right regularizer for b in $\partial\Omega$. Analogously one can construct a left regularizer. □

b) The case where k is odd and $c > 0$. In this case we consider the equation

(67.86) $$b(x', D')v = f \quad \text{in} \quad \partial\Omega$$

subject to the boundary condition

(67.87) $$P(x', D')v|_\gamma = g.$$

We assume that P is a ψdo of order $\alpha < \frac{1}{2}$ and

$$P_0(x', \xi''_{x''}, \lambda_1(x', \xi''_{x''})) \neq 0 \quad \text{for} \quad x' \in \gamma,\ \forall \xi''_{x''} \neq 0, \tag{67.88}$$

where $\lambda_1(x'', \xi'')$ is the same as in (67.73).

Theorem 67.7. *Suppose k is odd and $c > 0$ in (67.73). Suppose the condition (67.88) holds. Let $D_{b,P}$ be a subspace (not closed) of $H_1(\partial\Omega)$ consisting of all $v \in H_1(\partial\Omega)$ such that $bv \in H_{1-r+\frac{k}{k+1}}(\partial\Omega)$, $Pv|_\gamma \in H_{1-\alpha-\frac{1}{2(k+1)}}(\gamma)$. Then the operator $Bv = (bv|_{\partial\Omega}, Pv|_\gamma)$ is a Fredholm operator from $D_{b,P}$ to $H_{1-r+\frac{k}{k+1}}(\partial\Omega) \times H_{1-\alpha-\frac{1}{2(k+1)}}(\gamma)$.*

Remark 67.3. Note a particular case of Theorem 67.7 when $P = 1$, i.e., we have the boundary condition $u|_\gamma = g$ on γ.

As in the case a) of subsection 67.4, Theorem 67.7 implies that the boundary value problem $Au = 0$ in Ω, $Bu|_{\partial\Omega} = f$, $Pu|_\gamma = g$ determines a Fredholm operator from D_A to $H_{1-r+\frac{k}{k+1}}(\partial\Omega) \times H_{1-\alpha-\frac{1}{2(k+1)}}(\gamma)$, where D_A is defined similarly to D_A in Remark 67.2.

Proof of Theorem 67.7. The beginning of the proof is the same as the proof of Theorem 67.6. Let $\psi_j b_{1j}^{(-1)} \Phi_j w$ be the same as in (67.82). Then

$$b\psi_j b_{1j}^{(-1)} \Phi_j w = \psi_j \Phi_j \Big(-i\frac{\partial}{\partial y_{n-1}} - i y_{n-1}^k c_{j1}(y', D_{y''}) \Big) w + T_0 w, \tag{67.89}$$

where k is odd and $c_{j1} > 0$. Substituting $v = \psi_j b_{j1}^{(-1)} \Phi_j w$ into the boundary condition $Pv|_{x_{n-1}=0} = g$, we get:

$$\psi_j P_j b_{j1}^{(-1)} \Phi_j w|_{x_{n-1}=0} = \psi_j \Phi_j \hat{P}_{j1} w|_{x_{n-1}=0} + T_{\alpha-r}^{(1)} w|_{x_n=0}. \tag{67.90}$$

In (67.90) we used Theorem 62.7 again and the fact that the restriction to $x_{n-1} = 0$ commutes with Φ_j (cf. (67.81)).

Note that the principal part of $\hat{P}_{j1}(y', \eta')$ is equal to $P_j(x', \xi') b_{j1}^{-1}(x', \xi')$ when (y', η') and (x', ξ') are related by the canonical transformation (67.78). In particular, $\xi_{n-1} = \lambda_{j1}(x', \xi'')$ corresponds to $\eta_{n-1} = 0$. Therefore the condition (67.88) implies

$$P_{j1}(y', 0, \eta'', 0) \neq 0, \tag{67.91}$$

i.e., (67.91) coincides with the condition (67.52). Define $w = K_{j2}\Phi_j^* \varphi_j f + K_{j20} Q_j \Phi_j^*|_{y_{n-1}=0} \varphi_j g$, where K_{j2} and K_{j20} are the same as in (67.59) and $Q_j(y'', \eta'')$ is a ψdo in $\mathbb{R}^{n-2}$ with the principal symbol $P_{j1}^{-1}(y', 0, \eta'', 0)$. We get (cf. (67.59)) that

$$\begin{aligned} b_j \psi_j R_j \varphi_j f &= \varphi_j f + T_{j,-\frac{1}{k+1}}(f, g), \\ P_j \psi_j R_j \varphi_j (f, g)|_{x_{n-1}=0} &= \varphi_j g + T_j(f, g), \end{aligned}$$

where

$$\psi_j R_j \varphi_j(f,g) = \psi_j b_{j1}^{(-1)} \Phi_j K_{j2} \Phi_j^* \varphi_j f + \psi_j b_{j1}^{(-1)} \Phi_j K_{j20} Q_j \Phi^* |_{x_{n-1}=0} \varphi_j g.$$

Taking the sum over j, $1 \le j \le N$, we get a right regularizer of the operator $(bu|_{\partial\Omega}, Pu|_\gamma)$. Analogously one can construct a left regularizer.

c) The case where k is odd and $c < 0$. In this case we consider the equation with a "coboundary" condition:

$$b(x', D')v + G(\rho\delta(\gamma)) = f, \quad x \in \partial\Omega, \tag{67.92}$$

where G is a ψdo in $\partial\Omega$, $\rho(x')$ is an unknown function on γ, and $\delta(\gamma)$ is the delta-function of γ. We assume that

$$G_0(x'', 0, \xi''_{x''}, \lambda_1(x'', 0, \xi''_{x''})) \ne 0, \quad x'' \in \gamma, \quad \xi''_{x''} \ne 0, \tag{67.93}$$

where G_0 is the principal symbol of G in the $(x'', x_{n-1}, \xi''_{x''}, \xi_{n-1})$-coordinates in $T_0^*(\gamma_\delta)$, $\lambda_1(x'', x_{n-1}, \xi''_{x''})$ is the same as in (67.73), and $\operatorname{ord} G_0 = \beta$, $\beta < -\frac{1}{2}$.

Theorem 67.8. *Denote by $D_{b,G}$ the subspace of $H_1(\partial\Omega) \times H_{1-r-\beta+\frac{2k+1}{2k+2}}(\gamma)$ such that $bv + G(\rho\delta) \in H_{1-r+\frac{k}{k+1}}(\partial\Omega)$. Assume that (67.93) is satisfied and $\beta < -\frac{1}{2}$. Then (67.92) defines a Fredholm operator from $D_{b,G}$ to $H_{1-r+\frac{k}{k+1}}(\partial\Omega)$. The boundary value problem*

$$\begin{aligned} &Au = 0, \quad x \in \Omega, \\ &Bu|_{\partial\Omega} + G(\rho\delta) = f, \quad x \in \partial\Omega, \end{aligned}$$

also defines a Fredholm operator.

Proof: As in the proofs of Theorems 67.6 and 67.7, we use the operators $b_{j1}^{(-1)}$ and Φ_j to reduce the construction of a right parametrix to the case similar to (67.61). Note that (67.62) is satisfied if the condition (67.93) is satisfied. Then, constructing a right parametrix as in (67.71), (67.68) and using Φ^*, we get a right parametrix for (67.92).

The construction of a left parametrix is similar.

Remark 67.4. Differentiating the operators K_1, K_0 of the form (67.53), (67.54) with respect to x_{n-1}, we get an estimate (cf. (67.55))

$$\begin{aligned} &\|K_2 f\|_m \le C\|f\|_{m-1,\frac{k}{k+1}}, \\ &\|K_{20} f\|_m \le C[g]_{m-\frac{1}{2(k+1)}}. \end{aligned} \tag{67.94}$$

Assuming that $\alpha < m - \frac{1}{2}$, we can change 1 to m in Theorem 67.7; i.e., we can prove that $(bv|_{\partial\Omega},\ Pv|_\gamma)$ is a Fredholm operator from $D_{b,P}$ to

$H_{m-r+\frac{k}{k+1}}(\partial\Omega) \times H_{m-\alpha-\frac{1}{2(k+1)}}(\gamma)$, where $D_{b,P}$ consists of $u \in H_m(\partial\Omega)$ such that $bu \in H_{m-r+\frac{k}{k+1}}(\partial\Omega)$, $Pu|_\gamma \in H_{m-\alpha-\frac{1}{2(k+1)}}$.

Having estimates (67.94), the proof of Theorem 67.7 proceeds without changes. The same is true for Theorem 67.6. To extend the proof of Theorem 67.8 to the case $m > 1$, one needs to assume that $\beta < -m + \frac{1}{2}$ and to take more terms in the modification $K_{\text{mod}}f$ of Kf (cf. (67.69)) to make $K_{\text{mod}}(f - G(\rho\delta(x_{n-1}))) \in H_m(\mathbb{R}^{n-1})$, provided that $\rho(x')$ has the form (67.68). For example, when $m = 2$, we take, instead of (67.69),

$$\begin{aligned}K_{\text{mod}}(f - G(\rho\delta)) &= K_3(f - G(\rho\delta)) + \theta(x_{n-1})e^{-x_{n-1}\Lambda''}e(f - G(\rho\delta)) \\ &- x_{n-1}\theta(x_{n-1})(-\Lambda'' + x_{n-1}^k c(x', D'') + c_1)e^{-x_{n-1}\Lambda''}e(f - G(\rho\delta)).\end{aligned}$$

68. Problems

1. Let $A(x,\xi) = x_n - g_1(x',\xi) - ig_2(x',\xi)$, where $x' = (x_1, \ldots, x_{n-1}) \in \mathbb{R}^{n-1}$, g_1, g_2 are homogeneous in ξ of degree zero and smooth when $\xi \neq 0$. Find a canonical transformation that takes $x_n - g_1(x',\xi) - ig_2(x',\xi)$ into $y_n - ig_3(y',\eta)$, where $y' = (y_1, \ldots, y_{n-1})$, and g_3 is homogeneous in η of degree zero.

Construct a FIO corresponding to this canonical transformation.

2. Let Φu be a FIO of the form (65.8) with nondegenerate phase function $\psi(x, y, \theta)$. Let Λ be the Lagrangian manifold corresponding to Φ and let $(x,\xi,y,\eta) \in \Lambda'$ iff $(x,\xi,y,-\eta) \in \Lambda$. Suppose that Λ' is a graph of a canonical transformation $\alpha(y,\eta) = (x,\xi)$. Extend the proof of Theorem 63.1 to show that $WF(\Phi u) \subset \alpha \circ WF(u)$, where $\alpha \circ WF(u)$ is the image of $WF(u)$ under the canonical map α.

3. Let $\{f(x,\xi), g(x,\xi)\}$ be the Poisson bracket of f and g, that is, $\{f(x,\xi), g(x,\xi)\} = \sum_{j=1}^n (f_{x_j} g_{\xi_j} - f_{\xi_j} g_{x_j})$. Let $(x,\xi) = \alpha(y,\eta)$ be a canonical transformation. Prove that α preserves the Poisson brackets, i.e.,

$$\{f(x,\xi), g(x,\xi)\} = \{f_1(y,\eta), g_1(y,\eta)\},$$

where $f_1(y,\eta) = f(\alpha(y,\eta))$, $g_1(y,\eta) = g(\alpha(y,\eta))$.

4. Consider a Hamiltonian system

$$\frac{dx}{dt} = \frac{\partial H(x,\xi)}{\partial \xi}, \quad \frac{d\xi}{dt} = -\frac{\partial H(x,\xi)}{\partial x}.$$

Let $(x,\eta) = \alpha(y,\eta)$ be a canonical transformation. Prove that

$$\frac{dy}{dt} = \frac{\partial H_1(y,\eta)}{\partial \eta}, \quad \frac{d\eta}{dt} = -\frac{\partial H_1(y,\eta)}{\partial y},$$

where $H_1(y,\eta) = H(\alpha(y,\eta))$, i.e., $(x,\eta) = \alpha(y,\eta)$ transforms the given Hamiltonian system to a Hamiltonian system.

5. Let $x_1 = x_1(y_1,\eta_1)$, $\xi_1 = \xi_1(y_1,\eta_1)$ be an arbitrary canonical transformation of $\mathbb{R}^1 \times \mathbb{R}^1$ onto $\mathbb{R}^1 \times \mathbb{R}^1$. Prove that the Jacobian $\det \frac{\partial(x_1,\xi_1)}{\partial(y_1,\eta_1)} = 1$. Prove the same result for any $n \geq 1$.

6. Landau levels. Consider the Schrödinger equations describing an electron in a constant magnetic field in $\mathbb{R}^2$:

$$Hu = \lambda u,$$

where

$$H = \frac{1}{2}\left(-i\frac{\partial}{\partial x_1} - \frac{eB_0}{2}x_2\right)^2 + \frac{1}{2}\left(-i\frac{\partial}{\partial x_2} - \frac{eB_0}{2}x_1\right)^2 \pm \frac{1}{2}eB_0.$$

For simplicity, assume that $eB_0 = 1$.

a) Prove that the transformation $(x,\xi) \to (y,\eta)$, where

$$\eta_1 = \xi_1 - \frac{1}{2}x_2, \quad y_1 = \xi_2 + \frac{1}{2}x_1,$$

(68.1)

$$\eta_2 = \xi_2 - \frac{1}{2}x_1, \quad y_2 = \xi_1 + \frac{1}{2}x_2,$$

is a canonical transformation.

b) Show that

$$S(x,y) = x_1y_2 - \frac{1}{2}x_1x_2 - y_1y_2 + x_2y_1$$

is a generating function for (68.1), i.e., $\xi = S_x(x,y)$, $\eta = -S_y(x,y)$.

c) Denote by Φu the following operator:

$$\Phi u = \frac{1}{2\pi}\int_{\mathbb{R}^2} e^{iS(x,y)}u(y)dy.$$

Let Φ^* be the formally adjoint to Φ:

$$\Phi^* u = \frac{1}{2\pi}\int_{\mathbb{R}^2} e^{-iS(y,x)}u(y)dy.$$

Prove that

$$\Phi^*\Phi u = u.$$

d) Prove that

$$\hat{H}u = \Phi^* H \Phi u$$

can be written as

$$\hat{H}u = -\frac{1}{2}\frac{\partial^2 u(y_1,y_2)}{\partial y_1^2} + \frac{1}{2}y_1^2 u(y_1,y_2) \pm \frac{1}{2}u(y_1,y_2).$$

e) Finally, prove the L. D. Landau theorem:

Theorem 68.1. *The spectrum of H consists of the eigenvalues*

$$m + \frac{1}{2} \pm \frac{1}{2}, \qquad m = 0, 1, 2, \dots .$$

Each eigenvalue has an infinite multiplicity.

Bibliography

[A1] S. Agmon, *Lectures on elliptic boundary value problems*, Van Nostrand, New York, 1964.

[A2] S. Agmon, *Spectral properties of Schrödinger operators and scattering theory*, Ann. Scuola Norm. Sup. Pisa 2 (1975), 151–218.

[ADN] S. Agmon, A. Douglis, and L. Nirenberg, *Estimates near the boundary for solutions of elliptic differential equations satisfying general boundary conditions*, CPAM 12 (1959), 623–727; II, CPAM 17 (1964), 35–92.

[AH] S. Agmon and L. Hörmander, *Assymptotic properties of solutions of differential equations with simple characteristics*, J. Analyse Math. 30 (1976), 1–38.

[AK] S. Agmon and Y. Kannai, *On the asymptotic behaviour of spectral functions and resolvent kernels of elliptic operators*, Israel J. Math. 5 (1967), 1–30.

[Ar1] V. Arnold, *Mathematical methods of classical mechanics*, Springer-Verlag, New York, 1978.

[Ar2] V. Arnold, *On the characteristic class entering in the quantization condition*, Funct. Anal. Appl. 1 (1967), 1–13.

[At] M. Atiyah, *K-theory, Lecture notes*, Benjamin, Inc., New York-Amsterdam, 1967.

[AtB] M. Atiyah and R. Bott, *The index problem for the manifolds with boundary*, 1964, Differential Analysis, Bombay Colloq. 1964, pp. 175–186.

[AtS1] M. Atiyah and I. M. Singer, *The index of elliptic operators I*, Anal. of Math. (2) 87 (1968), 484–530.

[AtS2] M. Atiyah and I. M. Singer, *The index of elliptic operators III*, Anal. of Math. (2), 87 (1968), 546–604.

[B] R. Beals, *A general calculus of pseudodifferential operators*, Duke Math. J. 42 (1975), 1–42.

[BF] R. Beals and C. Feferman, *Spatially inhomogeneous pseudodifferential operators I*, Comm. Pure Appl. Math 27 (1974), 585–639.

[CH] R. Courant and D. Hilbert, *Methods of mathematical physics*, vol. II, New York, 1962.

[CK] D. Colton and R. Kress, *Inverse acoustic and electromagnetic scattering theory*, Springer-Verlag, 1992.

[ChP] J. Chazarain and A. Pirion, *Introduction à la théorie des équations aux dérivée partielle linéaire*, Gauthier Villars, Paris, 1981.

[D] J. J. Duistermaat, *Fourier integral operators*, Birkhäuser, Boston, 1996.

[DH] J. J. Duistermaat and L. Hörmander, *Fourier integral operators II*, Acta Math. 128 (1972), 183–269.

[DL] G. Duvaut and J.-P. Lions, *Les inéquations en méchanique et en physique*, Dunod, Paris, 1972.

[DSj] J. J. Duistermaat and J. Sjostrand, *A global construction for pseudodifferential operators with non-involutive characteristics*, Inv. Math. 20 (3) (1973).

[E1] G. Eskin, *Boundary value problems for elliptic pseudodifferential operators*, Translations of Mathematical Monographs, 52, AMS, Providence, R.I., 1981.

[E2] G. Eskin, *Asymptotics near the boundary of spectral functions of elliptic selfadjoint boundary problems*, Israel J. Math. 22 (1975), no. 3-4, 214–246.

[E3] G. Eskin, *Degenerate elliptic pseudodifferential equations of principal type*, Math. USSR Sbornik 11 (1970), no. 4, 539–582

[E4] G. Eskin, *Elliptic pseudodifferential operators with a degeneracy of the first order in space variables*, Trans. Moscow Math. Soc. 25 (1971), 91–130.

[Eg1] Yu. V. Egorov, *On canonical transformations of pseudodifferential operators*, Uspekhi Math. Nauk 25 (1969), 235–236.

[Eg2] Yu. V. Egorov, *Linear differential equations of principal type*, Nauka, Moscow, 1984. (Russian)

[EgK] Yu. V. Egorov and V. A. Kondratiev, *On the oblique derivative problem*, Mat. Sb. (NS) 78 (120), 1969, 148–176. (Russian)

[EgS] Yu. V. Egorov and B.-W. Schulze, *Pseudodifferential operators, singularities, applications*, Operator theory: Advances and Applications 93, Birkhäuser-Verlag, Basel, 1997.

[Ev] L. Evans, *Partial differential equations*, Graduate Studies in Mathematics, AMS, Providence, 1998.

[EIO] G. Eskin, H. Isozaki, and S. O'Dell, *Gauge equivalence and inverse scattering for Aharonov-Bohm effect*, Comm. in PDE 35 (2010), 2164–2194.

[ER1] G. Eskin and J. Ralston, *Inverse scattering problem for the Schrödinger equation with magnetic potential at a fixed energy*, Commun. Math. Phys. 173 (1995), 199–224.

[ER2] G. Eskin and J. Ralston, *Inverse backscattering problem in three dimensions*, Commun. Math. Phys. 124 (1989), 169–215.

[ER3] G. Eskin and J. Ralston, *Inverse scattering problems in several dimensions*, Journées EDP St. Jean de Monts, 1993, Exp. 17.

[F1] B. V. Fedosov, *A direct proof of the formula for the index of an elliptic system in Euclidean space*, Funct. Anal. Appl. 4 (1970), 339–391.

[F2] B. V. Fedosov, *Theorems on the index*, Modern problems in mathematics, Fundamental directions, VINITI, 1991, vol. 65, pp. 165–268. (Russian)

[Fo] G. Folland, *Introduction to partial differential equations*, Princeton University Press, 1995.

[FrW] L. Frank and W. Wendt, *Coercive singular perturbations II, Reduction and convergence*, J. Math. Anal. Appl. 88 (1982), no 2, 464–504.

[GS] A. Grigis and J. Sjöstrand, *Microlocal analysis for differential operators*, An Introduction, London Math. Soc. Lecture Notes Series 196, Cambridge University Press, 1994.

[GSh] I. M. Gelfand and G. E. Shilov, *Generalized functions*, vol. I, Academic Press, New York and London, 1964.

[GSt] V. Guillemin and Sh. Sternberg, *Geometric asymptotics*, Math. Surveys, vol. 14, AMS, Providence, R.I., 1977.

[GT] D. Gilbarg and N. Trudinger, *Elliptic partial differential equations of second order* (2nd. ed.), Springer, 1983.

[H1] L. Hörmander, *The analysis of linear partial differential operators*, Springer-Verlag, Berlin, vols. 256 (1983), 257 (1983), 274 (1985), 275 (1985).

[H2] L. Hörmander, *Fourier integral operators I*, Acta Math. 127 (1971), 79–183.

[H3] L. Hörmander, *Linear partial differential equations*, Sringer-Verlag, 1963.

[H4] L. Hörmander, *Pseudodifferential operators and non-elliptic boundary problems*, Ann. of Math. (2) 83 (1966), 129–209.

[I] V. Ivrii, *Precise spectral asymptotics for elliptic operators*, Lecture Notes in Math., vol. 1100, Springer, Berlin, 1984.

[Isa] V. Isakov, *Inverse problems for partial differential equations* (2nd. ed.), Applied Mathematical Sciences, vol. 127, New York, 2006.

[Iso] H. Isozaki, *Asymptotic properties of solutions to 3-particle Schrödinger equations*, Commun. Math. Phys. 222 (2001), 371–413.

[KN] J. J. Kohn and L. Nirenberg, *On the algebra of pseudodifferential operators*, Comm. Pure. Appl. Math. 18 (1965), 269–305.

[KS] D. Kinderlehrer and G. Stampacchia, *An introduction to variational inequalities and their applications*, Academic Press, 1980.

[KV] R. Kohn and M. Vogelius, *Determining conductivity by boundary measurements*, Comm. Pure Appl. Math. 37 (1984), 289–298.

[L] J.-L. Lions, *Quelques méthodes de résolution des problèmes aux limites non lineaires*, Dunod, 1969.

[LM] J.-L. Lions and E. Magenes, *Problèmes aux limites non homogenes et applications*, vols. 1–3, Dunod, Paris, 1968.

[LU] J. Lee and G. Uhlmann, *Determining anisotropic real-analytic conductivities by boundary measurements*, Comm. Pure Appl. Math. 42 (1989), 1097–1112.

[M1] V. P. Maslov, *Perturbation theory and asymptotics methods*, Moscow State University, Moscow, 1965. (Russian)

[M2] V. P. Maslov, *Operator methods*, Nauka, Moscow, 1973. (Russian)

[Ma] V. Mazya, *The degenerate problem with an oblique derivative*, Math. USSR Sbornik 16 (1972), 429–469.

[Mal] M. B. Malyutov, *Oblique derivative problem in three-dimensional space*, Soviet Math. Dokl. 8 (1967), 87–90.

[Me] R. Melrose, *Geometric scattering theory*, Cambridge University Press, 1995.

[MF] V. P. Maslov and M. V. Fedorjuk, *Semi-classical approximation in quantum mechanics*, Math. physics and applied mathematics, Reidel, Dordrecht, 1981.

[MaP] V. Mazya and B. Paneah, *Degenerate elliptic pseudodifferential equations*, Uspekhi Math. Nauk 25 (1970), 193–194. (Russian)

[N] L. Nirenberg, *Lectures on partial differential equations*, Proc. Reg. Conf. at Texas Tech., May 1972, Conf. Board of AMS.

[Na1] A. Nachman, *A reconstruction from boundary measurements*, Annals of Math. 128 (1988), 531–576.

[Na2] A. Nachman, *Global uniqueness for a two-dimensional inverse boundary value problem*, Annals of Math. 143 (1996), 71–96.

[No1] R. Novikov, *Multidimensional inverse spectral problem for the equation* $-\Delta\psi + (v(x) - Eu(x))\psi = 0$, Funct. Anal. Appl. 22 (1988), 263–272.

[No2] R. Novikov, *The inverse scattering problem at a fixed energy for the three-dimensional Schrödinger equation with an exponentially decreasing potential*, Commun. Math. Phys. 161 (1994), 569–595.

[NoKh] R. Novikov and G. Khenkin, *The* $\overline{\partial}$*-equation in the multidimensional inverse scattering problem*, Russ. Math. Surveys 42 (1987), no. 3, 109–180.

[R] W. Rudin, *Real and complex analysis*, McGraw-Hill, New York, 1974.

[Sch] L. Schwartz, *Théorie des distributions*, I, II, Hermann, Paris, 1950–51.

[Se1] R. Seeley, *Complex powers of elliptic operators*, Proc. Symp. in Pure Math., vol. 10, pp. 288–307, AMS, Providence, RI, 1967.

[Se2] R. Seeley, *The resolvent of elliptic boundary problems*, Amer. J. Math. 91 (1969), 889–920.

[Se3] R. Seeley, *Integro-differential operators on vector bundles*, Trans. Amer. Math. Soc. 117 (1965), 165–204.

[Sh] G. Shilov, *Mathematical analysis* (second special course), Nauka, Moscow, 1965. (Russian)

[Shu1] M. A. Shubin, *Pseudodifferential operators and spectral theory*, Springer series Soviet Math., Springer, 1987.

[Shu2] M. A. Shubin, *Pseudodifferential operators and spectral theory*, Dobrosvet, Moscow, 2005. (Russian)

[Si] B. Simon, *Functional integration and quantum physics*, Academic Press, New York, 1979.

[Sj] J. Sjöstrand, *Operators of principal type with interior boundary conditions*, Acta Math. 130 (1973), 1–51.

[St] E. Stein, *Singular integrals and the differentiality of functions*, Princeton Univ. Press, Princeton, N.J., 1972.

[SU] J. Sylvester and G. Uhlmann, *A global uniqueness theorem for an inverse boundary value problem*, Ann. Math. 125 (1987), 153–169.

[T1] M. E. Taylor, *Pseudodifferential operators*, Princeton Univ. Press, Princeton, N.J., 1981.

[T2] M. E. Taylor, *Partial differential equations I. Basic theory*, Applied Math Sciences, vol. 116, Springer, 1996; II. *Qualitative studies of linear equations*, Applied Math. Studies, vol. 116, Springer, 1996; III. *Nonlinear equations*, Appl. Math. Sciences, vol. 117, Springer-Verlag, 1997.

[Tr] F. Treves, *Introduction to pseudodifferential and Fourier integral operators*, vols. 1, 2, Plenum Press, New York, 1980.

[W1] R. Weder, *Global uniqueness at fixed energy in multidimensional inverse scattering theory*, Inverse Problems 7 (1991), 927–938.

[W2] R. Weder, *Completeness of averaged scattering solutions and inverse scattering at fixed energy*, Comm. PDE 32 (2007), 675–691.

[Ya] D. Yafaev, *Mathematical scattering theory: General theory*, Translation of Math. Monographs, vol. 105, AMS, Providence, RI, 1992.

Index

Titles in This Series

124 **David A. Cox, John B. Little, and Henry K. Schenck,** Toric varieties, 2011
123 **Gregory Eskin,** Lectures on linear partial differential equations, 2011
122 **Teresa Crespo and Zbigniew Hajto,** Algebraic groups and differential Galois theory, 2011
121 **Tobias Holck Colding and William P. Minicozzi II,** A course in minimal surfaces, 2011
120 **Qing Han,** A basic course in partial differential equations, 2011
119 **Alexander Korostelev and Olga Korosteleva,** Mathematical statistics: asymptotic minimax theory, 2011
118 **Hal L. Smith and Horst R. Thieme,** Dynamical systems and population persistence, 2010
117 **Terence Tao,** An epsilon of room, I: pages from year three of a mathematical blog. A textbook on real analysis, 2010
116 **Joan Cerdà,** Linear functional analysis, 2010
115 **Julio González-Díaz, Ignacio García-Jurado, and M. Gloria Fiestras-Janeiro,** An introductory course on mathematical game theory, 2010
114 **Joseph J. Rotman,** Advanced modern algebra: Second edition, 2010
113 **Thomas M. Liggett,** Continuous time Markov processes: An introduction, 2010
112 **Fredi Tröltzsch,** Optimal control of partial differential equations: Theory, methods and applications, 2010
111 **Simon Brendle,** Ricci flow and the sphere theorem, 2010
110 **Matthias Kreck,** Differential algebraic topology: From stratifolds to exotic spheres, 2010
109 **John C. Neu,** Training manual on transport and fluids, 2010
108 **Enrique Outerelo and Jesús M. Ruiz,** Mapping degree theory, 2009
107 **Jeffrey M. Lee,** Manifolds and differential geometry, 2009
106 **Robert J. Daverman and Gerard A. Venema,** Embeddings in manifolds, 2009
105 **Giovanni Leoni,** A first course in Sobolev spaces, 2009
104 **Paolo Aluffi,** Algebra: Chapter 0, 2009
103 **Branko Grünbaum,** Configurations of points and lines, 2009
102 **Mark A. Pinsky,** Introduction to Fourier analysis and wavelets, 2009
101 **Ward Cheney and Will Light,** A course in approximation theory, 2009
100 **I. Martin Isaacs,** Algebra: A graduate course, 2009
99 **Gerald Teschl,** Mathematical methods in quantum mechanics: With applications to Schrödinger operators, 2009
98 **Alexander I. Bobenko and Yuri B. Suris,** Discrete differential geometry: Integrable structure, 2008
97 **David C. Ullrich,** Complex made simple, 2008
96 **N. V. Krylov,** Lectures on elliptic and parabolic equations in Sobolev spaces, 2008
95 **Leon A. Takhtajan,** Quantum mechanics for mathematicians, 2008
94 **James E. Humphreys,** Representations of semisimple Lie algebras in the BGG category $\mathcal{O}$, 2008
93 **Peter W. Michor,** Topics in differential geometry, 2008
92 **I. Martin Isaacs,** Finite group theory, 2008
91 **Louis Halle Rowen,** Graduate algebra: Noncommutative view, 2008
90 **Larry J. Gerstein,** Basic quadratic forms, 2008
89 **Anthony Bonato,** A course on the web graph, 2008
88 **Nathanial P. Brown and Narutaka Ozawa,** C^*-algebras and finite-dimensional approximations, 2008
87 **Srikanth B. Iyengar, Graham J. Leuschke, Anton Leykin, Claudia Miller, Ezra Miller, Anurag K. Singh, and Uli Walther,** Twenty-four hours of local cohomology, 2007

86 **Yulij Ilyashenko and Sergei Yakovenko,** Lectures on analytic differential equations, 2007
85 **John M. Alongi and Gail S. Nelson,** Recurrence and topology, 2007
84 **Charalambos D. Aliprantis and Rabee Tourky,** Cones and duality, 2007
83 **Wolfgang Ebeling,** Functions of several complex variables and their singularities (translated by Philip G. Spain), 2007
82 **Serge Alinhac and Patrick Gérard,** Pseudo-differential operators and the Nash–Moser theorem (translated by Stephen S. Wilson), 2007
81 **V. V. Prasolov,** Elements of homology theory, 2007
80 **Davar Khoshnevisan,** Probability, 2007
79 **William Stein,** Modular forms, a computational approach (with an appendix by Paul E. Gunnells), 2007
78 **Harry Dym,** Linear algebra in action, 2007
77 **Bennett Chow, Peng Lu, and Lei Ni,** Hamilton's Ricci flow, 2006
76 **Michael E. Taylor,** Measure theory and integration, 2006
75 **Peter D. Miller,** Applied asymptotic analysis, 2006
74 **V. V. Prasolov,** Elements of combinatorial and differential topology, 2006
73 **Louis Halle Rowen,** Graduate algebra: Commutative view, 2006
72 **R. J. Williams,** Introduction the the mathematics of finance, 2006
71 **S. P. Novikov and I. A. Taimanov,** Modern geometric structures and fields, 2006
70 **Seán Dineen,** Probability theory in finance, 2005
69 **Sebastián Montiel and Antonio Ros,** Curves and surfaces, 2005
68 **Luis Caffarelli and Sandro Salsa,** A geometric approach to free boundary problems, 2005
67 **T.Y. Lam,** Introduction to quadratic forms over fields, 2004
66 **Yuli Eidelman, Vitali Milman, and Antonis Tsolomitis,** Functional analysis, An introduction, 2004
65 **S. Ramanan,** Global calculus, 2004
64 **A. A. Kirillov,** Lectures on the orbit method, 2004
63 **Steven Dale Cutkosky,** Resolution of singularities, 2004
62 **T. W. Körner,** A companion to analysis: A second first and first second course in analysis, 2004
61 **Thomas A. Ivey and J. M. Landsberg,** Cartan for beginners: Differential geometry via moving frames and exterior differential systems, 2003
60 **Alberto Candel and Lawrence Conlon,** Foliations II, 2003
59 **Steven H. Weintraub,** Representation theory of finite groups: algebra and arithmetic, 2003
58 **Cédric Villani,** Topics in optimal transportation, 2003
57 **Robert Plato,** Concise numerical mathematics, 2003
56 **E. B. Vinberg,** A course in algebra, 2003
55 **C. Herbert Clemens,** A scrapbook of complex curve theory, second edition, 2003
54 **Alexander Barvinok,** A course in convexity, 2002
53 **Henryk Iwaniec,** Spectral methods of automorphic forms, 2002
52 **Ilka Agricola and Thomas Friedrich,** Global analysis: Differential forms in analysis, geometry and physics, 2002
51 **Y. A. Abramovich and C. D. Aliprantis,** Problems in operator theory, 2002
50 **Y. A. Abramovich and C. D. Aliprantis,** An invitation to operator theory, 2002

For a complete list of titles in this series, visit the AMS Bookstore at **www.ams.org/bookstore/**.